Health Care in America: A History

JOHN C. BURNHAM

Johns Hopkins University Press

Baltimore

© 2015 Johns Hopkins University Press
All rights reserved. Published 2015
Printed in the United States of America on acid-free paper

2   4   6   8   9   7   5   3   1

Johns Hopkins University Press
2715 North Charles Street
Baltimore, Maryland 21218-4363
www.press.jhu.edu

Library of Congress Cataloging-in-Publication Data

Burnham, John C. (John Chynoweth), 1929– author.
Health care in America : a history / John C. Burnham.
   p. ; cm.
Includes bibliographical references and index.
ISBN 978-1-4214-1607-6 (hardcover : alk. paper) —
ISBN 1-4214-1607-7 (hardcover : alk. paper) —
ISBN 978-1-4214-1608-3 (pbk. : alk. paper) —
ISBN 1-4214-1608-5 (pbk. : alk. paper) —
ISBN 978-1-4214-1609-0 (electronic) — ISBN 1-4214-1609-3 (electronic)
   I. Title.
[DNLM: 1. Delivery of Health Care—history—United States.   2. History, Modern
1601—United States. WZ 70 AA1]
   RA445
   362.10973—dc23   2014018369

A catalog record for this book is available from the British Library.

*Special discounts are available for bulk purchases of this book. For more information,
please contact Special Sales at 410-516-6936 or specialsales@press.jhu.edu.*

Johns Hopkins University Press uses environmentally friendly book materials,
including recycled text paper that is composed of at least 30 percent
post-consumer waste, whenever possible.

CONTENTS

ACKNOWLEDGMENTS

This book is designed as an introduction to the field of the history of health and health care in America. As such, it depends upon the scholars who have created and contributed to that field. The book itself is an acknowledgment of gratitude to them, as well as recognition of their work. The notes indicating specific items of indebtedness also serve to thank at least some of these fine scholars.

This book was written and revised over several years, and in the process I received an extraordinary amount of encouragement, assistance, and correction. I should and do recognize and thank in particular two colleagues, Stephen Casper and Christopher Hamlin, who have sustained my efforts.

Moreover, several colleagues who also read the entire manuscript in one draft or another deserve special recognition for service above and beyond the usual collegiality. On this honor roll are Gerald Grob, Susan Lawrence, Tamara Mann, George Paulson, and John Sauer, plus three anonymous readers. Major sections were very kindly reviewed, to my great profit, by Daniel Fox, Guenter Risse, Wolfgang Sadee, Dale Smith, and the late Harry Marks. In addition, three separate chapters were vetted by members of the Ohio State University Medical Heritage Center Writing Workshop/Seminar. Helpful suggestions were made by them and many other individuals, including a graduate class of Christopher Hamlin's at the University of Notre Dame, who reviewed one draft of the book.

In the course of writing, I received many kindnesses from many librarians. For facilitating the illustrations, I owe thanks to a number of helpful permissions officers and archivists. Substantial efforts were made in all cases to locate and acknowledge possible copyright holders.

I also received notable assistance and support from personnel in the Department of History and the Medical Heritage Center at Ohio State University. I offer particular thanks for invaluable assistance offered by Kristin Rodgers in the Medical Heritage Center.

And always there was the constant, understanding support of Marjorie Burnham.

As these words are written in the second decade of the twenty-first century, health efforts are consuming between 15 and 20 percent of all expenditures in the United States. How could such a staggering volume of medical and health activities come into being in a society that originated with a mere handful of European settlers in a virtual wilderness four centuries earlier? This book offers a narrative of how that happened. It starts with the customs and problems that European settlers brought with them and then describes successive small steps followed by a series of new eras in which Americans developed their health and medical concepts, practices, and institutions. It was in the course of those transformations that health care changed from a personal or at most neighborhood concern into a gigantic system that engaged a sixth of the productive effort of the nation's 300 million people.

Medicine and health activities are people's responses to times when human bodies suffer illness, malfunctioning, and accidents. Typically people try to assist or cure themselves and others. Or they call in a healer (in this history, most often a physician) and employ customs and distinctive institutions to assist recovery. This book is designed to introduce the reader to the history of that kind of activity in America, from 1607, the date of the first permanent European settlement, to the opening decade of the twenty-first century.

As is appropriate in a book exploring a broad sweep of history, I focus on major currents in the mainstream. There are many side eddies and countercurrents along the way, explored in other publications by other historians. In the narrative that follows, however, I attempt to identify and suggest obvious main directions of the flow of historical developments. Moreover, the book reflects the reality of its subject: small beginnings (seventeenth century) end up in extreme complexity (twenty-first century).

Modernization

It is not just the volume of recent health care that is astonishing. Over the centuries, *how* Americans devised health care changed profoundly. For convenience,

I frame the major developments as a story of modernization, beginning with traditional health care and ending with the health care of a substantially postindustrial society. Much of the fundamental change began only in the late nineteenth century. It was at that point that residents of the United States undertook to institute what they later designated "modern" medicine and health care.[1] Modern health care was marked by *organizational* and *technological* innovation, both of which followed in the wake of industrialization. It is this process that historians, looking back, have described as modernization. I have used a general framework of modernizing as a convenience, not a theoretical statement, to suggest some of the coherence and continuity in this many-faceted history.[2]

The book starts with a description of traditional health care in contrast to and as a point of departure for modernization. Immediately, basic elements in the story of medicine and health appear. There was of course a sick person, but a second set of figures was already in place in society in 1607 and afterward: healers and caregivers. Being sick meant that the ill person related to other figures in society, the most notable of which, beyond family and neighbors, was the healer or doctor. As time passed, traditional belief and knowledge changed. So did customary ways of dealing with sick people. Indeed, the book ends just when the doctor-patient relationship shifted again, at the turn of the twenty-first century. And all the while, influencing and interacting with health care, changes were accelerating in American society, not just in science and industrialization but in those fundamentals of modernization, technology and organization.

The term *modernization* appropriately highlights the epochal transformations in the United States. A society and culture that in the very early nineteenth century was still traditional, local, and deferential gradually gave way to a dynamic, cosmopolitan, manipulative one in the twentieth century. In one view, the Confederacy, defeated at the end of the Civil War in 1865, represented the last stand of traditional society in the United States before modernity became dominant.[3] Moreover, in that same year Joseph Lister, in Scotland, began to use antiseptic means specifically to control infection in surgical wounds, a landmark event in medical history signifying the beginning of the end of a traditional framework for thinking about medicine and health.[4]

It is true that in the United States, health care initially was unusually deeply shaped by traditional ideas and customs, and so major modernization came somewhat later to medicine and health than it did to other areas in American society. The historian Michael Bliss, for example, writes about "the coming of age of modern medicine between 1885 and 1922." A model in which modernization displaced tradition nevertheless produces an informative fit.[5] And scholars have recognized that in medical history, technology and organization worked together, whether or not the term *modernization* was used.[6]

Epochs and Eras

I have arranged events after the mid-nineteenth century into a sequence of eras, each one characterized by a set of major new developments recognized by well-informed people at the time. These eras are of course not clear cut; in each new era much of the old continued. The narrative therefore begins with the traditional medicine that came into place in the New World beginning in the early seventeenth century. Tradition, ironically, has never been static. Yet traditional medicine, in which people in health activities looked back in time even as they made various additions and transfigurations, endured until the sequence of modern eras began in the late nineteenth century. Only with the traditional and local in mind do later events—including the persistence of familiar content and customs in succeeding, modernizing eras—make sense.

Part I opens with the first permanent European settlers in North America in 1607 and proceeds to show how two things happened. In the nineteenth century, the European medicine that Americans continually imported began to depart from tradition in one limited area and then another. Simultaneously, the institutions, beliefs, and folk customs that Americans had clung to evolved, persisting but also, here and there, beginning to adapt. Nevertheless, basic patterns of thinking about disease and therapy lasted for 250 years, from the seventeenth century until the mid-nineteenth century. It is at that point in the story that it becomes clear that a major break was impending. In a new epoch, which begins with part II and the 1880s, one distinctive era of modernization in the history of medicine and health began to succeed another, chronologically, into the early twenty-first century. For each era, I sketch the overall patterns of change in health care and the adaptations that healers, patients, and social movers found necessary as circumstances changed.

The concept of eras within larger epochs should be a familiar one, but I have given a special reading to the process of how one era succeeded another.[7] That is, my narrative spans three grand epochs: traditional health care, modernizing health care, and, at the very end, a new epoch in health care distinguished by a shift in the doctor-patient relationship. Thus it is within the second, modernizing epoch, from roughly the 1880s to the 1980s, that the series of eras particularly clarifies what happened.

My idea of an era differs somewhat from the customary. In this book, I refer to an *era* as a period when contingent events caused people at the time to reorient their medical and health efforts because they had a sense that something new was occurring. Yet, ideas, practices, and institutions from previous eras and epochs persisted within the new orientation. Each new era was more complex than the preceding one. The era of surgery and germ theory, for example, contained and preserved content from the epoch of traditional medicine and care. The next era, physiological medicine, not only had distinctive new emphases and elements but also continued much of the old, namely, traditional ideas and customs, plus, now, the innovations that had accompanied surgery and germ theory. And so on through the eras of antibiotics, technology, and environmentalism. This additive model of succeeding eras explains how the story became so complex by the end of the twentieth century: each new era contained much from all the preceding eras.

Overview of the Book

The book is divided into four parts. Part I, as I have indicated, introduces the diseases, treatments, beliefs, and institutions that constituted traditional medicine and health in the United States from colonial days until about the 1880s. Slowly, early European settlers in America and their descendants and successors set up, expanded, improvised on, and then tried to adapt common patterns of health care known in many cases since ancient times. Over time, however, imports, local adaptations, and pressures of social change set the stage for successive eras of modernization even as people of that time were living fully in their own "present."[8]

Part II takes the main narrative into a new epoch, showing how modernization and the rise of biomedicine shaped medicine and health. It was at this point that the history of modern health care not only began but started to work out in successive eras. The eras in part II are defined by what people in each period perceived as the most conspicuous innovations in health care: exciting new developments in surgery, germ theory, and then physiology and, ultimately, antibiotics.[9]

After World War II, however, the character of change shifted. Part III covers the second half of the twentieth century, when the scale and complexity of the health care enterprise modified the nature of the eras and reframed the history. The story of modernization becomes more general with the overwhelming impact of technology and then environmental thinking on patterns of medical and health efforts and, simultaneously, the greatly expanded role of the federal government and other sociocultural shifts.

At the very end of the story, part IV explores a disjunctive turn and introduces a whole new epoch. Part IV contains a summary of one more era, genetic medicine, and then a short conclusion suggesting how a new version of the doctor-patient relationship had begun to transform the meaning of medicine and health in the United States at the end of the twentieth century. At that point, the narrative disappears into the obscuring mists of the present.

Emphases and Themes

In each era, recognizable health care institutions and practices and beliefs provide signs of change or familiar elements in the continuity of the narrative. They include sicknesses current at any time; people labeled "doctors" or "healers"; the popular image of "the physician"; nurses, trained and untrained; other paramedical personnel, such as midwives and laboratory technicians; medical treatment, including surgery and technology; folk medicine; commercial health services and goods; medical education; hospitals; public health efforts; medical research; books and journals communicating knowledge about health and medicine; payment systems and the economics of medicine and health; popularization of medicine and health beliefs; and political interactions with medicine and health care. In addition, there were common types of belief about how the body functions and the best ways to preserve one's health.

Not all of these institutional arrangements and common beliefs and expectations were significantly present, changing, or important at all times. Indeed, major change could be marked by the appearance, disappearance, or alteration of any one or any combination of them.

Trying to Recapture the Past

As a history written for the general reader, this book assumes that the reader wants to know not only how the present medicine and health apparatus evolved out of the past but why health care had, and has, a continuing, growing, and special place in American history. Most people will know at least fragments of the story. They are part of our culture. Moreover, many Americans are now relatively sophisticated in matters dealing with health and illness. For any reader, however, the story that follows should both answer questions and raise new questions about how Americans tended their sick and sought health over the past four centuries. Any national history carries with it an implicit comparison with the narratives from other nations.[10] Such comparisons constitute important histories in themselves, filled with wisdom. Unfortunately, they have to remain merely implicit in this book, in which I focus on a particular set of activities in a limited geographical and sociocultural area.

I have tried to show not only what happened but how people at the time were thinking about health and health care. Much can be learned from struggles against disease and accident in another time. The events and struggles that I describe were not the inevitable outcome of some forces or scheme. Unforeseen, accidental factors, particularly the courses of diseases and the introduction of new technologies, redirected the flow of events. Long-term pressures of many kinds, from economic development to religious, intellectual, and scientific activities, could profoundly affect the institutions of medicine and the ways of doing medicine. People at any time spoke and wrote about what they witnessed and perceived. Their voices can sometimes evoke the realities of that time in the past.

Finally, this history of medicine and health in America has a dimension that is usually greatly underestimated. For centuries, as I have pointed out, Americans borrowed and imported from Europe most of what they knew and did about health and medicine. Beginning in the 1920s, however, and especially in the second half of the twentieth century, many aspects of American medicine and health services served as models for health care elsewhere in the world, ultimately even in Europe, with its many powerful legacies and established institutions and regulations. Increasingly, students and professionals from around the world made pilgrimages to the United States and took home versions and at least parts of American practices. Over the years, models from North America often affected health care work everywhere also by way of the World Health Organization and other international organizations. Even a history of medicine and health limited to the United States, therefore, has another layer of significance for problems of sickness and health everywhere on the planet in the twentieth century and after— another story that I do not explore here but one that the reader of the narrative may infer.

In this book, the story cuts off, perhaps abruptly, four hundred years after the first British settlement in Virginia. From one point of view, the early twenty-first century is only a stopping place in a story the conclusion of which still lies in the future. My hope is that my sudden stop will serve as a beginning: that readers, trying to understand their own "present," will be moved to make further explorations into the past records of people who attempted to provide health care for themselves and, directly or indirectly, for other human beings.

PART I

Establishing and Nurturing Traditional Medicine and Ideas about Health

Around the turn of the twenty-first century, historians were taking great interest in how and why people living in Europe five hundred years earlier had developed sociocultural institutions and technologies that led them to explore and try to exploit much of the rest of the world. The debates among the historians have not yet been settled. But the fact remains that in the Early Modern period, just as Europeans were developing rationalism and what became science, a small number of them, beginning in 1607, succeeded in establishing permanent colonial settlements on the east coast of North America, joining a series of other New World colonies that stretched from the Caribbean to Canada.

Those settlers, and the growing stream of mostly British migrants who followed, brought with them the cultures of Europe, including traditions of learning, a rich and varied array of folk beliefs and superstitions, and a set of customary social roles and ways of carrying out the usual human functions that sustain life and society. The Europeans even brought with them a physical legacy of diseases and disabilities, accompanied by various kinds of healers who by custom helped families attend anyone struck by illness or accident. The first part of this book is therefore a history of the global transfer of biology and culture and, simultaneously, an account of the development of local and even provincial customs, institutions, and illnesses as one generation succeeded another and tried to replicate traditional beliefs and ways of doing things. Calling on the authority of tradition was not always easy, as local communities and conditions changed and as European models and ideas also changed from the seventeenth century into the nineteenth century.

By 1700 the traditional medicine that had come from Europe to the New World already had three elements. The most basic was a set of miscellaneous, unorganized recipes for "cures" for all of the pains and miseries that constituted sickness. Such folk practices often functioned as superstition or simple commercial purchases. The second element was a growing tendency of educated healers to think in abstract or systematic terms. In the eighteenth and nineteenth centuries, this habit of generalization produced a variety of medical systems that explained, often in imaginative ways, why people got ill and what illness was. The third element in traditional medicine was an attitude of practical skepticism that seemed to pop up now and again in human societies. These three elements—practices, theories, and practicalities—operated as only one part of a pattern of health care in which family and friends did the actual physical work of taking care of someone who was ill or injured.

In the period before the 1880s, on either the learned or the folk level people looked to the past for authority on what to do for a person who needed care. Social leaders all had received a classical education, and so it seemed natural to venerate traditional medicine proper, as it had derived from teachings of the ancients in the Greek and Roman worlds.[1] In medicine, the ultimate authority was Hippocrates, to whom all authorities afterward referred, along with other ancients such as Galen and the commentators on all of them. In the seventeenth century, when Europeans began to travel to North America, medical writers were also citing later writers who had elaborated on the classical texts. Eventually the most important was Thomas Sydenham (1614–1672), who because of his emphasis on clinical knowledge and experience (as opposed to abstract theory) was known as the English Hippocrates—a complimentary designation that emphasized deference to established lines of thinking.

Thus, in the early nineteenth century medical knowledge was still traditional in that teachers looked back to earlier authorities, from the ancients to more recent writers who extended and expanded the teachings of venerated predecessors. Even when writing about "new" knowledge or "improvements and modifications" of medicine or "the advancement of medical knowledge" in the mid-nineteenth century, American medical writers repeatedly referred back to a traditional body of learning.[2] For the bulk of Americans, however, the effective authorities were family or other associates who passed on to them common sense and techniques and wisdom. Eventually those traditions too were perpetuated through publications and commercial marketing.

Part I ends when a critical number of people in medicine and health began to look to a new kind of authority and to new or changed institutions in health care. Even though Americans were often noisily independent and rebellious, especially in the early nineteenth century, in the areas of medicine and health they tended to hold on to older viewpoints and forms. Over a period of decades, however, both institutional changes in medicine and intellectual shifts on the part of some leaders opened the way for a different kind of belief in the science in medicine. Those figures mostly reacted in their own local ways to European developments. But that they did react makes them interesting and illustrates how quickly tradition eventually, in the late nineteenth century, gave way, at least in part, to other bases for health activities. Recognizing the power of tradition and the many paths of escape from tradition is essential to understanding that a new age in medicine and health began toward the end of the nineteenth century, without in any way devaluing the earlier efforts to seek and preserve health.

The society that initially got transplanted to coastal settlements in North America was aristocratic, paternalistic, deferential, heavily religious, and largely family-centered. People in populated areas knew their place in society. In that New World colonial society, it was literate and learned elites who carried formal, traditional medicine across the Atlantic. But, confusingly, they and everyone else carried the folk practices as well. At the same time, settlers had to confront the wilderness and the people who were already living there, who had very different cultures.

In 1776, thirteen of the colonies separated from Britain and won independence in a war that ended in 1782. After a second war, from 1812 to 1815, the residents of the new United States turned markedly less often to Europe and more to looking after their own expanding country. That country was meanwhile developing a distinctive, relatively prosperous society with substantial regional differences. In the early nineteenth century, physicians and folk healers alike were encountering enormous changes in trade, transportation, and material circumstances as the population spread across the continent toward the Pacific coast. People became less deferential and more self-confident. Middle-class, bourgeois standards and leaders came to dominate the growing urban centers, as well as the still largely traditional rural countryside.

The best way to understand what happened in health care is to follow those whom local populations accepted in a healing role, whether formally educated physicians, generally educated people who knew the learning of the day, mid wives, or just wise women and men in the neighborhood. But everyone visited the sick. At such times, which were frequent, they might try to help but also prayed and offered advice, a key means of perpetuating and spreading folk knowledge and belief.

As the colonies turned into political territories, states, and a nation, some formal institutions appeared. Here and there, physicians formed organizations that over time spread locally and, by the 1840s, nationally. Occasionally, citizens founded a charitable hospital. In the late eighteenth century the first medical schools appeared. Doctors started publications, and in the nineteenth century, the number of journals devoted specially to medicine grew rapidly. As commercially marketed medications gained powerful traction after 1820, pharmacy and dentistry became separate fields, parallel to medicine. Nursing gained an identity particularly during the Civil War (1861–65). By the time medicine and health entered a new epoch around the 1880s, basic institutions were in place that would still be familiar, at least by name, more than a century later. Moreover, most of the population had concluded that physicians with a medical degree were an important, distinct group in society.

For more than 250 years after 1607, before the late nineteenth century, members of the rapidly growing population perpetuated their beliefs and healing practices. At the same time, social leaders, including an elite among the healers who were recognized as "doctors," continued to introduce new medical ideas and practices from abroad. And sometimes ideas and techniques were modified as they slowly percolated to various geographical and social groups in North America.

The political and economic independence and growing commercial and military power of the United States were not matched by intellectual independence. Very little that later would look fresh and interesting in medicine and health came out of the new nation. The transit of health care practices followed a model familiar from colonialism everywhere: local residents accepted new ideas from Europe but simultaneously adapted them to meet local circumstances. Early nineteenth-century British books reprinted in the United States were heavily edited or annotated to make the content conform to local botanical and social resources.

In medicine, one can define shifts in the transatlantic flow of ideas by tracking which European universities elite American doctors chose to attend when seeking advanced training not available at home. In the eighteenth century, they usually traveled to Britain, and especially Scotland. Beginning around 1820 there was a French period, and then after 1850 a German period. Only a small part of the population, however, consulted the practitioners with European training. Instead, facing the challenges of sickness and injury, ordinary inhabitants depended on themselves, on their families and neighbors, on advice books, and on a wide variety of healers and doctors—who had much or little or no training, typically drawing on traditional learning, custom, and lore. The generations of the 1850s and after, however, came to know a level and type of science that was substantially different from that of their predecessors of even a century earlier. But it would be another generation or two after the 1880s before intellectual dependence on Europe finally tapered off.

North America changed greatly from the seventeenth or even the eighteenth century to the mid-nineteenth century. All along, people could see that material alterations were taking place, signaling a less obvious but perhaps more pervasive transition in society, culture, and belief. Most articulate Americans eventually developed some optimism about the future. In various areas of health and medicine, a sense of inevitable progress became a rhetorical commonplace. Yet actual changes from before the Revolution to after the Civil War did not always validate such an interpretation. Events did, however, show how Americans of all varieties struggled with both natural and social factors that affected their health and how those attempts led to the perceptions and actions of later generations in matters that were indeed of life and death. Very few, however, were prepared for the onslaught of modernization even as they set the stage for it.

Chapters 1 and 2 introduce traditional practices and beliefs from the seventeenth and eighteenth centuries and the specific environmental and health challenges with which the American colonists and their descendants had to deal. Chapter 3 explores how the traditional ideas and institutions changed and, most importantly, did not change in a time of tumultuous social and economic transformation. And chapter 4 explores the persistence of tradition in medicine even as the forces of modernization began, here and there, to break in and facilitate the transition to a whole new epoch.

That new epoch could come, however, only after what K. Codell Carter characterizes as "the collapse of traditional medicine" in the middle decades of the nineteenth century. As a first step, leading practitioners slowly stopped using strong interventions and just let nature take its course. Other scholars describe how most of the population tended to embrace and defend traditional health and healing practices in either commercial or folk versions in the first half of the nineteenth century. Ordinary practitioners, in the face of attacks from skeptical citizens and from popular medicine and alternative practitioners and doctrines, resisted new ideas and techniques. Yet among growing sectors of well-educated physicians and other elites, slowly and with hardly anyone's noticing, the whole idea of disease and the cause of disease shifted from symptoms and multiple possible causes to many single diseases and, finally, each with a single cause, in line with new scientific thinking.[3] It was in this way that around the 1880s, in a new epoch, Americans unmistakably abandoned substantial parts of traditional thinking and therapy and came under the influence of modernization, an extended process that operated especially as society in general was being transformed by industrialization and urbanization.

Landmark Dates

1492	Columbus makes first voyage to the Western Hemisphere
1607	Europeans establish first permanent settlement in North America: Jamestown, Virginia
1620	European settlement in Massachusetts begins
1708	First American newspaper ad for a commercial medical preparation published
1721–22	Boston controversially initiates inoculation for smallpox
1734	First American medical-advice book published: John Tennent, *Every Man His Own Doctor*

1736	First law regarding physicians passed (Virginia; lapsed after two years)
1730s	Cinchona bark imported to North America
1751	Pennsylvania Hospital founded in Philadelphia
1765–66	First medical school founded: University of Pennsylvania
1775–82	Revolutionary War
1788	First book of original medical contributions printed in the United States
1789	Washington inaugurated as first president
1797	First U.S. medical journal founded: *The Medical Repository*
1799–1800	Introduction of vaccination
1812–15	War of 1812
1820s	Some American physicians start going to France for advanced study
	Accelerated economic development and transportation revolution (canals and railroads) begins
1832	First cholera epidemic
1846	Use of surgical anesthesia introduced
1847	American Medical Association founded
1849	A regular medical school graduates a woman: Elizabeth Blackwell
	Virtual ending of physician licensing laws, leaving medicine to the free market
1850s	American physicians turn to German-speaking universities for advanced study
1861–65	U.S. Civil War
1866	Last great cholera epidemic in the United States
1871	German-style laboratory of physiology founded at Harvard
1872	American Public Health Association founded
1872–73	First formal nursing schools founded
1878	Yellow fever epidemic takes a significant toll
1879	*Index Medicus* begins
1886	Association of American Physicians founding signals new research elite

CHAPTER 1

Health and Disease in a Land
New to Europeans

For thousands of years North America was inhabited solely by Amerindians. Then, beginning in the seventeenth century, Europeans settled on the eastern shore of what would become the United States. They increased in numbers at an astonishing rate and began to spread across the continent. Nature furnished them a rich but often deadly environment, for diseases carried away large parts of the new population. This chapter describes how the first European settlers and the first generations of their descendants and successors viewed their bodies, encountered and thought about different kinds of diseases, generally suffered ill health, and tried to protect their settlements from sicknesses.

The Shock of First Arrival

The Europeans who came to eastern North America immediately interacted with the countryside—the physical environment—and the native inhabitants. The geography of the new land varied widely, from rocky New England to the swampy coastal plains in the South. Even the climate was different from what they had known. Europeans were accustomed to moderate weather patterns, held relatively steady by the ocean currents. In the New World, however, they found the weather shockingly more extreme and changeable, and a large number of them died from what was later called "exposure."

When the new settlers appeared and survived, the consequences for the health of the Amerindians were enormous. Those populations presented the Europeans with a great contrast in culture and custom and also differed widely among themselves.[1] No one, however, expected the general and disastrous health effects that followed the arrival of the colonists, who brought diseases with them from Europe. Scholars now estimate that in the whole Western Hemisphere seven-eighths or more of the Native American population perished.[2]

For generations before and after the first permanent settlement at Jamestown, Virginia, in 1607, European promoters of emigration described the new land and climate as healthful. In fact, however, settlers faced great dangers to their health. It was not just heat and cold that caused suffering. The first colonists learned to associate health problems with swamps and areas of bad air and water, associations that had origins in Europe and persisted for centuries afterward, as newcomers moved into western areas and met with the same health problems encountered on the coast.[3] One of the traditional themes in American history has been how the frontier experience was replicated anew each time people moved west. So it was with coming to terms with health issues as the population spread across the continent.[4]

In the seventeenth century, settlers believed that each person had a bodily "constitution." Illness was not so much a specific entity as an alteration in one's inner constitution.[5] That individual constitution was like the land, responding to changing seasons, producing different products, experiencing times of want and plenty. "[Being] close to nature" and "[identifying] with nature" were not just expressions. Many European (and, later, American) thinkers came to believe that it was good to get close to nature and follow "natural ways." Others, particularly by the eighteenth century, held that it was nature that was the source of disease and dangerous things. For centuries, this tension continued between the idea that nature was good and the assumption that nature was dangerous, something that needed to be tamed and controlled.

The Seasoning

From the first settlements, Europeans' initial contacts with the new country brought on a common health experience known as "the seasoning." Upon landing, newcomers got sick. Later, as people moved west, they had the same experience in any new geographical setting. In the nineteenth century, people still spoke of the relationship a person's body had with a "particular atmosphere, and the various objects which surround" it. A change in one's setting, in the seventeenth century or the nineteenth, would likely affect one's health.[6]

In 1610, when Lord De la Warr arrived in the new Jamestown colony, he suffered the usual seasoning, starting, he reported, with "a hot and violent ague," or fever. Then, amidst recurrences of the fever, he had a debilitating "flux," or diarrhea, with cramps and what he believed to be gout and scurvy, all of which left him weak.[7] More than a century later, in 1723, another Virginia newcomer described a similar experience in a letter back to Scotland (read *y* as *th*): "We had no sooner landed in this Country, but I was taken immediately with all ye most common distempers yt attend it, but ye most violent of all was a severe flux of which my uncle died. . . . All that comes to this country have ordinarily [*sic*] sickness at first wch they call a seasoning of wch I shall assure I had a most severe one."[8]

From the seventeenth century to the nineteenth, each time mobile elements in the population met new and challenging climates, many of the newcomers did not survive the encounter. In the early nineteenth century, the famous author Timothy Flint could still note laconically that "emigrants generally suffer some kind of sickness, which is called 'seasoning.'"[9] Historians have guessed that the most common recognizable diseases in the seasoning were typhoid, dysentery, and malaria.[10] A substantial percentage of newcomers did not survive the new conditions they encountered. The cause of death mattered little if at all.

The Amerindian Depopulation

Settlers also created new conditions for the people who had already been living in North America for untold generations, the Amerindians. Almost all scholarship on Native American peoples has become extremely controversial, but some basic facts seem to have survived. The various Amerindian tribes, or nations, already were living close to nature in 1607, many picking up the increased vulnerability to disease that comes with the transition from hunter-gatherer societies to settled agricultural communities, where pathological organisms found more chances to thrive and spread. Moreover, the Amerindian societies were already undergoing other major changes. Some groups were shifting their internal social arrangements, and some were conquering or displacing or trading with others. To all of these struggles was now added the sudden intrusion of Europeans, who brought not only guns and destructive alcoholic beverages but also diseases.

Much is unknown about exactly what happened. It is clear, however, that Native Americans who encountered Europeans suffered a stunning depopulation. In a period of less than fifty years in the eighteenth century, the interior tribes of the South below Virginia declined from 197,000 to fewer than 68,000, and their numbers continued to diminish after that.[11] Historians often quote William Bradford's vivid account of the impact of smallpox on a community near Plymouth in 1634:

> the pox breaking and mattering and running one into another, their skin cleaving by reason thereof to the mats they lie on. When they turn them, a whole side will flay off at once as it were, and they will be all of a gore blood, most fearful to behold. And then being very sore, with what cold and other distempers, they die like rotten sheep. The condition of this people was so lamentable and they fell down so generally of this disease as they were in the end not able to help one another, no not to make a fire nor to fetch a little water to drink, nor any to bury the dead.[12]

The immediate cause of the depopulation of the Americas was infectious diseases of many kinds. While Amerindians already suffered from many infectious diseases, such as gastrointestinal inflammations, tuberculosis, and pneumonia, enormous numbers of Native Americans were carried off by the whole host of

unfamiliar illnesses brought in from Europe: smallpox, measles, malaria, diphtheria, scarlet fever, typhus, and whooping cough in the seventeenth and eighteenth centuries, with more to come later. As late as the 1830s, malaria, probably
introduced by European traders, devastated the populations of Amerindians in
California. Of course these same diseases also reduced European populations on
both sides of the Atlantic, but not nearly to the extent that the Native Americans
were affected.[13] In addition, aggressive colonial policies and warfare against Amerindians caused many deaths and weakened the social structures that provided nutrition and care for the ill. A few intact communities socially isolated
from Europeans managed to maintain their populations despite the incursion of
epidemics.[14]

Experts differ regarding how isolated and genetically homogeneous the pre-1492
or pre-1607 Amerindian population was, decreasing their resistance to infection
compared with the more diverse European populations. There may also have been
levels of malnutrition and, particularly as the disruptive Europeans moved in, social stressors that weakened the immune systems of large numbers of people. The
fact remains that once diseases (especially European virus diseases) started affecting the Amerindian populations, social organization periodically broke down,
leading to further opportunistic disease, lack of care, and malnutrition. People who
became ill with smallpox and did not receive adequate water from their kin or
others often died. Meanwhile, severely ill people could not muster the strength to
produce adequate nutrition for what might be left of the community. A letter
from South Carolina reported that as smallpox and yellow fever decimated the
European settlers and their slaves in 1698–1700, one Indian nation was virtually
wiped out by the diseases, leaving only a few, who "ran away and left their dead
unburied lying on the ground for vultures to devour."[15]

British colonists did not at first recognize the vulnerability of the Amerindian
population. Many Europeans believed that Indians enjoyed unusual, even remarkable levels of health and physical well-being. This initial impression persisted as
colonists tried to learn about the Amerindians' medicines and health practices,
which the newcomers believed might help them adapt to the land. But soon the evidence of depopulation complicated the picture, as colonists witnessed the results of the terrible diseases that followed the Amerindians' contact with the
Europeans.[16]

Disease and Death in the Colonies and the New Nation

The progress of diseases in North America was often obvious and impressive. At
the end of the colonial period, between 1775 and 1782, for example, a great smallpox pandemic spread across North America and other parts of the Western Hemi-

sphere. First noticed in Boston, the disease spread through the armies fighting the Revolutionary War. What is most remarkable about this pandemic is that it affected both European and Native American populations. As in all visitations of disease, some populations suffered more illness and death than others, but clearly all the people, even in the Louisiana Territory and extending to the far West Coast, no matter how long they or their ancestors had been in the New World, could and did suffer from this wave of an obvious, dangerous, highly infectious disease.[17] It was a remarkable symbolic manifestation of the changing social environment and the persistent threat of infectious disease. Of course there were always exceptions to any patterns of death and disease based on individuals' genetics and life histories ("constitution" again), exposure, and social circumstances and stressors. The fact was that people in both Europe and America in the seventeenth and even eighteenth centuries lived in physically stressful conditions and typically were disabled frequently by illness and miserable with cold, malnutrition, and sores.

Deaths occurred very frequently. Even after 1800, people did not live long. Conditions in New England were the most favorable, but even there as late as the eighteenth century the life expectancy at birth was only nineteen to thirty-three years owing to the usual high rate of infant mortality. At the age of five, one could typically expect to survive to the age of thirty-eight to forty-nine. If one reached thirty, one might live to be fifty-two to fifty-seven. Those who survived to age forty could expect to live into a decent old age, and those few who made it to seventy could often reach eighty. Everywhere, infectious diseases carried away children and young adults at a very high rate, while older adults died from a wide variety of diseases and conditions. The death rates were much higher in the southern colonies.[18]

In the colonial period, the main diseases were two: "fevers" and "fluxes," or diarrheas. People of that time did not see many "diseases" that were distinctive in the modern sense. Instead, they saw these two large categories of disease: fever and flux. They witnessed how they and others developed symptoms that they believed depended upon their interactions with the land and climate. By the seventeenth century, Europeans could in a few instances discern groups of symptoms occurring together in a sufficiently distinctive way that they constituted particular, named diseases. Moreover, between 1600 and 1800 the number of such groups of symptoms increased, so that very sick people could complain that they were suffering "a complication of disorders." The most feared were spectacular epidemic diseases, like smallpox and, in the nineteenth century, cholera, which came unannounced and affected many people. But many diseases were quiet, chronic, or endemic (i.e., always present in the local population).[19] In New Hampshire in the eighteenth century, "consumption," or tuberculosis of the lungs, was the most common single

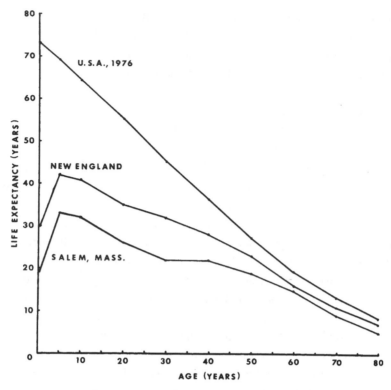

Life expectancy for people of different ages in late eighteenth-century (the two lower lines) New England (rural) and Salem, Massachusetts (relatively urban). The added upper line, from the late twentieth century, suggests the contrast between life courses and expectations in colonial America and today. J. Worth Estes, "The Practice of Medicine in 18th-Century Massachusetts: A Bicentennial Perspective," *New England Journal of Medicine*, 305 (1981), 1040. Reproduced by permission, copyright Massachusetts Medical Society.

attributed cause of death, although that diagnosis could sometimes cover other respiratory ailments.[20]

Fevers were at first the most common and most often fatal obvious diseases, and they could occur in either epidemic or endemic forms. Most common contagious diseases, such as those of childhood, induced a fever that was evident to anyone who touched the patient. Hence the category "fever" made sense (it would later be thought of generally as infectious disease). There were different kinds of fevers, and a person's individual constitution could incline him or her toward certain types of fevers.

Many a patient could detect a fever based on feeling chilled. Today, because of common drugs like aspirin, few people in developed countries have witnessed the full course of a serious fever. In addition to manifesting a hot skin and a flushed countenance, the patient can become delirious. Delirium typically involves loss of accurate perception and reason. Sometimes the patient shows a characteristic picking at the bedclothes and often a thrashing about. If the patient survives, he or she afterward feels exhausted and weak.

Common Contagious Diseases

From Europe the colonists brought measles. An epidemic of the disease was recorded in Boston in 1657, but it could appear anywhere. Indeed, it was one of the most contagious diseases. The patient first showed symptoms resembling those of a cold, then "fever, headache, thirst, and restlessness. On the fourth or fifth day, the characteristic eruptions," or spots, appeared on the skin. A child affected at age eleven in 1719 recalled later that the disease had been serious: "I was visited with a Low fit of sickness beginning with a fever and attended with the Meazells, and after that with great weakness and infirmities as also great pain."[21]

As with other diseases, when settlements were scattered and transportation difficult, outbreaks were often limited. With the growth of population and improved transportation in the eighteenth century, epidemics of measles were more likely to spread throughout the colonies. Measles later seemed a largely mild, childhood disease, but especially in adults some strains had many complications and a high mortality rate. In 1729, measles was the leading cause of death in Boston. In New London, Connecticut, in 1714, forty-five out of one thousand people with measles died.[22]

A nother disease that affected mostly children was whooping cough. Residents of the colonies began recording lamentable attacks in the mid-seventeenth century, and it was common, but apparently not very alarming, throughout the eighteenth century—even though it carried off many children and could last for several weeks in the many more who recovered. The distinctive "whoop" in the cough was often followed by vomiting. Children with this highly contagious illness became worn down by weeks of coughing. Mumps, another disease affecting chiefly children, was also recorded throughout the colonies. Although mumps could produce serious complications, the colonists seemed to take it in stride.[23] The reason that these distinctive but common varieties of "fever" caused so little comment was simply that other illnesses were much more serious.

Diseases such as measles and whooping cough have always been relatively distinctive, but historians have great difficulty in understanding what most diagnoses of that time might mean. The throat distemper, or putrid sore throat, appeared relatively late in the colonies (probably described first in 1659), and people came to fear it greatly in the eighteenth century. Later scholars have identified diphtheria and scarlet fever as the probable culprits, and some physicians at the time seemed to distinguish between the two. Scarlet fever was a very serious sore throat (much later traced to strains of streptococcus) producing scarlet-colored infected areas that could become ulcerated. Diphtheria, which had a dramatic later history, was an infection producing a membrane that could block the windpipe and choke the victim to death. Both diseases could easily be confused with other throat infections, but as early as 1736 one physician distinguished between "Cankery Ulcers in Glands or throat and mouth, with a glandulous fever," and an "inflammatory eruptive fever." Both versions of the throat distemper may have evolved between the seventeenth and nineteenth centuries to become more serious, but each, in whatever form, consistently affected children primarily. Diphtheria especially was highly fatal. Even in the colonial period, diphtheria carried off half or more of the children in many families. In Kingston, Rhode Island, in 1736, more than a third of the children in the community died in an epidemic.[24]

The most terrifying epidemic disease, however, was smallpox. In the eighteenth century, at least some strains of smallpox became markedly more virulent in both Europe and the Americas.[25] One historian described the classic symptoms as "a temperature of 103 degrees or higher, a quick pulse, an intense headache, vomiting, and pains in the loins and back. . . . On the third or fourth day the typical eruptions appear, usually coming first on the forehead and at the roots of the hair and gradually spreading over the body. The eruptions are dark red spots which eventually develop into papules, or pimples." These eruptions caused terrible scarring.[26] What such a description omits is the reality of the sickroom: "The patient often becomes a dripping, unrecognizable mass of pus by the seventh or eighth day of eruption. . . . The putrid odor is stifling, the temperature often high, and the patient in a wild state of delirium."[27] Smallpox came on without warning and had a high death rate. In 1731 a terrible epidemic of smallpox brought business in New York to a standstill and took the lives of 547 people out of a population of 8,620.[28] It is no wonder that people feared this disease.[29]

Two other major fevers, typhoid and typhus, were endemic among Europeans. Typhoid, it was later concluded, was spread largely by consuming human fecal material, and typhus by lice, but into the nineteenth century the two diseases were often confused with each other as well as with other fevers. Typhoid, characterized by violent gastrointestinal symptoms and body pains and sometimes a mild rash, had one distinctive symptom: a persistent fever. Hence it was called "slow fever," "burning fever," or "long fever." Stupor was also common. The death rate was about 20 percent among those afflicted. Typhus had a sudden onset of feeling ill, followed by a rising fever (hence the confusion with typhoid). After four to six days a rash

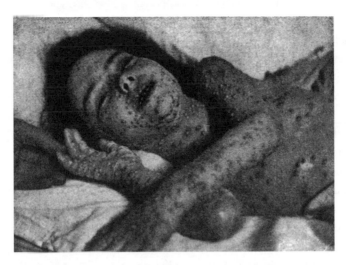

A very sick smallpox victim, six years old, photographed in St. Paul, Minnesota, c. 1900. She had running sores all over her body and face. She was fortunate enough to recover, but surely her face and body bore scars from the sores ever afterward. She was, in the language of the times, "pock marked." Frederick Leavitt, "Smallpox in Children," *Pediatrics*, 11 (1901), 92.

appeared. Although the role of lice in carrying typhus was not understood until much later, from the beginning of the colonies, the names given to typhus indicated that it spread in crowded conditions: "gaol fever," "military fever," "hospital fever," "camp fever." Except in such special locations, typhus was not conspicuous among Americans, although during the Revolutionary War town dwellers feared that "camp fever" would spread from the camps to the towns.[30] Typhoid, by contrast, held a significant and growing place among illnesses in the nineteenth century, especially after the Philadelphia physician William Gerhard helped distinguish between the two fevers in 1837.

Malaria and Yellow Fever

Two other epidemic fevers caused great concern, because they took many lives, especially among children, and often seriously impaired the health of victims who survived. What malaria and yellow fever had in common was marked seasonality and—as became clear only in 1900—contagion by means of mosquitoes (so-called mosquito vectors).

Malaria, which was endemic in both Europe and Africa in two major varieties, arrived in the colonies in the seventeenth century. Before 1800 the most virulent variety, which flourished in warm climates, was being reported in such places as the Carolinas. Americans came to know both varieties, although there were many exceptions to the typical patterns. In both strains, chills (often accompanied by intense headache and cramping, nausea, and diarrhea) were followed by several hours of high fever and sweating. Then the paroxysm would end. One kind of malaria had a typical seventy-two-hour cycle (quartan), the other a forty-eight-hour cycle (tertian). Some episodes would end after two weeks, but another episode could occur after that. Malaria could weaken many organs in the body, and, most significantly, if the malaria did not kill the patient, other infectious diseases typically attacked the now-vulnerable body—again often with fatal consequences for the patient and marked effects on death statistics.[31]

Malaria was a continuing problem, as people complained of their exhausting chills and fever ("ague") and other symptoms. Often victims would feel better, only to suffer another series of attacks later. Looking back, historians can see that some people connected their bad health to swamps and to summertime. One late eighteenth-century observer, for example, commented that "Carolina is in the spring a paradise, in the summer a hell and in the autumn a hospital."[32] But the patterns of incidence remained baffling. Over many years in the colonial period, the disease tended to leave the northern colonies and become more common in the South. Later it spread with the frontier far up the Mississippi River valley as the population there expanded.[33]

Yellow fever probably arrived in the Caribbean area from Africa in the seventeenth century. Many ships traveled from the Caribbean territories to North America, and there are records of yellow fever epidemics in trading cities, possibly Boston in 1693 and definitely Charleston and Philadelphia in 1699. Thereafter, Charleston, New York, and Philadelphia especially had violent epidemics until the mid-eighteenth century, after which few cases were reported until the 1790s, when epidemics started again, most notably striking Philadelphia in 1793. The combination of epidemics and fears of epidemics led later commentators to refer to the years 1793–1800 as the yellow fever era. In the early nineteenth century, yellow fever moved to the Gulf Coast port cities, especially New Orleans, and in 1878 one terrible epidemic spread to Memphis and St. Louis and may have taken twenty thousand lives or more. But the disease was sporadic and ended in North America around 1900. Although effects spread unevenly in the population, yellow fever caused terror, suffering, and disruption.[34]

The illness began with sudden chills, fever, and pain, particularly headache. The patient might recover after two or three days, or the temperature would fall below normal and the skin would take on the yellow cast that gave the disease its name. Vomiting followed, typically bringing up partially digested blood, which gave the disease another name, "the black vomit." Internal hemorrhaging and widespread internal damage sometimes brought death within days, sometimes more slowly.

The case fatality rate was often 20 percent and could run much higher. A letter from 1897 describes what it was like to care for a dying young woman:

> The poor girl's screams might be heard for half a square and at times I had to exert my utmost strength to hold her in bed. Jaundice was marked, the skin being a bright yellow hue: tongue and lips dark, cracked and blood oozing from the mouth and nose. . . . To me the most terrible and terrifying feature was the "black vomit" . . . . By Tuesday evening it was as black as ink and would be ejected with terriffic [*sic*] force. I had my face and hands spattered but had to stand by and hold her.[35]

Dysenteries and Other Diseases Known to Europeans

In the colonies in the eighteenth century, dysenteries, rather than fevers, became the most commonly recorded American diseases. The shift reflected in part a growing and more concentrated population who contaminated water supplies and strained sanitary systems, causing frequent infection of the digestive system. Walter Jones, of Virginia, wrote to his brother in 1769 from the University of Edinburgh that the subject of his medical thesis "is the Bloody Flux, which is not a splendid one, but as it is amongst the worst Disorders with which our Country is often afflicted, I thought it my Duty to Study it particularly." Later scientists found that a number of different biological pathogens could inflame the lower part of the gastrointestinal tract and cause sustained violent diarrhea, cramps, and mucous and bloody discharges (the "bloody flux"), all leading to dehydration. Malnutrition can also cause dysenteries. The death rate from any of these dysenteries was often very high, perhaps 10 to 20 percent of those affected in an epidemic.[36]

From infants to those in the prime of life, everyone suffered from dysenteries. The records and complaints began with the first settlers in 1607 and continued relentlessly into the nineteenth century. In 1618 in Virginia, there was "a most pestilent disease (called the bloody flux) which infected almost the whole colony," as one inhabitant noted at the time. Year after year, community after community suffered from a "bloody flix," or flux, although in one year it might cause more deaths than in another. The afflicted colonists told of being unable to function, in addition to suffering extreme discomfort and weakness. A missionary in 1763 reported that he had "not had one day, or one Night's Ease or rest these four Months past & . . . no less than fifteen times a night I am oblig'd to get up; and that accompanied by the Most excruciating pains in my Bowels, my back, my Loins."[37]

In addition to the fevers and dysenteries, colonists as well as Native Americans suffered from respiratory diseases, particularly pneumonias, colds, influenzas, and other afflictions often labeled at the time "pleurisies." These chiefly seasonal

This report of medical cases from a unit of the Continental army in the Revolutionary War from 31 July to 22 September 1779 illustrates the relative frequency of diarrhea and dysentery, as well as arthritic complaints, in the colonial population. The largest number of diagnoses were of "biliary fever" (a catchall category), 31; dysentery and diarrhea combined, 30; and syphilis (lues) and rheumatism, each 14. The number wounded was 33, all just in the first few days. Historical Society of Pennsylvania, Francis Alison Papers, DAMS #11679, by permission.

maladies were a familiar source of mortality. Pneumonia in particular came to be known as "old man's friend" because a victim could so easily and quickly sleep away into death.

While leprosy and plague did not make it from the Old World to the New in the colonial period, most European diseases did. As population increased, some other European afflictions, such as syphilis (which was reported in American seaports as early as the mid-seventeenth century), appeared and became serious threats.[38] Indeed, influenzas and other epidemics followed travel of any kind. Despite their distance from Europe, from the beginning Americans shared in world epidemics.

The most notorious was cholera. Cholera first came to North America in 1832 and visited again as epidemics in 1849 and 1866, with isolated local outbreaks occurring throughout the midcentury period. The epidemics constituted a widespread disaster that intensified nineteenth-century concern and debates about the causes and mechanisms of disease. The first cholera epidemic was already raging in Europe by 1831 and then came to North America via Canada as well as independently via ship to ports such as Baltimore. Like the other great terror of the period, yellow fever, cholera was not clearly transferred from person to person, and traditional quarantines, which were introduced everywhere and sometimes brutally enforced, were not notably effective. Even small settlements along transportation routes were affected, and before the epidemic died out the next year, from 50,000 to 150,000 people had died.[39]

A terrifying disease, cholera killed more than half of those who became infected in the nineteenth century. The patient developed violent gastrointestinal symptoms and was soon dehydrated and otherwise very sick. The suddenness with which the symptoms came on was terrifying. A man in New York in 1832 felt no premonition of the disease "until he pitched forward in the street, 'as if knocked down by an axe.'" And as in other great epidemics, bodies piled up in city streets, reminding people of the fate that awaited them.[40]

At the end of 1848, cholera came again and began to spread up the Mississippi River valley from New Orleans. After a brief earlier outbreak, by May 1849 the epidemic hit the East Coast. This time the growing railroad network helped spread cholera across the country, and wagon trains carried it even across the deserts to California with the Forty-Niner gold seekers.[41] Not until after 1854 did the disease leave the country again.

Season and Geography as Special Causes of Illnesses

From the beginning, settlers could see how illnesses derived from the countryside. John Tennent wrote in his 1736 guide to health:

Our Country is unhappily subject to several very sharp Distempers. The Multitude of Marshes, Swamps, and great Waters, send forth so many Fogs, and Exhalations, that the Air is continually damp with them: This, in Spight of all our Precautions, is apt to shut up the Pores at once, and hinder insensible Perspiration. From hence proceed Fevers, Coughs, Quinsies [sick sore throat], Pleurisies, and Consumptions, with a dismal Trail of other Diseases, which make as fatal Havock here, in Proportion to our Number, as the Plague does in the *Eastern Parts of the World.*[42]

This traditional belief in the causal power of the atmosphere carried over into the nineteenth century, and it underlines how early Americans connected common diseases to their local geographical environments, a theme that would emerge again in the late twentieth century (see chapter 11).

In the early nineteenth century in particular, observations and beliefs connecting illness to seasonal and geographical variation were not without some empirical justification, and they could appear to be commonsensical. The best example was a new, terrible disease, the milksick. Abraham Lincoln's mother died tragically of the milksick in 1818, for example. As a Lincoln family member wrote at the time, "We war perplext by a disease cald milksick." The illness was first described in North Carolina in the Revolutionary period but more definitely in 1809 and 1811 further west. Exactly as an environmentally based medicine would predict, the disease was known in only a limited geographical area, from the western Appalachians to eastern Missouri and from southern Michigan to northern Alabama and Georgia. It was a terrible and often epidemic disease that came only in the summer. It was so devastating that at one point one-fourth of the deaths in Medina County, Ohio, were from the milksick. The milksick indeed caused an entire community, Darlington, Indiana, to be permanently abandoned.[43]

The milksick affected mainly humans and cattle. In livestock it was called "the trembles" or "the slows" because of obvious signs of the animal's weakness, in addition to an alarming loss of appetite and abdominal distention. The animals could lose consciousness and die, especially if exercised. In humans "the symptoms include loss of appetite, listlessness, abdominal discomfort, severe constipation." Seriously lowered blood sugar levels and coma could follow.[44]

The milksick gradually diminished in incidence in the late nineteenth century, but only in 1928 did the cause become clear in the medical literature: acute acidosis from ingestion of a common weed, the white snakeroot. People who drank milk or ate butter derived from poisoned cows took in often fatal amounts of tremetol. Over time, with fencing and clearing of land, milk cows' access to white snakeroot decreased dramatically, and so did the incidence of the disease.[45]

The geographical distribution of the mysterious, devastating milksick of the nineteenth century, a seasonal disease occurring in specific geographical areas. Based on William D. Snively Jr. and Louanna Furbee, "Discoverer of the Cause of Milksickness," JAMA, 196 (1966), 1056.

In the meantime this important disease was produced by geography, season, and weather (in dry years cattle would wander into weed patches in search of food)— exactly what people of the colonial and early national days believed about many diseases.

Everyday Diseases

In addition to epidemic and contagious illnesses, malignant and degenerative diseases were always present in all populations, particularly among the aged. "Superficial cancers were recognized," as historian Richard Shryock remarks, "but heart and vascular conditions, nephritis, and so on were hidden behind such terms as fits, dropsies, and decay."[46]

Travel to the New World was itself often fatal. The accommodations on ships for both passengers and crews were crowded and foul, and provisions were typically inadequate or contaminated. Both biscuits and water were often filled with worms or insects or were otherwise unfit for consumption. On some ships, many people died, and on others, few. Fifteen ships arrived in Philadelphia in 1738. On only two of the ships were passengers in reasonably good condition; on the others, sixteen hundred passengers had died.[47] Ships often landed with almost all aboard seriously ill and weakened, their bodies unprepared for the seasoning experience that was to follow in the new land. The experience was intensified for people captured in Africa, who were enslaved and had to endure the notorious Middle Passage as, beginning in 1619, they were transported by force to the North

American colonies. In just the two to four months of that voyage, 20 percent on average died. That was, however, many fewer than died in African ports waiting to be transported. Many were weakened before and during the voyage by traders who fed them little to save money on food. And disease flourished in conditions even more crowded than on other kinds of vessels. The South Carolina physician Alexander Garden in the 1750s reported slavers' ships that "have had many of their cargoes thrown overboard; some one-fourth, some one-third, some lost half; and I have seen some that have lost three-fourths of their slaves." In general in the eighteenth century, the mortality rate on slave ships went down to perhaps 10–12 percent, by one count. The passage remained traumatic for the uprooted people and left them vulnerable to disease upon landing. In South Carolina, of those who survived the voyage, about one-third died within the first year.[48]

Women and Children

Among the diverse people who lived in the North American settlements, two population groups had special vulnerabilities: childbearing women and the infants whom they bore. Rates at which children survived to one year (infant mortality) or to adolescence varied widely, but at best children were always in grave danger from infectious diseases as well as from the usual accidents and other hazards of childhood. Infants, especially, often perished. In Virginia in the seventeenth century, half did not survive their first year. Later, in most areas a majority did make it past the first year, but recent studies show that infants in seacoast settlements did not survive as often as those inland, where perhaps only 6–9 percent of the children died in the first year. In one Chesapeake population, 20 percent of the infants did not get past the first year. The mortality rate for infants of nearby enslaved mothers, however, was perhaps 25 percent.[49] Of the privileged 808 children of those graduating from Harvard between 1658 and 1690, 162, or 20 percent, died in childhood.[50]

The rates seem higher when one reads the lamentable records of some families, sometimes silently displayed on gravestones. One Plymouth, Massachusetts, woman and her twenty children were buried together in the same grave. Cotton Mather recorded a mother with twenty-two children, "whereof she buried fourteen sons and six daughters."[51] The mournful list goes on and on. In some families many children survived. In others, most or all perished, virtually all from infectious disease, before reaching adolescence. Parish bills in Philadelphia for the 1780s show that about half of all deaths recorded were for children under ten.[52]

It is difficult for later generations to understand the centrality of childbearing to society in general and to women in particular in colonial and early national America. Even in relatively healthy New England, one out of five women died as

Glover children, Salem, 1784. A late eighteenth-century gravestone recording the deaths of three children from the same family suggests the constant presence of death in those times. The children were aged seven, three, and two years when they died, two of them in the same year, 1776. Harriette Merrifield Forbes, *Gravestones of Early New England and the Men Who Made Them, 1653–1800* (Boston: Houghton Mifflin, 1927), opp. 128.

a consequence of childbirth.[53] Much of an average woman's life was spent either preparing for the birth of a child or dealing with the immediate consequences. Judith Walzer Leavitt has used diaries to chart the lives of married women, such as Mary Vial Holyoke in New England. For twenty-three years after her marriage in 1759, Holyoke was pregnant perhaps 40 percent of the time and nursing another 25–30 percent; in only a third of the time was she free from childbearing. She had twelve children, of whom only three lived to adulthood.[54] In seventeenth-century Hingham, Massachusetts, Sarah Cushing bore twelve children and died at age thirty-eight; Sarah Hawke had seven children and died at thirty-six; Deborah Hobart Lincoln had ten children and died at thirty-seven. Other women in the community had different experiences. Ruth Andrews bore ten children and did not die until she was ninety-seven.[55] For a long time, however, women of childbearing age had a significantly elevated risk of dying. Many men buried more than one wife.[56]

Although women's place in society from the colonial period into the nineteenth century was quite restricted, they joined together as best they could to support one another in childbirth and childrearing. They gathered with the midwife at birth times, and they advised, assisted, and comforted especially during confinement and birth. To childbearing they brought customs, rituals, superstitions, and cures—herbs, potions, teas, whatever constituted the local practice. In populated areas, women typically developed mutual support systems that included nursing at times of illness as well as assisting in childbirth.[57] Their actions reflected a special world of women as well as the harsh social hierarchy in which, for example, in many areas men took their choice of food, while women had only what was left (perhaps sharing with children).

Life Expectancy and Population Growth

All the diseases that afflicted the population took a terrible toll in the colonial period, even if it was not comparable to that suffered by the Amerindians. There is a record of the experience of missionaries sent out from Britain by the Society for the Propagation of the Gospel All were young and apparently healthy. Between 1700 and 1750, sixty-two went to the Carolinas. After five years, twenty-seven (almost half) had either died or resigned because of poor health. In the perhaps relatively more healthful middle colonies, nearly a third were lost after only five years. In the South, by the early eighteenth century, the illness and death of clergy, most of whom had come from England, had completely undermined the government's attempt to make the Church of England the established church. Dissenting preachers filled the vacuum that disease created.[58]

Historians have found innumerable records of settlements in which a substantial percentage of the population died from just a single disease in a single year. Familiarity with death and expectation of death were part of life. Children then and later learned to recite each night the equivalent of "If I should die before I wake . . ." In the Chesapeake region, where children often lost, besides siblings, one or both parents, a planter recalled in 1698, "Before I was ten years old . . . I look'd upon this life here as but going to an Inn, no permanent being."[59] In Virginia, of the six thousand settlers who arrived between 1607 and 1625, forty-eight hundred, or 80 percent, did not survive their early encounters with fevers and other infectious diseases. Even though the chances of survival became much better within two or three generations, asking after a person's health was not just a politeness.[60]

Historians have figured that death rates in British cities by the eighteenth century were so high that only migration of population into the cities could sustain the communities. Similar analysis suggests that only the constant addition of migrants sustained parts of the North American colonies at different times. Even Philadelphia at times before 1760 did not produce births sufficient to sustain population numbers. In lowland South Carolina in the eighteenth century, the toll from malaria especially reached a point that the population was not self-sustaining. Indeed, the lower death rate for enslaved laborers from Africa, many of whom, it was later found, carried the sickle-cell trait that for some years helped them resist some types of malaria (but ultimately brought on a number of pains and weaknesses and early death), was a rationale for bringing in more enslaved people. In fact, the death rate among the enslaved, who were often malnourished, was frequently extraordinarily high.[61] In the seventeenth century, a major cause of death among all parts of the population was starvation or at least deficiency diseases.[62] Even in later, more plentiful times, diets were often wholly inadequate. Moreover, malnutrition could manifest itself in indirect ways, with serious symptoms such as swelling, pain, eye and skin problems, and loss of appetite.[63]

Altogether, however, residents in the North American colonies became more and more numerous. In the late seventeenth and eighteenth centuries their diets improved in quantity and variety. In addition, most of the population was rural and spread out far more than in Europe, so that, as noted above, people as a whole suffered relatively fewer urban or crowd diseases than in the Old World. Figures from the eighteenth century show that death rates were inversely proportional to the size of towns. In rural New Hampshire infant deaths were only 11–16 percent of all deaths.[64] Often, as suggested above, colonists had large families. Except for the rigors of climate and infectious diseases, plus the usual high costs of accidents and fighting, there was every reason for the population to increase. And increase it did, even as the population of the British Isles was actually declining in the early to mid-eighteenth century.

By the end of the eighteenth century, members of the Revolutionary generation, such as Benjamin Franklin, had already figured that after including the addition of immigrants and the subtractions effected by disease, the number of inhabitants was doubling approximately every twenty years.[65] Standard population estimates show the increase. In the British North American colonies that ultimately became the United States, the population went from about 2,500 in the 1620s to 52,000 in the 1650s and 155,000 in the 1680s. There were 655,000 colonists in 1734; 1,207,000 in 1754; and 2,205,000 in 1774. The first U.S. census, in 1790, showed a population of almost 4 million, almost half the number in the British Isles. Before the Revolutionary War, Philadelphia was the second largest city, after London, in the whole British Empire.

While the vast majority of American colonists lived in a rural setting, the growth of population did mean the growth of some towns and a few small cities. By the mid-eighteenth century, then, the disparity between health in the country and health in towns became significant. The death rate in Boston was perhaps three times that of the rapidly increasing but mainly rural population in New Hampshire.[66] In addition, there were remarkably higher survival rates for many northern communities compared with similar communities in the southern colonies.

The middle colonies of course showed a middle profile. Particularly in the South, people in settlements further west, toward the frontier, lived longer. Moreover, while death rates in Europe often reflected epidemics and starvation, the figures in rural northern America were usually more regular and lower.[67]

The people who lived a settled life in such rapidly changing social conditions not only had to deal with death and disability from disease but also had to establish everyday conditions of living, including what later generations included as part of health under the label "quality of life." In fact, the colonists' expectations were low. Everyone before the mid-nineteenth century expected to suffer from cold and heat. Even in young, healthy people, arthritic and rheumatic afflictions were remarkably common. In the northern climates especially, people kept the same clothing (wool and leather typically in colonial times) on for most of the year, with interesting biological effects. Bathing was long considered dangerous, although in the summer some people might go into local waters. Only at the end of the eighteenth century did baths become more common and personal cleanliness take on an additional dimension. One historian quotes a Quaker lady's

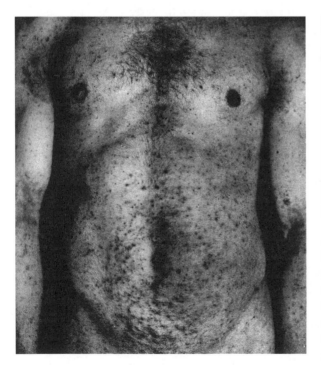

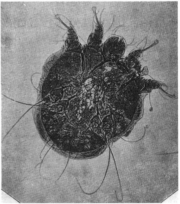

Left, a man with scabies. The inflammation is caused by an allergic reaction to a mite, *right* (greatly enlarged), burrowing under the skin. Scabies is a skin infection that causes nearly intolerable itching. Hence, "the itch." Henry G. Piffard, *A Practical Treatise on Diseases of the Skin* (New York: D. Appleton, 1891), 100, 101.

diary from 1799: "Nancy came here this evening. She and self went into the Shower bath. I bore it better than I expected, not having been wett all over at once, for 28 years past."[68]

Living in often dark, crowded quarters and never bathing led to any number of problems, most notably "the itch," or scabies, caused by a fecund, tiny insect that burrowed under the skin. Other parasites, including intestinal worms, were common, as were fungal skin diseases. The colonists always had an intimate relationship with flies, gnats, and mosquitoes, along with lice and other organisms. Everyone expected to be in discomfort to some extent most of the time, if not all the time. In addition to body pains, for example, a person's teeth were usually in very bad condition. In the late seventeenth century, a visitor to New England reported, "The women are pitifully tooth-shaken." And then there was constant intestinal upset from bad food and contaminated water. Lack of sanitation in any part of one's existence intensified everyday bad health and meant that those who abused their bodies with alcoholic beverages rather than contaminated water may have survived better, at least in the short run. Altogether, to have "the miseries" was a decidedly normal part of everyday life well into the nineteenth century.[69]

Changing Human Factors in Health Conditions

From at least the mid-seventeenth century, health and survival in the colonies substantially exceeded that for comparable populations in Europe. All testimonies indicate that diet provides the major explanation. There were flocks of wild fowl, abundant fish, and many game animals, in addition to much land for domestic animals, so that in contrast to European conditions, protein resources were available in all parts of the colonies from the earliest times on. Settlers also commented often on the wild berries, but scurvy and other deficiency diseases were still to be found not just on ships at sea but also on farms.

By the mid-eighteenth century, however, the growing population and better living conditions reflected a decided improvement in diet. One factor was probably the ever-increasing planting of orchards. Even vegetables came more frequently into the diet in different places and different times; overall, what people ate was decidedly better (from a later point of view) in the late eighteenth century than earlier, although typically heavy, greasy, and badly cooked, as in the hog-and-hominy diet that became common in many parts of the South. The great constant of ill health, malnutrition, which stunted development and reduced immunity to disease, was still found everywhere, especially in the winter, but substantially less so at the end of the eighteenth century.[70]

Geography and climate did not change over the generations. Yet as the population figures indicate, society did change, often in ways that affected the health of the inhabitants. Particularly as agriculture spread and towns grew, settlers changed the environment in which they lived. The most dramatic example was the clearing and damming of the swamps in the southern coastal low country to grow rice and indigo beginning in the late seventeenth century. The newly created wetlands were perfect breeding places for mosquitoes, just when people coming from Africa brought with them the more virulent malaria, along with its carrier mosquito species.[71]

Some Beginnings of Community Health

As they grouped together, the colonists not only attempted to cure any illnesses they suffered but also took steps to prevent and ward off disease. Many of the steps consisted in personal actions, such as swallowing tonics or undergoing seasonal bleeding just on general principles. Not least, in those days, was praying for one's own health or that of another. At the same time, communities took steps collectively to stem the spread of disease, particularly epidemic disease. Such actions were those of traditional European public health, based on the belief that some diseases were contagious and that others were caused somehow by local unsanitary conditions.

The historian John Blake has identified different periods in Bostonians' concern about epidemic diseases. From 1722 to 1775, they feared smallpox. The Revolutionary War created another era, when smallpox inoculation became common. And the years from 1793 to 1800, unsurprisingly, constituted the era of yellow fever. Smallpox and yellow fever instigated standard quarantine measures. In ancient times quarantine lasted forty days, the period in which any communicable disease was supposed to expire, although in practice there were many exceptions. Ships that might be carrying disease among the crew or passengers were forbidden to dock and were, for example, held in isolated places in the Boston harbor. In 1721, after rumors that plague had broken out in Marseilles, the Massachusetts General Court (legislature) ordered a quarantine of forty days for any vessel from the Mediterranean. Penalties for violating the quarantine could include death. Philadelphians and authorities in other port towns instituted similar measures to quarantine ships that might bring disease.[72]

Most of the towns depended upon trade, and cutting off transportation by means of quarantine was economically damaging. Therefore, authorities often denied that an obvious epidemic was present. Acknowledging the presence of disease could result not only in quarantine but also in much of the population's leaving the area.

Most people before the mid-nineteenth century believed that God sent epidemics to punish inhabitants for their wickedness. The clergyman Michael Wigglesworth wrote as early as 1662:

> Our healthfull dayes are at an end,
>> And sicknesses come on
> From yeer to yeer, because our hearts
>> Away from God are gone.[73]

In the face of an epidemic, fear for one's life and one's livelihood therefore made prayer the first line of defense. Public days of fasting, prayer, and humiliation frequently accompanied quarantine in a community's response to an epidemic.

By the middle of the eighteenth century, two developments had begun to modify and focus public health efforts. One was simply the more frequent concentrations of population, which gradually led to community regulations to make inhabitants take sanitary measures such as had been established in Europe, especially in crowded cities. The second was more concern about natural, as opposed to supernatural, causes of disease. In particular, people became increasingly concerned about the health effects of nasty and putrefying matter, a concern that continued into the nineteenth century.

It was the custom for people to throw all kinds of garbage and trash into yards and streets. Cleaning one's property and the area in front of one's property was the responsibility of the property owner. In the eighteenth century and after, in some towns scavenger services were available to haul garbage away. But inhabitants seemed unconcerned when rains caused privies to overflow, if indeed they bothered to use privies. Ponds, streams, and streets all served as sewers in villages and towns. Animal and sometimes human manure sat everywhere waiting for flies and rain to break it down. In New York City, some slums had lots so small that the people used basements in their houses for privies. Commercial establishments, particularly slaughterhouses, used public areas for discarded materials from butchering and created what everyone agreed were "nuisances." Visitors to early American towns commented on the awful sights and smells. Philadelphia for a while was nicknamed "Filthy-Dirty." And when an epidemic hit, it became customary, as in Europe, to order a general cleaning of the streets. When Nathaniel Potter, of Baltimore, in 1802 wanted to explain the "epidemic distempers" of the winter just past, he noted that it had been warm and that that had intensified the connection between filth and disease: "The gutters, docks, and other repositories of *filth*, emitted an effluvium little less offensive than that which is exhaled during the summer and autumnal months."[74]

Gradually, after the Revolutionary War, voluntary committees were formed to meet health needs (including burying the dead in an epidemic). Then officials in some larger urban areas acquired health-policing responsibilities. And officials also instituted economic regulations with health side effects, such as forbidding the sale of unclean and adulterated food. All such attempts at organized public health, however, tended to be local, sporadic, and ineffective. The very attempts, however, indicated the conditions that explained why concentrations of population brought relatively high death rates.[75] How Americans of the colonial period reacted when their health and that of their neighbors failed, however, did not change greatly, although, as will be shown in the next chapters, a few small innovations suggested the potential for powerful changes in the future.

CHAPTER 2

Traditional Treatment and Traditional Healers

When a person in colonial America became ill or injured, he or she would look to the past to try to find traditional means of treatment or cure, whether administered by oneself, by one's family and neighbors, or by a person who was designated as a healer, most typically a physician. Rationales for the treatment could vary, but recourse to traditional procedures persisted to a remarkable extent well into the nineteenth century.

Traditional Medical Practices and Thinking

Already in ancient times, a number of medical procedures were in place. Ways to patch up those injured in war or those who had broken or torn their bodies were conspicuously traditional, and classical-era healers used some surgical instruments of a kind still in use today. Standard procedures, such as mechanical repair of a dislocated joint, are also still part of medical practice. For other types of ailments in the seventeenth and eighteenth centuries, the legacy of the ancients and one's family and community was much more complex. Oral and written traditions shaded into folk beliefs and superstitions. Everyone, from the most learned to the most superstitious, had "recipes" for treating various ailments. Often, too, people improvised ways to prevent and cure disease. Occasionally some procedure or dose appeared to work and became labeled as traditional.

It is true that over the centuries, ideas, theories, and knowledge about the body had gone through different stages and formats. Regardless of rationales, however, basic methods of treating illness persisted as venerated tradition for two millennia not only in Europe but, for two and a half centuries after 1607, in settlements on the western edge of the Atlantic world.

In practical terms, then, a colonist who was ill would seek an appropriate medicine. Lists of "cures" came from learned treatises or from other writings, folk wisdom, or tradition in any form. Most typically, the format for medical treatment was the recipe, often recorded, for example, in a family Bible.

Some of the records we have include also the advice sent incidentally in letters. In 1643, Massachusetts governor John Winthrop received from Dr. Stafford, in England, medical instructions for the benefit of the people of the colony. Among the instructions for treating madness, burns, yellow jaundice, skin disease, and other afflictions were the following (read *ye* as *the*):

> For disease of ye Bladder,—Give ye partie to drink (if it be an inflammation heat of Urine;) emulcions made with barlie, huskt almonds, and ye 4 great cold seeds, if his drinke hath been strong before; but if small drinke and water, give him old Maligo and Canarie [wine], such to drinke Warme either by itselfe or mixt with Water; And applie to the region of his bladder, a poltis made with barlie meal, and ye rootes of Arum; make Injections of ye decoction of Hypericon, ye barke of a young Oake (the Outward black skin being taken off) and linnseede; and by God's grace he shall find present ease and cure with continuance. . . .
>
> For ye Bloodie Flix [flux, or diarrhea]; Purge first with Rhubarbe torrified; and give the partie to drinke twice a day a pint of this caudle following: Take a dragme of ye best Bole-Armoniak, a dragme of Sanguis draconis; and a dragme of ye best terra Segillata of a yellow colour . . . ; Make these into a fine powder, and with a quart of red stiptick Wine, the yolks of halfe a dozen eggs, & a quantitie of sugar.

These instructions sent from the Old World to the New illustrate the pattern of the practice and beliefs of traditional medicine. To modern eyes, the recipes were in no logical order. Each one stood on its own. They were simply in a list or collection of knowledge passed from one person to another. Finally, an escape clause was often included: these cures worked only with God's grace, or with God's help.[1]

Most, if not all, were biologicals, chiefly plant products but occasionally a "boyled Toade" or some such. Winthrop's son, John Winthrop Jr., who as an educated person in the mid-seventeenth century acted as physician without the title, often recommended, in addition to biological substances, many mineral ingredients, such as saltpeter, antimony, mercury, tartar, acids, sulfur, and iron.[2]

Applying Medications to the Body

Europeans and colonists alike believed that medicine should go into the gastrointestinal system for absorption into the body. Indeed, the first cookbook published in the colonies, a 1742 reprinting of a volume originally published in Britain and written by an English woman, Mrs. Smith, contained "above three hundred family receipts [recipes] of medicines; viz. drinks, syrups, salves, ointments, and various other things of sovereign and approved efficacy in most distempers, pains, aches, wounds, sores, &c."[3]

Colonials mostly took medicines orally. The "Injections" in the 1643 instructions, however, referred to a different avenue of administration, that is, injecting the medicine into the gastrointestinal system from the lower end. The usual term then was *clyster*. Clysters therefore offered an alternative route for getting medicines into the body. A favorite such medication of the time was cayenne pepper in oil. The clyster to deliver drugs was not the same as the later water enema, which was used to stimulate the bowels, although the goal of many, if not most, traditional medicines was to bring about copious bowel movements and sometimes vomiting. How else would one know if a medicine was efficacious if it did not produce such visible effects?

A number of other types of treatments were in use. Medicines were often applied to the skin not only in ointments but by other means. Soaking a hand or foot in water containing a medication would, it was hoped, permit a substantial absorption of the solution. Plasters and poultices (like the one of barley meal and arum root noted in the 1643 letter) could be applied anywhere. Often irritants were applied, such as cantharides (Spanish fly, prepared from certain dried beetles), which could raise a blister on the skin. Or people created artificial ulcers ("issues") by placing threads or a pea under the skin so as to cause presumably healthful drainage.

This early nineteenth-century pewter bleeding bowl, with a handle for convenience, was used to catch the blood in a bloodletting. Gradations for measuring the amount of blood taken were scratched into the sides of the bowl. Medical Heritage Center, Ohio State University. Photo by Kristin Rodgers.

To increase that drainage, the thread under the skin could, for example, be rubbed with an irritating ointment and pulled back and forth. Other ways of raising blisters or sucking imagined "peccant [unhealthy] matter" out of the body through the skin included a traditional therapy, cupping, in which paper, perhaps soaked in brandy, would be set afire and placed in a cup that was applied to the skin, creating a vacuum; and sweating was induced also to rid the body of bad substances. But above all, physicians and others who followed traditional ways recommended bleeding a patient.[4]

The attending physicians' account of the treatment of George Washington's last illness in 1799 is notable in part for showing that almost nothing had changed in the two centuries since the first colonists landed. Washington suffered an inflamed sore throat accompanied by chills and fever. He had himself bled by a local bleeder, but only after repeated entreaties did he call a physician. When the first one arrived, "foreseeing the fatal tendency of the disease," he summoned two consultants, one might think to share the blame. Meantime, the physician ordered

two copious bleedings; a blister was applied to the part afflicted, two moderate doses of calomel [a violent purgative] were given, and an injection [clyster] was administered, which operated on the lower intestines—but all without perceptible advantage; the respiration becoming still more difficult and distressing. Upon the arrival of the first of the consulting physicians, it was agreed, as there were yet no signs of accumulation in the bronchial vessels of the lungs, to try the result of another bleeding, when about thirty-two ounces of blood were drawn, without the smallest alleviation of the disease. Vapours of vinegar and water were frequently inhaled, ten grains of calomel were given, succeeded by repeated doses of emetic tartar [usually to induce vomiting] . . . with no other effect than a copious discharge from the bowels. The powers of life seemed now manifestly yielding to the force of the disorder. Blisters were applied to the extremities, together with a cataplasm of bran and vinegar to the throat. Speaking, which was painful from the beginning, now became almost impracticable; respiration grew more and more contracted and imperfect, till half after eleven o'clock on Saturday night, when, retaining the full possession of his intellect, he expired without a struggle.

He was fully impressed at the beginning of his complaint, as well as through every succeeding stage of it, that its conclusion would be mortal, submitting to the several exertions made for his recovery rather as a duty than from any expectation of their efficacy. He considered the operations of death upon his system as coeval with the disease, and several hours before his decease, after repeated efforts to be understood, succeeded in expressing a desire that he might be permitted to die without interruption.[5]

Washington underwent what was later called "heroic treatment," in which the more serious the disease, the more extreme the measures to cure it. At the time, many physicians thought the amount of blood Washington's physicians let was excessive. Regardless, some bleeding was expected.

Bloodletting

Bloodletting requires special comment, for it has been greatly misunderstood.[6] Bleeding a patient was standard therapy from ancient times, often undertaken by the patient or relatives without bothering with the intervention of a physician. Bleeding was done also for one's general health, for example, in the springtime.[7] In the case of illness, the bleeding ritual had an empirical, practical base. Someone with an infectious disease usually ran a fever. Anyone could see that the face flushed with fever indicated that the ill person had too much blood. Therefore, sufficient bleeding could reduce the delirious patient's frantic movements and cause the face to become less flushed, indeed ultimately quite pale.

Most good clinicians employed bleeding with restraint, and only in select cases, but even those practitioners swore by it as the physician's "sheet anchor." The usual technique was to tie a ribbon or cloth tightly around the arm about two inches above the elbow. When a vein became visible, one cut into it lengthwise with a lancet. The blood then ran down the arm into a bowl, often a special vessel with markings on the side to show how much blood was being drawn. Often blood was taken from the site of an inflammation or sore, such as the ankle. Things could go wrong. One could go too deep and cut the bottom part of the vein, so that it bled internally. Or if one were particularly incompetent, one could choose an artery instead of a vein. In ordinary practice, when enough blood had been taken, the cut was closed with pressure and a square compress of soft linen placed over it. If bleeding continued, cold water or astringent vinegar could be used to try to unite the two sides of the cut. When long, slow bleeding was needed, such as in rheumatism, leeches could be placed over the area afflicted.[8] Using leeches was in fact a major mode of bleeding.

Later generations have looked back at the practice of bleeding with some horror. Yet one should not be too quick to condemn the loss of some blood. Experiments at the end of the twentieth century suggested that a person could lose up to a third of the volume of his or her blood before the immune system was affected or cellular shock set in. Indeed, a small amount of bleeding was found to have an

This mid-nineteenth-century photo of bloodletting (venesection) shows the blood flowing off the arm from a vein that has been opened. The procedure remained the same over the centuries. Stanley Burns, MD, and The Burns Archive, New York.

anti-inflammatory and soporific (sleep-inducing) effect, not very different from that of aspirin and similar compounds consumed in great amounts by later generations. The effects of bleeding may explain in part the successes of earlier physicians who bled patients but were restrained in their use of the procedure.[9]

Therapeutic bloodletting was deeply embedded in European culture and health beliefs. As we shall see, the practice faded away in the mid-nineteenth century, but physicians and patients still swore by it.[10]

Judging Effectiveness

Did any of the traditional cures or therapies actually affect the course of disease? From the colonial period into the nineteenth century, people believed so, enough that they kept trying them. But they had formidable problems in judging medical measures other than seeing if a person vomited or moved his or her bowels or became less delirious. First of all, physicians and other healers depended on their impressions. Practitioners and ordinary people did have valuable experience from simply observing; none of them, however, ran controlled experiments. Much later generations, for example, knew a common ailment, the twenty-four-hour flu. If that disease existed in colonial times, and if when one first got sick, one took any medication, within twenty-four hours the person would be effectively cured, and the medication gained a good reputation. No one in that period did statistical tests to chart the patterns of illnesses that were relieved without medication.

Moreover, people then, as now, could also be misled by the placebo effect. Investigators have shown that if a physician (an authority figure) administers a medicine, *any* medicine, about one-third of the patients will report improvement. Indeed, the placebo effect is so powerful that only elaborate controls can even begin to filter it out. Both before and after the American Revolution, physicians and lay people did not have that modern perspective, and it is no wonder that effective therapies, in their eyes, existed all about them.[11]

There was a tendency to describe medications as specific cures for specific diseases, or at least for specific symptoms. Yet people presumed that general factors of constitution and environment always were operating. Most of the time, dosing of any kind did not work except to induce action of the gastrointestinal system. And of course in the age of fevers and fluxes, few specific diseases were identified. At the beginning, therefore, there were virtually no medical "specifics," one medicine for one disease.

What medical substances could produce were clear effects on the body, although not necessarily on the symptoms. Again, most of the effects involved action in the GI system (diarrhea or vomiting). People believed that some substances affected various organs, such as the liver, the eyes, or the lungs. Some dosing was designed to promote secretions or discharges, such as salivation, menstrual flow, mucous discharge, or sweating. Other treatments instigated drainage through blisters or running sores. At that time, medications that produced clearly detectable reactions were considered effective as such.[12] Any effect on the disease, however, was more complicated.

Prescribing by educated physicians in the colonial period and after is best understood by viewing it in terms of concurrently operating understandings of illness. Part of the medical approach was based in theory, part on the individual doctor-patient relationship. And nonphysicians adopted parts of educated physicians' ways of working and imitated parts or all of the physicians' style.[13]

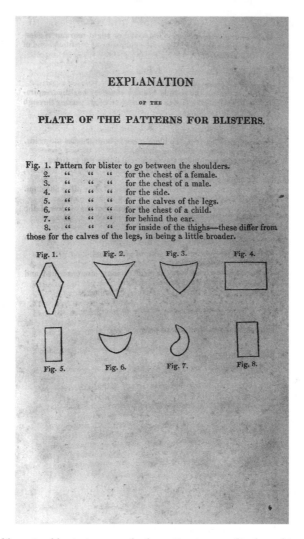

Blisters could be raised by irritants or by heat. In 1830, medical teaching included maps of the exact shape of blisters to raise on different parts of the body, as demonstrated on this page from a medical textbook. William P. Dewees, *A Practice of Physic*, 2 vols. (Philadelphia: Carey & Lea, 1830), 2:818.

Internal Balance in the Body

Much of the time, physicians and others were attempting to read the internal balance of the patient's constitution, or "system." They saw sickness as something personal and internal.[14] In traditional, classical medicine, the body had four humors (corresponding to the elements fire, water, air, and earth): blood, phlegm, black bile,

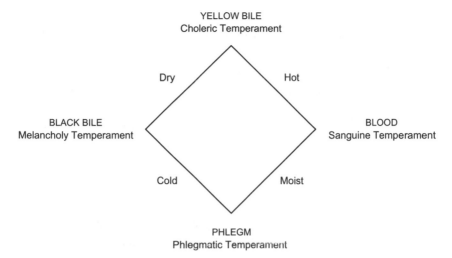

The ancients taught that there were four humors, each of which had two qualities. When a person was ill, the lack of balance between the humors (deduced from the symptoms) indicated the type of treatment required. Hot and wet symptoms, for example, required treatment to cool and dry the body, such as with astringent substances.

and yellow bile. When there was an excess or deficiency in any humor, a person was ill. Various types of disease reflected various kinds of imbalance. In the case of a common cold, there was, obviously, too much phlegm (the body was too cold and too wet and should be treated with heat and dryness). An excess, or plethora, of blood in a fever brought a flushed countenance (sign), sweating (too much wetness), and elevated temperature (too much heat). Reducing the excess of wet, hot blood by bleeding was therefore an obvious treatment. Even unlearned practitioners and laypeople thus consciously or by custom were following ancient, traditional teachings.

The second kind of adjustment to the patient's constitution that physicians attempted came from various Enlightenment theorists in the eighteenth century. They contended that healthy people had a proper balance in the flexibility and reactions of the blood vessels and nerves. If the vessels and nerves became too tense or too loose, the natural motions of the body were disturbed (as in bowel irregularities). The pulse could indicate whether the fibers in the blood vessels were too loose or too tense, or breathing could suggest whether the nerves were carrying tension or looseness in the windpipe and lungs. If the patient's system was weak, a stimulant was called for; if the system was overactive, a sedative. A person could die either from too much irritation and tenseness or from an ultimate relaxation and lack of irritability.

Each physician preferred to use certain substances to restore the natural balance of the humors or the correct tenseness and nervous reactivity. Records of practicing physicians in the late eighteenth century show, however, that despite their individual preferences, they tended to use many of the same old and new drugs much more often than they did other items in the lists of recipes available to them. Rhubarb and mercury preparations predominated among the many cathartics, and there were favorite additional purges and emetics. Peruvian bark (see below) and Virginia snakeroot were used for general tonics. Narcotics and sedatives included camphor and opium. Blistering (usually by applying cantharides or by cupping, as explained above) served as a counterirritant when there was too much irritation.[15]

By the seventeenth century, some drugs that did have clear effects assumed a central place in medical treatment. Perhaps the most notable was opium, which was used to fight pain, induce sleep, and slow (or stop) the actions of the gastrointestinal system. Another was ipecacuana, or ipecac, which became a standard emetic (to induce vomiting) and in smaller doses helped relieve coughing.

People also thought in terms of a "specific," that is, a substance that countered a particular condition. When a virulent disease most scholars believe to be syphilis swept across Europe after Columbus's sailors returned from the New World a second time, mercury, taken externally and/or internally, appeared to be a specific for that disease. Current medical authorities still disagree about how efficacious mercury was or is, but common wisdom from the sixteenth century reflected the belief that there was this one specific cure: "One night with Venus, and a lifetime with Mercury."

New Developments in the Eighteenth Century

During the eighteenth century, with new cures and preventives, the apparent efficaciousness of standard medicine actually began to shift slightly. To begin with, a few developments in Western medicine that we now think must have had some effect on some roughly identifiable diseases came to the colonies. The first was cinchona bark, often called Peruvian bark or Jesuits' bark, after the men who had found it effective against malaria in South America in the seventeenth century. There is a record of cinchona being used in Virginia in the 1680s, but general use did not begin until into the eighteenth century. So the tertian and quartan and other fevers of malaria could give way to a substance the active principle of which was in fact quinine (first purified in 1820). Of course it often was not understood as a specific, and many practitioners tried it on all "fevers" and then held it useless because it did not cure every one.[16]

The second new development was inoculation for smallpox, an innovation that very much involved the colonies. On 27 April 1721, a small British naval vessel from the West Indies, the *Seashore*, put in at Boston. Not until May did the selectmen of the town learn that a diseased person from the ship had come ashore. A case of smallpox showed up at the Paxton house, and two men were sent to stand guard over the house so that no one could leave or enter—a formal quarantine. A special cleaning of the streets was ordered, and the ship was sent to Bird Island to keep it isolated. By 27 May there were eight known cases in the town, and by mid-June there were so many that house quarantines were abandoned.

Early in June, Cotton Mather, who with his father, Increase Mather, was a chief minister of the town, circulated among the local physicians a scientific account of the process of inoculation that had been used in the Middle East to prevent the spread of smallpox. Inoculation consisted of placing into a cut in the skin a thread or a bit of cloth that had been soaked in a running sore of someone suffering from smallpox. The idea was that smallpox taken artificially would induce a mild case and render the person immune forever to the disease. Mather also knew that people in Africa resorted to such a practice. He had resolved to advocate it if smallpox ever came to Boston.

None of the doctors responded to Mather's plea until Zabdiel Boylston inoculated his six-year-old son and two servants on 26 June. After several anxious days, they all recovered. Soon Boylston inoculated three more people, and then ten more. But by that time a terrible controversy had arisen. Some physicians asserted that inoculation had bad effects on the patient. Others, and not just physicians, objected violently because the inoculated people were spreading the disease. Still others had religious objections. Mather, who was a highly educated person, was able to answer both religious and scientific arguments, but he came to believe that many of the "Enemies of Inoculation" were inspired by the devil.

Boylston took up inoculating again in August, and he continued into November, when the epidemic was beginning to decline anyway, even though the controversy did not. Eventually, from the smoke of the public arguments (a lighted "grenade" was actually thrown into Mather's home at one point) emerged an arithmetical test of this medical innovation. Out of almost 6,000 cases of smallpox in Boston, 844 people died, or about 1 in 6. Out of 286 people inoculated, only 6 died, or 1 in 46. These and later figures from other places finally resulted in a figure of about 1 in 50. Clearly, in one of the very first numerical tests of a medical procedure, inoculation offered substantial but not perfect protection from a greatly feared disease. It is easy to figure that Boylston had demonstrably (or perhaps luckily) saved more than forty lives in Boston.[17]

During the rest of the eighteenth century, other colonials saw and understood that the smallpox figures from Boston and elsewhere suggested that getting inoculated was a good idea, although opposition continued. In 1738 the Boston experience of innovation, opposition, and substantial acceptance was largely repeated in the Charleston, South Carolina, area. Eventually the fears of many informed people that inoculation would spread the disease led to a sensible social arrangement. A number of physicians set up inoculation institutions in isolated places, such as a country inn or, famously, a small island in Boston Harbor. A boat would carry the patient to this isolated site, where he or she would be inoculated and then spend several days getting sick and recovering. Patients passed the time reading and playing games, treating it as later people would a vacation at a lodge. Some young people met patients of the opposite sex whom they married. In any event, the practice of smallpox inoculation was widespread among some groups by the 1760s. This exotic procedure had taken a place alongside other measures from standard medicine and also local folk recipes. Instructions were printed in the newspapers, and on occasion ordinary people carried out the procedure independently. During the Revolutionary War, Continental troops were inoculated when smallpox appeared in their area. In 1777, George Washington ordered that all troops and all new recruits be inoculated. "Scarcely one who was inoculated died; whilst almost none who took the disease by contagion got well," reported one contemporary. Indeed, in the last years of the war Washington depended for medical advice on a physician who had earlier run an inoculation institution.[18]

Vaccination

The inoculation story ended, however, in 1799–1802. In 1799 Benjamin Waterhouse, a Boston physician and professor at the Harvard Medical School, read Edward Jenner's account of a new preventive measure against smallpox. Jenner, an English country practitioner, found that people inoculated with cowpox, an animal disease that had only the mildest effects on humans, were thereafter immune against attacks of smallpox. This relatively harmless form of inoculation represented a great improvement over artificially giving people real smallpox. After learning that leading British physicians supported the finding, Waterhouse arranged for some cowpox, or "kine pox," infectious matter to be sent to him. It arrived in 1800 by ship on threads in a sealed bottle. He immediately tried it out by vaccinating (from the Latin word *vacca*, for "cow") his five-year-old son and six servants. Then he tried to give them ordinary smallpox from patients who clearly had the disease. All were immune. The kine pox worked just as Jenner said it would.[19]

Waterhouse proceeded to publicize his willingness to provide the new procedure, but he did not distribute the vaccination material to other physicians—except some at a distance who paid a liberal licensing fee to have a local monopoly. Eventually other New England physicians obtained kine-pox "matter" from their British correspondents. Meanwhile, physicians elsewhere had taken steps similar to Waterhouse's. As early as 1799 a successful vaccination had taken place in South Carolina. Many practitioners were soon publicizing the procedure. Despite some controversy and some failures, by the mid-1800s the procedure was generally available, at least in urban centers, across the United States.[20]

Foxglove

Meanwhile, another specific drug had come into the new United States just before the end of the eighteenth century. A New Hampshire physician, Hall Jackson, read a new book (1785) by the distinguished British physician William Withering in which Withering reported on his work showing that a common plant, the purple foxglove, whose active ingredient turned out to be digitalis, could dramatically reverse a familiar, life-threatening condition, dropsy. Dropsical patients had accumulations of fluid that appeared chiefly as swelling in the limbs and abdomen. Twentieth-century experts have determined that such patients typically died from heart disease.

In 1786 Jackson, in Portsmouth, obtained seeds and preparations of purple foxglove directly from Withering, and he reported his clinical success to other American physicians. Jackson's success rate in relieving the dropsy was eight out of eleven, a most impressive record. "Mr. Joseph Norton of Kittery," for example, "had great oppression and difficulty of breathing with swelled oedematous legs and thighs, was relieved with the foxglove." Jackson was indeed dealing with a specific, although its full power did not become clear until a century later. But as a grateful correspondent wrote after his wife was cured by Dr. Jackson, "It appears to be provided and designed by the great author of nature as a remedy exactly fitted for such a disease." Withering's work also generated trials and adoptions in the middle and southern colonies. For a generation, foxglove was successful, but because practitioners of the nineteenth century demanded that it also cure tuberculosis of the lungs, the medicine's reputation diminished until modern digitalis was indeed confirmed as a specific.[21]

By the opening of the nineteenth century, then, two new drugs, cinchona bark and foxglove, plus vaccination, were taking physicians beyond the traditional therapies and preventives such as dosing, bleeding, and purging. Any new ideas or techniques, however, had to compete not only with recommendations from the traditional medical literature (particularly the doctrines of Galen) but also with an enormous number of other traditional and folk therapies and superstitions.

MEDICAL REPOSITORY.

VOL. I.—No. II.

ARTICLE I.

ON THE DIGITALIS PURPUREA.

By JAMES MEASE, *M. D. Refident Phyfician of the Port of Philadelphia.*

THE *Digitalis Purpurea*, or Purple Fox-Glove, is a medicine which, for fome time paft, ftood high in the lift of the Materia Medica, but, for various reafons, appears to have, in fome meafure, loft the character once formed of it. The frequent trials made by the judicious Dr. Withering, and the repeated fuccefs that has attended its ufe in his hands, certainly fhew, that the opinion he entertained of it, as a valuable addition to the ftock of medicinal plants, was well founded. Many other phyficians have likewife fpoken favourably of its good effects; but many inftances of failure have alfo been recorded, and in cafes, too, apparently favourable to its operation. I have feen it exhibited both in public and private practice, and have given it myfelf, fometimes with advantage, and at other times with little benefit, or obvious injury. The caufes of this different fuccefs have never been fully afcertained; but the refult of the prefent experience of this medicine, I believe, is rather unfavourable to its character. From my own reflections and obfervations, I am difpofed to attribute the want of fuccefs of *digitalis* lefs to its own inert quality, than to the circumftances attending its exhibition, which I fhall point out in the following paper; and I fhall alfo endeavour to afcertain the manner moft proper to exhibit the medicine, and the cafes moft favourable to its operation.

The caufes influencing the fuccefs of the *digitalis* may be referred to the following heads:

VOL. I. No. 2.   A

An article on the medicinal qualities of the purple foxglove (*Digitalis purpurea*) appeared in 1797 in the second issue of the very first medical journal published in the United States, the *Medical Repository*, which began publication in 1797. From the original in the collections in the Medical Heritage Center, Ohio State University.

Traditional and Folk Treatments

In actual practice, physicians and ill people did not distinguish empirical treatments (those based on personal experience) from treatments that were traditional in Europe or from new substances and practices from the New World. In addition, the colonists borrowed freely from the Amerindians with whom they associated, especially in the Chesapeake and southern areas, where in the seventeenth century there were significant numbers of Indians among the enslaved laborers.[22] Moreover, all population groups tried to integrate practices the people from Africa brought with them (such as inoculation for smallpox).[23]

In the colonial and early national years, all local traditions came together, because all of the inhabitants, new and old, practiced polypharmacy; that is, they used a variety of substances to treat any one disease in any one patient. It was easy to add new materials to such a practice, especially for Europeans, who had to make do with materials they could find or, rarely, import. Thus in the absence of Peruvian bark for fevers, some North Americans used the root of the local tulip tree.[24] European polypharmacy sometimes meant using a standard mixture, such as mithridate, which contained fifty ingredients, and "Venice treacle," which contained more than sixty, ranging from French lavender, wine, and red rose flowers to opium and anise seeds.[25] In America, the residents did not so often have standard mixtures available, and so they employed a wide variety of different substances in their simpler forms or made rough-and-ready substitutions. By the late eighteenth century, however, standard practice among trained physicians included remarkably fewer attempts at polypharmacy.[26]

How quickly the use of new "cures" spread is suggested by the fact that the first export from New England (in 1602, even before permanent settlement) was the bark and pith of the North American sassafras tree (known to moderns as a flavoring in root beer). So popular was sassafras tea as a medicine that in just the year 1770 England imported nearly seventy-seven tons of sassafras. Obviously, many later migrants from England would already have known about sassafras when they debarked on the western coast of the Atlantic, where sassafras originated and was popularly utilized. The famous Swedish traveler Peter Kalm reported that when sassafras became widely recommended for syphilis in the colonies, some English people were afraid to drink the tea publicly for fear of being thought diseased![27]

In practice, then, colonists, Amerindians, and transported Africans all might use any substance or procedure that someone recommended either orally or in writing. Often, too, they used plants available to them in ways they had seen or heard about. The same could be said of the religious and superstitious aspects of healing.[28] All found some patronage in the marketplace.

European colonists believed that the movements of the stars and planets affected one's individual constitution, and many learned people used astrology. A 1715 almanac recommended that for general health and balance in the humors, a laxative be taken "when the Moon is either in Gemini, Libra or Aquarius. Take a vomit . . . when the Moon is in Aries, Taurus, or Capricorn."[29] Most European settlers prayed for divine intervention when someone was ill. They also sometimes ascribed illnesses to witchcraft—as notoriously in Salem in Massachusetts. As one historian comments, "While Indians used charms for their ailments, white men bought magic stones for treating the bite of the mad dog."[30] It was therefore not a big step for Europeans to treat respectfully, if not to adopt, the many beliefs in conjuring and conjurers that came from Africa or were encountered on the Amerindian frontier. The Huron and the Iroquois distinguished between conjurers casting spells and those who were herbalists, although the same person might fill both offices.[31] The conjurers of African background also employed vegetable and other compounds. All in all, the line between folk superstition and folk medicine was indeed hard to draw. The fact that surviving records were all written by Europeans increases the difficulty of knowing exactly how folk medicine spread among American communities before the Civil War.[32]

Communicating Medical and Health Information

One thing the Europeans brought with them was communication of medical knowledge by printed materials. In addition to oral traditions passed on through families and communities, then, ideas about treatment could be conveyed in an impersonal, public form. The colonists therefore had a whole range of sources of advice on how to care for themselves and others. The closest were family, neighbors, and others in one's environment, who communicated in person. By the eighteenth century, newspapers and magazines began to appear and then multiplied. Health advice and "recipes" appeared frequently in these publications and in the widely read almanacs that printers produced, which included not only advice on how to live healthfully but also recipes for cures, many submitted by regular readers—alongside astrological advice about possibly dangerous seasons and when to bleed or purge for the preservation of health.[33] One title page reads: *The Virginia Almanack for the Year of our Lord God 1771 . . . Wherein are contained. . . . Approved Receipts for the Whooping Cough and Rotten Quinsy, an infallible Cure for the Distemper that has so long raged among the Horned Cattle, a noted Preservative both for Man and Beast, to keep them from Infection, a List of Drugs and Medicines necessary to be kept in every Family in the Country. . . .*[34]

There were even the beginnings of advertising for "remedies" that merchants offered for sale. The first newspaper ad for a commercial medication appeared in 1708 in the earliest newspaper, *The Boston News-Letter*: "Daffy's, Elixir Salutis, very good, at four shillings and six-pence per half pint Bottle; And good Hungary Water, at one shilling and six-pence per Bottle. To be Sold by Nicholas Boone at the Sign of the Bible in Cornhill, near School-Street, where any that want a quantity of either, may be supply'd very Reasonably." In this early number of the paper, there were only three other ads, two for slaves for sale and one for a lost horse.[35]

Caring for Invalids at Home

In a book designed for professional physicians, unless it was in the original Latin, the content was such that any educated person—or perhaps less educated—could read and understand it. The category of strictly professional literature, intellectually accessible only to physicians, did not develop effectively until well into the nineteenth century. By the late eighteenth century, however, there were many books deliberately designed to convey medical advice to lay readers. Indeed, the late eighteenth and early nineteenth centuries constituted the high point for medical-advice books. At that time, the bulk of medical care, the authors of the books assumed, would be carried out not by physicians but by those who increasingly needed information from popular writings. The first book written by an American for the guidance of Americans appeared initially in 1734: John Tennent's *Every Man His Own Doctor; Or, The Poor Planter's Physician; Prescribing Plain and Easy Means for Persons to Cure Themselves of All, Or Most of the Distempers, Incident to This Climate, and With Very Little Charge, the Medicines Being Chiefly of the Growth and Production of This Country*. Tennent, although writing within the standard medical ideas, made such suggestions of native products as the following:

> For spitting, or Pissing of Blood, *bleed* 8 Ounces. The next Morning, *purge* with *Indian Physick*; and drink nothing but *Tea*, made of *Comfry Leaves* or *Root*, and sweeten'd with *Syrup of Quinces*. But whenever a *Fever* produces Loss of Blood, the Heat of that must be taken off by *cooling Medicines*, before the *Bleeding* will cease.[36]

Such medical-advice books were not just the general moral, diet, and lifestyle treatises that had been common for years in Europe. Instead, these books incorporated specific recipes for specific illnesses or complaints, plus instructions for binding up wounds and other practical matters—exactly as one would find in newspapers, magazines, and almanacs. But the books were explicitly designed for home use, including, in the American setting, use on an isolated farm where a physician was not available.[37] Not least was one widely used early British compendium composed largely while the author, John Wesley, the famous religious leader, was resident in Georgia in 1736–38 and saw the special need for medical advice. Wesley's *Primitive Physick, or an Easy and Natural Way of Curing Most Diseases*, was first published in 1747 and over many decades went through thirty-two editions. By 1780 the book contained more than 900 recipes for 288 conditions.[38]

Late in the eighteenth century yet another slight difference suggested how medical care was changing. Amidst all of the recipes for cures that had come from various sources, a note of skepticism was driving out some of the most outlandish prescriptions, such as "powdered peacock dung in succory water" for epilepsy.[39] In 1784, Abigail Adams recorded in her diary that she was reading "Buchan Domestic Medicine. He appears a sensible, judicious and rational writer."[40] No doubt the Age of Reason had some influence, but so, too, did shifts in health care as more educated, and then more professional, healers appeared in the growing population of the colonial and Revolutionary eras.

As reflected in the advice books, from the seventeenth century through the nineteenth, the customary site for actual care of the ill was a person's home. There members of the family nursed anyone who fell ill. Sometimes friends or neighbors would come in to help. And sometimes an outside party would be called in, someone recognized in the community for special knowledge or talents in the art of healing—beyond the oral knowledge or knowledge from advice books available to the patient and family.

As noted in chapter 1, one of the customs that came from Europe was the social duty of visiting the sick. In those times, no one wanted to be alone when disabled by disease. And privacy was not an issue. As one expert historian writes, "Modern conventions which protect the privacy of people excreting excrement or giving birth simply did not exist in seventeenth-century England."[41] So, too, in the New World visitors came and helped and nursed and prayed and of course gave advice. In less populated areas, people welcomed any company that would break the loneliness of being invalided.

Particularly in the colonial period, only on rare occasions did people employ a trained physician. Instead, people who were educated or wise filled the social role of healer. As folk wisdom passed on in Benjamin Franklin's *Poor Richard's Almanack* explained: "By the time a man is forty, he is either a fool or a physician." When, for example, William Byrd, the Virginia planter, in the early eighteenth century found the bloody flux rampant nearby, he immediately wrote to one of his supervisors with instructions to use Byrd's own treatment—a relatively mild course of cures followed by a soft diet—if the flux reached his property. "I order'd him to communicate this Method to all the poor Neighbours, and especially to my Overseers, with Strict Orders to use it on the first appearance of that Distemper."[42]

Healers in the "Medical Marketplace"

As the population increased, local men and women with knowledge and talent for healing often charged for their services. In turn, purchasers of healers' services could be discriminating or critical of the services that were offered in the marketplace. In frontier settings or when one lacked money to pay for medical advice and services, people had little choice; with their oral traditions and advice books in hand, they improvised. Only in population centers did colonials have much choice.

The British arrangement from which the colonists mostly took and adapted their practice was a very broad medical marketplace, offering a range of alternatives from neighborhood healers and outright frauds to fully certified physicians. In urban centers, the formally recognized healers were organized in guilds, much as in medieval times. There were university-trained physicians who were notable for their learning and had high social status. There were trained surgeons in another guild, who also had status. Alongside them were trained apothecaries, who not only sold medications but were able to prescribe them. In the British countryside, distinctions between physician, surgeon, and apothecary tended to melt away, and anyone with the qualifications of any of the three could undertake general practice, which included diagnosis and advice, surgery, and selling remedies. So, too, did the distinctions disappear in the New World.

Yet in every European society, those at the top recognized that even in the chaos of the marketplace some practitioners had what passed for expertise at the time.[43] The colonists, of course, in carrying over deferential social arrangements, had British and other social hierarchies as a distant model, and local "doctors" won some partial professional recognition or aspired to.

Until 1765, there were no medical schools in the colonies. And almost no university-trained physicians in Britain or on the Continent would undertake a risky voyage to an uncertain situation in the colonies. Some apprentice-trained surgeons and apothecaries occasionally did make the trip. On the whole, however, the colonists had to select from the lower end of the medical marketplace and utilize such healers as they could find. William Byrd described one sort of healer in 1706: "Here be some men indeed that are call'd Doctors: but they are generally discarded Surgeons of Ships, that know nothing above very common Remedys."[44]

As each of the colonies was founded, special circumstances determined the range of healers who might appear there. Where European commercial venturers sent settlers, such as Jamestown, New Amsterdam, and the Swedish Delaware settlements, the company typically hired an apprentice-trained surgeon or apothecary to care for the ill. Otherwise any educated person, such as the clergy in Massachusetts and elsewhere, could fill the office of healer. Increasingly, however, the local medical authority would be some self-appointed healer. When Adam Cunningham, a young Scotsman who was actually an educated physician, tried to set up practice in Williamsburg in 1730, he found not only the town but also the countryside so crowded with "phisitions" who had set themselves up that he was unable to make a living there.[45] In 1711, a woman in South Carolina wrote home to

Britain about how the conversion to practicing medicine was even planned ahead: "My husband having provided himself with lancets in England resolved upon practicing Physick and Surgery which he has done with success beyond his expectations."[46]

In 1775, some citizens of northeastern Pennsylvania sent a petition to the provincial assembly complaining that their area was "infested with a Set of Men, who taking upon them the Offices of Physicians and Surgeons (though in Reality no better than Empiricks or Quacks) administer Drugs so unskillfully and ignorantly, that some Persons have, in all Probability, thereby lost their Lives, and others been rendered Cripples." The petitioners, recognizing that there were skilful and well-regarded practitioners in the community, asked for the licensing of medical practitioners.[47] No licensing occurred there or in most other areas until almost the beginning of the nineteenth century.

The petitioners nevertheless showed that they knew about the different grades of practitioners in the medical marketplace. Lowest on the scale were the quacks, or those who, in order to make money, deliberately deceived people. Sometimes they claimed to have formal medical qualifications, when in fact they did not. Or they sold people worthless "cures" in the form of procedures or substances that the quacks knew were fraudulent, like the legendary "snake oil."

Empirics differed from quacks in that they actually believed in some cure or cures that they had discovered through their own personal experience. One of the most famous empirics in the eighteenth century was Elisha Perkins, of Plainfield, Connecticut, a regular physician who also boarded students and bred mules to support his large family. In 1795–96 he announced his discovery that passing pointed metal objects over afflicted parts of a patient's body would effect a cure. Perkins supposed that electricity was involved, which was not surprising because Benjamin Franklin's contrivance of pointed lightning rods was still news at that time. Perkins therefore devised, patented, and sold metal "tractors," round at one end and pointed at the other. He firmly believed that his tractors were effective, and followers worked with them even in Britain and Denmark. So convinced was Perkins of the healing power of "tractoration" that when a yellow fever epidemic hit New York in 1799, he traveled there to treat the victims. Unfortunately, he himself died in the epidemic.[48] Clearly, Perkins was an empiric, not a quack, because he believed in his remedy, right down to his death.

Unlike Perkins, most empirics were, as the petitioners in Pennsylvania asserted, more or less ignorant lay people, and it was hard to distinguish empirics from quacks. How could one tell? An ad in a Virginia newspaper in 1752, for example, announced that "Mr. Richard Bryan, living in King George County, is most excellent at curing the *Iliack Passion*, or the Dry-Gripes, the cure of which he is dextrous

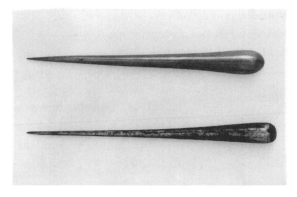

This set of metal tractors was used at the end of the eighteenth century, when Elisha Perkins was trying to cure people by passing the points of the tractors over the afflicted area of the body. The two pointed tractors were used together at the critical site. Dittrick Medical History Center, Case Western Reserve University.

in."[49] In Huntington, Connecticut, a shoemaker happened "to cure an old woman of a pestilent mortal disease, [and] he thereby acquired the character of a physitian [and] was applied to from all quarters."[50] Joseph Pynchon, born into a prominent family but without any training, simply started practicing in Boston, using various traditional cures, such as his own centipede soup. Before he retired in 1760, he had a substantial practice among leading citizens.[51] In such ways did many people find themselves offering medical care.

Empirics shaded over into the educated and wise people who were not physicians but who used a combination of regular medical knowledge and personal intelligence. Clergy in particular served as healers of the body as well as of the soul. Often educated, they were able to read medical works. Moreover, they often needed the fees. The Reverend Mr. Jacob Green, of Hanover, New Jersey, formally asked his congregation for permission to practice medicine. Replying that "country congregations could not have ministers unless ministers would help support their own families," they granted Green permission to "practice Physick if he can bair it."[52]

Probably the most important part-time (and sometimes full-time) healers were neighborhood women, who most typically served as midwives. The office of midwife was well understood in Europe and often licensed, and such was the arrangement, for example, in New Amsterdam under Dutch rule. On the famous voyage of the *Mayflower* in 1620, there was on board a midwife, Bridget Fuller, who was the wife of Deacon Fuller. She was still in practice in Plymouth in 1663.[53] In all the colonies, many women filled the role of midwife without supervision. As midwives, they testified in court and held official status unusual for women. They customarily also performed healing functions. Throughout the colonial period, visiting the sick, and of course helping and offering advice, was conspicuously a female duty. Moreover, the advice and recipes of many women could come from formal books as well as oral and family tradition.[54]

One late eighteenth-century midwife, Martha Ballard, not only delivered hundreds of babies but practiced healing very extensively in her area. Beyond nursing, she used local plants and common household substances such as salt and eggs and also store-bought substances such as licorice, pepper, vitriol, and myrrh.[55] Such women healers were also expected to perform midwifery. Until the end of the eighteenth century, in Cape May County, New Jersey, for example, women midwives/ healers provided virtually all of the medical care available there.[56]

In the seventeenth century, at least, being a woman practitioner could be a hazard even when the woman had grateful patients. Closeness to life, death, and illness could raise suspicions of witchcraft. In 1648, Margaret Jones, a midwife in Charlestown, Massachusetts, was hanged for witchcraft. Among other signs, her neighbors noticed that "she [was] practicing physick, and her medicines being such things as (by her own confession) were harmless, as Aniseseed, liquors, etc. yet had extraordinary violent effects."[57]

Sometimes communities formally recognized outstanding lay practitioners. For a century or more after 1607, such instances show up in the record. The Connecticut legislature issued medical licenses by special act to Thomas Lord in 1652 and Daniel Porter in 1654. New York did the same for Jonas Wood in 1677. The Rhode Island legislature by a simple law granted one citizen the title of "doctor of physsicke and chirrurgery" in 1654.[58] Mostly, however, the title "doctor" was awarded by usage, or a person just assumed it. Altogether the "doctors" in the marketplace included so many quacks and ignorant empirics that the whole category was often condemned. In 1764 one New York merchant, hearing of a proposal to found a medical school, commented that the community "had rather One half were hangd that are allready practicing, than breed up a New Swarm in addition to the Old."[59]

Systematically Trained Physicians

Early in the eighteenth century, a few regular physicians who were well trained, according to the standards of the day, began to appear in the colonies. At the time of the Boston inoculation controversy, in 1721–22, one of the local physicians was William Douglass, who had studied at Edinburgh, Paris, and Leyden and held a medical degree from Utrecht. He opposed Boylston and Mather on the grounds that they were spreading smallpox, although he soon accepted the practice of inoculation. In the following years, an occasional formally trained physician—such as the Edinburgh graduate Alexander Garden, of Charleston, South Carolina, who was also a naturalist and after whom the gardenia is named—might settle elsewhere in the colonies. But such well-qualified physicians were rare. Of the recognized practitioners in Virginia before 1800, only one-tenth held European medical degrees.[60]

From the beginning, any qualified or talented practitioner could take on apprentices, and almost all colonial physicians who had any proper training had served an apprenticeship to learn doctoring. The custom increased during the eighteenth century and continued for a long time in the nineteenth. Since the apprentices were boys or young men, the apprentice system helped exclude women from the practice of medicine.[61] An apprentice was indentured for five to seven years and lived as a member of the physician's family. In some cases, an apprentice was also an indentured servant paying off his transatlantic passage, such as one in Maryland in 1734 who agreed "to be an Apprentice and Servant to the said Doctor Gustavus Brown in his Business and Imployment of Physick Surgery and Pharmacy" for six years. In taking on an apprentice, the doctor was supposed to instruct the apprentice as he would a son, including looking after his morals and character. In turn, apprentices were to assist the physician in his practice. They curried the horses and performed personal service, gathered herbs and mixed prescriptions, ran errands, and dressed wounds.[62]

Out of the body of apprentice-trained physicians who served the population in the eighteenth century a few went on to study for a time in Europe. Some took degrees there, and some did not. This pattern of study with a preceptor in America and then further study in Europe continued for a century and a half and produced many, if not most, of the leaders in American medicine. In the colonial period, the main destination of ambitious Americans was the University of Edinburgh, although some studied in London hospitals or even in Paris. In 1765 Edinburgh awarded as many medical degrees to Americans as to Scots.[63]

The first American medical school was established in Philadelphia in 1765–66 as part of what became the University of Pennsylvania. By 1800 that medical school alone had graduated 179 students. Other medical schools followed, and at first all were affiliated with a college or university: in 1767, King's College (later known as Columbia University) in New York, stimulated by rivalry with Philadelphia; in 1782–83, Harvard; and in 1797, Dartmouth.[64] The schools had substantial entrance and graduation requirements, and the curriculum was demanding, including, at the University of Pennsylvania, clinical lectures at the Pennsylvania Hospital.

The Supply of Physicians

Historians have identified many, if not most, of the medical practitioners in the colonial period whom community leaders recognized as physicians. By

TABLE 1
Total population and numbers of doctors in Massachusetts, 1700–1790

Sample Year	Total Population	Number of Doctors
To 1700	70,000	70
1710	77,700	104
1720	89,000	130
1730	121,350	192
1740	153,000	244
1750	180,000	316
1760	215,000	388
1770	235,308	478
1780	268,627	643
1790	378,787	725

Source: Based on Eric H. Christianson, "The Emergence of Medical Communities in Massachusetts, 1700–1794: The Demographic Factors," *Bulletin of the History of Medicine*, 54 (1980), 66.

extrapolating from well-studied areas, it is possible to say that by 1776 about 3,500 people were serving as doctors. Only about 400 had had any formal training (which would include apprentice training), and only half of the 400, perhaps 5 percent of the total, held medical degrees. Given a population at the time of 2–3 million, there was about one practitioner for every 600 to 800 people.[65]

The historian Eric H. Christianson has laid out the numbers of doctors available for the changing population of Massachusetts during the whole eighteenth century. Although the population grew rapidly, the number of doctors grew even faster (table 1). Fewer than half of the doctors were even trained as apprentices, and fewer than a third held a degree (some doctors had done both). But over the decades, the younger men coming into practice were better trained than their predecessors. By the second half of the eighteenth century, there was one doctor for about every 500 to 600 inhabitants.[66] What was true of Massachusetts was also true of other colonies (after 1776, states). At the end of the eighteenth century, out of 500 doctors in Virginia, 55 had degrees.[67]

The Eighteenth-Century Legacy of Physicians

In the medical marketplace, the better-trained practitioners, often appealing to class prejudice, had always complained loudly about their competitors, the ignorant healers and frauds. Beginning in the mid-eighteenth century, however, a number of changes altered the landscape. Late colonial and early national society began to show some slight premonitory signs of modernization—economic development and social organization, with less sensitivity to tradition and more openness to new ideas and arrangements. Not least of the signs were the hesitant beginnings of increased medical professionalization. In a few places, the better doctors tried to group together in voluntary organizations to establish their qualifications, identity, and place in the social class hierarchy. Not only had they finally managed to establish medical schools in the New World but they had participated conspicuously in movements to found hospitals. In Philadelphia, the hospital came first, then a medical school and a medical society. In Boston, the order of appearance was reversed. The medical society came first, then a medical school and a hospital. But the ultimate outcome was the same.[68] Sometimes physician groups won recognition from governments. Over time, they managed to make it more difficult for less formally qualified people to assume the title "doctor."[69]

Physicians and members of the educated public perceived that medical knowledge was improving in the Age of Reason. Not just recognized practitioners were more intensely weeding out ridiculous and complex items from their customary recipes. Lay people too caught the spirit of skepticism. In Virginia, between 1739 and 1767 the legislature had given awards to people who claimed to have cures for pleurisies, dry gripes, and cancers. But beginning in 1770, the legislators uniformly turned down such claims and appeals.[70]

Another sign of the changing landscape for medicine was that occasionally doctors' sons were taking up medical practice, a sure sign of pride and confidence.[71] A marked, if subtle, sign of change in the late eighteenth century was that male physicians began to invade the practice of midwifery, the start of their century-long struggle to enter the birthing room. Customarily, male physicians were called in only when there was an emergency, but increasingly "midwifery" became part of medical education in Europe, where new techniques and instruments were appearing. As another sign of changing times, the attraction of expertise and "progress" in Enlightenment times overcame views that women were supposed to suffer during birth. But many decades would pass before the issues of women's sphere, modesty, and other gender and professional issues were resolved largely in favor of male physicians.[72]

At the end of the eighteenth century, the high point of the Age of Reason, advanced thinkers throughout the Western world tried to establish a world of rationality. In medicine, the result was a number of systematic theories explaining how the human body operated and how those theoretical systems dogmatically justified various types of therapy. Did the best understanding come from emphasizing chemicals or mechanics? Or was there too much tension in the fibers in the body? It is not easy to see how the pretentious, abstract systems of medicine of that day could lead to better medical practice. From one point of view, at least, they probably stimulated some practitioners to question tradition and perhaps simplify some of the recipes then in use, recipes that, as I have noted, functioned as superstition as much as learning.

Hospital and Charity Care

In Philadelphia, Thomas Bond, who had studied abroad at London and Paris hospitals, was able to mobilize support among the charitably inclined Quakers to create in 1751 the first major hospital in the thirteen colonies. The Pennsylvania Hospital was modeled after charitable hospitals in Britain that aimed to care for the "industrious" poor.[73] In the French colony of Louisiana, authorities had established small official and military hospitals, and as early as 1736 what became Charity Hospital, which was to have a long, distinguished history, came into being as a result of a private bequest and community support.[74] Before and after these permanent general medical hospitals, authorities occasionally set up temporary single-disease isolation hospitals during epidemics. By the early nineteenth century, there were occasionally almshouses that incidentally cared for poor and unfortunate ill and convalescing people, as well as outpatient charity dispensaries in the few urban centers.[75]

The hospital of the eighteenth and nineteenth centuries was not a medical institution. The hospital was a charitable institution, set up to care for persons who were so unfortunate as to be without a home, in which one's family would, as was customary, care for the ill person. Travelers or poor and isolated people were also the objects of hospital charity. In small towns, a family would take in a person who

Perspective created in 1768 of the Pennsylvania Hospital (*left*) in its geographical and social setting in late colonial Philadelphia, showing the placement of the hospital alongside other social services institutions, the workhouse (*center*), and the almshouse (*right*), with part of the city (*far right*). Thomas G. Morton, *The History of the Pennsylvania Hospital, 1751–1895* (Philadelphia: Times Printing House, 1897), 221.

had no home care. In New England, the selectmen of the town provided for some-one who, for a price, would look after an impoverished sick person without fam-ily. In New York between 1687 and at least 1713, the Common Council appointed a physician to look after the poor in their own homes.[76] In Virginia, church authori-ties acted for the government. In one vestry record, entries show attempts to find a doctor for ailing citizens. In 1745, 459 pounds of tobacco (the usual currency) was paid to "Sarah Broker for curing Jo'n Moone's Leg & Washing the surplice," and another sum was paid "to Capt. Doran for Salivating Eliz'th Taylor," presumably with calomel (a procedure discussed below).[77]

On rare occasions in a larger urban center, then, a hospital was set up to fur-nish keep and care, including the common household curative measures that any family would administer. As in a private home, a doctor might be called in. So it was that physicians volunteered to periodically see patients in the hospital. But phy-sicians were external to the hospital or, if there was one, the almshouse. Lay trust-ees administered the charity. In the case of military hospitals, for soldiers who ob-viously were away from home, military authorities provided the service.

Following the model of voluntary hospitals in Britain, founders of hospitals understood that taxpayers would benefit from the charitable institution. Benja-min Franklin, a major supporter of the hospital in Philadelphia, argued that car-ing for the sick poor in a hospital would cost taxpayers much less than hiring care for them.[78] And the Pennsylvania Hospital made special provision for an-other group of unfortunates, the mentally disturbed—who, again, would not have to be farmed out to local families. As an additional harbinger of the future, the citizens of Virginia in 1776 set up a special hospital in Williamsburg for the men-tally ill, recognizing that physicians traditionally treated mental illnesses. As the governor explained, there was

> a poor unhappy set of People who are deprived of their Senses and wander about
> the Country, terrifying the Rest of their Fellow Creatures. A legal Confinement,
> and proper Provision, ought to be appointed for these miserable Objects, who
> cannot help themselves . . . where they are confined, maintained and attended
> by able Physicians to endeavour to restore to them to their lost Reason.[79]

The founding of the Pennsylvania Hospital stimulated the founding of a sec-ond major general hospital, the New York Hospital (chartered in 1771, but use of it was delayed by the Revolutionary War). In both cases, the medical school used the hospital for teaching purposes, as was the custom in Europe. Over the years, as the population grew, hospitals appeared elsewhere to fulfill social and charita-ble obligations. Not until the late nineteenth century did the design, purpose, and governance of hospitals change substantially.

Beginning in 1775, as in the case of the New York Hospital, the American Revolution interrupted the functioning and growth of formal medical institutions. In the war, however, physicians received substantial social recognition for their service to the armies. The patriots organized their armies after the model of the British army, which many residents had served in during the earlier colonial wars. Thus the armies had medical services and medical officers, some recognized by appointment to a high rank.[80] After the war they constituted an important element in the Revolutionary generation that led American society, including elements of a medical establishment, into the nineteenth century.

CHAPTER 3

The Beginnings of Change
in Traditional Health Care

Between the American Revolution and the 1850s, traditional medical care continued throughout the expanding country. Meanwhile, changes in American society affected and even threatened customary practices and recognized physicians. This chapter covers how the health marketplace dealt with attempts to establish the "profession of medicine" alongside folk and commercial approaches to treatment. In these tumultuous decades, Americans in general, and physicians in particular, tried to preserve their customary ways of providing care in the face of attempts by new professional and commercial elements to modify how people thought about their bodies, about diseases, and about public health.

Framing and influencing all these changes were rapid, disorienting geographical expansion and economic development. Especially after about 1815–20, internal development and economic expansion dominated events in the nation. Not only did the population increase but that population became much more literate, more mobile in all senses, and socially less deferential. The population went from 4 million in 1790 to almost 13 million in 1830 and then to an astounding 31.5 million in 1860, by then substantially larger than that of the British Isles. By 1830 eleven states had joined the original thirteen in the union. In 1859 there were thirty-three. The extent of health activities was transformed; the basic patterns of health care not so much.

In the 1830s the railroad system started to grow. By 1869 rails connected the East to California. In the 1850s, industrialization in the United States "took off," accompanied by significant urbanization, although the country remained overwhelmingly rural, albeit now spread out from coast to coast. In that vast territory, which still had a moving frontier, there were by 1860 fifty-two medical schools and—something new, and very successful in the United States—three dental schools. There were also six institutions offering training in pharmacy, one sign of the beginnings of standardization of purity and dose. According to the 1860 census, 55,000 physicians of various kinds provided professional health care, about 1 for every 573 people.

The Medical Professionals

Medical professionals in general did not fulfill the promise that was present at the beginning of the nineteenth century, even though after 1830 the thousands and then tens of thousands of physicians could be classified in new ways, as will appear below. But according to a traditional classification, based on education or skill, some physicians were considered very good, many were very bad indeed, and the vast majority ranged somewhere in between. Differences among the doctors complicated medical care and helped generate in large parts of the population contempt for most people in the health business, especially physicians.

Under attack from competitors, from do-it-yourself advice publications, and from antielite intellectuals, most conventional physicians tended to hunker down and defend medical traditions and the authority of traditional medicine. In the face of doubt, they intensified an assertive style of practice that appeared extreme to later generations. When their authority was challenged, they did not innovate. Instead, the articulate ones, at least, invoked cultural nationalism and local customs and beliefs in order to defend customary actions.[1] Any time they did innovate, it was within some context of established medical thinking.

In 1820, with collaboration from both physician and pharmacist groups, a notable institution appeared, the *United States Pharmacopoeia*, a list of standard drugs and medical preparations with standard ways of preparing them, measuring them, and using them. The Massachusetts Medical Society had published a pharmacopoeia in 1808 based on one issued in Edinburgh, but physicians in New York in 1817 called for a national meeting to compile an authoritative national pharmacopoeia. The convention made a useful compilation, published in both Latin and English in the same book, and stipulated that a convention would be called to issue a revision every ten years. Remarkably enough, with much jockeying and rivalry, a revision did appear thereafter every ten years, providing material evidence that physicians and pharmacists were part of a progressive as well as traditional science. The pharmacopoeia was followed by a handbook dispensatory, which also went through many editions and was described by John S. Billings in 1876, in reference to sales, as "the most successful medical book ever published in this country."[2]

Trying to Be a Doctor and a Professional

In the health marketplace of the late eighteenth and early nineteenth centuries, people had some idea of how a doctor was supposed to act. As the decades passed in the new United States, without regard to the realities of health or to whether doctors could heal anyone in the decades before the Civil War, a stereotype of ideal physician behavior emerged in both medical treatises and popular and literary works. To some extent this ideal drew on foreign models and the experience of the better-off classes, who could afford doctors, as well as on the professional aspirations of physicians. In trying to upgrade their professional status, some physicians contributed to general publications read by the opinion makers, praising the professional abilities and character of "the doctor." Literary writers often reinforced that picture but sometimes showed physicians in a less favorable light.[3]

The status of the doctor was in fact still uncertain. One of the roles of the professional was still ceremonial: he was someone to call, along with a clergyman, when the patient's situation was dire. Also, as earlier, those who "sent for the doctor," or "summoned the doctor," in the mid-nineteenth century viewed him as a personal servant.[4] From the practitioner's perspective, professional status brought an added dimension to his services. The physician usually claimed to have a certain amount of medical knowledge, but in addition to the knowledge that came from an education, he had the wisdom that came from experience. In particular, the doctor—of whatever variety—knew his patients' life histories, their constitutions, and their circumstances. He was, above all, a good friend, one who would give sage advice but also one who would spend nights by the bedside of a sick child or adult, nursing and watching.[5] Reinforcing the image of wise family friend was the fact that ordinary medical practice was organized around families. Even the account books of physicians listed families, not individual patients.[6]

The family basis of medicine extended in a special way to the southern states, where the plantation and slave system had become well developed. Records show that members of the slaveholder families, as well as the family doctor, cared for the enslaved people as members of the family, as people of that place and time understood family. Like other residents of the United States, slave owners often believed themselves to be more successful practitioners than the physicians they hired. A Louisiana plantation owner who treated his slaves himself in one instance bled them and puked them because he noticed the first sign of an epidemic. All of those slaves survived. The planter claimed that area physicians had delayed treatment, causing many of their patients to die.[7]

Patients in the slave system differed from those of their free "white" neighbors in some special respects. Empirical observations recorded at the time suggested that "white" and "negro" bodies differed in their reactions to many diseases and conditions beyond malaria, including yellow fever and respiratory diseases. Later students concluded that in addition to the sickle-cell trait, there may have been some genetic variations in disease resistance, reinforced by different childhood exposures. Many people were bred for profit, which meant that the circumstances of their pregnancy, birth, and childcare could vary to an extreme with the local conditions. Almost all births were attended by midwives, who had a special place in slave society.[8] Enslaved patients and caregivers could oppose treatments, whether ordered by an owner or a physician. Slave patients could evade treatment, or they could be forced to undergo treatment. There was continual conflict between traditional and medical diagnoses and treatments, probably even more than in the rest of society. Meanwhile, slaves who medicated other enslaved people provided an arena of independence for themselves and fellow slaves.[9]

The motives of slaveholders usually comprised, in widely variant combinations, property protection, personal concern, and conventional paternalism.[10] Much of the history of health care of enslaved and free antebellum African Americans was at most just an exaggeration of general medical care in the United States, including care, or lack of respect and care, for those members of society whom dominant groups valued least. Indeed, physicians were often an essential part of the slavery system and the general social system.[11] In general, physicians with a plantation practice, whether on a fee or a retainer basis, found medicine a more profitable occupation than did most of their colleagues.

Everywhere in the country, at any time, collecting for services was the major preoccupation of practitioners of every variety. One eminent physician of the 1840s complained: "It runs me almost crazy to think that with hundreds upon hundreds due me professionally I find the greatest difficulty in raising a simple fifty dollars."[12] A Wisconsin physician of that time collected only 24 percent of his billings, and much of what he did collect was in kind. Sometimes a whole community would have a "bee," with all pitching in to furnish the local doctor with wood or corn. Or a yearly "doctor's dinner" brought all kinds of farm goods into a physician's yard. One historian has teasingly described such cooperative community efforts as "an early form of a group health plan." Physicians' incomes in the pre–Civil War era averaged about as much as a skilled artisan's (doctors averaged $600 a year, common laborers $300, and artisans $600). Lawyers held much more property than doctors, but doctors accumulated more than clergy. Southern physicians did better than northern physicians, but the latter included many immigrants from Europe, whose incomes kept the averages down. The range of incomes everywhere was extreme, but physicians' economic status was generally somewhat above that of other white males,[13] evidence that professional status was possible for doctors, albeit far from universal.

Trying to Establish a Medical Profession

In the midst of overwhelming social and economic changes, many people who thought of themselves as physicians did indeed attempt to use that identity to claim to be part of a medical profession, just as lawyers of that time were also trying, with some success, to professionalize. As noted earlier, eighteenth-century physicians who had a sense that they were different from the usual empirics and quacks began spontaneously to join together, and in the nineteenth century the pace of forming medical organizations increased markedly. These organizations were always based in an urban area. One such organization is notable for publishing, in 1788, the first original medical publication in the thirteen original states: *Cases and Observations by the Medical Society of New Haven County, in the State of Connecticut.* By 1800 not only were there a number of local medical organizations in the new nation but in nine of the thirteen states physicians had established state organizations, with varying degrees of success. Medical societies constituted an important means by which physicians tried to gain recognition as, collectively, a profession, indeed to define a profession of medicine.[14]

Not only did medical organizations publish transactions but independent medical journals started to appear in the medical community, beginning with *The Medical Repository* in 1797. The founding of new journals increased particularly in the 1840s. Two hundred journals had been founded by 1850, although most of them did not last very long.[15] There was, for example, only one issue of the *Medical Society of Maine Journal*, dated 1834, and the *St. Joseph (MO) Journal of Medicine* lasted only from 1859 to 1861. Much of the content of any journal was reprinted from other sources, and any original material in the journals usually consisted of physicians' reports of cases they had treated. Such reports extended the "experience" of the readers, who followed the narrative patient "history" as one followed any narrative in the storytelling culture of that day.[16]

With the development of a more distinctive medical literature in the nineteenth century, the general population was slowly excluded from medical knowledge and from practitioners' experience and learning as it was communicated in print and in meetings. Inexorably, albeit very slowly, physicians were developing special knowledge, often expressed in technical language. Eventually first one part and then another of this knowledge was no longer intellectually accessible to just anyone, and possession of it often gave educated physicians an advantage in the medical marketplace, even beyond their possible surgical skills. Journals generally repeated personal reports, opinions, and wisdom, but they also helped import and spread innovation. Daniel Drake, of Cincinnati and Louisville, observed sarcastically in 1852 that "even the village surgeon now cuts according to the newest fashion of some great transatlantic operator." Perhaps a majority or more of practicing doctors did not read any medical literature. Still, some did. It was they who noticed "the improvements and modifications incessantly taking place in the departments of Pathology and Therapeutics," as an 1840s textbook author put it.[17]

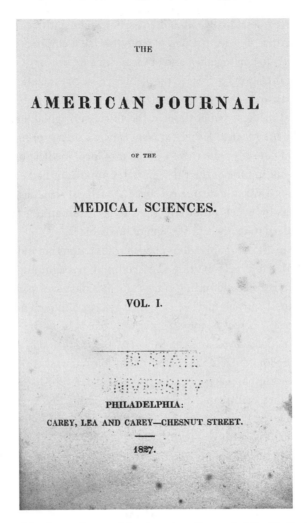

The title page of the first volume of the *American Journal of the Medical Sciences*, in 1827, the earliest U.S. medical journal that was still being published, under the same title, in the twenty-first century. It is a symbol of the centrality of medical journals as instruments to introduce and store "new" knowledge that could undergird the developing profession of medicine. From the original in the collections of the Medical Heritage Center, Ohio State University.

Goals of Organized Physicians

From time to time, beginning in the eighteenth century with the earliest organizations, scattered physician groups attempted to institute fee bills—a step toward claiming a monopoly on medical practice. The earliest attempts took the form of asking governmental authorities to set fees. In 1736 the Virginia House of Burgesses put some limits on physician charges, noting the practice of some practitioners who "demand excessive fees, and exact unreasonable prices for the medicines which they administer." For a visit and prescription in town, the limit was set at five shillings, with proportionately more for more distant calls. For a simple fracture, the maximum charge was two pounds, and for a compound fracture, four pounds.[18] This legislation soon lapsed, however.

By the 1790s, physician societies were trying to set fees themselves through group agreement to charge the amounts stipulated in fee bills rather than to ask for legislative intervention. The Medical Society of New Jersey had such an agreement as early as 1766, but only in the early national period did many more fee bills appear, such as a frequently reissued New York table of charges of 1798–1816, which included procedures such as cupping, dressing a wound, amputation, inserting setons, an operation for a fistula, repairing a prolapsed anus, treating fractures and dislocations, and of course bloodletting.[19] In 1837, the Washington County, New York, Medical Society adopted a fee bill that included:

Advice at office, 50 cents
Ordinary visit under one mile, 50 cents,
For each additional mile, extra 25 cents
Obstetrics, ordinary, not over six hours, 4 dollars
Catheter, single introduction, 2 dollars
Dislocation, hip, 10 to 25 dollars
Vaccination, single patient, 1 dollar.

Ordinary procedures such as pulling a tooth, bleeding, or giving a purge or a vomit were priced at twenty-five cents. The most expensive item, cutting for bladder stones, cost fifty dollars.[20]

After the coming of the railroads, it became feasible for physicians to organize on a national scale. In 1846–47 the American Medical Association was founded. By 1856 more than three thousand physicians were members—a significant but small segment of the regular medical practitioners, some of whom were attempting to bolster the professional status of physicians in the United States by setting standards for medical education. For many decades, however, those reformers were ineffective.[21]

Another sign of the professionalization of physicians at the turn of the nineteenth century was the continued efforts by various states to recognize qualified medical practitioners, efforts that intensified beginning in 1811. Indeed, by the 1830s, only three states did not have some kind of licensing system. Although it varied greatly by locality, the licensing procedure mostly required that an applicant obtain a passing mark from physicians who examined the applicant orally. In frontier states, enforcement of any kind was improbable; in any event, the licensing procedure usually certified the fitness of an applicant rather than penalizing uncertified practitioners.[22] Typically, the medical societies, whose leaders advocated licensing, participated. In Michigan, for example, a state medical society was established in 1820 in order to meet the requirement of the territorial government that such a body license physicians—and, it was hoped, discourage quackery.[23] Regardless of formalities, a continually increasing proportion of physicians in the United States were in fact trained, even if only through apprenticeships.

The Ironic Undermining of Professional Development by Medical Schools

With organization and the beginnings of licensing of physicians, the professional status of physicians appeared to have significant momentum. After 1810, however, another sign of professional achievement, the continual founding of medical schools, in the end undermined professional standards, although this result was not evident at once. A school chartered in Baltimore in 1807 was named in 1812 the University of Maryland, with authority to add other university faculties.[24] By 1830 at least eighteen more medical schools had started up. Several lasted for only a short time. Medical instruction at Brown University, for example, began in 1812 but was abandoned seventeen years later. New degree-granting institutions tended to follow the spread and growth of population, as in New England, the District of Columbia, Kentucky, South Carolina, Georgia, and Ohio. By 1860 there were forty-seven medical schools across the country.[25]

The record is only approximate because a number of medical schools did not begin as formal or, especially, academic institutions. One impetus for new schools came from local physician organizations. A second source was apprentice training. In one county in western Massachusetts, 152 out of 215 doctors in the fifty years after 1790 had only apprenticeship training. But when a preceptor had acquired a number of students, the group could morph into an informal or even formal medical school.[26]

A system of so-called proprietary medical schools was evolving, each set up and governed by a few practitioners who appointed each other professors and obtained a state charter—a step that was usually easy, given the politics of local patronage and the hope of the leaders of every village to turn it into a flourishing metropolis with an advanced collegiate institution. By the 1830s the establishment of proprietary medical schools had become a major concern of leaders in American medicine, who were working toward higher professional standards for physicians.

The multiplying medical schools competed for students by keeping entrance and graduation standards as low as possible. The result was that many graduates were not well trained and went out into the medical marketplace to compete with better-educated colleagues, making it ever more difficult for any one of them to obtain enough patients to make a decent living—or for standards to be raised. By 1850, about seventeen hundred MD degrees, on average, were being awarded each year.[27] In contrast to the perceptions of many people at the time, it is unlikely that business competition between "doctors" improved health care for anyone, although many practitioners may have shaped their treatment regimens to suit consumer taste.

For most of the nineteenth century, reformers in medicine focused on medical education as a source and sign of weakness and a disgrace to the profession. Medical schools were also the focus of much of the rivalry and name-calling within medicine. Alfred Stillé, of the University of Pennsylvania, spoke of "vulgar rivalry . . . in spirit and conduct similar to that displayed in the competitions of steamboats, and railroads. . . . Education is cheapened, the period of study abridged, or lightened—no irksome examinations are to be endured, and degrees

In 1838 the Medical College of Virginia began holding classes in a former hotel. Although quite respectable, and in a substantial building, the college, to compete for students, had to honor the same low standards for admission and attendance as the many other medical schools founded in the mid-nineteenth century. Wyndham B. Blanton, "Augustus Lockman Warner, 1807–1847, Founder Medical College of Virginia, First Professor of Surgery and First Dean," *Annals of Medical History*, 3rd ser., 4 (1942), 5.

acquired easily and assuredly." Whenever a school attempted to raise standards, students went elsewhere, and the reformers had to retreat. And since all of the schools except that at the University of Michigan (founded 1830) were totally dependent on student fees, the market ruled, with terrible effects.[28]

As early as 1832 Daniel Drake noted the deficiencies of the proprietary schools. He lamented that the sessions were extremely short: "In some, the session continues but two or three months, and four are the prevailing term. . . . The rule that a student may matriculate at any time within the first month of the course, leads to great procrastination in starting from home; and in its bad effects is only equalled by the absurd custom of leaving the university before the expiration of the lectures." He went on to list the scandalous shortcomings prevailing in most of the schools.[29] The general outlook was bleak almost to the very end of the nineteenth century. Lax state legislatures chartered medical schools wherever local citizens requested them. Generation after generation, reformers in medicine agreed that upgrading medical education was an important step in improving the profession and the delivery of medical care.[30]

One of the side effects of an increased number of medical schools was an increase in the demand for cadavers—for anatomical dissection, especially in the better schools. The importance of the emblematic ritual of doing a dissection of a cadaver was well established, and it persisted in the nineteenth century as a means of separating educated physicians from just clever or learned lay people. Inevitable grave robbing to obtain cadavers continued to damage the image of physicians, however, although authorities tried to turn a blind eye to a practice that most social leaders deemed necessary. Gradually over the second half of the nineteenth century, state legislatures passed laws to make it possible for medical schools to acquire unclaimed bodies from jails and other places. Especially after the rise of the railroad network, cadavers, usually preserved in whiskey (the cheapest form of alcohol) and placed in barrels, were shipped from areas of surplus to those in need. One historian, however, has found twenty-two recorded instances of grave robbing that set off local crowd action against U.S. medical schools between 1765 and 1884.[31] In 1879 one observer estimated that dissectors used about five thousand cadavers each year, half or more still obtained extralegally.[32]

Beyond grave robbing, doctors had another serious image problem that started with the medical schools. Rightly or wrongly, medical students generally had the reputation of being among the worst elements in any community. At Geneva College in New York, it was reported that "the rowdy element was so pronounced at times that there was a strong sentiment in the town to take stringent measures to suppress the school." In the 1830s, recalled one eminent graduate of the leading school of the time, the University of Pennsylvania, "the term 'medical student,' with

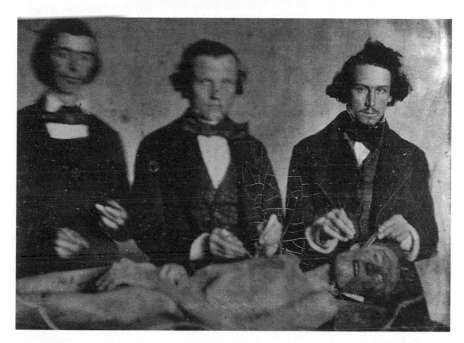

Dating from the 1840s, this photo of students posed with a cadaver is believed to be the first surviving photo of the dissection of a cadaver in the United States for the purpose of learning human anatomy. The practice was ancient, but photography was new. Stanley Burns, MD, and The Burns Archive, New York.

many citizens is intimately associated with 'roguery,' 'impudence,' 'lawlessness,' 'delicate sense of fashionable honor,'" the last because of dueling. Within the medical schools, students often ran wild, throwing things, fighting, and shouting during the lectures. The famous Doctors' Riot in New York City in 1788 allegedly began when anatomy students "impudently" and provocatively hung a cadaver's arm out a window, outraging local residents. In 1849 some medical students were observed treating a patient and then betting on "the number of hours the patient would live."[33]

The Tendency to Heroic Therapeutics

The physicians' public-image problems only began with medical students. In the competition among doctors and healers and commercial and home remedies, however, the issue was usually not so much professional standing as therapeutics. In therapeutics, traditional treatments did not always serve aspiring professionals well. Indeed, for decades, into the nineteenth century, physicians' collective reputation could be damaged by the then fashionable practice of aggressive "heroic medicine" (a label applied by later generations), such as that employed for Washington in 1799. Many later physicians, however, tended to modify their practices, particularly beginning in the 1830s and 1840s, when a new generation of young, enthusiastic practitioners, a number of whom had trained in France, took leadership roles and stimulated a faster pace of change away from traditional therapeutics and belief (as will be described below).

Heroic medicine was effectively spread by a charismatic, or at least politic, teacher at the University of Pennsylvania medical school, Benjamin Rush. A signer of the Declaration of Independence in 1776, Rush served in the Revolutionary army and taught until his death in 1813. He viewed heroic medical practice as another aspect of independence from Europe. Rush more strongly than most of his colleagues urged the physician to be active and to interfere with nature, which had brought the patient's disease. Rush repeatedly urged "efficient remedies" rather than "the supposed healing powers of nature." "The death of a patient," he noted shrewdly, "under the ill directed operations of nature, or what are called lenient and safe medicines, seldom injure[s] the reputation or business of a physician."[34]

Drawing on mainstream theory and practice, Rush was the most effective advocate of that heroic therapy in which patients underwent frequent and sometimes extreme measures, particularly bleeding and purging. Rush calculated that "not less than 6000 of the inhabitants of Philadelphia probably owe their lives to purging and bleeding" during the yellow fever epidemic of 1793. He himself "did not lose a single patient whom I bled seven times or more in this fever."[35] His students spread this heroic style of practice, notably in the South and West. At the same time, many physicians, especially in New England, advocated a milder style of practice.

In traditional medicine, what the physician knew as a basis for practice was the constitution and circumstances of the individual patient, a body "in a system of dynamic interactions with its environment." The body system, comprising parts or organs that affected other parts and organs, reacted with outside influences, such as mental upset, which could cause physical upset, and vice versa, just as each part of the body reacted to food and drink and light and dark. The physician was not concerned with specific diseases as much as with the total operation of the system. So one might vaccinate for smallpox but at the same time regulate the input and output of the GI system with diet and purges. The same watchfulness to assure the balance in a sick person's system was employed in natural developmental cases, such as teething in babies.[36]

By the early nineteenth century, physicians generally were utilizing three types of treatment to correct a person's unbalanced, diseased "system" or constitution: depletives, stimulants, and alteratives. Depletives, chiefly bleeding and purging, were well known (and were the most utilized by Rush and his followers). Stimu-

On 13 June 1791 Dr. John Foulke ordered that a "Negroe" patient be admitted to the Pennsylvania Hospital and then bled to treat his lunacy ("Phrenzy"). This page from the records in the hospital archives shows the several steps for admission and treatment, with orders directed to the steward and Dr. Cuthbush. Thomas G. Morton, *The History of the Pennsylvania Hospital, 1751–1895* (Philadelphia: Times Printing House, 1897), 125.

lants, which included alcohol, cantharides, arsenic, and sometimes opium and Peruvian bark, were used for diseases in which the patient showed weakness and flaccidity or was generally run down. And then there were alteratives, to shock and reset the functioning of the system, just as people subjectively sense after a violent encounter with vomiting and other symptoms of flu that their internal rhythms have been altered and reset. The medicines most commonly given as alteratives were emetics (to induce vomiting), especially tartar emetic (antimony and potassium tartrate). Even these three types of treatments could be confused in the minds of some practitioners. Nathan Smith, a moderate New England practitioner, recalled seeing "a written prescription, in which opium, wine, alcohol, cantharides and arsenic, were all directed to be taken several times in the course of twenty-four hours."[37]

Clearly, purgatives, useful as either depletives or alteratives, were the most frequently employed medications in the first half of the nineteenth century. For depleting, purgatives along with bleeding were effective. If there was too much stimulation, as particularly in a fever, a purgative could act as a counterstimulant or alterative.[38] Eventually the motto of rough-and-ready practitioners would be "Trust in God, and keep the bowels open!"[39]

One purgative, calomel (mercurous chloride), was symbolic of heroic practice. It dominated in the United States from the Revolutionary War to the Civil War and was particularly favored by Rush. Aside from acting powerfully on the bowels, calomel caused irritation and cramps in the GI system and sometimes vomiting. The most singular symptom was "salivation," or a copious production of saliva, as (appropriately) in acute mercury poisoning. Salivation was interpreted as a sign that the medicine was having the desired effect. With repeated doses, however, the mouth and pharynx could turn ashen grey, and the oral cavity could develop abscesses. The jaw could be affected so that the teeth would be loosened. Bloody diarrhea was taken as a symptom of the disease, not of the medicine. And with enough poisoning, there would be a sedative effect on the patient. "It is but trifling with the life of a man to give him less of a remedy than his disease calls for," noted one 1828 advocate of giving calomel in doses of 100 to 200 grains (3 to 6 grains was originally considered a normal dose, and Rush was criticized for giving as many as 10).[40]

Historians have uncovered endless records of patient treatments from the early and mid-nineteenth century that illustrate the heroic extremes employed by physicians and often imitated by citizens who dosed themselves and their families and friends. One midcentury Philadelphia physician boasted that he had never lost an adult patient with peritonitis. "The reason that he gave was that he always bled his patients from the arm until they fainted, and then put one hundred leeches on the abdomen." Scholars have further identified the high point of "active treatment" as the first quarter of the nineteenth century—and they have found that theory had little to do with practice. American followers of the French theoretician F. J. V. Broussais, for example, bled patients as enthusiastically as did followers of Rush. And any "active" practitioner, well trained or not, could prescribe amazing doses of calomel and other substances that caused later commentators to remark on the hardiness of a population that could survive such cures.[41]

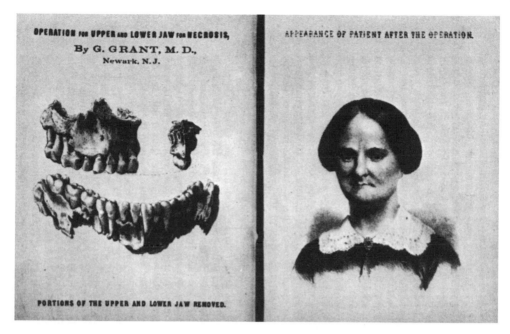

This woman took large amounts of the purgative calomel, which ate away her jaw. Such extreme effects often came of "heroic medicine" of the early and mid-nineteenth century. The picture on the right displays how well the surgeon reconstructed the woman's face after removing the destroyed jaw (*left*). David L. Cowen, *Medicine and Health in New Jersey: A History* (Princeton, NJ: D. Van Nostrand, 1964), 30, reprinting G. Grant, "Operation for Upper and Lower Jaw for Necrosis," *Transactions of the Medical Society of New Jersey*, 1859.

There were occasions when heroic therapy was demonstrably effective. A Boston physician reported in 1812 that he had used oil of turpentine in the case of "a lad of sixteen years of age, but considerably under the usual size." The boy had for six years "discharged" parts of a tapeworm. Under the physician's direction, the boy over two days took several one-ounce doses of turpentine "with the same ease that he would rum or gin." Sometimes the medicine was vomited up, but eventually some segments of worm came out the anus, and on the second day the boy "presented me with upwards of twenty feet of the Taenia lata; one end of which I found upon examination terminated in the head" (a sure sign of success of the treatment). The boy suffered minimal toxic effects from the turpentine: "a little warm in his stomach . . . a little dizziness in his head" and a hard pulse, in addition to nausea and diarrhea. It is easy to see how the experience of this physician and others at that time who also used oil of turpentine for intestinal worms (a common affliction) could confirm such practitioners in their inclination to strong dosing and heroic therapy.[42] Similarly, when large numbers of the better practition-ers used mild means, competing therapeutic activists could criticize them for fail-ing to take adequate steps to counter their patients' diseases.

Contention and Chaos in the Medical Marketplace

It was hard to discern best practice anyway amidst the noisy public controversies in which medical practitioners and even the public engaged. Doctors themselves participated in often extreme and vituperative arguments over medical and pro-fessional matters. As late as the 1850s, a Cincinnati medical editor observed that "it has become fashionable to speak of the Medical Profession as a body of jealous, quarrelsome men, whose chief delight is in the annoyance and ridicule of each other."[43] Those were often crude and rough times. In New Orleans, for example, physicians would occasionally feud to the point of fighting duels that were some-times fatal. In 1838 one eminent, if disagreeable, physician, Charles Luzenberg, chal-lenged a colleague, J. S. McFarlane. It was reported that Luzenberg was "in the habit of suspending the bodies of persons who had died under his care whilst House Surgeon of the Charity Hospital, and shooting at them as marks with pistols, in order to improve his skill as a marksman in his expected contest with Dr. McFarlane."[44]

Public distrust of doctors brought the ultimate breakdown of the promising start made in education and licensing at the beginning of the nineteenth century. Over several decades, any licensing or certifying of medical practitioners was repealed by various state legislatures. Between 1833 and 1844 state after state repealed or weakened penalties for unlicensed practitioners. As early as 1849, officials of the new American Medical Association found that only New Jersey and the District of Columbia were still regulating the practice of medicine, and the New Jersey law was made ineffective within five years. The laws that were repealed had not been very strong, and in Georgia, for example, anyone who failed the state board exami-nation could appeal to the legislature to overrule the board and grant a license.[45] The idea of a free people making choices in a free market was also the argument made for abolishing any legislation that would uphold "the medical monopoly." A Cincinnati journalist in 1849 wanted "reason and public opinion" to be "the sole legislators of the medical profession."[46] And so by the 1850s the United States was formally launched on an experiment to impose the free market on health care.

The marketplace for health care had another ironic effect. There may have been many badly trained physicians—one physician of the 1840s noted that "the pro-fession is not only crowded, but a large proportion of it is made up of unworthy and ignorant men," and an 1847 estimate was that there were forty thousand reg-ular physicians and another forty thousand "irregulars."[47] Yet the multiplication

of cheap, low-standard proprietary medical schools meant that apprentice training of physicians diminished dramatically. Obtaining the easy MD degree was
faster and possibly cheaper than putting in the usual seven years on an apprenticeship. True, many medical schools had a requirement, which might or might not
be enforced, that students work for two years with a preceptor who was a qualified physician. But even though doctors who attended medical schools might be
inferior, patients increasingly expected to have someone with a formal education
and a medical diploma. This ideal operated gradually in the marketplace to eliminate many self-appointed practitioners. Even on the frontier, there was often a surplus of degreed physicians. A person with an MD degree also was likely to identify with a professional medical community. Drastic upgrading of the profession,
however, awaited the full force of new scientific thinking and ideals that would
reorient actual practice.[48]

Meanwhile, the increasing recognition of formal education to qualify a physician inadvertently provided a pathway for some women to enter into medicine officially and become part of the medical community. Indeed, the significance of the
graduation of Elizabeth Blackwell with a medical degree from Geneva College in
1849, the first female to graduate formally from a medical school in the United
States, was recognized immediately, if often unhappily.[49]

The Advent of Medical "Sects" with the Thomsonians

Beginning in the 1820s and 1830s, the problem of physician professionalization and
image took on a whole new dimension when the medical sects joined in the medical marketplace and challenged the authority of regular, traditional doctors, either heroic or moderate. The new movements came to be called "sects" because
such practitioners professed an almost religious belief in the curative powers of a
set of dogmas, just as the "regular" physicians often did in theirs.[50] Indeed, these
distinctive sectarian practitioners established themselves as equals to traditional
doctors. Many Americans viewed the "regulars" as just one of the sects competing in a fluid, raucous marketplace.

The first medical sect to appear and enlist significant numbers of supporters was
the botanic physicians. These were initially the followers of Samuel Thomson, a
self-educated, itinerant healer who developed the only medical system that was ever
patented. He restricted himself to botanic preparations and prescribed most often
a common plant, lobelia, which operated as an emetic, and capsicum (cayenne pepper), which was well known in medical use, presumably to relieve gas and settle
the digestion. Eventually he assembled a number of plant products that in combination or singly covered all of the ailments of humankind, medications that he
believed would mechanically control body heat, the source of illness and health.

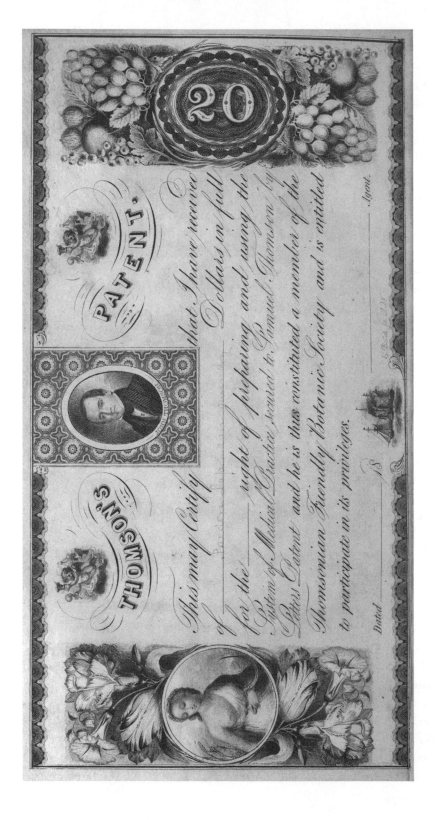

He also utilized steam baths and other forms of sweating. Thomsonians were commonly known as "puke doctors," because of the action of lobelia, or as "steam doctors." In 1837 a Michigan newspaper reported that a New York doctor had been accused of "steaming to death a Mr. French" in a "Thomsonian Infirmary." Doggerel from that day describes a Thomsonian physician's practice:

> I puke, I purge, I sweat 'em
> And if they die, I let 'em.[51]

Thomson began full-time practice in 1805. Soon he set up shops and organized Friendly Botanic Societies, whose members spread the use of his preparations. In 1813 he obtained a patent on his system, and he set about selling rights to use it. By the 1820s Thomson had developed an effective distribution arrangement. He or his agents would sell his book, *New Guide to Health* (1822), for twenty dollars. The book included a license that authorized the purchaser to practice Thomsonian medicine, that is, the "right of preparing and using" Thomsonian preparations, which Thomson sold.

Botanic medicine spread throughout the United States. Evangelistic agents repeated and elaborated on Thomson's rhetoric, emphasizing the democratic nature of the movement, which was open to, and easily understood by, everyone.[52] Indeed, a hidden agenda was authorizing women to practice medicine according to Thomsonian principles, at least in the home.[53] In the cholera epidemic of 1832 the Thomsonians did better than those who bled and purged, and the movement grew dramatically. By the 1840s, a large fraction of practicing physicians adhered to one form or another of botanic medicine (e.g., perhaps one-third of those serving Ohio and one-half of those serving Mississippi).[54]

Over the years, leaders in botanic medicine split off from Thomson in large numbers, often using parts of regular medicine. The most important rival group gathered originally around another botanic—but non-Thomsonian—physician, Wooster Beach, originally from New York. Beach opened the first formally chartered botanic medical school in Worthington, Ohio, in 1830. Beach's group, which took the name Eclectics, eventually drew in many splinter Thomsonians. In the pre–Civil War decades, Eclectics and splinter Thomsonians imitated more

Opposite, This license to practice Thomsonian medicine was received by the purchaser of Thomson's book, *New Guide to Health* (1822), which contained instructions for treating patients and oneself. The license also gave the purchaser membership in the society composed of all official purchasers of the Thomsonian system. From the original in the collections of the Medical Heritage Center, Ohio State University.

conventional medical forms, setting up medical schools, organizations, and journals that competed with both regular medicine and strict Thomsonianism. Some Eclectics strayed into using nonbotanic remedies. In 1839 a dissident group of Thomsonians became the core of Physio-Medicalism, which to outsiders appeared to resemble closely the Eclectics.[55]

The groups that came out of botanic medicine had a substantial popular base in the Midwest and the South. They opposed regular medical practice and often had close ties to contemporary reform movements such as the antislavery crusade.[56] They were distinctive in that they advocated specific medications for symptoms, and they eventually lost out only with the popularization of the germ theory of disease. Just before that began to happen, in 1873, by one count, the specifically Eclectic physicians constituted about 5–6 percent of American physicians.[57]

Homeopathy and Hydropathy

The second major school to enter the medical marketplace were the homeopaths. Homeopathy originated in Europe. Devised by a disillusioned German physician, Samuel Hahnemann, the approach would spread throughout the world. From one point of view, homeopathy was another eighteenth-century "system" of medicine. This one, however, had a special appeal in the United States, an overwhelmingly Protestant country where dissenters were encouraged.[58]

Homeopathy had three major aspects. First was the idea that the physician should assist nature in healing. Instead of countering the effects of an illness, treatment should assist those effects, which were nature's efforts to cure the disease. In the case of fever, for example, homeopaths utilized heating medications. The popular motto for homeopathic treatment was "Like cures like." Second, Hahnemann used simple pharmaceuticals that he tested on himself and others. The "provings" were empirical in format, but they had the potential to be more evidence-based and "scientific" than the therapeutic traditions that guided many treatments at the time. A typical homeopath carried a case with a set of numbered drugs that would be prescribed in different combinations for each individual patient. Third, homeopaths believed that medicines were more effective when they were diluted, which was in direct contrast to the aggressive dosing of many regular and botanic physicians. Indeed, enemies of homeopathy told stories of homeopaths who went to extremes, such as one who supposedly put a couple of drops of a substance into Lake Michigan and then waded out some distance and scooped up water, maintaining that it was an extremely dilute, and therefore effective, medication. Hence the expression, still used today, "homeopathic dose," meaning a possibly ineffectively weak version of a medication. The spirit was caught by a satirist in 1848:

The homeopathic system, sir, just suits me to a tittle.
It proves of physic, anyhow, you cannot take too little;
If it be good in all complaints to take a dose so small,
It surely must be better still, to take no dose at all.[59]

Highly dilute medications gave homeopathic physicians a substantial economic advantage, because their medicines cost them almost nothing and could be given to patients without additional charges. It was long believed that in the cholera epidemics, homeopaths' patients, by avoiding bleeding and purging, less often suffered fatal dehydration. Indeed, in the mid-nineteenth century, a belief that patients of homeopaths survived better than those of regular physicians led some life insurance companies to offer lower rates to followers of homeopathy.[60]

The homeopaths were educated physicians. For some time, graduation from a regular medical school, in addition to training in Hahnemannian principles, was

This classic homeopathic practitioner kit from the latter part of the nineteenth century shows numbered bottles of medications that were used in various combinations to treat individual patients' ailments. From the original in the collections of the Medical Heritage Center, Ohio State University.

a requirement for membership in the professional organization.[61] They not only constituted a significant element among degreed physicians (perhaps 5–10 percent) but drew most of their clientele from the upper middle class and elite groups who employed educated doctors. They consequently had a disproportionate cultural— and political—impact. They had close ties, for example, to advocates for women's rights, and a significant proportion of homeopathic practitioners were women.[62] The press of the day was often friendly or even favorable. One critic called home- opathy the "Aristocracy of Quackery." In time, the well-educated homeopaths drew patients away particularly from less-educated botanic physicians, as education came to count for more among those choosing medical assistance.[63] As many his- torians have pointed out, the homeopathic alternative caused many regulars to modify their treatment regimens, sometimes radically. At the same time, homeo- paths over the years came to accept much of modern science and even vaccination. Later they splintered over how much of "regular," or "allopathic," science (using medicines to counter symptoms, as opposed to homeopathic medications, used to reinforce) to accept.[64]

Still another alternative treatment system came in with hydrotherapy. The "water cure" was systematized and introduced in Austria in the 1820s. It soon swept across Europe, and the first American water-cure institutions opened in 1843. By the 1850s, water cure was a major force in American health care. Advo- cates proclaimed that pure water applied both internally and externally had ex- traordinarily beneficial effects and definite curative powers.[65] Across the country, people with various complaints went to water-cure establishments. A bland, cool diet—taken with water—and exercise rounded out the hydropathic regimen. Women especially found the hygienic ideals and valuation of the body inspira- tional. At one point in the early 1850s the *Water-Cure Journal and Herald of Re- form* had the astounding circulation of fifty thousand. In New York, one char- tered school awarded MD degrees not only to men but to women, some of whom stayed in practice for many decades.[66]

Another European import taken very seriously by educated people was phre- nology, or the attempt to identify brain function by head shape. In the United States, practitioners came to believe one could cultivate personality functions that exami- nation of the head showed to be deficient. Such thinking prepared people later in the century to accept scientific localization of brain functions and in other ways to break from traditional views of humans and human nature.[67]

Self-Help and Commercial Home Treatment

In addition to the sectarian and various faddist physicians, the pre–Civil War med- ical marketplace was complicated by self-help alternatives. There were the self-help,

self-doctoring books, which flourished long after their eighteenth-century beginnings. There were traditional inherited "cures" that people indulged in. And there
were the commercial self-medications, the category generally referred to as "patent medicines."

Why were Americans so extraordinarily devoted to medicating themselves? Beyond the tradition of self-reliance and growing democratic scorn for any aristocratic expert, the growing American tradition of common school education made
it possible for an ever-increasing proportion of the population to read self-help
books. Nor were they passive readers, as notes and recipes in surviving copies of
the books testify. Also, physicians often were often not available to large parts of
the widely dispersed population. On the frontier and in other rural areas, geographical isolation encouraged many Americans to turn to instruction books and to commercially promoted medications.

Self-medication was a habit everywhere, in towns and cities as well as across the
countryside. In 1847, a St. Louis physician, J. V. Prather, examined 469 deaths of
children aged five or under occurring over a four-month period. Only 133 had been
under the care of a physician. Another 100 had been treated by quacks or irregulars. But 205 had had no physician of any kind.[68]

For the families of such children, the major alternatives to consulting a physician were folk remedies, folk wisdom, and the patent medicines. Actually few of
the "patent" medicines were actually patented, for that would have required manufacturers to disclose the ingredients. Instead, these were supposedly secret "proprietary" medicines protected by a copyright for the name. The vendors of patent
medicines were a colorful lot, and their advertising copy is endlessly amusing, especially when one knows what was actually in the mixtures. Again, popular liter-
acy encouraged people to read the nostrum ads. Many ads featured young women
of attractive countenance and bearing, suggesting even to those who were not very
literate that the dose was pleasant. Some of the makers of these commercial preparations and contraptions pioneered national marketing and advertising and thus
played a role in expanding commerce in general in the first half of the nineteenth
century.[69]

Merchandisers of medicines for self-dosing continued to flourish as newspapers
and other printed materials multiplied in the decades after American independence. Yet there was a change. Whereas in 1771 all the proprietary remedies in one
catalog were imported, by 1804 a majority of the nostrums in a comparable catalog originated in the United States, and the trend was clear.[70] Newspapers continued to increase in number and circulation, especially after postal rates declined in
the 1840s. By 1860 there were almost four hundred daily papers and thousands of
weeklies. In the nineteenth century, ads for medications, no matter how false,

This tonic label from the 1850s was designed to draw attention (note the attractive female) and to suggest graphically how bottles of Kentucky Tonic Bitters, a patent or proprietary medicine, would do battle with various human pains and ailments. Courtesy of the National Library of Medicine.

misleading, or offensive, filled such periodicals, including religious periodicals. A typical local paper from Hamilton, Ohio, in 1840 comprised four pages, with seven columns per page. Ads for medications filled fully nine of the twenty-eight columns.[71] Given the economic power of this advertising, no politician could move against even blatant quackery and deception in a free market.

The number and variety of commercial substances and gadgets marketed to an eager public is testimony to human chicanery and hope. A large percentage of proprietary medicines secretly contained alcohol, which anyway continued for decades to be considered of medical benefit. One of the most successful, Pe-Ru-Na, was 28 percent alcohol. Others were, for example, Mrs. Winslow's Soothing Syrup, for teething children (later found to be a sweetened solution of morphine) and "female regulators," supposedly for birth control. Most amusing, albeit clearly profitable, were the substances and contrivances for men that cured "genital debility" or "seminal weakness."[72]

Advertisers' claims were extraordinary, if often vague.[73] Moreover, the market continued to grow. By the end of the nineteenth century, the U.S. Patent Office had issued patents for 321 disinfectants, 250 extracts, 375 internal remedies, 56 plasters, and 371 topical remedies. And trademarks for proprietary medical compounds already numbered almost 6,000![74]

Personal Health

In general, when individuals in the first half of the nineteenth century took responsibility for their own health, they had in mind chronic diseases and quality of life, that is, freedom from ordinary debilities, aches, and pains, of which there was a continuing abundance. As noted earlier, people ascribed acute or epidemic disease to a punishment from heaven over which they had no control, as one might be struck by lightning.[75]

From the 1820s through the 1850s many Americans, as part of evangelistic zeal, sought day-to-day personal health as an aspect of religious salvation, and many opinion leaders maintained that leading a traditionally healthful style of eating, drinking, breathing, evacuation and retention, and control of the passions was a moral and religious obligation. The usual rationale was that the body was the temple of the soul and should therefore be looked after. Elisha Bartlett, a physician in Lowell, Massachusetts, in 1838 wrote a tract explicitly titled *Obedience to the Laws of Health, A Moral Duty*. In 1856, Catharine Beecher described the rules of health and observed that "our Creator has connected the reward of enjoyment with obedience to these rules, and the penalty of suffering with disobedience to them." Health reform was a basic constituent of the general reform movement of the 1830s–1850s. Uplift leaders typically viewed physical care of the body as a fundamental part of living a moral life. Doing without at least alcohol, undue sexual stimulation, immoderate eating, and other excesses would result in both physical and moral-religious improvement.[76] Reform movements were very broad. When the vegetarians met in a national convention in 1850, the supporters present included not only abolitionists and hydropaths but Reuben Mussey, who was president of the American Medical Association in that year.[77]

The best-remembered health reformer (and vegetarian) was Sylvester Graham, whose prescription for healthful living still persisted in various forms many generations later, not least in the guise of graham crackers, made of graham flour, which Graham supposed was good for a person's body because it was "natural" (coarsely ground and including the whole kernel). Graham lectured widely to large audiences, proselytizing for his beliefs, which reflected the traditional teachings of health enthusiasts, to which he added a little physiological theory of his own time. Graham insisted that health depended most upon a person's body, not external miasmas (see below) or other uncontrollable factors. Moreover, as a health reformer, he held that everyone was capable of controlling the conditions of his or her own body, just as other reformers were stressing personal responsibility for moral and economic well-being. This emphasis on personal responsibility for health led to a widespread mind-set that combined prevention with various fads for curing. It also fed into the thinking that was changing regular medical therapeutics.[78]

Similar to Graham, and more successful, was the reformed physician William A. Alcott, a self-styled "medical missionary" who preached on the virtues of moderation, vegetarianism, cold water, and other aspects of healthful living. In embracing nature as the only true physician, Alcott was, he said, declaring his "medical independence."[79]

One of the social reforms of the day had consequences that later took on great importance in the profile of medical care. A stereotypical New England spinster,

The imposing Western Lunatic Asylum in Staunton, Virginia, c. 1860, typified the response to a powerful reform movement led by Dorothea Dix that created a system of state hospitals to provide safety and care to mentally ill people. Henry M. Hurd, ed., *The Institutional Care of the Insane in the United States and Canada*, 4 vols. (1916; reprint, New York: Arno, 1973), 3:723.

Dorothea Dix, discovered that people with mental illness were often kept in pitiful and inhumane conditions at home or in jails and workhouses. In 1841 she began a crusade to get states to establish asylums or hospitals where people suffering from a disease, labeled at the time "insanity," could be given shelter and a chance to recover. She succeeded beyond any expectation, became a heroic national and international figure, and initiated what in the United States grew into a system of state hospitals for the mentally ill.[80]

Ideas about People's Bodies

Health advocates of the antebellum (pre-1861) period conceived of following the rules of health as keeping "the system" operating properly. The system was the series of organs that worked by mechanical balance and sympathetic interaction. Coordinating the whole was a vital spirit that animated all of the organs; one historian described it as "an integrated package of inbuilt mechanisms." Since each part of the body had a use—the stomach to digest food, the eyes to see—the mechanisms in the body, the organs, had moral purposes and, if abused, moral dangers. To antebellum thinkers, their physical health was an individual matter, just as the obedience to the traditional laws of health was an individual matter.[81] Yet people at the time also feared that the social and technological changes taking place around them were bringing new dangers to their bodies. Civilization and bad habits, they believed, could bring "weakness" of both will and physique. Such factors all worked together in campaigns to enhance morals, promote uplift, and improve personal well-being.

Another new factor also entered Americans' quest for health: the growing role for women concerned about their own bodies on not only spiritual but also social or even political grounds. In the paternalistic order of that day, women's bodies were supposed to be special. For most articulate Americans, women's physical functioning centered on their reproductive function. Women were considered weak and susceptible to emotional expression. But in the early decades of the nineteenth century, economic changes began to move production out of the home, creating a growing population of middle-class women who were expected to focus on the family but also to play their part in society by shopping and doing charity work. In the general social disorder, many women applied the messages of health reformers as a way of making sense of changing circumstances.

In a society in which men were supposed to employ self-improvement for economic and vocational advancement, women were enjoined to believe that caring for their bodies was a duty. Thus did male chauvinism open a path to feminism. Significant numbers of at least middle-class women eagerly sought means to guard their health and were conspicuous figures in the medical marketplace, especially

as they increasingly patronized irregular physicians and healers. They also looked
after themselves medically. Beyond reading advice books, middle-class women,
particularly in the Northeast, gathered in groups to hear lectures and to perform
hygienic exercises to keep disease from their bodies. A group in Boston in 1857 was
devoted to "the study of the human body, —to learn its structure, its functions, its
derangements and to prevent suffering, —to learn the laws which govern life and
health." And these groups took on social and political significance as women discovered that they could influence the world around them. The laws of life and health
once again meshed perfectly with the temperance and other reform movements.[82]
An 1850s female physician opposed to tightly laced corsets and high heels and generally advocating healthful dress reform exclaimed: "How . . . glorious would it be
to see every woman free from *every* fetter that fashion has imposed! . . . a day of
'universal emancipation' of the sex."[83]

Alternative medicine and self-dosing flourished in part because physicians
disagreed with one another and in part because so many people believed in
the self-reliant "common man." Many Americans, as I have suggested, utilized
sectarian and self-devised medicine for religious, political, or philosophical
reasons, if not sometimes to reinforce their regional and social class identities.
Moreover, as marketing intensified, dosing according to advertising began not
just to supplement but to replace folk medicine. In an increasingly commercialized world, this transformation was appropriate. The traditional almanacs, for
example, continued to flourish. However, the editors dropped their astrological
sections but expanded their recipes for cures, which were now supplemented by
ads for patent medicines.[84]

Traditional Medicine Rooted in the Community

By the 1850s a pattern of medical practice by physicians trained either by apprenticeship or in medical schools was in place across the country. More than 80 percent of the population lived in rural areas or in scattered towns so small that residents tended to know everyone else there. These, along with occasional small cities,
were the "island communities" that, with the surrounding countryside, were largely
self-sustaining except for trading with urban centers with ties to the larger world.
Any physician living in such a community had to fit in with the community and
deliver traditional medical care if his or her practice was to survive. Medical practice therefore did conform to standard expectations. The doctor had to respect and
work with patients who were used to treating themselves and who would decide
just how much of any recommendation to follow.[85]

A historian who has examined numerous records of mid-nineteenth-century
physicians describes what he calls the "country orthodox style" of practice, a prac-

tice that "remained remarkably constant over time," into the late nineteenth century. Indeed, it obviously had origins in earlier doctors' customs, or at least those who could provide a dignified model. To some extent, practice was a ritual. The doctor had to offer a diagnosis and to prescribe a treatment, usually, if not a dressing, one of a limited number of available medications, among which each physician had favorites, and usually more than one, perhaps generalized as "sundry medicines." Medical innovations and theories seldom reached the countryside, where patients usually would not have welcomed them. What counted was the physician's experience.[86] Doctors in this comfortable, familiar system could indeed act the part of a wise friend and stay to look after the patient.

Continued Dependence on European Ideas

After the War of 1812 and the internal economic development that followed, social changes fed into determining not only the circumstances but the very ways in which people cared for their health. In contrast to the turn to inward development and independence in economic expansion, medicine as such was still essentially a derivative activity. Almost all changes in ideas and practices were imported. The most advanced practitioners looked to what was happening in Europe if they wanted to improve their practices. They could even turn their distance from Europe into a unique American advantage. John Syng Dorsey, in his pioneer textbook of surgery in 1813, wrote that "an American, although he must labour under many disadvantages in the production of an elementary treatise, is in one respect better qualified for it than an European surgeon. He is,—at least he ought to be,—strictly impartial, and therefore adopts from all nations their respective improvements." This deliberate eclecticism implicitly entrenched traditional medicine. By citing all customary authorities, medical leaders could avoid embracing fully any disturbing new idea. Like one Ohio textbook author in 1831, John Eberle, they could continue to depend on established "essential phenomena and principles" and familiar classics.[87]

Figures on medical book publication assembled by John S. Billings suggest the continuing but finally diminishing colonial, that is, derivative, profile of American medicine (table 2). As Billings points out, the numbers not only reflect increases in population and wealth but also show that reprints and translations of European publications (in a day when there was no copyright protection) declined proportionately from 1800–1809, when there were more than twice as many European books published in the United States as books by local authors, to 1850, after which the proportion evened up and began to reverse.[88]

TABLE 2

Categories of medical books published in the United States, 1775–1869

	1775– 1788	1800– 1809	1810– 1819	1820– 1829	1830– 1839	1840– 1849	1850– 1859	1860– 1869
U.S. authors, first edition	39	24	51	48	83	96	101	157
U.S. authors, all editions (vols.)	51	31	77	86	136	162	197	256
Reprints and translations (vols.)	49	76	111	135	192	214	184	160

Source: John S. Billings, "Literature and Institutions," in *A Century of American Medicine, 1776–1876*, by Edward H. Clarke et al. (Philadelphia: Henry C. Lea, 1876), 294.

The French Clinical School

The first major new ideas from Europe that led some American physicians to modify their thinking came from the French clinical school, carried to the United States by Americans who recognized that Parisian teachers were indeed improving on what physicians knew. The movement began in the 1820s, when at least 105 American doctors chose to go to Paris rather than Britain for further study. One American estimated that 40–50 of his countrymen were studying in Paris in December 1829. This group was the beginning of a small flood, and they carried back the word about what was to be learned in France. They translated French works. They also transported instruments. As early as 1822 John Bell, in Paris, was sending a stethoscope to Usher Parsons, in Providence, who had become an advocate of French training when he stumbled on the Paris medical scene while traveling for his health. In the 1830s, at least another 222 American physicians went to Paris, including the first African American MD, James McCune Smith, in 1837. Hundreds more physicians made the pilgrimage from the United States before 1850.[89] Already in August 1826, Elisha Bartlett wrote home with unusual understanding of the changes he was seeing: "The celebrated Laennec [who devised the stethoscope] died at his country residence on the 13th of the present month. The publication in 1819 of a new method of ascertaining diseases of the chest forms an era in the history of medicine."[90]

The Americans studying in France were selective, but they were impressed by "facts" that their French teachers demonstrated to them. Most basic was pathological anatomy, the idea that diseases had their seats in specific locations in the body, rather than being a general condition of the whole body. It was autopsies that allowed anyone to see, for example, the intestinal lesions typical in typhoid. The greatest Parisian teacher, Pierre Louis, in particular showed that treatments

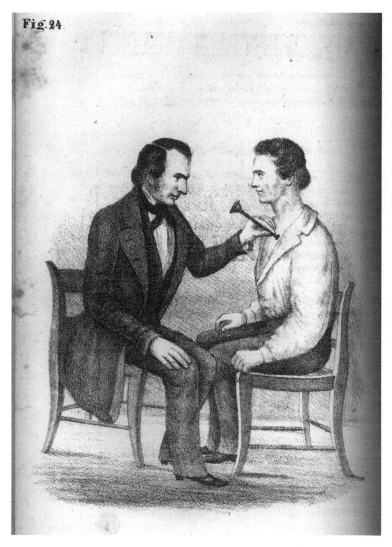

Illustration from a medical journal in 1854 showing a physician using a monaural stethoscope. Subsequently the instrument was modified to provide tubes for both of the listener's ears. M. Lawson, "Lectures on the Pathology, Diagnosis and Treatment of Diseases of the Chest . . . Mode of Auscultation—Stethoscopes," *Western Lancet*, 11 (1854), 137.

could be found effective or ineffective by counting the outcomes of patients rather than relying on selective clinical impressions. Most controversial was that persistent bloodletting could not be shown to produce favorable outcomes in pneumonia and some other diseases.[91]

Americans also brought new techniques from France. Some were surgical innovations, but use of the stethoscope, above all, marked the beginning of a significant shift in medicine. With a stethoscope, the physician gained information about a patient's body that the patient did not and could not know. Increasingly, this kind of technique would give the well-trained doctor a concrete advantage over just an ordinary wise person.[92]

Perhaps the most significant immediate legacy of American physicians' visits to Paris was the extent to which they rejected the great theoretical systems of Europe. Those systems had justified extremes in treatment, as in the case of Rush. In a traditionally American, practical way, the new generation of doctors tried to concentrate on the individual patient and on therapy.[93]

The Healing Power of Nature

The style of friendly, observant medical care could apply to physicians following the tradition of moderation as well as to those devoted to heroic treatment. But now it could apply also to physicians who returned from France with a somewhat different approach to healing. Critics of that fresh approach designated it suggestively "the nature-trusting heresy," that is, heretical to the traditional medical devotion to therapeutic activism.[94] The approach was soon carried to extremes by physicians in Vienna, where deference to natural processes was designated "therapeutic nihilism." The view that the physician should take advantage of the healing power of nature, as I have suggested, contrasted with, for example, that of the followers of Benjamin Rush, who believed that therapeutic intervention ("medical art") should counter the natural forces that were causing the disease.

One critical essay in the new nature-trusting heresy was published by Jacob Bigelow, of Boston, in 1835. Subsequently it was reprinted, and other medical writers across the European world cited his manifesto. Bigelow pointed out that many diseases were "self-limited." Such a disease, for example measles or chicken pox, could be identified, and

> after it has obtained foothold in the system, cannot, in the present state of our knowledge, be eradicated or abridged by art [medical intervention], but to which [i.e., to the disease] there is due a certain succession of processes to be completed in a certain time; which time and processes may vary with the constitution and condition of the patient, and may tend to death or recovery, but are not known to be shortened or greatly changed by medical treatment. . . . These are strictly self-limited diseases, having their own rise, climax, and decline, and I know of no *medical* practice which is able, were it deemed necessary, to divert them from their appropriate course or hasten their termination.[95]

Bigelow characterized the physician as "but the minister and servant of nature." And it was clear that in many cases the physician functioned merely to nurse the patient. So the doctor could be one "who turns a pillow or administers a seasonable draught of water" to palliate the suffering of the patient until the physician had seen the disease through to a natural termination.[96] Bigelow expressed skepticism about the usefulness of many medications, and he implicitly agreed with the notorious assertion made by the physician Oliver Wendell Holmes as late as 1860 that, leaving out opium and a few specifics, "if the whole materia medica, *as now used*, could be sunk to the bottom of the sea, it would be all the better for mankind,—and all the worse for the fishes." Indeed, Holmes in 1860 was launching an even more direct attack than Bigelow against heroic practice and against the faulty logic that justified powerful dosing. Holmes went on, for example, to assert that "the injuries inflicted by over-medication are to a great extent masked by disease.... How is a physician to distinguish the irritation produced by his blister from that caused by the inflammation it was meant to cure?"[97] At that time, just before the Civil War, the current against traditional heroic practice was growing ever stronger, particularly in light of another long, local tradition in New England.

Identifying Specific Diseases

The nature-trusting heresy was only one face of a major change that was affecting advanced American medical thinkers in the pre–Civil War period. Particularly under the influence of pathological anatomy brought back from France, they began increasingly to discuss the causation of specific diseases (etiology). Indeed, in American medical publications etiological discussions reached a high point in the 1840s. In that period, medical authorities were becoming more and more aware of specific diseases, with the implication that each disease had a special cause related to change (detectable at least in an autopsy) in some location in the body. By 1840 the old distinction between remote and proximate causes of disease was losing meaning as less speculative pathologies came into play, and people noticed immunity, which, as induced in vaccination, applied to a specific disease. Such imaginative diagnoses as "biliary affection," which could mean almost anything, began to disappear.[98]

In medical practice and especially in general public discussion, yellow fever, malaria, and the new threat, Asiatic cholera, were distinctive and in general baffling. Some diseases that generated popular apprehension, such as smallpox and rabies (typically caused by the bite of a "mad dog"), were traditional and obviously passed on by contact, just as syphilis was. The problem was with the old standards, fevers and fluxes.

In order to identify specific diseases, one had to answer two questions: (1) What in the external environment brought the disease to someone's system—which led to hygiene and public health questions; and (2) What was the peculiar nature of the disease process that made one disease different from another? From 1847 to the Civil War the standard U.S. textbook listed a number of specific diseases, such as scarlet fever, smallpox, measles, plague, and yellow fever. But the larger part of the work still was not organized by diseases but instead covered symptoms grouped as they occurred in various parts of the body, such as the mouth, the joints, the bowels, the urinary organs. For most physicians, any disease was still specific to the patient, and the patient's constitution and immediate environment, like the weather, should modify any therapy.[99]

Pathological anatomy clearly was becoming central to differentiating how one disease differed from another.[100] A major step came in the work of William Gerhard, of Philadelphia, who in the 1830s differentiated typhoid from typhus on the basis of pathological anatomy.[101] This rare contribution of an American to world medicine came not from a frontier post but from a practitioner doing clinical research in an urban setting. Indeed, already in 1829, the first American textbook of pathological anatomy, written by William E. Horner, of Philadelphia, had appeared.

Departing from Convention in Medicine

An occasional American physician in the first half of the nineteenth century, like Gerhard, did make a contribution to medicine that Europeans recognized—and that patriotic medical writers and speakers could then cite as evidence to bolster their argument that there was an admirable medical profession in the United States. Most such innovations, however, were in the surgical realm. The best-known U.S. contribution, copied in Europe within weeks, was devising ether anesthesia for surgery, which resulted in reducing pain, making surgery more acceptable, and allowing surgeons to work more slowly and more carefully. Even at the time, the first public trial of anesthesia by a physician was recognized as "one of the important discoveries of the age." Although the practice was introduced at Massachusetts General Hospital in 1846, physicians for years used anesthesia only selectively. They were slow, for example, to apply it to childbirth. When in 1847 Fanny Longfellow (wife of the poet) reputedly became the first American woman to use ether to ease the pains of childbirth, she was aware that already "such a thing had succeeded abroad."[102]

There were also accounts of innovative gynecological surgical procedures introduced by accomplished Americans such as J. Marion Sims, of Alabama, which led a number of midcentury physicians to specialize in gynecology. Oliver

William T. G. Morton, an inventor and promoter of ether anesthesia for surgery, used this model to apply for a patent at the U.S. Patent Office. The patient would breathe the fumes in air passing over the ether in the container and become insensible to pain. George B. Roth, "The 'Original Morton Inhaler' for Ether," *Annals of Medical History*, 3rd ser., 4 (1942), 391.

Wendell Holmes pioneered the idea that puerperal (childbed) fever, an infection that carried off many women just after they had given birth, was caused by physicians who somehow carried the disease on their hands and persons. And one frontier doctor, William Beaumont, who by chance acquired a patient with an external opening into his stomach, made major contributions to the physiology of digestion.[103] Such instances of American ingenuity, much publicized later in the nineteenth century and after, were remarkably few.

There was another level of innovation, one that is seldom noticed, but it was not trivial. In the United States, a lack of governmental or institutional regulation, together with the culture of individual independence, permitted practitioners to try out new ideas spontaneously. Such experiments usually did not end well, but occasionally something useful came from them. In one instance, Thomas Fearn, a doctor in Huntsville, Alabama, in 1831 was called to treat a local farmer's daughter, who was terribly ill with a seasonal "bilious fever" (a high fever; the very term was traditional, referring back to bile, one of the ancients' four humors, which could become out of balance). Fearn had some quinine sulfate, which had recently been extracted as the active principle in Peruvian bark. Since it seemed that the girl was going to die anyway, Fearn attempted a heroic and dangerous dose of quinine, ten times the usual amount. This was essentially a human experiment. To Fearn's astonishment, the girl's fever broke, and she recovered.

Fearn did not publish his account until 1851, but for twenty years he traveled the South and the West telling other practitioners about the nontraditional dosage with which he had obtained dramatically better effects. Others had similar experiences, and over time quinine became not just a tonic often used in fevers but a much more precise medication for periodic fevers, that is, malaria, fulfilling the empirical use of bark as one of the few specifics available.[104] Clearly, as in past centuries in other places, there were clinicians in America who were quietly extending or breaking away from tradition and instigating innovations. The source of variation in practice might be Paris or Vienna or, more rarely, the rural back country. Despite the tension between medical thinking from Europe and the local experience of practitioners, some Americans were beginning to distance themselves from at least some traditional therapies and thinking.[105]

Defending Tradition

Even as they strongly advocated the traditional therapies, leading American physicians in the early to mid-nineteenth century were quietly abandoning them. The historian John Harley Warner has shown vividly that the actual use of bleeding and purging declined decisively in medical practice from 1820 to 1880. And identifying specific diseases that were universal was rendering obsolete therapies aimed at general conditions and individual constitutions. But people at the time did not recognize how radically medicine was changing, because the process of departing from tradition was so slow and the rhetoric so misleading. Even late in the century it would take great courage for a physician to deny that "bloodletting is the sheet anchor" of medicine.[106]

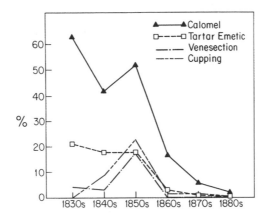

Graphic depiction of the variation and then decline in the actual use of traditional heroic therapies, particularly calomel and bloodletting ("venesection"), in practice at Commercial Hospital in Cincinnati, regardless of what writers at the time publicly recommended or defended. John Harley Warner, "Power, Conflict, and Identity in Mid-Nineteenth-Century American Medicine: Therapeutic Change at the Commercial Hospital in Cincinnati," *Journal of American History*, 73 (1987), 941. Reprinted with the generous permission of Oxford University Press and John Harley Warner.

Beginning in the 1820s, as orthodox physicians became aware that they were under attack from sectarian practitioners, the regulars defended the unity of their profession. Even as leading regular physicians were slowly moving away from traditional practices, especially bleeding and purging, in public debates they tended to emphasize how traditional they were—to distinguish themselves from the sectarians with their new-fangled therapies. In 1859, for example, in the St. Louis Medical Society, where a discussion showed that bloodletting was seldom used, no one would speak up against therapeutic bleeding, as the secretary reported, and "we must conclude that the lancet holds the same place as a remedial measure, among well educated medical men, that it ever did." In such ways, orthodox rhetoric honoring traditional medicine continued to disguise the extent to which physicians were moving away from traditional medicine.[107]

Changes in the Health Status of Americans

One question historians have asked is whether improved health care, especially the decline of heroic practice in the middle decades of the nineteenth century, had any effect on the health of Americans overall. Demographic studies show that life expectancy did not steadily "progress" in the nineteenth century, as one might expect.

Evidence suggests two substantial changes. The first one, noted above, came in the decades immediately after the American Revolution, a change for which physical evidence exists: Americans grew taller, markedly taller, than their colonial forebears and taller than their contemporaries in Europe. In an expanding economy, American children's diets had improved markedly.[108]

From the 1830s to the 1850s, however, something happened. People died earlier than had been the case before 1810. Indeed, life expectancy began to decline in 1810 and did not recover and begin to increase again until the 1880s. Living conditions apparently deteriorated for many Americans as industrialization and urbanization got under way immediately before the Civil War. Skeletal remains and other evidence show that in the period of economic expansion the average height of the population, the continuingly good measure of nutritional resources and health for children, began to decline substantially in the decades just before 1860, even though the U.S. per capita net economic production (a measure of wealth) increased by more than 40 percent between 1840 and 1870. Economic growth came at the expense of children's nutritional well-being as well as adults' health.[109]

There was yet one other aspect of medicine and health that did not change before the Civil War: public health and controversies over the nature of epidemic diseases. The cholera epidemics in particular precipitated public arguments over the causation of all specific diseases. By the mid-nineteenth century, physicians and

other informed Americans were carrying on an explicit debate (reflecting a similar debate in Europe) about whether specific diseases were contagious or noncon-
tagious. If specific diseases were contagious, they spread by personal contact. Perhaps in crowded, poorly ventilated quarters, poison spread from the "emanations"
of one person to another person. Or some mysterious poison was transmitted by
material objects, fomites—"substances which are supposed to retain contagious effluvia," as Dunglison's medical dictionary put it at the time—for example, the dirty
bedclothes of a smallpox or cholera patient.[110]

Those who believed epidemic diseases spread by means other than from person to person were the anticontagionists. They drew on a rich, traditional European literature, and they dominated American medicine. Already in 1827, a French
physician reported that he had found 568 American physicians who held yellow
fever not to be contagious, compared with only 28 who thought it was contagious.[111]
Most anticontagionists commonly invoked a traditional theory of *miasma*. The
miasma was an atmospheric emanation that came from decaying organic material
and stagnant water. Just as in the early nineteenth century, on a hot summer's day,
one can still sense what might be the miasmatic vapors rising from a swamp—
"febrific agents of a gaseous form," in the words of Eberle, the textbook writer.
And when miasma came, so did malaria, yellow fever, and the summer diarrheas
that killed so many babies. Moreover, "miasmata" could come from not only
swamps and ponds but also gutters, puddles, or close, damp air in cellars or
ships' below-deck quarters. Indeed, any decaying or putrefying animal or vegetable matter exposed to air could give rise to miasma, "vegetable and animal substances contained in the public filth of cities, in marshes and in other situations
furnishing these materials," as Eberle put it.[112]

Closely related were theories blaming atmospheric conditions. Many physicians
and public-spirited people recorded weather conditions conscientiously. When the
atmosphere became corrupted, it would act to transmit poisons that caused diseases. In 1849, a physician wrote of yellow fever that "the efficient cause . . . is an
aerial poison, probably organic, which requires a certain temperature for its generation and existence, and affects special localities and persons."[113] In the mid-
nineteenth century, physicians commonly used the term *zymotic* to categorize diseases that were "epidemic, endemic, contagious, or sporadic," that is, spreading by
some mechanism, as in a fermentation process. Such thinking assumed the existence of some "principle" or poison that traveled.[114]

There were social as well as economic issues in this medical debate. Quarantines, as noted earlier, were common and costly, whether because they interrupted
trade or because they isolated or confined sick persons. Yellow fever epidemics were
profoundly disruptive to businesses in the Gulf Coast states, even without the death

lull. At the end of the 1850s, following critical yellow fever epidemics in southern ports and a small outbreak in New York, there were three national sanitary conventions. One hope was to standardize local quarantine regulations. Instead, the delegates repudiated quarantine. At the third, held in 1859, the assembled health officials and experts resolved, 85 to 6, that it "is the opinion of this Convention that the personal quarantine of cases of Yellow Fever may be safely abolished, provided that fomites of every kind be rigidly restricted."[115]

The irony was that the anticontagionists were most strongly in favor of clean water, sewage and garbage disposal, and fresh air, and they opposed filth in general. Indeed, a few cities introduced large waterworks to bring in fresh water. Between 1830 and 1860, the number of waterworks in the country more than tripled. Unfortunately, however, the urban population increased by six times. A reform writer in 1851 declared, "Thousands die annually before their time, and tens of thousands waste much of their lives on beds of sickness, not by the inscrutable purposes of their Creator, but because the noisome atmosphere of uncleanliness, disease, and death has been allowed to gather and float about them, til the lamp of life has gone out." A writer advocating public sanitary measures to stop yellow fever in 1855 went so far as to declare that opposition to sanitary progress was "un American."[116] The contagionists, by contrast, advocated quarantine and concentrated, not on the environment, but on people who carried the disease. In the case of smallpox, individuals could be vaccinated, and as early as 1845 Cincinnati, for example, compelled unvaccinated people to be vaccinated when smallpox hit the city.[117]

Neither contagionists nor anticontagionists had a clear idea of how poisons or specific diseases traveled. In 1853–54 the idea of London's John Snow that cholera was borne by water entered medical discussion. The impact, however, was only to intensify calls for a clean water supply, which already were growing louder in urban areas. It would be another few decades before both contagionists and anticontagionists could unite around the new germ theory.[118] Meanwhile public health measures consisted still of sporadic prayer, cleaning of the streets, and quarantine.

The Persistence of Traditional Thinking

History generally is about change. But a more appropriate focus with regard to health care in the United States in the first half of the nineteenth century is how superficial the changes were. Folk beliefs could be repackaged as commercial nostrums. Traditional wisdom could appear in newspaper and journals produced by new steam-powered presses. There were medical schools and doctors and medical organizations in sectarian medicine just as in regular medicine. Physicians continued to treat individual constitutions, whether in heroic or nature-trusting styles.

The treatments were rationally organized—stimulants, depletives, alteratives—but later authorities believed the treatments were still limited in their effect on diseases. Surgeons had improved instruments and even anesthesia, but little changed in goals or outcomes. A few more specific diseases had been identified, but mostly disease still depended upon personal constitution, fevers, fluxes, atmosphere, miasma, and mysterious metaphorical poisons. It would be at least a generation later before modernization greatly modified traditional thinking in medicine and health in the United States.

Meanwhile, the pathological studies that connected localized bodily conditions to specific diseases were slowly redirecting traditional thinking among physicians and well-educated people. The traditional medical view was that doctors should treat the symptoms, which were caused by a combination of contingent factors. On the one hand were an individual's constitution and personal health regimen, such as indolence, drunkenness, and lack of exercise. On the other hand were external factors such as the weather, the climate, and the pressures of living. The multiplicity of factors gave a physician great leeway in treatment. But as pathological studies, or in such cases as Bigelow's, clinical studies, carved defined diseases out of the symptoms, physicians silently and even unconsciously used fewer traditional prescriptions and procedures, particularly bloodletting. They hung on to traditional rhetoric and formulations even as they gradually abandoned many traditional ideas and therapies.[119]

CHAPTER 4

Setting the Stage for Modern Medicine and Health, 1850s to 1880s

By the 1850s first one circumstance and then another began to set the stage for a new way of thinking about medicine and health. Most obviously, American medicine was still derivative, if not colonial, but ambitious physicians traveled less often to French medical schools and more often to those in Germany, where laboratories and new scientific studies were bringing far-reaching changes to medical thinking worldwide. In the United States, well-educated people's understanding of the healthy and the diseased human body was changing rapidly. In addition, urban centers, which were developing on the basis of industrialism as well as trade, became the site of social and especially intellectual and cultural change in the 1860s and 1870s and after. Population concentrations in cities, particularly in the East and the Midwest, could now sustain significant beginnings of a novel feature, medical specialization. Specialized functioning of course was the purest element in the "organization" of modernization.

Traditional Practice and Thinking under Siege

Industrialization, the underlying basis for modernization, came with a rush in the 1850s to 1880s. With industrialization came another kind of organization, national-scale businesses. And more than ever, opinion makers praised innovation, not tradition. Yet in medicine and health, tradition still shaped ideas about a person's constitution and environment, even as a number of leading intellectuals advocated a skeptical, laboratory-based view of health and disease. Step by step, and without really admitting it, leading regular physicians were abandoning traditional therapies and then traditional ideas about disease. With no full replacement for the traditional, it was sensible for them to adopt a strategy of therapeutic nihilism for therapy and shift their faith to new forms of science.[1]

Some historians of American medicine have characterized the 1850s to the 1880s as the "doldrum decades" because, in those years, U.S. physicians made remarkably few recognized contributions to the rapidly expanding body of medical knowledge.[2] Some members of the medical profession, particularly in the growing cities, sensed that they were doing important, highly skilled work. However, they were not contributing significantly to new medical thinking, and they regretted this shortcoming. In 1877 the first president of the American Dermatological Association quoted a colleague from 1871 who had deplored the lack of American publications on dermatology: "Can we wonder, then, that America has as yet contributed little to dermatology?" Years later, a colleague could still describe the literature in that specialty produced in the United States as "vast in yearly amount, meager in contributions."[3] The growing volume of all medical publications was in fact made up mostly of digests, textbooks, popularizations, and translations of European work, as suggested in Billings's statistics in table 2. Any claims to original work were very limited. American physicians did continue to share their clinical experience by publishing case reports. In the first issue of the *Journal of the American Medical Association,* in 1883, all three of the original contributions were standard case reports, although, at that late date, subsequent issues began to carry an occasional experimental or more complex clinical study.

At least part of the later perception of American backwardness derived from comparison with the growing flood of innovation coming from overseas. Just keeping up with new discoveries and ideas was a real achievement for a physician. The steamship, the railroad, and then the telegraph sped communication. The number of national and international medical meetings increased dramatically. Moreover, surgeons more frequently visited one another's clinics, a fundamental type of communication not directly recorded in print.

In the United States, laboratory and clinical studies coming out of German-language universities and health institutions simply swamped local islands of excellence. Not only were American physicians studying in German, Austrian, and Swiss medical schools in record numbers but after the revolutions of 1848 in Europe, many German-speaking physicians emigrated to the United States. Both print and personal communication with Europe, including Britain, kept ideas flooding in from Germany and from some French connections as well. American medicine, exposed to so much from outside, became perhaps even more intellectually derivative than earlier in the century.[4]

Organizing Medical Information as Progress

Ironically, this lack of originality in American medicine led to one of the most important contributions to world medicine: the *Index Medicus.* In the 1870s John Shaw Billings, then a young physician working in the office of the surgeon general of the army, began adding to the small library in that office all of the medical publications he could acquire—from all over the world. But anyone in that library, or anywhere else, attempting to find material on a particular subject had no guidance.

John Shaw Billings working in his study, presumably when he was gathering material for *Index Medicus*, which he created in 1879, bringing order to efforts worldwide to recognize discoveries in medicine and making modern "progress in medicine" visible in each successive issue. Courtesy of the National Library of Medicine.

One had simply to review likely publications one by one. Billings therefore got authorization to issue a printed catalog of the library with subject entries. As increasing numbers of publications poured into the library, he worked furiously to add to the subject-category index.

Publication of the catalog, however, lagged years behind the appearance of new articles and books, and so in 1879 Billings began to issue a monthly classified list of a surprisingly large percentage of all of the newest medical publications in the world, chiefly journal articles. Thus was born the *Index Medicus*, which flourished and grew and became an absolutely essential element in medical research in all countries for a century and a quarter or more. The *Index Medicus* may not have constituted an original research contribution, but it was a nice parallel to the also distinctively American exploitation of the interchangeable-parts system used in mass production in factories, which likewise multiplied "output" in a modernizing world. In addition, as one issue followed another, the index represented and implicitly emphasized newness in medicine, one way to define what was modern by the 1870s.[5]

In the midst of the takeoff in American industrialization and, of simultaneous changes in medicine that began with German imports in the 1850s, came a terrible interruption, the 1861–65 Civil War. But soon after the war, the momentum of change in medicine intensified. Moreover, in discussions of medicine, innovation became a dominating theme. Typically, physicians, and others as well, equated newness with progress and now spoke relentlessly of "the progress of science." Journals had sections titled "Medical Progress," containing abstracts of recent publications. As early as 1858 George B. Wood, revising his textbook of medicine for a new edition, complained of having "to keep pace with the rapid progress of theoretical and practical medicine in the last three or four years."[6] After 1865, sensitivity to innovation and progress pervaded medical publications, implicitly devaluing the old.

The Impact of the Civil War

Historians have so devotedly explored the medical history of the American Civil War that we have a remarkable number of personal stories and accounts of medical care in and around the armies and navies. This extraordinary detail reveals vividly what health care was like at that time. Unlike the usual published cases, in which the authors focused on their successes, which presumably symbolized progress, the war yielded an extraordinary abundance of descriptions of, and knowledge from, failed as well as satisfactory treatment. A young man died despite attempts at therapy, for example, but autopsy showed exactly how a musket bullet had damaged his liver and produced the symptoms he complained of before his death.[7]

In spite of all the historical attention to medical care in that war, the war did not shift the general course of medical knowledge or science into new channels. The official contemporary historians of the war scrupulously recorded in footnotes innumerable small claims of originality but could detect no particular breakthroughs. From the field, some surgeons reported ingenious operating techniques, such as an emergency direct puncture of the bladder, or some mechanical innovation, like a duck-billed speculum.[8] The most outstanding contribution from the war was research on traumatic damage to the nervous system, so that in the newly developing specialty of neurology Americans suddenly stood out among world authorities.

Major changes did occur, however, in the ways the armies from the North and the South organized and standardized medical care. The war's greatest effect on health care, then, came through the ideas that physicians and others took back from their exposure to military medicine, which during the course of the war improved markedly.[9]

Neither side was prepared for the massive conflict. Nor was the expectation that the sick and wounded would have regular medical treatment well established. Things began to change particularly after publicity about the disasters of the Crimean War (1854–56) depicted injured soldiers lying about untended and dying after battle. In the Civil War, Americans set an international precedent in developing an effective ambulance service to transport the wounded to places where they would receive medical treatment.[10] Moreover, in the North, a powerful voluntary civilian group, the Sanitary Commission, provided soldiers with basic preventive comforts such as vegetables, blankets,

Women from the Michigan State Relief Association providing supplies, cooking, and care for wounded soldiers, showing the important role of civilian groups in looking after soldiers in the Civil War. Francis Trevelyan Miller, ed., *The Photographic History of the Civil War*, 10 vols. (1910; reprint, New York: Review of Reviews, 1957), 7:341.

and instruction on cleanliness and also agitated for proper care of the sick and wounded.

The government of the Confederacy had to create many institutions from the ground up. Eventually the Confederate army set up a medical service and lines of supply for the medical care of their ill and wounded. The Union army started out with leaders who had no idea how to deal with the thousands and thousands of ill and wounded from the first battles of the war. Witnesses saw acres of suffering soldiers lingering helplessly on the battlefield without care or injured men filling towns near encampments, begging for food and assistance. With the appointment of a new surgeon general, William A. Hammond, in 1862, however, a whole system and organization slowly came into place that set a model for military medicine for generations.

Two features of the Union army medical reforms stood out: the organization of an official ambulance corps and military hospitals, which now reflected the belief of the sanitarians that fresh air would keep disease from spreading. Hospitals in both the North and the South were constructed on the pavilion plan, spreading out the patient beds in airy wards in detached wings connected by walkways. Moreover, the hospital patients were regarded not as charity cases housed out of pity but as medical cases, a point of view that would help prepare, as it turned out, for a reconceptualization of the hospital as a medical, not a charity, institution.[11]

Medical Problems in the Militaries

The realities of the war were appalling. Of a national population of 33 million, more than half a million men in the military died. Two-thirds of the dead were Union soldiers. The two hundred thousand Confederates who died constituted a full third of the Confederate army. In addition, an uncounted but very large number from both sides were left disabled. As was to be expected at that time, about 64 percent of the Union deaths and an even higher percentage of Confederate deaths were attributable to illnesses, not military action.[12] Raw recruits from isolated rural locations were crowded together and were first sickened and then frequently killed by "childhood" infectious diseases, such as measles and mumps. An appalling lack of sanitary measures accounted for many other fatal illnesses. Early in the war, the secretary of the Sanitary Commission, in a review of twenty military camps, reported that "a complete system of drains so essential to the health of the men, did not exist in any of the camps . . . the sinks were unnecessarily and disgustingly offensive, personal cleanliness among the men was wholly unattended to, and the clothing was of bad material and almost always filthy to the last degree."[13]

About half of the fatalities from illness were from diarrheas and malarial fevers. A recent historian listed the major killer diseases as "infectious diarrhea or dys-

A drainage trench used for toilet purposes at the Camp Morton prisoner-of-war camp in Indiana suggests the difficulty authorities had in maintaining sanitary conditions during the Civil War. At the same time, the effort shows that many leaders believed in the importance of sanitation. Francis Trevelyan Miller, ed., *The Photographic History of the Civil War*, 10 vols. (1910; reprint, New York: Review of Reviews, 1957), 7:71.

entery, pneumonia, typhoid fever, tuberculosis, measles, and smallpox." Scurvy affected African American troops, whose rations tended to be inferior. Soldiers' susceptibility to diseases varied according to their geographical origins. Among the Confederate forces, for example, pneumonia and typhoid were notable in causing deaths.[14] At any point in the war a sufficient proportion of the troops (typically 40 percent among the Confederates, fewer among Union troops) were so ill as to threaten or prevent any attempt at military action.[15]

Among the physicians who served at one time or another were a number of the better trained of the country doctors, as well as some from the urban elite. Sectarian physicians ultimately were eliminated from the Union corps, although a number were able to pass allopathic examinations and serve anyway. In addition to the fifteen thousand doctors who served on the two sides, there were stewards assisting the physicians by filling prescriptions and carrying out other duties. A significant number of the stewards later became physicians and pharmacists. And given the large number of illnesses (a total of 6 million were reported in the Union army alone), much of the medical service was indeed medical, as opposed to the more spectacular surgery.[16]

The Impact of the War on Medicine

Especially through medical inspections in both armies, medical staff were continuously learning about standards for good medicine.[17] In 1863 Surgeon General Hammond, in a move that reflected the best practice of medicine at the time, issued an order removing calomel and tartar emetic from the regular supplies of the Union army. Many physicians both in and out of the army objected vehemently to this rejection of traditional medicines, and Hammond, an often difficult person anyway, was forced out of office. Yet the order defying tradition was not revoked. Rather, it stood as part of the education of practitioners about what superior practice in medicine had come to be.[18]

Physicians in the field learned a great deal about surgery. They often used anesthetics, and they saved many lives by removing bullets, repairing wounds, and amputating when a limb had developed a localized infection. If the limb was not removed, the infection would prove fatal. Overall, there was a reduction of 50 percent in the amputation fatality rate from the beginning to the end of the war. Some common wound infections were uncontrollable. The most horrifying was hospital gangrene, described by one physician as follows: "A slight flesh wound began to show a gray edge of slough, and within two hours we saw this widening at the rate of half an inch an hour, and deepening." Or as the Confederate surgeon Joseph Jones reported, "It is not uncommon to see large surfaces of muscles and even of bones exposed, the skin and cellular tissue having been completely dissected away by disease."[19]

It is true that many physicians who had never seen much surgery got experience by working on wounded soldiers. And it is true that the conditions were often appalling and that infection from surgery killed vast numbers of patients. Yet the results improved during the war, and the statistics of survival were impressive indeed compared with those of earlier conflicts.[20] It is no wonder that the experience the thousands of physicians took back to their civilian practices led to a sudden improvement in American medical practice overall. In the army, formerly isolated practitioners were suddenly exposed to physicians who were treating a variety of difficult maladies, such as congestive heart failure, syphilis, and liver disease. S. Weir Mitchell, of Philadelphia, later realized that "the constant mingling of men of high medical culture with the less educated" had significantly upgraded the practice of medicine in the United States.[21]

Spreading Awareness of Sanitation

The most significant takeaway from military experience to civilian life, however, was the idea of sanitation. According to one army sanitary inspector, if one found "dirt at one end" of a soldier, one would find "cowardice at the other."[22] Recruits were in fact often dirty in their habits and indifferent to the importance of using a latrine even when one was available. It also took hard military discipline to get them to wash their clothes or themselves or to comb their hair. Medical officers' insistence on cleanliness, fresh water, and a decent diet sometimes brought them into conflict with less-concerned line officers. In the end, medical personnel, regular officers, and to some extent the troops all took back home the longstanding ideas of the sanitarians, who believed that environmental factors caused many major diseases and that cleanliness, fresh air and water, and a varied diet would prevent disease.

Many of those who served also carried back ideas about the relation between enforced sanitary standards and disease prevention. The memory of a clean camp was a powerful one for both soldiers and doctors. It should not have been surprising that public health efforts gained momentum throughout the United States after the Civil War.[23]

The public health movement had roots going back to the 1840s and 1850s, when cities in the United States were growing. City leaders slowly began to develop a sense of community responsibility for health conditions, particularly the dirty streets that many people associated with disease. In 1845 in Cincinnati, a city representative of the expanding country, for the first time health efforts took on the label "public" and thereafter became known as "public health." Under that heading, the citizens demanded social action targeting not only filthy streets but also water, sewage, milk, and smelly, noxious trades and activities. The Civil War experience, in which masses of men were gathered together, intensified leaders' often well-developed sense that health was indeed a public concern.[24]

In 1877, Billings reported as an eyewitness that "there are now scattered, all over the country, physicians who have had more or less army experience, and whose ideas about remediable causes of disease, and the best mode of dealing with them, have been, to a considerable extent, derived from that source."[25] But it was not only soldiers who experienced the sanitary measures and the physicians and officers who ordered them who learned personally about pure water, air, diet, and cleanliness. Anyone who volunteered or sent vegetables or heard tales of the camps acquired some impression of the value of organized and disciplined mass action and the importance of one's immediate environment and conditions in preventing illness.

A New Role for Women: Nurse

Still one more innovation fostered especially by the wartime experience was the entrance of women into the medical care of the sick and wounded in the role of paid nurses. In hospitals, nursing usually had been performed by recovering patients (male and female), who looked after the more disabled. Now specially designated and trained females began to take on that role, even in the all-male army. By the end of the nineteenth century a nurse typically was a woman.

Women came into service as nurses following the British example of Florence Nightingale in the Crimean War and afterward (an example well known in the United States by 1860). The idea of women's carrying out domestic roles for the ill was a familiar one. Moreover, the example of religious sisterhoods, which also gave care and comfort to the armies in the Civil War, was well known. Both trained and untrained sisters provided critical assistance to both sides throughout the conflict. As a Sister of Charity on a crowded hospital evacuation ship noted, "Our sisters shared with their poor patients every horror except that of feeling their bodily pains."[26]

When the war started, Dorothea Dix was asked to head up an official nurse corps for the North. She enlisted a number of middle-class women, who received some brief training. In spite of worries at that time about having women working among soldiers, elsewhere, on both sides, large numbers of women were enlisted as both paid and volunteer nurses looking after soldiers in hospitals and many other sites. "All our women are Florence Nightingales," noted an editorial writer in the *New York Herald*. The effectiveness of the women volunteers led to a growing tolerance and willingness to have women serve as nurses in public arenas like hospitals. At the same time, across the country women continued to answer the calls of "kinship and community" and take on most of the caring for those who were ill at home or in the neighborhood.[27]

The idea of professional (female) nurses and nursing schools continued and grew after the war. A nurse-training school opened in Boston in the New England Hospital for Women and Children in 1872, and the next year two more schools opened in New England and one in New York.[28] For a long time, graduates served as private nurses in patients' homes more than as hospital nurses, but clearly a new figure was present in model health care. The nurse was to be strictly disciplined and subordinate to the physician, but she could often provide the nursing care previously offered by the physician,[29] so that the physician's role could be solely to deliver scientific medicine.

Science rather than just caring propelled some women to follow Elizabeth Blackwell and become physicians. One model was Marie Zakrzewska, who founded the New England Hospital for Women and Children, an institution where women medical graduates could gain hospital experience. She started as a midwife in Germany and then, with the help of Blackwell, took a medical degree at Cleveland Medical College in 1856. Zakrzewska, who was determined not to let women

In the late nineteenth century, patients were often placed in hospital wards more crowded than this one, but the nurse was already a familiar, indeed, stereotypical figure in the ward and in medicine. A. B. Ward, "Hospital Life," *Scribner's Magazine*, 3 (1888), 706.

be excluded from medicine or be placed in a patronized, secondary position, introduced the latest methods and instruments to the New England Hospital.[30]

From the Sanitary to the Public Health Movement

In 1866 a chance event, another epidemic of cholera, made the sanitary experience of the Civil War suddenly more effectual. The cholera appeared first in New York

City, where, just a few months earlier, a number of physicians and local organizations had succeeded in persuading the state legislature to set up a board of health with real continuity and enforcement powers. Using two approaches, sanitation and isolation, the board appeared to succeed in keeping the number of cases small.

Subsequently, Chicago and other cities also established strong, permanent boards of health, and by the 1870s a whole public health movement had been launched in the United States. Americans had learned from European examples that powerful boards of health could be effective in their communities. Physicians and some public figures had taken up in North America the sanitary movement developed even before midcentury in Britain. This was an arena different from the individual patient therapies over which sectarians and regulars argued. One of the reasons the regulars ultimately won out was that they convincingly claimed the territory of public health.

Following European thinkers, many physicians had become convinced that even if not contagious per se, cholera was "portable." At the least, medical authorities believed that the means of portability was likely the feces of patients. In New York City, in the face of the cholera of 1866, not only was an enormous amount of filth removed from the crowded, dirty city but when a case of cholera was reported, within an hour Board of Health personnel arrived with a wagon to remove and burn the clothing, bedding, and other objects the patient had contaminated—the traditional "fomites." The sanitarians had developed a number of chemicals to disinfect places where sick people had been, so the dwelling areas of those patients were disinfected—even before there was a precise idea of "infection."[31]

During the Civil War, such powerful chemicals as iodides had been used on hospitals, barracks, and patients alike. Other strong disinfectants, notably lime and carbolic acid, had been used on walls and floors and other areas to achieve the ultimate in cleanliness, judged not least by the effect of rendering decaying material relatively odorless. Or one could view the process as destroying dangerous miasmas that one could detect from smells. Robley Dunglison, in his 1857 dictionary, defined a disinfectant as "any substance that will neutralize moribific effluvia,"— wording that would serve later when there were other ideas about what the moribific effluvia were.[32] Whatever the process, sanitation seemed to contain cholera and other diseases.

The other great scourge of the post–Civil War years was yellow fever. After Benjamin Butler, the occupying commander of New Orleans, in 1862 instituted a program of cleaning the city and imposing a quarantine rigorously enforced during the summer months, New Orleans became relatively free of the terrible annual visitations of yellow fever.[33] Eventually public health efforts became ineffective in New Orleans, and in 1878, as noted earlier, that city and the South in general suffered from a remarkable epidemic of yellow fever. In Memphis, of the 20,000 people who did not manage to flee the city, 17,000 contracted the fever, of whom 5,150 died. Only after a substantial delay did some sanitary measures and public health authorities come into place in the South, following the northern models.[34]

The sanitarians faced many obstacles, particularly political opposition based on the skepticism and traditional beliefs of many citizens. Before and after the war, such was the public's distrust of quarreling physicians that in many localities the law limited the number of physicians who could serve on the public health board—or even excluded them altogether. And so widespread was the attention paid to the rights of ordinary citizens to befoul their own property and public spaces and watercourses that political bodies hesitated to take any action to curb people's behavior or even clean up after them at taxpayers' expense.

Sanitary Conditions

People put up with conditions in the second half of the nineteenth century that later generations find hard to imagine. Industrialists were often ruthless in disposing of wastes and by-products of any kind, from the traditional offal from slaughterhouses to poisonous substances dumped into streams. And as industrialization begat further urbanization, the cities became sites of incredible public health problems, problems that seemed alien to the still majority rural dwellers, who anyway did not see anything wrong with the usual outhouse and barnyard smells.

Even in large cities, water came largely from wells, which, given the crowded conditions, were usually sunk close to privies, which often overflowed. It was wryly observed at the time that water from many city wells could serve as an effective purgative.[35] Sometimes human wastes were still dumped into the streets along with garbage, the carcasses of dead animals, and industrial waste, which mixed with huge quantities of horse manure and other animal droppings. In New York at mid-century, pigs roaming the streets were still a major means for disposing of garbage. Citizens hoped that rain would carry the waste away from public areas and streets and so remove the unbearable stench that was everywhere and which no words or pictures can convey to later generations.

Newark, New Jersey, for example, began to use the Passaic River as a source of water in 1870, just as new industries upstream began dumping waste into the river and municipalities such as Paterson were channeling sewage into the river. The river became unfit even for recreational boating, and the population along the banks was exposed to overpowering smells and the spectacle of sewage, trash, and dead animals floating by. Indeed, the river was so polluted that below Newark it became a virtual cesspool, unsuitable for use even as a sewer.[36]

Added to the many dangerous urban arrangements was the population density of the huge numbers of poverty-stricken, exploited, and despised city dwellers. Beginning in the 1840s, Irish migrants were a conspicuous segment of the ill and socially dependent. More than 40 percent of those who died from cholera in New York City in 1849 had been born in Ireland.[37] The Irish continued to arrive in waves until 1910, crowding into cities—whereas many other midcentury immigrants had the capital to move to the country and farm. Beginning in the 1880s, huge numbers of poor people from southern and eastern Europe joined the Irish in forming a vast urban underclass. People at the time recognized the increasing distance between the articulate medical elite and the growing numbers in the laboring classes, whose members suffered not only discrimination and exploitation but serious and continuous health problems.

An additional element appeared particularly in the South. Formerly enslaved African Americans and their descendants crowded into restricted urban areas. In Atlanta in the 1870s, for which we do have some statistics, the death rate for those classed as African Americans, including children under ten, was twice that for so-called whites.[38] Particularly across the countryside, the Civil War disorganized or destroyed the existing medical and folk-healer care that served the enslaved people who typically worked and lived on plantations. As the Union armies gradually moved in, large numbers of the new "freedmen" left and sought better conditions. At first, untold numbers contracted and perished from illnesses incident to deprivation, particularly lack of food and housing. In the immediate aftermath of war the suffering was terrible, exacerbated by famine and by pervasive racist assumptions that are shocking to people today. Both federal army efforts and local efforts were focused on the need for labor, leaving few social resources for people disabled either permanently or temporarily by illness. They were miserable and died under a system of neglect and occasional exploitation that soon extended to Amerindians being resettled (or exterminated) in the West. In the South, as populations stabilized into the twentieth century, a number of rural African Americans were nevertheless able to get around harsh segregated social arrangements and obtain some care from "white" physicians serving rural communities.[39]

Health Costs of the Social System

In the United States in the mid- to late nineteenth century, social leaders and even the unfortunates themselves often assumed that the poor deserved their bad conditions. One could point to their unclean habits, their drunkenness, and their presumed less than saintly behavior, in addition to their obvious place at the bottom of the social hierarchy. In that poverty-stricken world, disease was just one of the punishments that God or nature visited upon the wretched masses who destroyed themselves with unhygienic and presumably lazy lifestyles. Indeed, in cities anyone could see the connection between poverty and the main killer, tuberculosis (never mind that tuberculosis also thrived among rural populations). The mindsets of the leaders in that society were such that they defied humanitarian reformers and resisted spending any money to systematically assist unfortunate people of any kind.

The decline in the actual height of Americans in the midcentury and post–Civil War years, mentioned in chapter 3 as a sign of children's inadequate diet, was only one manifestation of poor health conditions in which those who made up the labor force for industrialization lived. Indeed, there is an unsettling correlation between maldistribution of wealth in the mid-nineteenth century and physical signs of stunted growth, even without factoring in the appalling environment of the slums. There is also evidence that Americans in the mid- and late nineteenth century suffered and died not only from infections but also from chronic conditions, such as heart disease, far more often than their long-lived descendants in the twentieth century.[40] The life expectancy of men at age 20 declined from 47 years in 1801–9 to 41 years in the 1850s. For women, the decline was from 48 years to 37.1![41]

Beyond Common Disease Patterns and Responses

Beyond the brief wave of cholera in 1866, experts tracking diseases noticed that malaria was disappearing from most northern areas of the United States, unaware, no doubt, of the consequences of draining swamps and (unwittingly) introducing cattle as an alternative target for mosquitoes. Yellow fever had for a long time occurred entirely in the South. Smallpox still appeared in various places, but vaccination could control major epidemics. Everywhere, people continued to suffer respiratory infections and numerous gastrointestinal difficulties, many of which were serious diseases. Malnutrition, hazardous living conditions, and common infections, including some that were very serious (many probably streptococcal), dogged inhabitants in both harsh rural and crowded city environments.[42]

The incidence of diseases and fatalities resulting from them varied. For example, apparently a combination of public awareness and active quarantine reduced the annual death rate from scarlet fever in Massachusetts (which had reasonably complete statistics) from 8.2 per 10,000 in the 1860s to 2.1 in the 1880s.[43] In Pittsburgh in the 1870s, among infectious diseases recognized then, diphtheria, typhoid, scarlet fever, whooping cough, and smallpox caused the most deaths. Within a few years, from the 1880s to the end of the century, smallpox almost disappeared, but diphtheria and typhoid rose to consistently high, dominating levels.[44]

Everywhere, sanitarians and public health authorities hoped they could control epidemic diseases. Moreover, they had substantial (albeit not unanimous) citizen support. Many Americans expected scientific information to empower community action. In the late nineteenth century, public health officials came to exercise remarkable authority. Courts did not intervene when these officials destroyed property or prevented the free movement of citizens. As late as 1900, officials in Honolulu who were doing selective burning in Chinatown to stop bubonic plague accidentally set the whole area afire and confined the fleeing inhabitants so as not to spread disease. Those same officials also burned mansions belonging to the well-to-do so-called whites where rats were found and confined the inhabitants of those houses as well. The plague, incidentally, soon came to an end.[45]

The Growth of Commercial Health Products

Comparable to the extremes of wealth and poverty, the medical marketplace included a few advanced medical thinkers alongside numerous less zealous practitioners and, in addition, the patent medicines and commercial quackery already established by the 1840s and 1850s. Not only self-help publications but also commercialized quackery and patent medicines reached a quantitative high point after the Civil War.

As communication, industrialization, and publication flourished, so did marketing, advertising, and gullibility. The number of periodicals published in the United States more than tripled between 1860 and 1900, and they expanded in size as well, substantially financed by ads for patent medicines. Color lithography became available for advertisers, and posters and signs adorned the out-of-doors. Huge announcements advertising bitters, regulators, and purges desecrated Niagara Falls and Yellowstone Park. The 1880 census identified 563 establishments in the United States devoted to producing patent medicines and nostrums.[46]

The legendary proprietary medicines of Lydia Pinkham were launched in the 1880s. By means of advertising, they became an American institution. Playing upon the many forms that "women's troubles" took, advertising drove sales of the 19 percent alcohol tonic and other preparations. Claims that Lydia Pinkham substances would cure fallen womb (a sometimes distressing result of inadequate care in childbirth) as well as tiredness soon evolved into general suggestions that such purchased medical means might help American women whose lives were unhappy and uncomfortable, not least with implied claims of contraceptive power. Indeed, eventually "Mrs. Pinkham" (clerks in a room at the company) answered all women's letters personally with advice, accompanied by a pamphlet touting how a company product would make the recipient feel better (as, indeed, the alcohol content might). Similar claims by still unregulated medical practitioners did not diminish the credibility of the advertising.[47] But the strategy of having a health care individual (even a fictitious one) look after the particular individual troubles of each person set a model that would have profound importance in the Progressive era (chapter 7).

Opposition to Aggressive Reformers and Sanitarians

Suspicion of physicians and even outright opposition to legitimate medical improvement continued to be commonplace among both educated and uneducated people in the second half of the nineteenth century. Political as well as popular opposition to medical authority grew also as sanitarians and public health officials expanded their efforts to prevent disease by advocating costly public works to provide clean water and dispose of sewage.

Furnishing public works was understandable and important. Public funding of the infrastructure, for example, was so widespread as to blunt the economic effects of the terrible depression that came after 1873.[48] Yet piped-in home water service came only slowly to many areas, and sewage systems still much later. The wealthy and progressive areas obtained service first. Impoverished areas and populations suffering social discrimination, notably African Americans, continued with contaminated wells and particularly lack of sewage systems well into the twentieth century. Sewers in Mobile, Alabama, were still open, not covered, as late as 1913.

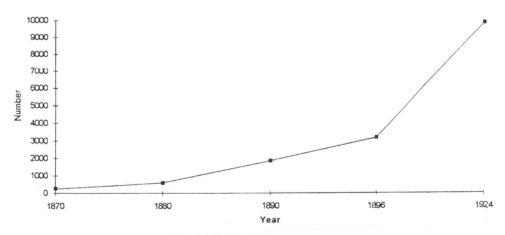

Number of waterworks in the United States. The increase in the number of waterworks in the United States from 1870 to 1924 suggests remarkable efforts by American communities to obtain a good water supply. A clean water supply became a standard part of the social infrastructure, with major effects on general levels of health. Martin V. Melosi, *The Sanitary City: Urban Infrastructure in America from Colonial Times to the Present* (Baltimore: Johns Hopkins University Press, 2000), 118. Reproduced with the generous permission of Martin V. Melosi.

While health was the issue (along with water for fighting fires), the ranks of organized sanitarians for decades after the Civil War included engineers and particularly plumbers alongside physicians. Americans were quick to improve on European designs for sanitation works and eventually took a special interest in sewerage projects. A physician in the 1870s commented, "There seems to be arising among the citizens a kind of panic relative to the drains of their own houses, and they have a great horror of the least odor of sewer gases," which many supposed carried disease.[49] In the 1880s alone, Americans patented 1,175 drainage and sewerage devices.[50]

Late nineteenth-century sanitarians also carried over from the 1840s an intensifying concern about correctable personal behavior that they believed brought on ill health. Whole books have been written about reformers' campaigns against tightly laced corsets and other items of female apparel, such as high-heeled shoes. Getting affluent people to take a bath once a week was a major accomplishment, and in a number of cities paternalistic leaders actually set up public baths to enable poor people to protect their health by getting clean. The sanitarians were aided by new advertising campaigns by soap manufacturers, who raised the consciousness of members of the public concerning the virtues and healthfulness of cleanliness.

Diet continued to be a major concern of health reformers among the sanitarians. They condemned heavy foods and fried foods. They also decried lack of fresh air, crowding on railways and the new urban streetcars, and, as always, using tobacco and alcohol. One sanitarian wrote in 1874, "Healthy life is tantamount to social and political prosperity, and it is the sure support of morality and religion."[51] The sanitarians' self-righteous preaching generated resistance and political opposition that played out in every locality—just one more response to the profound technological and social changes that crowded on and confused Americans in the post–Civil War decades.

Physicians in Transition

Among those most beset by change were medical practitioners. The 55,000 reported in the 1860 census increased to 82,000 in 1880 (the population increased from 31 million to 50 million).[52] Many, perhaps most, physicians continued in the old ways, little affected by the startling laboratory reorientation coming out of German-speaking areas of Europe. Most doctors were more concerned with just seeing patients and the continuing worry of collecting fees. In Spring Green, Wisconsin, as late as the 1880s, a normal birth was worth a five-dollar pig; minor surgery, several hams; and a fifty-cent office call, eight dozen eggs, a bushel of potatoes, or three haircuts.[53]

The elite, however, and particularly the younger elite, saw in medicine progress, innovation, the excitement of science and discovery, and realistic hope of defeating the diseases that caused so much human suffering and death. Physicians who shared this viewpoint communicated with one another and became a significant factor in shaping the profession.

In 1886 a number of elite practitioners in a significant act formed a small group, the Association of American Physicians.[54] At the first meeting, the president, Francis Delafield, of New York, recognized the transformation that had brought the group together, and he distanced it from the ceremonial, political/professional, and social activities that marked the meetings of the American Medical Association (AMA):

> We all of us know why we are assembled here to-day. It is because we want an association in which there will be no medical politics and no medical ethics [discussed]; an association in which no one will care who are the officers and who are not; in which we will not ask from what part of the country a man comes, but whether he has done good work, and will do more; whether he has something to say worth hearing, and can say it. We want an association composed of members, each one of whom is able to contribute something real to the common stock of knowledge, and where he who reads such a contribution feels sure of a discriminating audience.[55]

Delafield was recognizing the ongoing shift in the public identity of the physician, from the family friend and counselor to the bearer and agent of the nonlocal science that had to be applied to the illness at hand. By the 1880s, such "scientific" physicians increasingly viewed the world of illness as one in which patients were suffering from definite, separate diseases, not maladies peculiar to the particular patient's constitution and disposition.

More than ever, the physician could know more about the patient's body than did the suffering patient, and a small but growing number of practitioners broke with traditional practice by employing new instruments. Beyond the stethoscope, the thermometer came into more general use. The pulse was timed and counted, and temperature charts appeared in hospitals in the 1860s. Some advanced practitioners were using ophthalmoscopes and laryngoscopes in the 1870s and 1880s. And still more authority and knowledge came from the microscope, a device that further distanced what a trained physician could know apart from information available to either a patient or a merely personable practitioner. As early as the 1840s, many Americans had been learning about the new achromatic microscope, which was opening up new fields of investigation. Oliver Wendell Holmes learned to use one and taught his anatomy students at Harvard how to use it. Others did likewise,

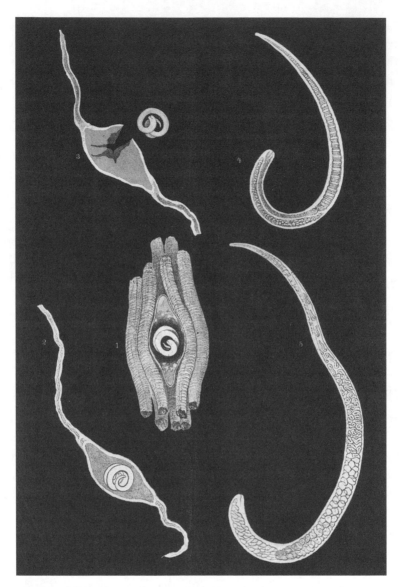

These images of the life stages of the tiny but devastating parasite *Trichina spiralis*, which invaded human bodies and caused trichinosis, a terrible and painful illness, were made in 1869 by a New York City teacher and investigator. Such investigators were using microscopes and creating a mind-set that would be receptive to the germ theory of disease. Like germs, trichinae were parasites found in humans and were virtually undetectable without magnification. John C. Dalton, "Trichina Spiralis," *Medical Record*, 5 (1869), 79–83.

and an editor in 1847 could say that "a good microscope has become . . . essential to the physician," an idea that did indeed spread, albeit slowly, in succeeding years. Two American textbooks on microscopy appeared in the 1850s, and after the Civil War many physicians were using a microscope to examine not only anatomical and pathological specimens but even the blood, always trying to keep up with the flood of discoveries and publications coming out of European laboratories.[56] In 1878 young William H. Welch returned to New York from study in Europe. In order to make a living in the United States, he gave lessons on the use of the microscope. At that late date he was still a pioneer in the use of that instrument in the United States.[57]

In 1887, after hearing a paper at a meeting of the Association of American Physicians on the traditional supposed disease typho-malarial fever, a longtime favorite of clinicians who compromised in the face of superficial symptoms, William H. Draper, of New York, noted that the speaker had shown "that this hybrid disease does not exist." It had to be either typhoid or malaria. In a further comment, the young William Osler, then of Philadelphia, told of using the microscope because "in these doubtful cases, a careful histological examination of the blood will determine whether the affection is malarial or not."[58] Such certainty about specific diseases indicated the distance that medicine had come from the 1850s to the 1880s.

The Vision of Bringing Laboratory Science into Medicine

It was still possible for only a very few of the medical elite to carry out laboratory research part time, much less full time. And when they did do research, their contributions, with few exceptions, continued to appear relatively insignificant compared with those of European investigators. And no wonder: any time that they devoted to research was carved out of their active practice. When W. W. Keen and Silas Weir Mitchell were working on poisons in the 1860s, as Keen recalled, "the front office was full of patients and the back office of rattlesnakes and guinea-pigs."[59]

It is true that the few scientific investigators in medicine in the United States were ambitious, yet in the mid-nineteenth century they still found that no institutions existed to support their research. American medical schools were totally preoccupied with turning out practitioners. Moreover, colleagues in medicine actively discouraged research as a self-indulgent waste of time when there were patients who needed care. When young S. Weir Mitchell, just beginning his medical career in the early 1850s, suggested to his physician father that he would like to spend some years doing research, his father exclaimed, "Weir, if you do that, people will look upon you as they would if you joined a circus."[60]

Where anatomy and pathological anatomy had been important in localizing and identifying diseases and making surgery more useful, a new kind of science was making a dramatic entrance with the development of a single discipline, physiology. Physiology was based primarily on experimentation, not just static description as in anatomy. In the 1850s John C. Dalton, who had studied in Paris with the advocate of "experimental medicine," Claude Bernard, pioneered teaching using live animals and vivisection at Buffalo, Vermont, and the College of Physicians and Surgeons in New York.[61]

Physiological experimentation with animals in investigation and teaching signaled a new intellectual orientation in medicine. Traditional medicine was based on clinical observation or at best on numerous listed observations of patients. Using animals to test physiological reactions suggested that human beings too were animals, a notion that already was causing trouble at the time in the form of Darwinism. In medicine, moreover, suggesting that physiology was universal undermined the long observation of human patients that was the very basis of

Using an animal in an experiment on respiration (a process humans share with animals) shows how easily physiologists and other scientists of the mid-nineteenth century generalized from observable events in animals' bodies to physiology in humans. Such experiments with animals were becoming customary by 1856, when this image was published. John William Draper, *Human Physiology, Statical and Dynamical; Or, The Conditions and Course of the Life of Man* (New York: Harper & Bros., 1856), 170.

medical knowledge. Many leading practitioners denounced the idea that what happened in a dog had any relevance for the health of a human's unique individual constitution. Physiological "universalism" was another heretical attack on tradition.[62]

Teaching with vivisection became the hallmark of the new scientific medicine. Vivisection came in not only with Dalton but elsewhere at midcentury, not least with one brilliant, creative, and important scientist, Charles Édouard Brown-Séquard, who moved back and forth between France, Britain, and the United States. In the 1860s he taught at Harvard, of course using animal vivisection.[63]

Meanwhile, some physicians did do part-time investigation. One was S. Weir Mitchell, who became internationally prominent in neurology. He started out as a physiologist, experimenting in a laboratory as he practiced medicine. He too had studied with Bernard. Even before the Civil War, experts in Europe cited Mitchell's work on the physiological and toxicological effects of snake venom and curare,[64] but he never was able to obtain a regular teaching appointment.

In 1871 a landmark event occurred: a thirty-one-year-old independently wealthy physician and scientist, Henry P. Bowditch, founded a well-equipped laboratory at Harvard Medical School, where he taught physiology full time using vivisection. Bowditch, who had begun his studies with Bernard in Paris, had ended up in Leipzig with Carl Ludwig, who had a better-equipped laboratory. Through his influence and his students, Bowditch would redirect the orientation of American physiology not only to science but to German methods and traditions of laboratory research. As he wrote even before taking up the post at Harvard, physiology, by which he meant "the patient, methodical and faithful way in which the phenomena of life are investigated by the German physiologists," was "the only true foundation of medical science." As Oliver Wendell Holmes observed in 1884, "Physiology, as now studied, involves the use of much delicate and complex machinery," including "the cardiograph, the sphygmograph, the myograph, and other self-registering contrivances."[65] Many MDs, however, regarded such "contrivances" with suspicion and preferred traditional types of observation.

Slowly, other physiologists joined American medical schools and universities and made not only physicians but the educated public in general aware of the ideal of experimental science and technology in medicine. In 1882 D. Webster Cathell, of Baltimore, advised his medical colleagues: "If, at your office and elsewhere, you make use of instruments of precision—the stethoscope, ophthalmoscope, laryngoscope, the clinical thermometer, magnifying glass and microscope, make urinary analyses, etc., they will not only assist you in diagnosis, but will aid you greatly in curing people by heightening their confidence in you and eliciting their co-operation."[66]

Demystifying the Body

The new medical science of the mid-nineteenth century and after did more than give confidence and hope to patients and doctors in a professional relationship. Educated Americans in general developed a new view of the human body. For them, the body became a material machine that operated mechanistically—without any mysterious "vital principle." Moreover, the cells, visible under the microscope, could be understood in terms of reductionism. That is, enthusiasts held that all life processes could be reduced ultimately not just to tissue and protoplasm but to mere chemistry and physics.[67]

For most American physicians and intellectuals, reductionism did not translate into philosophical extremism (atheistic materialism), but many were enthusiastic about studying the body, and ills of the body, "scientifically." In 1856 the New York physician John W. Draper wrote a textbook of physiology that would inspire technical and other writers for decades. Many of the illustrations were based on Draper's internationally recognized pioneer work in adapting the microscope to take photographs. And he spelled out his purpose, to write a truly scientific textbook: "It was chiefly, indeed, for the sake of aiding in the removal of the mysticism which has pervaded the science that the author was induced to print this book. Alone, of all the great departments of knowledge, Physiology still retains the metaphysical conceptions of the Middle Ages from which Astronomy and Chemistry have made themselves free."[68]

The new view of the body was much more mechanistic than the earlier view of a congeries of organs animated by a vital force. Now writers employed the familiar steam engine as a metaphor in books for children. Authors enjoined children to shovel the food in, as one would coal under the boiler, to create energy for their bodies. As the authors of a textbook on "the human body as a living machine" wrote, "By a machine we mean an apparatus, either simple or complex, and usually composed of unlike parts, by means of which *power* received in one form is given out or applied in some other form."[69] Such metaphors sometimes helped more than scientific studies to shape the course of medical thinking.[70]

Most telling was the use of the telegraph and, later, the telephone as models for the nervous system. For now, instead of the vague sympathies between parts of the body that many writers had described earlier, late nineteenth-century Americans found in the signals sent by the nervous system a mechanical explanation for not only reflex actions but all sensory and motor functions of the body.[71] George F. Barker in 1872 explained that "chemistry teaches that thought-force, like muscle-force, comes from the food; and demonstrates that the force evolved by the brain, like that produced by the muscle, comes not from the disintegration of its own tisʃuɛ, but Is the converted energy of burning carbon. Can we longer doubt, then, that the brain, too, is a machine for the conversion of energy?"[72]

How far physiological research was carrying such ideas was demonstrated dramatically in 1874, when a major figure in American medicine, Roberts Bartholow, of Cincinnati, famously contributed to world medical knowledge experimental confirmation that the human brain was not a general organ but was divided into areas having specific functions. Earlier experimenters in Europe had stimulated the brains of animals and shown localization of function, but their results had not yet been accepted, and certainly not as applicable to humans. Bartholow had a patient in the Good Samaritan Hospital who suffered from a terrible cancer that was eroding her scalp, so that much of her brain was exposed. The doctors had done everything they could, but she was dying. With the patient's consent, Bartholow stimulated her exposed brain with electricity, eliciting specific motor responses first on one side and then on the other, confirming definitively the localization of function in human brains.[73] By showing this reaction in the organ that was supposedly the seat of a person's soul, Bartholow gave the materialistic and mechanistic view of the body a resounding reinforcement, as well as moving medicine toward later brain surgery, which, however, awaited other developments.

The Shift from Natural to Normal

Beyond the machine-telegraph model of the body and the new emphasis on the nervous system as the regulator of bodily, that is, human, actions, there was a third, even more fundamental transition in the decades around the Civil War. American medical thinkers were moving away from the idea of health as *natural* toward the idea of health as *normal*.[74]

Early in the nineteenth century, as noted earlier, medical thinkers had struggled with the idea that health was natural and sickness unnatural. Illness was an imbalance that was unique for each person. In the second half of the century, the use of instruments brought a shift in conceptualization. Physicians were increasingly able to measure with precision any number of functions of the human body. And once determined, these measurements could be placed alongside readings from other bodies. It remained only to gather enough numbers to be able to say what was typical (the mean) in a human being and to try to decide the point at which an instrumental measurement indicated a divergence from the "normal." Thus disease could be defined in terms of deviation from the normal, as, most typically, how high or low the patient's temperature was. In the 1870s and 1880s, instrumental temperature in fact became of particular concern to many American physicians. In the 1865–66 records of the Massachusetts General Hospital,

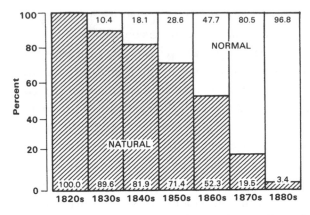

This remarkable graph showing the percentage of Massachusetts General Hospital case histories "in which terms *natural* and *normal* appear (out of all cases in which one or both terms appear)" illustrates how the concept "normal" replaced the concept "natural" in physicians' and educated Americans' understanding of what was and was not disease and pathology. John Harley Warner, *The Therapeutic Perspective: Medical Practice, Knowledge, and Identity in America, 1820–1885* (1986; reprint, Princeton, NJ: Princeton University Press, 1997), 89. Reproduced with the generous permission of John Harley Warner.

temperature recorded by observation was labeled "temperature," but that by instrument was labeled "thermometer" or "ther."[75]

By the late nineteenth century, physicians were serving in factories and administering examinations for life insurance—an increasingly important source of regular income for a practitioner working in an overcrowded profession. In an industrial accident, a dynamometer, for example, could measure the extent of damage to the muscle. For insurance companies, a physician could go beyond oral, visual, and tactile evidence to report whether a person's physiological functions, such as blood pressure, were abnormal enough to suggest that the person would not be a good risk for life insurance. In both industry and insurance, paper forms based on standardized instruments began to be used during the 1870s and 1880s. On the forms, the limits of normality were indicated numerically. Altogether, practical medicine became more precise, more useful, and more focused on abnormal functioning.[76] The very term *abnormal* became an indication of medical concern, even without the social and organizational implications that "abnormality" also acquired. The ideal human body became much more extensively defined, especially as, over the years, physicians devised or adopted more and more measuring devices to apply to their patients. Not least in clinical medicine were the scales, which doctors used to determine babies' normal weight and growth.[77] Rather than by a unique constitution, a person was defined by a battery of measurements. The reconceptualization was well under way by the 1880s.

Competing and Professionalizing by Exclusion

All this exciting, reorienting science before and particularly after the Civil War affected American doctors, who had day-to-day practical concerns, not least of which was defending the image and social value of physicians. For twenty years after the Civil War, leaders in the profession did not have realistic hope that medicine could be improved by licensing, although beginning in the 1870s one state or another attempted to establish nominal licensing through the state medical organization.[78] These sporadic efforts reflected the success of physician leaders whose private medical organizations had given them public standing and had brought some improvement to the image of the medical profession. Of course still only a small minority of practicing physicians moved beyond their local communities and rivalries and bill collecting and joined an organization.

The economics of medical practice, more than any other factor, divided doctors. As noted above, early local medical organizations had often tried to impose standard minimum fee bills, and some still appeared here and there throughout the nineteenth century.[79] Most practitioners, however, charged each patient individually, knowing that poor people would not pay. Following both ethics and community expectations, the physician furnished medical care without charge to those who could not afford the fees. For ordinary patients, the country doctor customarily still charged for travel to rural areas, for decades typically a dollar a mile. Indeed, a significant portion of a *country* doctor's time was spent simply traveling by horseback or buggy.[80]

Even though leaders in regular medicine still had not succeeded in raising educational standards, medical organizations at every level were often effective in excluding from their ranks quacks, irregular and incompetent physicians, and, increasingly, sectarians. By exclusion, the educated and organized allopathic practitioners found a basis on which to make claims of competence. Moreover, rules against physician advertising, removing medicine from "commercialism," provided another badge of professionalism on which to base a claim of elevated social status. Indeed, campaigning against patent medicines as well as irregular physicians was an important element in establishing and defining a modern medical profession.[81]

In the case of quacks and many irregulars, professional leaders' efforts at exclusion made sense, but regulars' campaigns against well-educated eclectic and homeopathic physicians were not always sensible or just. Attacks on homeopathy in particular had a secondary agenda beyond attacking economic

Even late in the nineteenth century many, if not most, practitioners spent a great deal of time traveling and were expected to visit patients in their homes. For a very large percentage of Americans, a visit from a doctor in the humble circumstances depicted here constituted their total experience with formal medical care. "Remarks on the Doctor's Office in America," *Ciba Symposia*, 2nd ser., 1 (1939), 289, from *Harper's Weekly Magazine*, 1890.

competitors: physicians who wanted to base medicine on laboratory experiments feared successful practitioners of homeopathy because they were part of the traditional world of systems and empiricism. In everyday practice, regulars were strictly banned from consulting or associating with sectarian practitioners, a rule that was not useful when there were well-respected homeopaths or eclectics in a community. In one instance in 1878 a medical society in Connecticut ejected a colleague because he was associating with a homeopathic physician—who, it turned out, happened to be his wife![82]

Many Americans still decried attempts by physicians to organize and to exclude anyone who wanted to practice on the public. The Jacksonian-era outcry against an established "medical monopoly" intensified over the years as physicians became better organized and tried to contain or manipulate market forces. Thus alongside the new image of the scientific physician, the old antimedical, free-market rhetoric continued.

In 1870 the AMA voted to exclude an African American local organization and in effect all African Americans. Between 1868 and 1900, fourteen schools were established to train African American physicians. In 1895 the excluded African American physicians formed their own organization, the National Association of Colored Physicians, Dentists and Pharmacists, which grew slowly and eventually became a physician group, the National Medical Association.[83]

Winning Public Recognition

For decades, into the twentieth century, the AMA and state and local societies were devoted to local, state, and occasionally national campaigns to obtain favorable legislation and other forms of public recognition in direct competition with sectarians, irregulars, and quacks. Leaders disagreed among themselves over strategy and tactics and the details of ethical standards. It was no wonder that AMA meetings could consist primarily in politicking, jockeying for representation, or socializing and touring rather than the scientific interchange that some leaders desired.

Yet by the 1880s, the urban elite physicians were enjoying some success. Signs of the growing prestige of scientific medicine were everywhere. There was favorable legislation of many kinds. Physicians were recognized leaders in the sanitary and public health movements. Indeed, success in public health areas, such as holding off an epidemic of smallpox with isolation and vaccination, was a potent instrument that medical leaders used locally to promote physicians as representatives of science.[84] Moreover, leaders in society tended to recognize members of the elite in scientific medicine.

Ironically, in the 1880s AMA regular physicians gained a remarkable legal victory that guaranteed a basis for serious and consequential licensing legislation by the states. A faction of regulars in West Virginia managed to get a law passed giving the state board of health authority to license educated physicians. That meant granting an exclusive and enforceable monopoly to qualified practitioners. Against all expectations, the U.S. Supreme Court in 1889 confirmed the power of the states to recognize the special position of the medical profession, a status previously granted only to lawyers and military officers. The decision, *Dent v. West Virginia*,

" Keen, swift, sensitive R., the surgeon."

A surgeon described as "keen, swift, sensitive" embodied one version of the ideal physician in the 1880s, an ideal that journalists as well as physicians tried to cultivate among middle-class audiences. A. B. Ward, "The Invalid's World," *Scribner's Magazine*, 5 (1889), 60.

encouraged and facilitated a movement to license and privilege physicians in the United States whether or not their practice did anyone any good. This licensing movement was based on state agencies such as boards of health rather than on medical societies as earlier.[85]

As noted earlier, historians have observed that it was particularly by picturing themselves as representatives of the new and the modern—"science" and laboratory—that regular physicians, now with a powerful boost from the Supreme

Court, ultimately won out over the sects. For sectarians—homeopaths, eclectics, botanics, hydropaths—continued to be devoted to systems and dogmas in the realm of therapeutics. A new sect, osteopathy, was founded between 1874 and 1892 by Andrew Taylor Still. Still's magnetic and bonesetting talents relieved increasing numbers of chronic, noninfectious complaints. But osteopathy also was marked at that time by dogmatic therapeutic beliefs.[86] Finally, chiropractic, a competing sect very similar to osteopathy, was effectively established between 1895 and 1906. Chiropractors retained dogmatic doctrines almost to the end of the twentieth century. The late dates of the successful new sects indicate how long the public continued to have enough doubts about regular doctors that significant numbers turned to alternatives.[87]

Regular physicians, in contrast to sectarians, increasingly claimed to be dedicated to following new discoveries and discarding old practices—traditional or dogmatic—that laboratory testing did not confirm.[88] With their at least rhetorical openness to change, they were able to mobilize credibility on the basis of medical and related sciences coming out of Europe in the 1850s and after. A critical group of leaders in science—including medicine—took on the identity "men of science" and set about making science and the descriptive *scientific* desirable and popular. It was they who embodied the dynamic element in modernization that underlay the appearance of a new epoch around the 1880s.[89]

S. Weir Mitchell, a close observer of the profession, in 1878 summed up how support for scientific medicine grew:

> Fifty years ago the public concerned itself little as to any form of scientific progress. . . . [Now] the profession of Medicine as such, no longer lives a life of intellectual seclusion. The increase in the number of scientific men, not physicians; the diffusion of knowledge as to Anatomy and Physiology; the ever-increasing interest in all forms of scientific activity; the growing value attached to Hygiene and to large measures of sanitary use; and what I might call the secularization of every addition to medical knowledge, by its instant record in newspapers, popular science journals, and reviews, have combined to give us, as a profession at least, what we once lacked,—a court, where we are heard with respect, and, for the most part, judged with fairness and interest.[90]

Signs of Departure from Tradition, Even in Medical Education

For a long time the generality of practitioners, particularly country practitioners, did not change their ways or join organizations. The better country doctors after the Civil War still deplored the large numbers of their colleagues who did not read or meet with other physicians to learn how to improve their practice. More than ever, the proprietary schools in the 1890s were still turning out many practitioners who, although degreed, were nevertheless unqualified and contributed to over-crowding and cruel competitiveness in the profession.

Meanwhile, from the 1860s through the 1880s, members of the public ever more intensely demanded that a physician have a degree, which served as at least some evidence of professional capability. Given the number of degree holders turned out by medical schools each year, an MD degree was not an unreasonable requirement. But with little or no regulation, the demand for degrees produced "diploma mills," from which anyone with enough money could purchase a degree. And even the bet-ter medical schools still competed for students by keeping terms short and fees low. There were no effective admission requirements other than the ability to pay the fees. The professor of surgery at Harvard famously objected to instituting written examinations in 1871 because, he claimed, more than half of the students could just barely write.[91]

Parallel to the distancing of well-educated physicians from their lower-status counterparts, some medical schools began to change. Although national efforts to reform the medical curriculum still continued to fail repeatedly in the decades after the Civil War, here and there, without fanfare, one school or another added an extra course or a summer course. It was also possible in some urban areas that a hospital now might provide an appointment that would give the student clini-cal experience—something increasingly necessary because by the 1870s the cus-tomary two- or three-year apprenticeship or preceptorship with a physician had fallen into disuse or been replaced by just a purchased certificate.[92]

In 1871 a medical journal editorialist could note that it was still necessary to award students degrees for the usual two-year course. But now the out-of-term in-struction available at the University of Pennsylvania—and open to students from other Philadelphia schools—made it possible "to extend to those who are willing to choose a wiser course . . . every opportunity for acquiring as high and complete an education as can be obtained in any medical centre in the world." And the edi-torialist recognized that those courses could constitute training in medical specialties.[93]

Leaders in medical education at the time were aware that changes had begun to take place.[94] During the 1870s first Harvard and then Penn and Michigan moved to create medical schools much more closely affiliated with the university and of-fering longer terms and a graded curriculum (one had to complete the first year before moving on to the second year, which was different and more advanced, and so on). The university affiliation signaled that medical education focused on con-veying new information and discoveries, not just recitations of traditional knowledge.

The changes were pioneered by a growing body of medical educators who wanted to teach and to upgrade the practice of medicine with new knowledge. These leaders were not interested in practice per se, nor in the average income of physicians. They were inspired by German-style laboratory investigation and other aspects of German medicine to supplement or replace the close observations earlier leaders had learned in France. In the United States, medical students more and more frequently had to learn by doing, including using the microscope and performing other experiments as well. By the 1880s the medical education reform movement had become a force in itself.[95]

The power of that force came substantially from the elite physicians who had taken further training in German-speaking countries, most often at the University of Vienna. Between 1870 and 1914, fifteen thousand American physicians—an astounding number—carried out further study at German, Austrian, and Swiss universities. By the 1880s a new generation of leaders was bringing back an ideal based on what they had seen and experienced: university-model medical education, research as part of medical education, and expertise developed through specialization of practice, teaching, and research. By the 1920s, as we shall see, those leaders had put into place a whole new system of medical education in the United States.[96]

Meanwhile, in some urban centers another institution, the dispensary, appeared as the result of philanthropic efforts to provide some rudimentary medical service to the many helpless and poverty-stricken people. By the 1870s there were twenty-nine dispensaries in New York and thirty-three in Philadelphia. There the deserving poor—not drunks or prostitutes—could see a doctor or apothecary, who would make a diagnosis and dispense free or very low-cost medicine. In the dispensary, one could see what most people, urban or rural, at and after midcentury believed was the heart of health care: a professional who dispensed a medication. By the late nineteenth century, dispensaries were generally staffed by young physicians who could gain experience and advance their careers, often in a specialty, by practicing on the poor.[97]

Specialists

As noted earlier, the appearance of specialists at midcentury was still another threat to the traditional doctor. Because specialization in medicine embodied both technological expertise and that fundamental principle of bureaucratic systematization, division of labor, even a very few specialists in American cities constituted a significant sign of the organizing that was to come. The first model for physicians came from French hospitals and clinical practice. The next was the strong specialization in German-speaking countries, in which laboratories reinforced the hospital and university specialty units that many Americans admired.[98]

One view of specialization was that it was a natural development as physicians came to conceptualize diseases, especially those that might yield to surgery, as manifestations of localized, distinct pathologies.[99] Yet in the United States even the most able surgeon would have considered it unethical to turn away a patient who had a purely medical problem. As late as 1876 the famous surgeon Samuel Gross wrote that "it is safe to affirm that there is not a medical man on this continent who devotes himself exclusively to the practice of surgery. . . . American medical men are general practitioners, ready, for the most part, if well educated, to meet any and every emergency, whether in medicine, surgery, or midwifery."[100]

Nevertheless, specialists like those known in Europe began to appear in the few very large urban areas in the United States. In a list of the country's elite physicians published in 1878, fewer than 20 percent were listed as having a special interest of any kind.[101] Eventually the obvious place for specialized practice was confirmed by the appearance of specialized hospitals—at first for women and children and then typically for eye disease. The Civil War called into existence hospitals for the eye and ear, nerve injuries, and orthopedic problems.[102] Hospitals and specialists for mental illnesses had now been around for decades, but they were working in isolated, often rural sites and were not part of the same kind of private practice found in cities. Slowly physicians on medical school faculties and in practice began to take on identities as experts in surgery, obstetrics, gynecology, and ophthalmology.

A physician who announced that he had special knowledge or abilities that his colleagues in medicine did not possess offended those colleagues, and officially any such claim was unethical. Identifying oneself as a specialist, as Billings observed in 1876, served "as a respectable means of advertising, and of obtaining consultations."[103] As early as the 1850s, a group of specialists in Cincinnati tried to advertise their expertise explicitly, but the AMA denied their application for an exception. The test case involved an ophthalmologist in New York who turned out to be mentally unstable. He made extravagant claims in 1865–68 and finally forced state and national groups to condemn him and use him as a symbol of why claiming specialized knowledge was unacceptable.[104]

In 1874 the AMA Judicial Council made a full statement that, as one historian points out, "virtually accused specialists of dereliction of duty":

> The title of Doctor of Medicine covers the whole field of practice, and whoever is entitled to that appellation has the right to occupy the whole or any part of the field, as he pleases. The acceptance of this honorable title is presumptive evidence to the community that the man accepting it is ready to attend practically to any and all duties [to] which it applies. As all special practice is simply a self-imposed

limitation of the duties implied in the general title of doctor, it should be indicated, not by special or qualifying titles, such as *oculist, gynecologist,* etc., nor
by any positive setting forth of special qualifications, but by a simple honest notice appended to the ordinary card of the general practitioner, saying, "Practice
limited to diseases of the eye and ear," or "to diseases peculiar to women," or "to
midwifery exclusively," as the case may be. Such a simple notice of limitation,
if truthfully made, would involve no other principle than the notice of the general practitioner that he limits his attention to professional business within
certain hours of the day.[105]

And so the "practice limited to" form of claiming a specialized practice (but not
special knowledge) remained standard well into the twentieth century.

To general physicians, successful specialists represented not only economic competition but a deeply divisive element in the profession. M. H. Henry, of New York,
for example, in 1876 spoke of "the injuries to the profession, committed by ill-
cultured and irresponsible specialists within our own ranks," although even Henry
recognized the legitimate place of a specialist who was actually well trained and
experienced and who, in the standard formulation, "decides to treat only a certain
class of diseases."[106] Unfortunately, contributions to world medical knowledge
tended to be based on specialization, which joined with science in bringing professional prestige.[107] Well-informed physicians and members of the public knew
who the leaders were in each specialized field.

The appearance of well-qualified specialists intensified the problems of physicians and public-spirited citizens in the 1870s and 1880s. Many people offering medical care were unqualified to begin with: quacks, traditional healers, and degreed
individuals who were scandalously ill informed. Even good doctors could be accused of meddling in areas in which their expertise was decidedly less than that
of the better specialists, operating, for example, with ghastly results for the patients.
The better physicians worried about competition and even engaged in price-cutting.
Now, however, when they spoke up as professionals against colleagues who cheated
and actually harmed patients, the physicians' concern took on a whole new dimension and spread to leading laypeople who worried about the social problem of
incompetent practitioners who tried to drum up business by claiming to be
specialists.

Change and Lack of Change in Medical Care

The institutions of medical care that had held so much promise at the end of the
eighteenth century were clearly in an unsatisfactory state by the late nineteenth century. Perhaps without being fully aware of it, citizens of the United States in the

mid-nineteenth century had tried a great social experiment, applying an un-regulated free market to the area of health care. The historical verdict from later in that century is remarkably clear: subjecting health care to free-market forces ultimately does not work.

Regardless of economic arrangements, the dynamic of what was becoming bio-medical science continued to influence what both well-educated patients and their physicians believed medical care to be. Everyone's understanding of the body and of disease was shifting dramatically. Diagnosis in particular became more precise, based on more knowledge of pathological processes and conditions.

Even the best-qualified physicians, however, had little to offer by way of ther-apy outside of being present and showing concern. Enormous numbers of children under five died each year from diarrheal diseases. The most scientific specialists, the neurologists, could locate pathological conditions in the brain with uncanny accuracy and often could predict how a tumor or a hemorrhage would develop, but they had almost no therapy. They could only stand by helplessly and offer the com-fort of a certain, but gloomy, prognosis.

Physicians continued to claim a place in the sickroom, however. Watching the patient and devising ways to strengthen and "preserve the vital forces," as the great clinical authority Austin Flint wrote, was more than ample justification for the

THE MODEL SICK-ROOM.

A physician's idea of an ideal sickroom in 1875. The sickroom was of course located in a private home and was essentially an ordinary, well-furnished bedroom that could accommodate visitors as well as the doctor. Geo. H. Napheys, *The Prevention and Cure of Disease: A Practical Treatise on the Nursing and Home Treatment of the Sick* (Spring-field, MA: W. J. Holland, 1875), frontispiece.

physician's presence. And indeed, attention to regimen and hygiene became markedly more conspicuous in the medical literature.[108] A Boston physician well known for devising a narrow needle for draining accumulations of body fluids wrote in 1881 of the physician's duty "to watch with care the symptoms, to prevent their aggravation by intercurrent affections [additional diseases] or external physical agents, to regulate the temperatures and supply of pure air and remove impurities from the body and its surroundings. To supply proper food; to relieve pain and provide sleep."[109]

Strikingly, the few new drugs introduced from the European laboratories and clinics did not offer direct treatment of diseases. What they did offer was some control of physiological processes. Notable innovations included bromides, aspirin, and other new substances with narcotic properties, which could calm patients. Electrical stimulation made a comeback. Nitroglycerine was being tried to control heart malfunctions. *Physiological* referred to the fact that that the body processes worked regularly, scientifically predictably, and that drugs, not least bowel regulators, were designed to assist functioning. Diet was especially important in regulating patients' bodies.[110] But most physicians relied on traditional means.[111]

Or, following the lead of Bartlett (chapter 3) and of physicians in Vienna, American physicians simply waited for nature to take its course as they practiced ever better informed therapeutic nihilism. From one point of view, following the lead of the surgeon general in the Civil War, getting rid of traditional medications could represent genuine progress. A Wisconsin physician in 1881 declared that half of the then current medications had come in only in the preceding thirty years. And indeed, in the 1880s the presence in an up-to-date doctor's office of not only different medications but, as noted above, microscopes, thermometers, and other technologies such as ophthalmoscopes did in fact begin to transform even the local image of the American physician.[112]

In 1880 about three-quarters of the population still lived in rural areas or small towns and villages. The bulk of medical practitioners continued to offer largely traditional care to people in their homes, where family and friends provided nursing services. This health care was largely unaffected by elite practices in the cities. And even in larger urban areas, physicians maintained the rituals that they had observed in rural general practice. Yet as transportation and industrialization began to impinge on more and more areas of life, the idea of being up-to-date began to infiltrate everywhere into American culture. Then new technology and ideas about the body and disease transformed medical ideas and practice.

During the 1880s, traditional theories of disease and such practices as bloodletting continued to fade away gradually.[113] Many Americans in medicine were converting to a "scientific" basis for health care based on physiological universalism and laboratory procedures. They were in effect abandoning well-worn pathways even before they had new ones to replace them. Therapeutic nihilism therefore signaled that leading physicians were waiting for new theories and practices. Indeed, it turned out that therapeutic nihilism was a holding operation, until new ideas of disease and therapy could replace the old traditional medicine. But no one before the 1870s and 1880s really foresaw the new, which arrived dramatically in the form of germ theory and antiseptic surgery.[114]

PART II

Medicine and Health in the Age of Science and Modernization

The period from the 1880s to the 1930s brought overwhelming modernization to the United States. Many people who came of age in the 1880s were still in their sixties in the 1930s. A physician could have started practicing before the automobile and general use of the telephone and still be practicing when the first antibiotics came in. In the 1880s the railroads were only beginning to bring the ultimately massive breakdown of the isolation of the local island communities that made up most of the United States. By the 1930s, still within a single lifetime, there were national radio networks and dominant national businesses and other organizations. In parallel with other revolutionary technological and organizational changes, American medical thinking underwent profound changes in a remarkably short period.

Medical and health efforts in particular shifted further from focusing on the traditional to searching for the new. Health care in the time of the 1890s depression and the Spanish-American War of 1898 was in many ways an activity different from health care during the Great Depression of the 1930s and World War II. Germ theory, physiological investigations, and new therapeutic means transformed curing and care. Only in their most basic elements did doctors, hospitals, and "calling on the sick" remain the same.

This part therefore introduces a fundamentally new epoch in the history of Americans' struggles with disease. Modernization in medicine brought with it not just industrialization, and organization in the form of bureaucratization, but that special aspect of bureaucratization, the recognition of expertise.[1] *Expertise* could refer to the physician's specialized knowledge or to specialization in medicine. Also in the late nineteenth century, *science*, a term often taken to mean experimental laboratory science originally imported mainly from Germany, continued to mean "modern." Laboratory medicine was modern because, as one historian explains, it "was implicated in challenging modes of thought both within medicine and without," including prevailing cultural norms and the personal authority of previously dominant opinion makers. Indeed, at the end of the nineteenth century major changes were occurring simultaneously throughout the world in physics, chemistry, and biology.² These great intellectual transformations affected how people conceptualized their bodies and medicine.

How educated people and less educated people understood disease changed decisively. Instead of seeing diseases as resulting from a series of vague causes, such as personal constitution, geography, and health regimens, Americans, following European scientists, began to see each disease as the result of the action of a specific agent, typically at first a microscopic "germ" that caused that specific disease.³ The implications of this new knowledge reverberated through medicine and society in many ways. Moreover, the modernization of medicine in this new epoch, along with other modernization, meant more than newness or improvement or routine advance. As noted in chapter 4, innovations and changes in this epoch had the dynamic quality that came to be referred to as "modern."⁴ People sensed that their identities were not fixed by the past. They set up organizations, common in health care as well as typical of the late nineteenth and early twentieth centuries, to try to reshape at least parts of the world.⁵

In the 1880s, then, establishment medicine in the United States entered into a new phase.⁶ For the next century, innovations in technical and applied medicine, with powerful inputs from disease incidence, technology, and social change, defined a series of identifiable eras in the history of health care and medical thinking.

These eras were not general periods like the earlier Enlightenment, French, and German periods. These were eras marked by obvious innovations in medicine that had ramifications in special areas of practice, medical and public attitudes, and the relations of sick people with both science and society. In each era, writers in medical journals and popularizers in trendsetting magazines identified a theme of something new, different, impressive, and characteristic of the day.

Institutions persisted, responding to change but surviving the coming of each new era. Most, like the fee system of payment and medical schools, had origins in earlier periods. Medical organizations and journals also were in place, although now they often served to introduce specialization and sometimes radical new ideas. Likewise, from the mid-nineteenth century the profession of pharmacy and the beginnings of the pharmaceutical industry were in place. And continuing from previous times were self-treatment, folk medicine, patent medicines, alternative practitioners, and quackery. But in or after the 1880s, institutional changes crystallized modern versions of specialization, public health, and state licensing. Hospitals and nursing appeared in altered forms, and the mass media, based in a surprisingly educated public, began a period of increasing dominance in American life. At the same time, the way both professionals and lay persons thought about "the body" and their own bodies changed.

Physicians after the *Dent* decision rapidly formed a profession. Providing fellow citizens with medical services gave them a privileged status. Many physicians embraced specialization, identifying with the scientific disciplines that increasingly constituted medical knowledge. For most of the twentieth century, a physician had often conflicting social roles as a professional, a student of a specialized scientific discipline, and a fee-collecting businessperson.

Meanwhile, over a very few generations, industrialization and urbanization transformed the society repeatedly. Anyone could notice in the late nineteenth century that business was continuing to become more national in scale. Only in the 1930s, however, did the federal government, that is, national authority, become part of everyday life. The middle-class leadership in the social hierarchy in place at the turn of the twentieth century did not last much past World War II, when, among other things, a significant redistribution of income took place. Meanwhile, in the 1930s, pressure groups became the major means by which powerful social elements shaped political decisions, including those that affected health care.

The history of health and medicine from the 1880s through World War II presents a remarkable record of the ways leaders in health and healing efforts tried to deal with new ideas and ways of looking at humans and their illnesses. At the same time, social change and some intense modernization, both technological and organizational, now framed traditional and lasting ways of doing things—and also helped inspire and develop radical new approaches to diseases.

The deliberate innovators in medicine represented themselves as the embodiment of inevitable progress.[7] A focus on science, modernization, and anticipations of progress captures some of how health care in the decades from the 1890s to the 1940s became oriented to the future as well as the present of those Americans. And in such a narrative, the social context kept furnishing disconcerting realities, and the past kept providing a context for even the most startling innovation.

Landmark Dates

1876	Joseph Lister, founder of antiseptic surgery, speaks in the United States
1882	News that Robert Koch, in Germany, identified the microbial cause of tuberculosis
1886	Aseptic surgery begins to come in from Germany
1892	William Osler's *Principles and Practice of Medicine* published
1893	Johns Hopkins University Medical School opens
1893–97	Diphtheria antitoxin comes into use
1894	First large polio epidemic in the United States breaks out
1895–96	X-rays introduced

1896	Virtual completion of licensing of physicians in all states
1898	Spanish-American War
1901	American Medical Association reorganized
1903	Rockefeller Institute founded
1904	National Tuberculosis Association founded
1906	Pure Food and Drugs Act passed
1909–10	First chemotherapy (606/Salvarsan) imported from Germany
1910	Flexner Report published
1913	Pasteurization of milk introduced
1915	National Board of Medical Examiners established
1916	American Board of Ophthalmology (first specialist certification board) established
1917–18	American participation in World War I
1918–19	Influenza pandemic kills a significant fraction of the population
1919	American College of Surgeons begins rating hospitals
1922	Insulin announced and used
1924	World War Veterans Act provides federal government treatment for veterans
1927–32	Committee on Costs of Medical Care evaluates health care availability
1929–41	Great Depression
1933	Specialist board system decisively put into place
1935	Prontosil (sulfa) begins the antibiotic era
	Social Security Act passed
1938	Food, Drug, and Cosmetics Act passed
1941–44	Penicillin introduced
1941–45	American participation in World War II
1944	Paul de Kruif's *Kaiser Wakes the Doctors* published
1946	Hill-Burton Act passed
1947	National Institutes of Health funding begins to increase dramatically

CHAPTER 5

The Age of Surgery and Germ Theory, 1880s to 1910s

Specifics underline the extraordinarily rapid change in health care in the decades around the turn of the twentieth century.[1] As surgeons employed techniques of antisepsis and expanded what they could achieve, they rose to the top of the medical hierarchy. Their rise was accompanied by a modernized hospital. Almost simultaneously, germ theory provided a new understanding of infectious and contagious disease, filling the gap left by the erosion of traditional ideas of disease. Therapy, diagnosis, public health, and science were all transformed. Both hospitals and research suddenly became major components of medicine.

Physicians still collected fees, but they also were beginning to work miracles with surgery, diphtheria antitoxin, and techniques for containing yellow fever. There were new diseases and disease categories, not least polio. Ultimately, as we shall see, especially in chapter 7, the tumultuous and uneven change played into the interactions between health care and social reform.

Health and Diseases

Beginning in the 1880s the general health of the population (measured in terms of longevity) began to improve rapidly. The steep decrease in mortality rates could be attributed to a decline in deaths from infectious diseases, not least typhoid and diarrheas, which were strongly affected by water and sewer projects, particularly water filtration after the turn of the twentieth century (see, e.g., table 3).[2] Also in the 1880s, the average height of Americans began to increase again after a long period of decline or stasis. Among other likely factors, transportation developments, including the refrigerator railroad car, made a variety of foods more available and cheaper for city dwellers. One historian summarizes the factors behind the shift in health and longevity as "sanitation, nutrition, and birth rates."[3]

Meanwhile, disease patterns continued to change. As the western frontier closed, and transportation kept growing at a frantic pace, local geographical factors in disease distribution diminished. With the rise of bacteriology, which made infectious diseases each appear to be distinct and identifiable, Americans increasingly viewed diseases, especially infectious diseases, as operating in a realm of their own,

TABLE 3
Typhoid death rates before and after the addition of chlorine (hypochlorite)
to the municipal water supply in selected U.S. cities

City	Before (1900–1910)	After (1908–1913)	Change
Baltimore	35.2	22.8	35%
Cleveland	35.5	10.0	72
Des Moines	22.7	13.4	41
Erie	38.7	13.5	65
Evanston	26.0	14.5	44
Jersey City	18.7	9.3	50
Kansas City	42.5	20.0	53
Omaha	22.5	11.8	47
Poughkeepsie	54.0	18.5	66

Source: Adapted from John W. Alvord, "Recent Progress and Tendencies in Municipal Water Supply in the United States," *Journal of the American Water Works Association*, 4 (1917), 284, reprinted in *The Sanitary City: Urban Infrastructure in America from Colonial Times to the Present*, by Martin V. Melosi (Baltimore: Johns Hopkins Univ. Press, 2000), 144.

independent of local circumstances. Yellow fever, for example, had become distinctively located in the South, near the Gulf. But even that disease continued to retreat to the point that no cases were recorded within the United States after 1905, when any local occurrence there or anywhere in the world could with new knowledge be traced to a specific, controllable mosquito population.[4]

Infectious diseases continued to kill and disable people at a very high rate in the decades before and after the turn of the twentieth century. In 1900 the leading listed causes of death were influenza and pneumonia, tuberculosis, and gastroenteritis, which together accounted for 31 percent of all deaths. Heart disease, stroke, and kidney disease followed, with accidents and cancer trailing (cancer accounted for only 3.7 percent). Infant diseases and diphtheria ranked ninth and tenth, respectively, among the top ten killers. By 1914 the major killers were heart disease, tuberculosis, pneumonia and influenza, liver disease, and stroke. Infectious diseases in general were receding as a cause of death—down to about 40 percent in 1900 and rapidly declining much further in the 1920s, after the great flu pandemic of 1918–19.[5]

In an industrializing country, waves of illness became more conspicuous. In Muncie, Indiana, at one point in 1890, for example, 25 percent of the local factory workforce was laid up with "the grippe," presumably influenza, and the local newspaper reported that "schools and churches are both crippled . . . , and the flour mill has been forced to close because it has not enough experienced men able to work."[6] At a time when workers were not paid when they did not work, this was a remarkable testimony to the socially disruptive power of infectious disease.

In general, infectious diseases became less virulent or less destructive. No one knows why. Some, like smallpox, almost certainly changed biologically. Cholera too was no longer a major threat. In 1892 prompt and effective quarantine procedures, aided by the new laboratory bacteriology, kept the notorious Hamburg cholera epidemic in Germany from spreading to New York City (and the whole country). Other infectious diseases may have yielded to improved living conditions and water and sewer systems. That would include particularly typhoid fever, even though it continued to be the single most important disease in ordinary medical practice; indeed, typhoid was for most progressive practitioners at the time the model of all diseases. Moreover, it took decades for urban authorities not only to filter water supplies but also to separate sewer effluent (very seldom treated) from water intake—a delay that had continuing deadly effects. Between 1870 and 1920 the number of communities with some sewer system increased from one hundred to three thousand.[7]

Poor people continued to die at a higher rate than did rich people. The pioneering public health physician S. Josephine Baker in 1902 was discouraged by the defeatist attitude of Irish mothers in the New York City slums. They "seemed too lackadaisical to carry their babies to nearby clinics and too lazy or too indifferent to carry out the instructions that you might give them. I do not mean that they were callous when their babies died. Then they cried like mothers, for a change. They were just horribly fatalistic about it when it was going on. Babies always died in summer and there was no point in trying to do anything about it."[8]

Jacob Riis in 1890 published an account of illness in the lamentably bad living conditions of those New York slums. In a tenement on Cherry Street:

"Listen! That short hacking cough, that tiny, helpless wail . . . will have another story to tell . . . before the day is at an end. The child is dying with measles. With half a chance it might have lived. . . ."

"It was took all of a suddint," says the mother, smoothing the throbbing little body with trembling hands. There is no unkindness in the rough voice of the man in the jumper, who sits by the window grimly smoking a clay pipe, with the little life ebbing out in his sight, bitter as his words sound: "Hush, Mary! If we cannot keep the baby, need we complain—such as we?"[9]

Location had a significant impact on death rates. Urban areas were still deadly places in which to live, in part because contagious diseases spread more easily there. Yet very large urban areas were the first to receive the water and sewer improvements that were having such a powerful effect in reducing death rates. In 1900, sewers were available to only 29 percent of the population, and it was 1907 before sewers came, for example, to New Orleans.[10]

Even within urban areas, public health and civic authorities identified dirty and diseased areas associated with one ethnic group or another. Health policy could

This photo, taken in the slums of New York City, shows children from the "Lung Block," a group of tenements in which tuberculosis infected virtually everyone who lived there. Most, if not all, of these children must have died of the disease sooner or later after the photo was taken. John Spargo, *The Bitter Cry of the Children* (New York: Macmillan, 1907), 171.

reinforce the crude racism of that period, as authorities used the customary characterization of infectious disease itself as an unwelcome alien invader and easily assumed that unclean ethnics (who could be American-born) were a source of danger. So, for example, officials in Los Angeles pathologized Mexican, Chinese, and Japanese neighborhoods as both different and health hazards.[11]

Changes in Diseases

Historians have to contend with an intimidating variety of factors that were operating to curb the toll taken by infections. Medicine, with a yet very limited list of medications, was still seldom effective. Tuberculosis, the great killer of the nineteenth century, continued to flourish and to cause an enormous number of deaths (each year one in five hundred Americans died from the disease), but more patients than earlier were finding the disease survivable, if debilitating.[12] The familiar measles, mumps, and scarlet fever continued, but they killed fewer children. And people noticed that the great epidemics of diseases of the nineteenth century were largely absent after about 1890.[13]

At the same time, physicians in everyday practice were concerned about some diseases that within a few decades had disappeared—but disappeared by reconceptualization, not cure. There was chlorosis, so named because the adolescent females who suffered from it took on a greenish pallor (later believed to be caused by anemia from iron deficiency). Or rheumatic fever, typically in children, in which there was a combination of arthritic symptoms and heart disease (later believed to be the result of streptococcus infection, as will be discussed below). Or pink disease (so called because the palms and soles turned pink), later known as acrodynia, which appeared to be infectious because it sometimes appeared as an epidemic in children's institutions; it turned out to be mercury poisoning. And the disease of masturbation was a very frequent concern, with well-known signs such as paleness and shifty eyes, a classic example of a folk and medical belief that only very slowly yielded to skepticism and empiricism.

One new disease was just becoming established and recognized, and it turned into an epidemic disease. First reported in Europe, polio, or "infantile paralysis" (named for the most feared form), appeared in local epidemics in the United States from 1894 to 1910. Polio became especially notorious only with a 1916 epidemic that generated much publicity, especially in New York City, where it led to twenty-four hundred deaths. The public fear of polio was constantly growing, and polio became, in people's minds, menacing in the same way cholera had been. Some people of course blamed filth and the poverty-stricken immigrants who lived in degraded slums. But the disease, particularly the paralytic variety, also hit exclusive suburbs and the rural countryside. Everywhere, social functioning was disrupted when polio appeared in the summer. Any facilities where children might gather together, whether schools, swimming pools, churches, or the new movie theaters, were in danger of being closed down. Parents kept their children at home to avoid any possible exposure.[14]

Urbanization and "Progress"

The spectacular growth of big cities increased crowding even as infrastructures made urban living easier for many. Smaller cities and even towns across the countryside also grew rapidly and brought the experience of some urban living to large parts of the population. For physicians, urban populations and urban ways of life continued to stimulate the growth of specialization and even some medical research.

By 1890 the United States had more industrial production than any other nation in the world. Other signs of modernity appeared in a startling array, from business and social organizations to electric power and communication. Most of the changes involved not only shifts in viewpoints and perspective but also technological and scientific innovation.

Health care professions provided both symbol and substance for modernization. It was not just in larger urban settings with laboratories and specialists that medicine functioned to advance modernization. In local communities too physicians served as the face of modernization. A practitioner, even one who was not doing the most innovative procedures, within his or her community still was acting as a symbol and as an actual educator to move people away from folk beliefs and patent medicines. Nowhere was this educative role better acted out than in rural communities. Or in towns in the South, where scattered African American practitioners were markedly close to their communities, and their status was recognized, even though they did not network with advanced, big-city specialists.[15] Physicians modernized even though it was doubtful, as late as the 1880s, that there were treatment modalities that greatly differentiated the practices of the most scientifically advanced from those of the less modern physicians.

Just a few years later, descriptions of what up-to-date physicians could actually do for patients demonstrated the rapid changes. Medical leaders began to speak openly of convincing evidence of "progress" and to defend medicine aggressively.[16] Regular physicians' attacks on sectarian and traditional practitioners took on a new edge. It has been customary to picture physicians as economically motivated, for example, in their attacks on the large number of midwives still assisting in most births in rural and ethnic settings (as late as 1930, even after powerful medical campaigns, still 15 percent of the births). Market forces explain some actions, but the evangelical overtones came not just from greed and misogyny but also from the drive for science or modernization. Why did physicians criticize midwives? Because they were unhygienic and "ignorant," and they represented obsolete social arrangements. Educated physicians, by contrast, could furnish birthing that was "modern" and protected against dangerous dirtiness.[17]

Surgery

By the 1880s an essential change in physicians' therapeutic effectiveness was under way. It came by way of surgery.[18] In 1880, if a patient had complained of fever, gastrointestinal dysfunction, and acute pain in the lower abdomen, a physician ordinarily would have treated the patient, probably with a purgative and diet and possibly opium for the pain. Before long, the patient probably would have died of "inflammation of the bowel," a common diagnosis of that day, or peritonitis, infection in the abdominal cavity. By the late 1890s, however, the up-to-date physician would have recognized the existence of an infection, probably by localizing the pain through the McBurney point (a technique first recommended by an American physician, Charles McBurney, of New York, in 1889). Immediately, a surgeon (or even the country doctor) would have operated and removed the infected appendix, and in 80–95 percent of cases the patient would have survived, probably survived very well. Thus, within a remarkably short time, a major realm of general practice, internal pain and upset, previously treated by medical means, was in part taken over and treated by surgeons—with dramatically better results.[19] A sample of original articles in the *Journal of the American Medical Association* in 1885 shows that 26 percent of the articles were on surgical topics; ten years later, in 1895, the figure had increased to 68 percent.

The details of events suggest the nature and extent of the change. Particularly after the introduction of Listerian antisepsis (described below), surgeons entered the abdominal cavity more frequently and would sometimes try to drain an abscess or clean infections in the lower abdomen. But for many years they disagreed about what processes, even what pathology, they were dealing with. Under such circumstances, conservative physicians continued to invoke the healing power of nature and tried to let the patient recover naturally. Many patients did.

In the 1880s, autopsies suggested that a number of fatal cases had an origin in the appendix, that is, in appendicitis (effectively named in the United States by Reginald Fitz, of Harvard, in 1886). At the same time, exploratory operations became much safer. Therefore a number of elite surgeons, doubting the healing power of nature, began aggressively to insist on immediate exploratory surgery rather than waiting until the possibly infected appendix wreaked havoc on nearby tissue. The surgeon, with ever safer techniques and more certain outcomes, could replace even a good physician invoking therapeutic nihilism.[20] What moved surgeons into a commanding position in medicine was the advent of antiseptic and aseptic techniques that finally held in check the appalling consequences of infection.

American surgeons had long had unusually good results with operations, probably because they generally operated in cleaner settings than their European counterparts. Yet the toll was still so awful that, even with anesthesia, they operated only with the greatest restraint. By the early twentieth century, however, surgeons were repairing not only the interior of the body but even the brain.

Antisepsis

Antisepsis was originated by the Scottish surgeon Joseph Lister after reading a report in the 1860s by the French innovator Louis Pasteur, who had found that bacterial infection (usually in the form of inflammation, suppuration, or putrefaction) was causing some diseases. Lister decided to try to counter the inflammation of surgical wounds that was killing his patients. He turned to dilute (10 percent) carbolic acid, which was used in the Glasgow sewage works to suppress odors. Soon stories of his success in producing infection-free healing spread across the globe.

Americans in Boston, Chicago, and New York heard about Lister's work from British sources as soon as it was published. Soon some Americans were visiting him personally in Glasgow. They also heard opponents, who believed that Lister was just another upstart claiming to have invented a new kind of surgical dressing. A number, however, tried to follow his techniques. One young Boston surgeon successfully employed the method in 1869 on a case of breast cancer, but he had to do so surreptitiously since senior surgeons in the hospital, who were at best ambivalent, might object.[21]

Lister's procedures were complicated, and when they did not succeed, clinicians could claim that experience, the traditional test, justified rejecting the whole idea of antisepsis. William Goodell, of Philadelphia, reported using Listerian techniques in a case of ovarian tumor in 1878. The patient died of subsequent infection. "I was very much disappointed," wrote Goodell,

> at the issue of this case. It was my first one performed under the spray, and every detail of Lister's method was scrupulously carried out. . . . The room had been thoroughly cleansed, and a spray of carbolic acid kept up for many hours. The smell of this acid was indeed quite over-powering during the operation. The students present had, at my request, all bathed their persons that morning, and had put on clothes which they had never worn in the hospital wards and in the dissecting rooms. . . . No one but myself touched the peritoneum, and I am sure that no poison was lurking about my person. . . . I am at a loss to account for the presence of poison germs. My confidence in antiseptic surgery was somewhat shaken.

Yet Goodell persisted and could report a series of very successful recoveries from such operations.[22] It is easy to understand, however, how a clinician could reject Lister's techniques after an experience such as Goodell's.

In the United States, Lister's personal visit and lectures at the time of the 1876 centennial celebrations won a substantial hearing for the complicated procedures by which he sprayed and soaked the surgical wound with carbolic acid and antiseptic dressings. Gradually, American surgeons began sterilizing not only the wound but their instruments and hands, trying to cut down on postsurgical infection and extend what they could do using surgical procedures. Lister's method remained controversial, however. As late as the early 1880s at meetings of the new American Surgical Association antisepsis provoked contentious discussion, with the "Anti-Listerians" in the majority.[23] Yet as B. A. Watson, of New Jersey, observed in 1883, "I know that Listerism in America has made but little progress, but, nevertheless, the present system of surgical practice has been modified to a very great extent by the introduction of the Lister treatment, and we find scarcely a wound treated in the United States today but what some part of Listerism is adopted."[24]

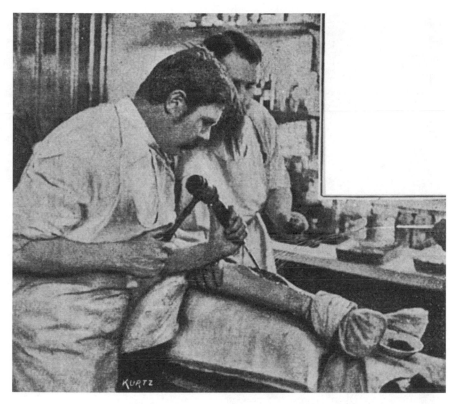

Surgeons in the antiseptic era removing a piece of dead bone so that a tibia can heal after an infection. Playing in from the right, onto the wound, is a hand-held spray of antiseptic, almost certainly dilute carbolic acid. The operator pictured can be identified as Arpad Gerster, of New York. Arpad Gerster, *The Rules of Aseptic and Antiseptic Surgery: A Practical Treatise for the Use of Students and the General Practitioner,* 2nd ed. (New York: D. Appleton, 1888), 195.

Shifting from Antiseptic to Aseptic

For some years, American practitioners tended to understand Lister's innovation as only an addition to surgical technique—even though Pasteur's ideas about germs had stimulated Lister to try his experiment. Gradually the idea spread that preventive cleanliness in and around the operation was the goal. Stephen Smith, of New York, as early as 1885 described the new procedure. In the old days a surgeon had not cleaned his instruments.

> During the operation he laid them down, or dropped them, and without cleaning applied them again to the wound. Now instruments . . . long before the operation . . . are placed in a carbolic solution, in order that any possible

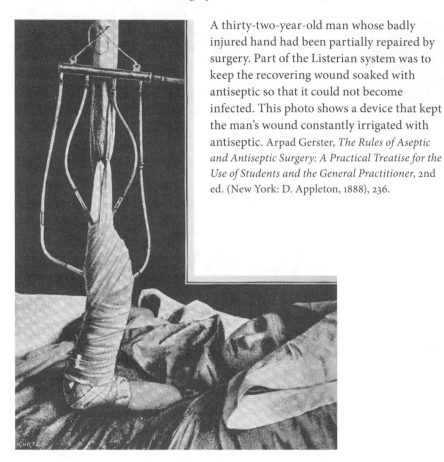

A thirty-two-year-old man whose badly injured hand had been partially repaired by surgery. Part of the Listerian system was to keep the recovering wound soaked with antiseptic so that it could not become infected. This photo shows a device that kept the man's wound constantly irrigated with antiseptic. Arpad Gerster, *The Rules of Aseptic and Antiseptic Surgery: A Practical Treatise for the Use of Students and the General Practitioner*, 2nd ed. (New York: D. Appleton, 1888), 236.

septic matter on them or their handles may be destroyed. During the operation one assistant devotes himself entirely to the duty of handing the instruments to the operator, and of receiving them from him and at once submerging them in the disinfective liquid. To avoid the possibility of laying an instrument down on an unclean surface, . . . towels wrung out of the antiseptic fluid are spread around the wound.[25]

Less than a year after Smith wrote, from Germany came the idea of disinfecting with heat as well as with chemicals, and the steam sterilizer was born. Indeed, Americans were relatively quick to adopt German methods of preventive cleansing of not only the instruments, the patient, and the room but also the surgeons and the operating room staff. As the logic of a completely germ-free environment spread over the years, the sterilized gowns, caps, and masks familiar a century later came into use. The effects of chemical cleansing agents on the hands of the personnel were often devastating to them, and, in an incident famous because he later married the lady, William Halsted, at Johns Hopkins Hospital, in 1890 introduced sterile rubber gloves so that a nurse could protect her hands from the corrosive disinfectives. Soon he and other surgeons also took to wearing rubber gloves, to have what were equivalent, as was commented upon at the time, to a "boiled hand," that is, germ free.[26]

The extent to which the antiseptic and then the new aseptic surgery was understood at the time to be merely an empirical modification of regular surgical procedures is evident in the fact that sectarian physicians readily adopted the new anesthesia and antisepsis/asepsis, but only as techniques, not because they were committed to a theory of disease. J. H. McClelland, of Pittsburgh, for example, in 1882 commented at the meetings of the American Institute of Homeopathy that "all

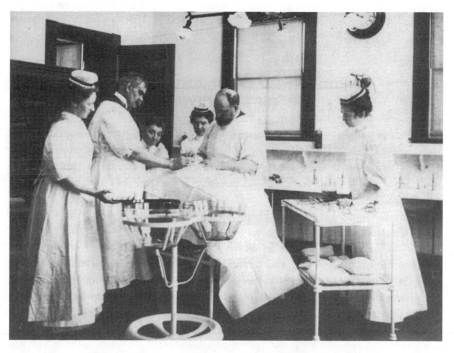

An operating room in 1901, showing the conversion to aseptic surgery. Street clothes are now covered by white gowns. There may even be rubber gloves in use. The antiseptic spray of an earlier day is gone. The masks and caps of a later period have not yet arrived, but the direction of change is clear. Joseph C. Merriam, *Framingham Medical Reflections* (Framingham, MA: Cesare George Tedeschi Medical Library, Framingham Union Hospital Library, 1975), 20. Generous courtesy of the Cesare George Tedeschi Library, MetroWest Medical Center, Framingham, MA.

good surgery in all time, wherein was provided thorough drainage, cleanliness and good wholesome dressings, was antiseptic surgery; and when we add to the means of cleansing and dressing wounds, certain chemicals thought or known to possess germicide powers, we complete the sum of an antiseptic dressing such as is accepted by a vast number of modern practitioners." Writing in a time of great change, McClelland noted that "in Vienna and Berlin they out-lister Lister," meaning that they used aseptic rather than antiseptic practices. His reference to "germicide powers" suggested acceptance of a microbial agent in putrefaction, but he was referring to surgery and wounds and did not necessarily mean to deviate from homeopathic therapies for nonsurgical afflictions.[27]

The X-ray

In 1896 came the x-ray, a technical innovation that almost immediately increased the effectiveness of surgical intervention. Surgeons had long depended upon pathological anatomy to localize injury and disease that might be corrected by surgery. The x-ray stood as an endpoint of providing exact localization by surgeons. Wilhelm Roentgen, in Germany, had identified the x-ray at the end of 1895. And now the advantage of having a medical school connected to a university became clear. Because the Crookes vacuum tube, which Roentgen used, was already standard equipment in advanced physics departments, the instrument quickly became available to alert physicians. The first x-ray photo published in the United States appeared within eleven weeks of Roentgen's announcement, and the first book on the subject within nine months.[28] E. A. Codman, a beginning physician at Harvard, in early 1896 received instruction from the physicists there and began extensive work on x-ray imaging:

> It would be impossible to give the reader an idea of the thrill experienced by those of us who did the early X-ray work. We each made weekly discoveries, only to find that our fellow workers in the same city and in all other cities had made the same ones at the same time. . . . Each of us had the self-importance to think that we were the first to show fractures of various types, to diagnose bone tumors or to locate foreign bodies in new parts of the anatomy.

Codman recalled showing the physiologist Henry Bowditch a bullet in his (Bowditch's) ulna, which Bowditch had carried since the Civil War. Codman also had a patient, an old woman, who for sixteen years had insisted that "she had a needle in her foot, when, after I had located and removed it, she shook it in the face of her doubting family."[29] One of Codman's young colleagues wrote to his mother: "Imagine taking photographs of . . . stones in situ—stone in the bladder—foreign bodies anywhere—fractures etc etc. . . . Its [sic] fearful uncanny."[30] Once again the

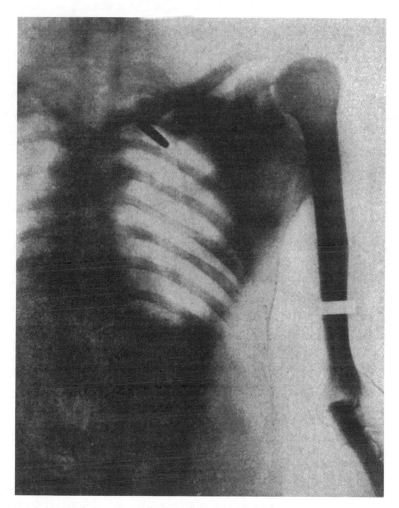

Just three years after the first report of an x-ray, at the time of the Spanish-American War in 1898, the pioneer radiologist Elizabeth Fleischman, of San Francisco, was already using the process to visualize a bullet lodged in the chest, as is perfectly obvious in this photo. Percy Brown, *American Martyrs to Science through the Roentgen Rays* (Springfield, IL: Charles C Thomas, 1936), 46. Courtesy of Charles C Thomas Publisher, Ltd., Springfield IL.

physician had a tool, in this case the x-ray, with which he or she could know still more about the patient than the patient knew. And by this time, in many cases surgery could demonstrably work a cure—shown, not least, by x-ray of, for example, a healed fracture.

Surgical Enthusiasts

The new safety and effectiveness of surgery made physicians and the public alike more accepting of surgical intervention. Surgeons saw new frontiers opening for their work, with the result that many acted, as historians have remarked, like soldiers assaulting a position or like heroic, daring explorers of the male variety. One historian has characterized this major change in surgeons' attitude as taking responsibility, asking what would happen if they did *not* operate and extirpate the enemy.[31]

Venturesome surgeons in fact undertook many dangerous or unnecessary operations, often on the basis of dubious diagnoses, as later surgeons pointed out. Particularly notorious was the wholesale, sometimes misogynistic removal of women's ovaries to cure supposed diseases based on female sexuality. One later specialist recalled tales from old practitioners of "the early days of abdominal surgery when a snowball in hell was almost as secure as an ovary in New York."[32] More benign, perhaps, was an increase in cesarean delivery of babies, traditionally very dangerous.

By the second decade of the twentieth century, however, the evidence encouraging surgery was convincing. A prominent surgeon writing for the public did not think it was too much to describe a whole range of the "marvels of modern surgery" that he himself had witnessed. Surgery, he wrote, "has explored the remotest regions of the body and carried a healing power to every part. It has even crossed the line between life and death, and brought back from the farther side the means of alleviation and recovery."[33]

The new image of surgery as miracle cure wrought powerful social change that transformed the entire configuration of health care. This transformation appears most clearly in the saga of the Mayo brothers, William (Dr. Will) and Charles (Dr. Charlie). Behind the appeal and drama of their incredible success story, however, lie extremely significant historical developments.

The Story of the Mayo Brothers

Dr. Will and Dr. Charlie were the sons of an English-born, largely apprentice-trained physician, W. W. Mayo, who in 1865 had settled in Rochester, Minnesota, a town in a remote prairie area south of Minneapolis. W. W. Mayo was a moderately successful general practitioner, and his boys grew up as typical hometown youngsters. He sent them both off to medical school, Will to the University of Michigan, Charlie to what became Northwestern University Medical School. In 1883, just as Will was finishing medical school, a tornado struck Rochester. As a consequence, the local Sisters of Charity decided to found a hospital to serve the community, St. Mary's Hospital. W. W. Mayo became surgeon to the hospital, and he was soon joined by Will and Charlie.

The Mayos of course took any kind of patient, but the boys, like many ambitious young MDs at the time, began to specialize in surgery. As surgeons to a local railroad, they had rail passes and so were able to travel. They did specialty training in New York and regularly attended clinics in Chicago. And they went to meetings, where their disciplined presentations and the large number of cases they reported soon made a powerful impression. When they could report a hundred cases of some procedure, others were reporting only a handful. As early as 1896 an operation performed on infected knees was known as the Mayo procedure. In 1893, when he was just thirty-one, Dr. Will was elected president of the state medical society, and in 1905 he was elected president of the American Medical Association. Citizens of the town of Rochester (with a population by then of six thousand) had never dreamed that one of their own would have such fame.

The media of that day took up this American success story and made these small-town kids into celebrities. But as the Mayo brothers added staff to what became the Mayo Clinic, their medical institutions also won recognition from their best-informed colleagues around the world and became synonymous with the ultimate in medical diagnosis and treatment. As early as 1901 the local Rochester paper reported:

> A number of the eminent surgeons of the country were in the city Friday. They attended clinics at St. Mary's Hospital, where several operations were performed. The gentlemen are Drs. J. B. Murphy of Chicago, one of the greatest surgeons of the United States; Robt. Weir, Professor of Surgery, College of Physicians and Surgeons, New York City; C. A. Powers, Professor of Surgery, University of Washington, Seattle; [Arthur Dean] Bevan, Professor of Surgery, Rush Medical College, Chicago. . . . The unanimous opinion of these famous surgeons is that St. Mary's is one of the best hospitals in the country, and all were delighted with their visit.[34]

This report, which sounds like puffery, was factual. The leading surgeons of the country and the world did come to remote Rochester, Minnesota, to see for themselves, and they ended up believing. The surgeon Owen Wangensteen said that when he traveled in Europe in 1927–28, his letter of introduction from Dr. Will would bring the head of any clinic rushing out, saying "Ah, Dr. Mayo!" Wangensteen observed that "the Mayo Clinic was appreciated in Germany and German speaking countries far more than it was here [in the United States]."[35]

The Mayo brothers' major impact, however, came through basic changes in patterns of providing medical services. The impact came through two institutions, the brothers' partnership practice and the hospital with its symbolic value. From the beginning, the two brothers covered for each other and kept each other up to date. As their practice expanded, they had to take in more partners, whose work they implicitly monitored and inspired. This model ultimately became the private group practice of the twentieth century, which succeeded the jealous individualism that had been altogether too characteristic of medical practice until then. And there were by-products of their group practice. Patient clinical and laboratory records were kept together in file folders, not recorded in great bound ledger books as had previously been the universal custom. The now-familiar examining room system was introduced, so that the doctor came to the patient in a separate room rather than having the patient come to the doctor's office and waste his or her time undressing, dressing, and so on. This meant that efficiency increased greatly, and one physician could see many more patients per day.[36]

The Transformation of the Hospital

What really changed the course of events, however, was how St. Mary's Hospital affected the imaginations of Americans of that day. The result was more profound, and different, than anyone might have anticipated. By the early years of the new century, the conclusion people were drawing from media stories was that if an isolated prairie community like Rochester, Minnesota, could have spectacular, world-class medical care, so could any town or city neighborhood in the United States. "The hospital" thus came to embody nonpartisan, cross-class hope. Consequently, there sprang up across the United States untold numbers of community hospitals and, along with them, expectations that modern medicine would work miracles for even the humblest citizen. Those expectations, often unrealistic, reverberated through the twentieth century into the twenty-first as a major aspect of American society and even politics, as we shall see.

At the beginning of the twentieth century, hospital care was increasing spectacularly primarily because surgeons, delivering most of the miraculous therapy of that day, needed sites for antiseptic and aseptic surgery—and also controlled areas for recovering patients. Even before 1900 a few surgeons practiced exclusively in hospitals. The technology and organization of the hospital, in addition to aseptic facilities, became standard within a remarkably short time. In 1877 the New York Hospital boasted that surgery done there was as safe as that done in the "most luxurious home." By 1907 the artificially lighted, elaborately equipped, and aseptically designed operating suite was clearly superior to care in any home.[37]

The change at the turn of the century was extremely rapid. A survey of hospitals in 1873 located 178 (although some additional institutions did not reply). The *American Medical Directory* of 1909 listed 6,152 hospitals, including 543 mental and tuberculosis specialty institutions. The figure was so high because it included many that were surgeons' small private hospitals for their own patients.[38] In 1903, an article in *Good Housekeeping* magazine was still asking, "Hospital or Home?" By 1910, articles in the same magazine were not mentioning the home option but bore titles such as "The Hospitalized Child."[39] Already in 1896, according to one report, in New York "even millionaires, having their own luxurious private homes, find in the hospital surroundings that are more favorable for a prompt and sure recovery than they can get elsewhere." They found the rooms comfortable but plainly furnished, with no draperies because, it was explained, "draperies encourage microbes."[40]

Hospitals at the turn of the twentieth century were also changing from charity institutions to institutions in which one paid for "care," which previously by custom had been provided by female members of the family. For hospitals, there was a growing tension between tending to the poor (the obligation of social leaders) and selling a service.[41] Social change also stimulated the growth of hospitals. In the urban areas, large numbers of both single persons and families in cramped quarters turned to hospitals for care that was not easily delivered at home.[42]

A second, related hospital struggle was between lay control, which one expected in a charitable institution, and medical control, which was appropriate for a medical institution. The lay trustees originally oversaw every detail of hospital operation. Because of the effectiveness of surgery in particular, physicians could demand conditions and control that would have been impossible earlier. Moreover, physicians had professional goals, including research and teaching, that they brought into hospitals. As early as 1900 the standard hospital admitting form impersonally separated the patient's own account of his or her illness from the results of the physical examination by the physician and the reports of laboratory tests. Patients often sensed that their bodies and diseases were more central to hospital operation than were their personal identities and agendas, exemplified in the morning report of an enthusiastic clinical assistant: "A pretty good lot of material. There's a couple of good hearts, a big liver with jaundice, a floating kidney, three pernicious anemias, and a flat-foot."[43]

And with technology, size, economic thinking, and organization came another consideration: management and efficiency. A third force, beyond charity and medicine, therefore, came into hospitals: professional administrators, to whom trustees delegated authority, just as they did in comparable situations in business, where managers were gaining power.[44] Not surprisingly, the institution of the hospital reflected the factory system and the organizational society of the time. In fact, as early as 1899 hospital administrators organized the American Hospital Association.[45] Meanwhile, the nurses and orderlies and maintenance personnel, including maids and cooks, were paid the lowest wages possible.

A Medical Care Hierarchy

The hospitals in particular brought with them two institutions that set the pattern for a whole social hierarchy in medical care for the twentieth century, a hierarchy that elevated the social position of physicians in particular. One institution was postgraduate physicians, typically interns and residents, although the terminology varied. Sometimes they were called "clerks" or "walkers" (in the wards). These medical school graduates performed a year or more of service in the hospital to learn the details of the practice of medicine in general, or of some specialty. Although hospitals exploited interns and residents and paid them little or nothing, they often served under expert practitioners and expected the experience to help their careers in medicine.[46]

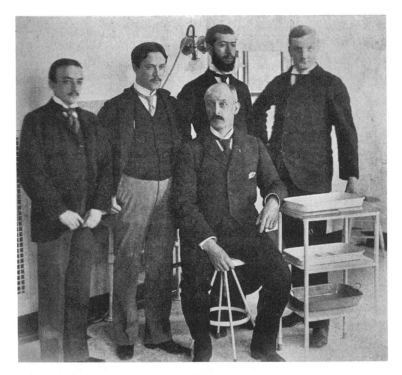

By the early 1890s the custom of newly graduated physicians' working on hospital wards to learn better how to practice medicine was common. The interns often posed with the chief surgeon in a formal portrait, as in this one at the Chambers Street Hospital in New York. J. C. Cutter, "Intern Service in the Old Days at Chambers Street Hospital, N.Y.," *Annals of Medical History*, 2nd ser., 3 (1931), 41.

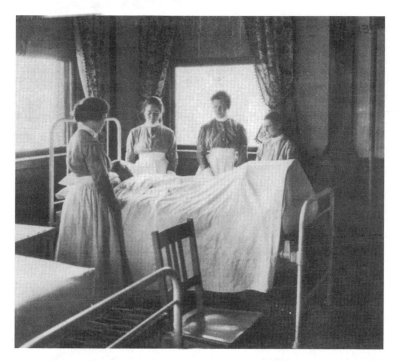

Students in the Framingham Hospital Training School for Nurses in a practical instruction area in a classroom, c. 1905. This Massachusetts hospital nursing school was typical of those that sprang up around the turn of the twentieth century. Such students furnished essential labor in American hospitals everywhere while providing a vocational opening for women. *Framingham Medical Reflections* (Framingham, MA: Cesare George Tedeschi Medical Library, Framingham Union Hospital Library, 1975), 84. Generous courtesy of the Cesare George Tedeschi Library, MetroWest Medical Center, Framingham, MA.

The other new institution was the nursing school, which was integrated into the hospital. By the 1890s larger hospitals, at least, became heavily dependent on the labor of nursing students—under the guise of training the students in nursing procedures. The students not only paid fees but served as unpaid workers, thus subsidizing health care with their labor. This arrangement worked because nursing provided women with one of the few opportunities they had to enter an established, respectable occupation. Indeed, in those years, few graduate nurses stayed to work in the hospital except in special positions. Not only were graduates too expensive for the hospitals but a private duty nurse could make perhaps four times what a hospital nurse was paid.[47]

By the opening of the twentieth century, then, a pecking order was coming into place, as exemplified in the hospitals. Earlier, as noted above, physicians had done nursing care and, for example, changed the dressings on patients' wounds. Now they could delegate those responsibilities to residents and interns. The residents and interns, in turn, were doing less of the manual labor in patient care, and so eventually they could assume that the trained nurses would change the dressings, assist surgeons, or, later, insert catheters. And the graduate nurses would turn housekeeping chores and jobs such as emptying bedpans, changing bedclothes, or bathing patients over to the nursing students as well as to the hospital maids and waitresses. Increasing numbers of nurses, interns, residents, and attending physicians thus steadily moved away from the manual labor of patient care. Nurses in particular found roles—in public health, for example—that moved them in other ways toward professionalization.[48]

Social Segregation and Discrimination

As hospitals shifted away from being charities, in a variety of additional ways they reflected the society in which they functioned. No one then thought it strange that wealthy patients had comfortable accommodations, services, and privileges or that the poor were crowded into wards and kept under strict discipline. One poor middle-class teacher wrote to the superintendent of Johns Hopkins Hospital to ask how she could "have the advantages of the Hospital . . . without going into the free ward."[49]

Especially in the late nineteenth century, charity often failed patients from groups suffering overt discrimination, in particular those with lower-class identities. Separate institutions therefore sprang up, primarily in the growing cities into which migrants from southern and eastern Europe were crowding, to serve groups discriminated against on the basis of identity regardless of wealth and education. Catholic hospitals multiplied, with labor and care based on the service—and administration—of sisters in several orders. Jewish hospitals attempted to accommodate both charity and middle-class patients. Even charity and sectarian hospitals competed for patients and community support as they opened everywhere across the country—and often failed and closed as well, as when, for example, a mine with ethnic laborers closed.[50]

Even as growing bureaucratic organization as well as depersonalizing scientific medicine shaped hospitals, the institutions perpetuated the conventional discriminations of those times. African Americans received care not only in separate wards but sometimes in separate hospitals—when they were not denied local hospital care altogether. Other kinds of segregation assumptions led to women's hospitals, which could also reflect the rise of specializations, as did pediatrics in children's hospi-

talo. Indeed, as all segregated groups produced more middle-class activists and ambitions for equality, the question became, would the separate hospitals and medical staffs, which had achieved their own status, "integrate" into the "white" male
medical establishment? By as early as 1882 seventeen state medical societies had admitted a total of 115 female physicians as members.[51] In 1900 there were seven thousand women doctors, about 5 percent of all doctors. Women practitioners were often irregulars, as noted above. As times changed, an increasing number, but still
only a trickle, of women graduated from good regular medical schools. They often had confidence in what they were doing because they had grown up in a social system that emphasized the moral superiority of women, and a few were in fact
outstanding national figures. Yet most women physicians had to put up with a great
deal of hostility and discrimination from male colleagues, not to mention patients.[52]

This photo of a class in the University of Michigan Medical School in the early
1890s shows some women students alongside the male students, as well as teaching
materials leading to a departure from the simple, rote, ceremonial lecture method.
Image bank of the University of Michigan Bentley Historical Library Collection.

Regardless, they and African American doctors, along with doctors from various national ethnic groups, suffered great indignities in a society that was not kind to people who did not fit certain superficial stereotypes.

Sponsorship aside, some hospitals were very good. With the latest technological advantages, such as operating tables and ventilation and lighting, as well as opulent furnishings, leading hospitals could attract both patients and medical staff. But there were other hospitals that were dirty, disorganized, and administered with shocking casualness and neglect. The range of quality and care paralleled the incredible distances that separated the good medical practitioners from the bad, a distance that could grow as medicine became more scientific and more effective. There were now not only good doctors but increasingly expert doctors. The surgeon George W. Crile, of Cleveland, recalled from his younger days that "surgery was in transition, textbooks became obsolete as soon as they were published. As there was no recognized authority, I relied upon the clinic [for teaching material] . . . and tried to keep in touch with the individual work of outstanding surgeons. . . . I visited the clinics of Bull and McBurney in New York, Senn in Chicago, Morris Richardson in Boston, Halsted and Kelly at Johns Hopkins and Gross and Keen in Philadelphia."[53]

The Struggle over Germ Theory

It is true that changes in surgery had profound professional and social effects from the 1880s to the second decade of the twentieth century. But those changes accompanied and interacted with the general intellectual revolution that came with the germ theory of disease—symbolized, for example, by two journals that changed their titles in the mid-1890s. One openly abandoned an older title, *Sanitary Era*, to become *Modern Medical Science*. The other changed from the *Battle Creek Sanitarium and Hospital Bulletin* to *Modern Medicine* explicitly on the basis of bacteriology introduced by the medical and religious reformer John Harvey Kellogg.

For many late nineteenth-century physicians, it was difficult to give up long-established ideas of a pathology based on the patient's constitution, environment, and life course. William Allen Pusey recalled that his father, a country doctor, could understand antiseptic surgery and even understand the importance of bacteria in surgery, but to the end of his life "he found it hard to see the new point of view in disease."[54] From 1876 until as late as 1881 most medical journal articles on antiseptic surgery actually ignored the germ theory of disease.[55] Physicians who employed antiseptic technique did not necessarily have to believe that bacteria caused ordinary diseases.

Older physicians who were slow to understand the basic change in framework had a point: it was mere theory. Bacteriology was very slow indeed in producing really effective therapies that could not be explained in other ways. As the editor of a medical journal wrote in 1885, "It has long been the reproach of the medical profession that therapeutics is not an exact science. In the light of the germ theory it might virtually become such." But he then turned quickly, and significantly, to the possible applications of germ theory to prevention, not treatment, of disease.[56]

Like the surgical journals, the leading general medical textbook did not introduce even the idea of bacterial disease until 1881. And the one medical area besides surgery in which germ theory proved of most immediate importance, public health, was an area low in prestige in medicine as well as controversial in social and political life. Good medical scientists, hardheaded materialists, could scoff at the idea that, as in witchcraft, a mystical entity that one could not see might come from the outside and enter the body and make a person sick. Or they granted that there were bacteria in diseased areas of the body but said that the bacteria developed only after the disease process started. They did not cause it. Moreover, there were problems with the theory. Scientists had to sort out the helpful or neutral bacteria and the dangerous bacteria in the body. They had to see that spores as well as adult bacteria could spread disease. They had to discover asymptomatic carriers of disease. And they had to have a good system of proof, which they did not have until Robert Koch, in Germany, in 1876 publicized his work showing beyond reasonable doubt that a bacterium caused anthrax.[57] The germ theory of disease had begun to come together in Europe in the 1860s, and educated Americans knew something about it by the 1870s, even as physicians were gradually learning about antiseptic surgical techniques. The 1880s were years of great controversy in medicine, as students fully converted to the germ theory of disease returned from Europe and confronted established colleagues who were not. In 1884 William Belfield, of Chicago, wrote that in the United States there were "at present perhaps a score of men who have given abundant evidence of competence in bacterial investigation," and it was to them, "not to dermatologists, surgeons, or pathologists, [that] we must look for facts upon this subject." He went on to cite sarcastically an eminent practitioner who had asserted "that bacteria, so-called, are in his opinion fibrin threads . . . and there is said to be a man in Virginia who insists that the earth is flat."[58] Not until the 1890s was the idea of microbial disease generally accepted in the field of medicine, and physicians no longer called it a theory but instead spoke of the facts of bacteriology and pathology.

Later generations may have believed that doctors and the public accepted bacterial explanations for diseases only very slowly. But in fact, beginning especially in the 1880s, adoption of the new viewpoint was, all things considered, remarkably rapid. It required seeing each infectious disease as a separate phenomenon that was defined by a particular kind of microbe and that did not vary in cause from person to person, a viewpoint that decentered the patient.[59] What American physicians still demanded was facts. In the French period earlier in the century, Americans wanted observable facts. Late in the century they still rejected theory, including germ theory. But as noted in chapter 4, gradually a new kind of evidence commanded authority: laboratory experiments.[60]

On 3 April 1882 news came from London via the Atlantic telegraph cable that Robert Koch had identified by rigorous experimentation a microbe that caused tuberculosis, that chronic, wasting disease that each year carried off a significant part of the population. TB was the model disease for which physicians offered constitutional explanations. Indeed, statistics showed that the disease often ran in families and so was probably hereditary.

The idea that tuberculosis was an infectious, contagious disease threatened traditional medical thinking as perhaps no other common fact could. And the idea also threatened the authority of physicians. In 1883 a physician in Wisconsin declared that despite Koch's work, "we certainly do *not know* that consumption [tuberculosis of the lungs] is an infectious disease. On the contrary, the vast majority of the profession . . . are pretty well satisfied that it is not. . . . To say that the profession has, all these years[,] been making a great and fatal mistake in regarding tuberculosis as a hereditary malady . . . is to make a very grave charge indeed."[61] But by the mid-1880s the idea of bacterial infection was flooding into American medical discussions, typically first in meetings of local medical societies and then in the journals. In 1884, in a new edition, the author of a leading medical textbook inserted a special appendix reporting Koch's work on tuberculosis.[62]

Tuberculosis did present great problems of evidence because the infection was not simple and direct but could produce very different kinds of symptoms in various parts of the body. Moreover, the disease was a chronic disease and so did not fit the pattern of "contagious" diseases like smallpox or childhood diseases. Finally, the germs turned out to be everywhere, and still most people did not have symptoms. That raised the question of what was later called resistance—as in—"resistance of the tissues" or "resisting power"—a concept new in late nineteenth-century medical practice.[63]

Germ Theory in Public Health

Even as a general, abstract idea, germ theory had remarkable immediate application in areas of public health. As early as 1885 Lewis H. Taylor, who was practicing in Wilkesbarre, Pennsylvania, was called upon to investigate an alarming epidemic of a virulent fever in nearby Plymouth, a town with a population of eight thousand. "So sudden was the onset," wrote Taylor, "that within a very few days

A TREATISE

ON THE

PRINCIPLES AND PRACTICE

OF

MEDICINE;

DESIGNED FOR THE USE OF

PRACTITIONERS AND STUDENTS OF MEDICINE.

BY

AUSTIN FLINT, M.D., LL.D.,

PROFESSOR OF THE PRINCIPLES AND PRACTICE OF MEDICINE AND OF CLINICAL MEDICINE IN
THE BELLEVUE HOSPITAL MEDICAL COLLEGE, ETC.

FIFTH EDITION,

REVISED AND LARGELY RE-WRITTEN.

WITH AN APPENDIX

ON THE

RESEARCHES OF KOCH, AND THEIR BEARINGS ON THE ETIOLOGY, PATHOLOGY,
DIAGNOSIS, PROGNOSIS, AND TREATMENT OF PULMONARY PHTHISIS.

PHILADELPHIA:
HENRY C. LEA'S SON & CO.,
1884.

Less than two years after the revolutionary demonstration in Europe that tuberculosis was an infectious disease caused by a microbe, the leading U.S. medical textbook appeared in a new edition in which the author, Austin Flint, included a special appendix conspicuously announced on the title page to report "the researches of Koch, and their bearings on the etiology, pathology, diagnosis, prognosis, and treatment of pulmonary phthisis [tuberculosis]." Austin Flint, *A Treatise on the Principles and Practice of Medicine: Designed for the Use of Practitioners and Students of Medicine*, 5th ed. (Philadelphia: Henry C. Lea's Son, 1884).

nearly a thousand people were stricken with the dread disease. The ravages were not confined to any class of people, nor to any section of the town, but the dwellers in the mansion as well as in the hovel were alike attacked." Typhoid fever was the diagnosis. Taylor continued: "That the mountain stream supplying the town with water might have become polluted by fecal matter was first suggested by Dr. R. Davis, of Wilkesbarre, in an article published in the *Record of the Times*. . . . The publishing of the article excited considerable interest, and, as great consternation prevailed among the people upon the subject of drinking water, a committee of physicians [was appointed]." The physicians found that "the stream supplied [residents] with an abundance of pure water." They also found that there was one house in the watershed and that a person recovering from typhoid, contracted in Philadelphia, had been living there. During the cold weather the "dejecta" from the patient had been thrown on the frozen ground or put into a faulty privy that drained onto the surface ground, and they surmised that with the spring thaw, infected matter polluted the water supply—the weather records exactly confirming the dates. The committee also used other evidence. The most compelling was that people who obtained their water from private wells rather than from the town water supply had not come down with typhoid fever. Taylor and his colleagues were satisfied that they had found the source of a "specific typhoid fever poison," a concept that could quickly translate into germ theory or might actually have derived from some version of it.[64] In this case the epidemic could be traced to one source of infection, indeed, a single person, which people could, and did, read as confirming the germ theory of disease.[65] They also understood that public health had a new strategy: to find the specific source of any infectious disease.

Familiar Ideas in Germ Theory

The germ theory spread as rapidly as it did because lay people as well as physicians already knew and believed in basic elements that went into it. Particularly important was the sanitary movement. The sanitarians emphasized cleanliness and fresh air and water to fight against the "diseases of filth" and drew on the long "cleanliness is next to godliness" tradition. As an Ohio sanitarian wrote in 1885, without referring to the germ theory of disease, "Special, intelligent, and careful investigations have demonstrated the connection between decomposing, foul-smelling, filthy human excrement and various filth diseases, of which typhoid fever is the type."[66] This sanitarian was concerned with "foul privies" everywhere, but it had long been customary to sanitize housing and other places in the environment. Lister, as noted above, had learned about carbolic acid because it was used on sewage in Glasgow to reduce the smells from festering organic material. Particularly in the growing urban environments in the United States, reformers concerned with filth in streets and homes connected dirt with disease as well as degradation.⁶⁷

The germ theory merely intensified the quest for cleanliness. Indeed, in the 1880s sanitarians and their campaigns against filth diseases were losing support. They could not show convincingly that their sanitary programs were successful. Some of them emphasized moralistic behavior and ever more intensely condemned not only dirt and "crowd poisons" but, as noted earlier, high heels and fried foods.⁶⁸ Indeed, the movement to open public baths in the cities, especially in slum areas where housing did not include bathing facilities, was aimed at moral uplift as well as health. As the New York State Tenement Commission reported in 1895, "The cultivation of the habit of personal cleanliness [has] a favorable effect . . . upon the character, tending toward self-respect and decency of life." Such moralism was, however, not saving sanitationism—and then the germ theory came along and restored credibility to the sanitarians and their campaigns against the filth diseases.⁶⁹

In 1886 one of the sanitarians' journals continued listing the traditional environmental concerns of their movement, with a section of the journal devoted to each: "Air and Ventilation; Water: its contamination, dangers, and purification; Waste: its drainage, sanitation, and utilization." And then there was the usual miscellaneous "General Sanitary Interests: Vital Statistics, Hygiene, Misc." Vital statistics, including epidemiology, was a still developing major element in public health. Hygiene was personal health. But quite suddenly in that year such articles as "The Elimination of Bacteria from Water" made it clear that readers were now expected to accept germ theory.⁷⁰ As William Henry Welch declared in 1889, "We are evidently at a great advantage when we can study the epidemiological facts with a knowledge of the substances which actually cause infection."⁷¹

Even among those outside the sanitary movement, specific elements in germ theory, beyond the value of cleanliness, were also familiar and made the idea easy to absorb. Most well-informed physicians after midcentury believed that diseases were specific entities, even if they had fluid boundaries and multiple vague causes. With germs understood to be corresponding specific causes, this common belief simply became much more precise. In zoology and medicine alike, parasites, even human parasites such as intestinal worms, were well known. The study of minute organisms, parasitical and otherwise, also had become well developed by the late nineteenth century with the widespread use of the improved microscope. The idea of poison was traditional, and, as I noted earlier, the special cases of transmission of some kind of poison as the cause of a disease, as in syphilis, were familiar. The former Confederate physician Joseph Jones, for example, writing in 1867, concluded "that typhus and typhoid fever are dependent upon the action of special poisons."⁷²

Physicians were familiar with the idea of contagion by contact, not least from the insistent writings, for decades, of Oliver Wendell Holmes and others about the spread of childbed fever on the hands of physicians.[73]

To put all of these elements together in microbes that entered the body, and poisoned it, challenged traditional ideas of disease. But separately these were all familiar ideas, and they opened the way to the new explanation: germs. As the years passed, investigators added a series of new elements, such as staining techniques for microscopic investigation and the idea that bacteria produced toxins, all of which made the mechanics of germ theory even more convincing. Indeed, as early as 1886 Boston City Hospital had a special laboratory for the microscopic identification and diagnosis of diphtheria.[74]

Building on the sanitary movement, then, physicians and public health officials tried to destroy dangerous germs. One obvious model was the operating theater, in which germs in the environment met antiseptic sprays and dips and heat, while the individual patient was also cleansed and disinfected as far as possible. For purposes of either prevention or cure, however, it turned out to be impractical to administer carbolic acid and other disinfectants inside an individual's body, whether one's own or someone else's. Because germ theory did not contribute directly to therapy, many physicians did not take it seriously.

Rabies and Diphtheria

Germ theory did, however, have practical applications in two major diseases. Both applications were based on the strategy of establishing in a person's body resistance to disease by using the disease microbes, on the model of smallpox vaccination, that is, a very specific injected serum for a very specific disease. The two practical examples, however, were based on countering, not the microbes themselves, but toxins the microbes produced.

The first was Pasteur's empirical treatment for rabies, or hydrophobia. The popular fear of the bites of mad dogs was longstanding, for the death rate in rabies was very high. In 1885–86 American newspapers and medical journals carried stories of Pasteur's experiments with an inoculation that appeared to prevent the development of rabies after a person was bitten. The newspapers sensationalized the first American children treated by Pasteur, who were sent across the Atlantic from Newark, New Jersey, to France. On their return, three of the children were put on display in New York and other cities and viewed by hundreds of thousands of paying customers. A pattern had been set for "the medical breakthrough," a pattern that was intensified by pictorial material in newspapers and illustrated magazines showing images of physicians giving hypodermic shots or of gleaming, complicated laboratories. Pasteur could not identify the microbe (later identified as a virus) involved in rabies, but he and others believed that the success of the treatment rested on germ theory.[75]

The second practical application of germ theory was the diphtheria antitoxin developed by Emil von Behring and Shibasaburō Kitasato in Berlin. Diphtheria, unlike rabies, was not an unusual disease. It continued and increased notably as a common and feared visitation in the late nineteenth century. Untold numbers of parents were still holding children in their arms and watching them choke to death as the infection created a false membrane across the windpipe. As late as the 1880s one small-town physician in the midst of a diphtheria "terror" that was taking many children in the community away prayed helplessly with his family, "Protect us, oh God, from diphtheria!" The first application of germ theory to diphtheria was diagnosis. The bacillus was definitively identified, and by the late 1880s and early 1890s American physicians could be expected to use the microscope to confirm the clinical diagnosis. Then Von Behring and Kitasato used serum taken from animals artificially infected with diphtheria to convey the resulting immune factor (a new concept) to a human patient.[76]

Isaac Abt, a young physician pioneering the specialty of pediatrics in Chicago, recalled his experience:

> One of my colleagues, a general practitioner, wanted to talk to me at once. When I finally got him on the phone, his voice was sharp with anxiety. He told me that he had a patient with a virulent diphtheria, and he feared the child's chance of recovery was very slight, unless there might be some good in von Behring's new antitoxin. He wanted a child specialist to administer it, and asked if I could go to the house immediately. There was no question about that; but antitoxin was not a thing that I carried around in my bag. . . . All that we had in this country was then imported from Koch's laboratory in Berlin, where von Behring had only recently discovered it. I knew of only one pharmacist in Chicago who might have a supply. Luckily, he had just received a small shipment, and I was able to obtain the serum without too great delay; but by the time I reached the patient he was very near death.
>
> "You can see why I sent for you," the attending physician said, "The serum may not help, but it's our only chance now."
>
> I gave the antitoxin and we waited. To my knowledge, this was the first time the serum had been administered in Chicago. Neither of us had ever seen its effect.
>
> The child lay still; his skin was blue, his pulse very weak; but he lived through the night. In the morning, his temperature was lower, he was breathing regularly, and color had returned to his lips. Twenty-four hours later, all immediate

danger was past; and in the days that followed, we saw him make a quick and uneventful recovery.[77]

As Abt made clear, the effect on the patient was remarkable. Science had worked a miracle comparable to the miracles of the Bible. But the effect on the physicians was equally remarkable. They had been more than helpless counselors and nurses: using laboratory science, it was they who had worked the miracle. No physician of that period who had a similar experience failed to mention the memory of it. They tended to forget occasions when antitoxin had been of no use or had had terrible side effects. And there were outright failures and improper uses and a great deal of confusion about diagnosis. Several New York

By 1902 antitoxin was a commercial product, obtainable from sizable factories. Here workers in the H. K. Mulford Company plant in Philadelphia are filling aseptic bulbs with antitoxin in preparation for shipment. H. K. Mulford Company advertisement, in *Pediatrics*, 8 (1902), unpaginated.

physicians expressed noisy skepticism and opposition based in part on the earlier failure of a serum against tuberculosis. But that was not what physicians remembered or what newspaper writers touted. As a pediatrician who had gathered 850 reports from physicians across the country declared, "Nothing has been so convincing as the ability of Antitoxin, properly administered, to check the rapid spreading of membrane downward in the respiratory tract."[78] The doctors believed that they had seen with their own eyes the new model of modern medicine: physicians saving lives by using technologies based on laboratory experiments.

It is hard to exaggerate the impact of the diphtheria antitoxin in convincing both practitioners and the public that germ theory and the laboratory did indeed represent a new kind of medicine that would displace much of the wisdom and experience of traditional practitioners. President Charles Eliot of Harvard recalled that the senior pediatrician there had strongly opposed the "fancy notion" of setting up a laboratory of bacteriology. But one day during a diphtheria epidemic Eliot encountered the pediatrician coming out of that lab, and Eliot asked him, "Doctor, isn't this an extraordinary place for you to be?" The pediatrician replied, "Yes, but nowadays you can't get your damned diagnosis anywhere else, and you can't get your damned treatment anywhere else." By 1913 all of the states but two had public health laboratories.[79]

The Public Impact of Germ Theory

Members of the general public in the United States were faster than physicians to try to apply germ theory. Information about microbes, along with fear, quickly began to appear in popular publications, school curricula, and other general media. Educators, and then the marketers of soap and disinfectants, first targeted middle-class homes, in which women's domestic role as health preservers took on unprecedented importance. The housewife became less a decoration and more an authority on modernization. And very quickly the target audience extended to include populations marked by social discrimination whose circumstances made them vulnerable to infectious diseases, particularly "new immigrant" groups and African Americans.[80]

One of the most dramatic signs of change was the disappearance from public spaces of the common public drinking cup, typically chained to a public water faucet. Instead, within a very few years, there were bubbling fountains everywhere. As late as 1909 the Pennsylvania commissioner of health was calling on school doctors and nurses to educate children about "the dangers of the common drinking cup and the promiscuous towel and of the lack of personal cleanliness generally." The results, he believed, would be immediate.[81]

This cartoon, complete with social class stereotypes, shows a public faucet with a cup conveniently chained to it. Public drinking water was widely available in this way. As germ theory came to be accepted, opinion leaders were horrified when they realized that "the common drinking cup" could spread germs from one user to another. In an amazingly short time the faucets were converted to the now familiar bubbling variety of drinking fountain, in which one's lips did not touch anything but a fresh stream of water. *Kansas State Board of Health Bulletin*, Feb. 1912, reprinted in *The Kansas Doctor: A Century of Pioneering*, by Thomas Neville Bonner (Lawrence: Univ. of Kansas Press, 1959), 140.

In addition to doctors mired in "tried-and-true" traditional medicine, two important groups in American society resisted or ignored the germ theory of disease. One was the military, where the lessons of the Civil War did not stick and line officers ignored advice from medical officers. In the Spanish-American War of 1898 the lack of sanitation and the amount of illness, particularly typhoid, among the troops created a national scandal. Later official inquiries led to a restructuring of military medicine and to research on yellow fever and other tropical diseases, such as dengue.[82]

The second major source of resistance was based on class and ethnicity. When the actions of elite, often racist health authorities appeared arbitrary and unfeeling to populations operating under various traditional ideas about illness, sometimes the people at the bottom of the social hierarchy resisted actively. During a smallpox epidemic as late as 1894–95 in Milwaukee, thousands of mostly German and Polish immigrants rioted against the health department vaccination and isolation campaigns, which were sometimes carried out in a heavy-handed manner. As one journalist reported, "Mobs of Pomeranian and Polish women armed with baseball bats, potato mashers, clubs, bed slats, salt and pepper, and butcher knives lay in wait all day for . . . the Isolation Hospital van."[83]

Because the germ theory drew on common ideas, it was often difficult to see where older concepts of dirt and infection left off and new ones began. Americans were already concerned about sewage and sewer gas. The so-called triumph of the white china toilet was in part the result of increased concern brought on by the germ theory. The plumbing manufacturers joined the makers of cleansers and disinfectants in employing noisy and somewhat educational advertising and claims about the dangers of germs. In general, members of the public learned to fear dust (which might contain spores as well as active germs) and watch out for other carriers of microbial infection: air, water, feces, and "flies, fingers, and food." Over the years, what began as private and voluntary actions became public and imposed, whether filtering water, substituting sewers for privies, disinfecting sickrooms, or establishing building codes. This striking social change, with women now as major actors, showed the impact of scientific information that became widely available at the end of the nineteenth century.[84]

Reconceptualizing Medicine and Prevention of Illness

Eventually Americans began to make significant contributions to ideas about microbial infections. In 1896 Theobald Smith, extending the new idea that mosquitoes carry malaria, showed that there was a tick vector (carrier) for Texas cattle fever. Shortly afterward, as a result of investigations initiated during the Spanish-American War, the U.S. Army physician Walter Reed led a team that definitively confirmed that mosquitoes were also responsible for transmitting yellow fever. The public health and social effects of this information, which led to prevention of the disease, were substantial. Members of the team, who provided conclusive evidence but turned into martyrs when they suffered or died from the disease, became public heroes. In follow-up studies, members of the team showed that yellow fever was caused by an agent that (unlike most pathogenic bacteria) could pass through a filter; that is, it was a filterable "virus." This was the first human disease to be traced to a "virus" (on the model of the tobacco mosaic virus, demonstrated in Europe in the 1890s), and slowly virology developed within biomedicine.[85] Meanwhile, public health workers had gained substantial control over a major, dangerous disease.

Another area of science, evolutionary biology, provided an additional model involving germ theory. Educated Americans were acutely aware of Darwin's ideas as soon as he published them beginning in 1859: humans, as part of nature, were not

The antituberculosis crusade continued to reverberate after World War I, when chewing tobacco was still common. Health authorities in Philadelphia, drawing on germ theory, placed this striking sign on streetcars to remind citizens to cultivate sanitary habits. "Presenting Truths to the People," *American Journal of Public Health*, 9 (1919), 207.

only animals but participated in struggles for survival in which only the fit survived. In 1883 Henry Gradle, a Chicago physiologist, repeated the obvious: "Diseases are to be considered as *a struggle between the organism and the parasites invading it.*"[86] The idea of "fitness" to resist germs thus flourished in the biological view of the human body. Indeed, physical fitness became a goal for many people, and gyms sprang up across the country where people could seek "physical fitness" if they had enough money. Even at home some Americans used machines and nostrums to try to improve their fitness.[87] But the idea that some humans were more fit than others complemented a European finding that hereditary mechanisms often determined a person's health—in part a continuation of the idea of a human's unique "constitution" but now in a more specific form. For some Americans, ill health as well as moral failings represented evolutionary failure, hereditary lines of unfitness certain to die out.

Sometimes undesirable heredity—like susceptibility to developing tuberculosis or nervous diseases or just having an extra digit on the hand or foot—ran obviously in large Victorian families. And sometimes members of dominant U.S. populations perceived other national or "racial" ethnic groups, especially those conspicuous in the "new immigration" at the turn of the century, to be carrying biological inferiority as well as disease. At the same time, business leaders used these people from other countries as cheap labor to fuel industrialization, and the immigrants endured not only deadly working conditions but often crowded, miserable living conditions and poverty that in itself probably caused ill health. All in all, immigrants, who were typically grouped and labeled in mean ways, came to constitute a "social problem" for dominant groups.[88]

Life insurance figures showed that one particular population group, African Americans, typically perished early and at a very high rate. This made it possible for at least some racist scholars to predict that African Americans would die out naturally in the cold Northern Hemisphere and the "Negro problem" would fade away with them. When, early in the new century, medical and health measures suggested that African Americans would be present permanently in the United States, not only did questions of social accommodation arise but African Americans' health care became a pressing concern for their fellow citizens.

Despite advances in the laboratory and the physicians' appropriation of the mantle of science, nonsurgical medical interventions based on new ideas of disease came very slowly. Knowing the germ that caused a disease did not immediately bring a cure for the disease. Thus public health took on a new momentum. When one cannot cure, one turns to prevention. To accompany the new ideas of disease, a regular group of experts, many in the American Public Health Association, tried to shape the national public health agenda. Gradually the plumbers and engineers dropped out, and a new kind of professional, often a public-spirited physician, staffed public health agencies and led public health efforts against specific diseases.[89]

A New Model of a Physician's Work

At the same time, the laboratory and the clinic were producing a new kind of doctor. Although physicians still could not cure, what they could do in a spectacular new way was to define and diagnose diseases, even hereditary diseases. The best sign was a massive textbook of more than a thousand pages, *The Principles and Practice of Medicine,* first published in 1892, by William Osler, a professor of medicine at Johns Hopkins. In successive editions, Osler's text dominated the field for more than a generation and taught innumerable medical students and practitioners a medicine based on physiology, bacteriology, pathology, and long clinical experience. To an astonishing extent, the book standardized modern medicine in the United States.[90]

What was most striking about the text was that it contained an enormous amount of wisdom about how to diagnose various maladies, specific diseases. By contrast, medical treatments accounted for a very small portion of the book. Osler would discuss the definition, etiology, morbid anatomy, symptoms, diagnosis, and prognosis of a diseases, ending with only a tiny paragraph titled "Treatment." Sometimes even that heading would be omitted. Osler held that pneumonia was a "self-limited disease," the language used by Bartlett a half century earlier. And he noted laconically that "many specifics have been vaunted in scarlet fever, but they are all useless."[91]

This dilemma of turn-of-the-century physicians, who could offer impressive surgery, diagnosis, epidemiology, and prevention, but not medical therapy, was evident as well in another famous teaching innovation in the United States. Richard Clarke Cabot, of Harvard Medical School, began to present to students, and then publish, a series of "case conferences." He was adapting a teaching procedure used in the local law school, the case method, to medical teaching, in which the facts of the case were presented and diagnoses offered. At the end, the pathologist reported definitively on what the autopsy showed to be the known cause of death— mirroring the French school of the early nineteenth century in attempting to correlate symptoms with localized pathology. Cabot's case conference model was widely copied in medical schools and clinical settings across the United States. It represented a twentieth-century shift from a chronological, often subjective history of the patient to a consideration of the presenting clinical signs assembled as a problem.[92]

One of Cabot's differential diagnosis cases from 1906–11 illustrates how far medical examination had advanced:

> A boy, 14 years old, of gouty family history, complains for a year of frontal headache, not very severe but persistent and wearing. Appetite excellent, but digestion not as good as it has been. Has grown suddenly very irritable, having been previously sweet-tempered. He has lost flesh during the year and seems listless and weak. Sleeps well. Bowels somewhat costive. Getting pale. Heart, lungs, and abdomen negative. Knee-jerks not easily obtained, but gait shows only weakness.

Urine normal color, acid, 1028, no albumen. Sediment negative. Temperature 98 [degrees], pulse 96. No oedema [swelling]. Blood negative.

1. What possible causes for the change in disposition? Masturbation [considered a distinct disease at the time], psychosis of puberty, brain tumor, diabetes.

2. Causes of frontal headache commonest at fourteen? Eye-strain, adenoids, frontal sinus disease, malaria, pubescence.

3. Significance of pallor both in general and in this case? Pallor may mean anaemia. . . . No diagnosis of anaemia is justified until the physician has seen the color of a drop of blood on filter paper (Talquist scale) or on a handkerchief. In this case no anaemia was present.

4. Diagnosis? Prognosis? Treatment?

Diagnosis: The careful student of this case will notice first of all the loss of flesh despite good appetite. This, with persistent headache and diminished knee-jerks, are the obvious physical signs. Eye-strain, adenoids, malaria, masturbation were excluded by examination and watching. Further questions revealed the fact that micturition was frequent and copious. This, with the loss of flesh despite good appetite, suggested diabetes, and the urine was found on examination to contain sugar (a point omitted in the examination of the attending physician . . .). As the glycosuria proved persistent, the diagnosis of diabetes mellitus was made. . . .

Prognosis: In a thin boy of 14, a few years of life is all that can be expected.

For treatment Cabot recommended, perhaps futilely, watching the diet. And the recommended treatment did not modify his gloomy prognosis. The account we have of the case shows extensive use of laboratory tests as well as physical examination. Not only was the sugar count of the urine (5%) noted in the discussion, but the quantity of urine was measured (3 quarts in 24 hours). The blood color was also put into a scale. In other cases Cabot mentioned leukocyte counts and other laboratory tests. But in most of the cases treatment was superficial at best, and prognoses were often unhappy. This constituted medical care as good as any at that time.[93]

In 1911, Cabot published a list of the medications actually used at Massachusetts General Hospital, which he found similar to lists of leading clinicians elsewhere. He found only a handful of specifics: quinine, mercury and Salvarsan (to be taken up below) for syphilis, iron in chlorosis (later believed to be female adolescent anemia); diphtheria antitoxin; tetanus antitoxin; thyroid and pancreatic extracts in hormonal diseases; and preventive vaccinations for smallpox and typhoid, as well as possibly two special infections that did not work out. The list shows that by 1910 physicians could directly address tetanus and typhoid, as well as the syphilis and diphtheria from an earlier time. Otherwise the list is remarkably like that available a century before, including digitalis, which Cabot put in another category, "drugs used to improve circulation and remove oedema." This second category of Cabot's included a wide variety of substances, from magnesium sulfate, strychnine, and calomel to adrenalin, caffeine, nitrites, and atropine. There were also drugs used for pain, for sleep and sedation, and for "supposed" action on the gastrointestinal system.[94]

Altogether, Cabot's list was extremely short compared with lists of drugs available half a century later, which included antibiotics, hormones, vitamins, antihistamines, steroids, and tranquilizers. In the chapters that follow, I shall show some of the ways in which physicians added new drugs and identified diseases ever more precisely. Yet it is possible to understand how the eminent practitioner Woods Hutchinson, writing in 1911 for a popular audience, could trumpet "the conquest of the great diseases." What is striking about Hutchinson's description is that almost all of the recent conquests had depended upon prevention and public health, such as a skin test for early detection of tuberculosis or the discovery of the mosquito vector in yellow fever. But on the basis of such scientific work, Hutchinson could envisage far more victories for scientific medicine in the coming years.[95] Many other writers carried this same message to both physicians and "the public" in general. From either viewpoint, the "progress" of medicine was remarkable and promised much more for the future. Biomedical scientists were confronting mysterious diseases and demystifying them despite not curing them.

Other Technologies Affecting Medical Practice

The routines of ordinary physicians across the country changed between the 1880s and the second decade of the twentieth century not only because of the science and technology of medicine but also because of other technologies. Increasingly, the horse-and-buggy doctor let the patient come to his or her office. That, after all, was where the medical equipment, including anesthesia and antiseptic devices, was located. But now so were the laboratory equipment, blood pressure devices, and even an x-ray machine. Medical instruments of all kinds came along quickly. An American had, for example, introduced the lighted cytoscope, for urethral examinations, in 1898. Osler in 1907 believed that the best tool for a doctor was "the little laboratory room attached to the office of the general practitioner."[96] Regular physicians were gradually making their scientific image materially evident.

The technological transformation went further, however. Two devices especially, the telephone and the automobile, changed medical practice. The telephone permitted the physician to avoid some travel, dispensing advice at long distance. And the automobile meant that a physician making rounds to patients could travel much

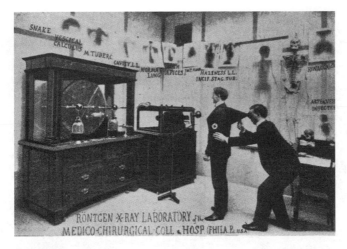

For a number of years after the x-ray machine was put into use, physicians and scientists who worked with the new technology did not know about the danger of exposure to the rays. This photo shows Mihran Krikon Kassabian working in Philadelphia c. 1900. He suffered from the exposure he received while pioneering in medical x-ray imaging. Percy Brown, *American Martyrs to Science through the Roentgen Rays* (Springfield, IL: Charles C Thomas, 1936), 90. Courtesy of Charles C Thomas Publisher, Ltd., Springfield, IL.

more rapidly and see many more patients. The ordinary doctor's routines and life rhythms were not the same in 1910 as in 1880.[97]

Alongside the medical practitioners came a number of pharmaceutical firms that furnished technologies. Some of these firms were dealers in patent medicines. Others were retailers and wholesalers supplying the market for drugs. Many small firms also did manufacturing. Mostly they produced pills, lozenges, powders, syrups, and tinctures. Beginning in the 1890s, tablet-making machines, in particular, brought mass production to the industry, and some large firms began to emerge. Gradually a number of pharmaceutical firms took on the identity of "ethical" drug houses attempting to meet the standards of the privately published *United States Pharmacopoeia*.[98]

In practice, purveyors did much to educate physicians and the public alike about any new or useful drug that came along, but if the "manufacturing chemists" employed scientists, it was almost always for the purpose of testing the purity and consistency of products. Only after the sensational introduction of diphtheria antitoxin in the 1890s did a few firms see how profitable innovation in pharmaceuticals might be. In 1902, even before the famous food- and drug-labeling acts of 1906, the federal government began inspecting and licensing the production of serums, toxins,

Dr. Scott, of Del Rio, Texas, posed in his new Maxwell. Physicians were often among the first to use such commercial technology as telephones and automobiles, and as roads improved, automobiles became more necessary for independent practitioners. The words on the door in the background read "DOCTOR SCOTT." Rose Papers, Western History Collections, University of Oklahoma, Norman, by permission.

and antitoxins (e.g., for diphtheria) because these substances were a new kind of medication, introduced at full strength into the body through hypodermic injection and not intermediated by the digestive system. The 1906 labeling requirements used for standards the publication already used by ethical manufacturers, the *United States Pharmacopoeia*, making a private formulation official. As a New York pharmacist noted at the time, "Standardization has become an absolute necessity, whether the average apothecary is competent or willing to assay all his drugs and extracts or not."[99]

Tough Times for Doctors

For the practicing physicians who were dispensing medical progress, the decades around the turn of the century were often extremely difficult. Even those with the best educations found the field overcrowded as proprietary schools churned out ever more graduates. Practice was hard to come by not only in the cities but on the frontier. In 1894 one witness reported:

As a result of our rapid railroad building, new villages are being located every few days, and it is not the rarest of sights to see one or two physicians hovering like shadows around a town which for the time being exists only on paper. These "too previous" aspirants frequently board at country houses, daily visiting the patch of waving prairie grass, the site of the prospective metropolis, and waiting sometimes for weeks before buildings are erected in which offices can be had.[100]

A medical reformer in 1889 reported that he had followed the careers of several hundred physicians who had taken not just the usual two years of lectures but four years of training, often with clinical and hospital experience. "These are, with few exceptions, the successful and prominent members of the profession in the different communities in which they reside. . . . They are successful, as a rule, because they have fitted themselves to command success."[101]

Degreed physicians had to compete economically not only with one another but also with quacks and the ever-expanding commerce in patent medicines. In 1897 the catalog of the mail-order pioneer, Sears, Roebuck, contained a twenty-page section listing more than four hundred proprietary nostrums. The last decades of the nineteenth century also gave rise to a remarkable increase in spiritualism and faith healing—particularly Christian Science and other groups that attracted a middle-class following, thus directly cutting into the most profitable segment of patients.[102]

Physicians' Fees and the Pressure of Contract Medicine

In that same turn-of-the-century period, well-established physicians introduced professional standards regarding fees and, here and there, business methods of collection through agencies run by medical organizations. They could justify collections because of the *sliding scale of fees* that had been and was still the standard in country practice. That is, a professional provided service to everyone and charged patients based on their ability to pay. Well into the twentieth century, physician groups claimed that medical services were available to everyone because ethical physicians did not charge the poor but made up for it from fees charged the more affluent.

In 1892 F. E. Waxham billed a wealthy Chicago man the then enormous sum of two thousand dollars for successfully intubating the man's son, who had diphtheria; that is, Waxham inserted a tube to break the membrane that was choking the child. Waxham explained, "I have never yet refused a case, however poor or destitute. You have had the benefit of this experience which has cost me so

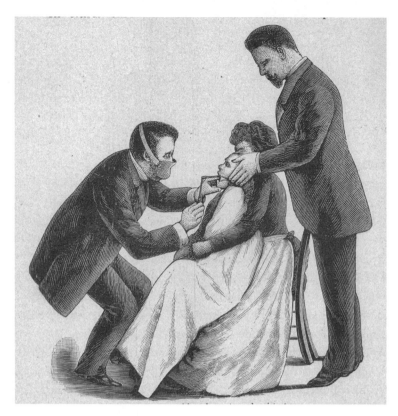

A physician demonstrates how to insert a tube into the windpipe of a child who has diphtheria to break through the membrane preventing the child from breathing. The physician shown is F. E. Waxham, of Chicago, whose bill for such a procedure is discussed in the text. F. E. Waxham, *Intubation of the Larynx* (Chicago: Charles Truax, 1888), 40. James M. Edmonson identifies Waxham in his introduction to *Mechanics of Surgery*, by Charles Truax (1899; reprint, San Francisco: Norman, 1988), xxxii–xxxvii.

dearly. . . . The skill which saved your boy's life is not a common commodity." The wealthy man objected to the large sum at first, but after a panel of physicians declared the charge just and reasonable given Waxham's expertise, he paid the bill.[103]

This was the sliding scale operating as it should, a progressive income tax enforced by individual physicians. But in a rapidly industrializing and unjust, segregated society, coverage was extremely incomplete. Workers and others with limited incomes therefore devised another way to protect themselves. They formed groups, and the groups contracted with physicians for medical care for all the families in the group, usually at a certain price, typically twenty-five to seventy-five cents or more per person covered for a year. The contracting groups were most often fraternal group lodges, like the Eagles or the Odd Fellows, or occasionally labor unions.

Physicians hated and feared the contract system because it oppressed them economically and professionally. How many patients could they see how fast and how cheaply? In the overcrowded medical profession, one could almost always be underbid by a desperate colleague. Physician organizations therefore opposed the contract system. In the 1890s, for example, the cigar makers of Tampa, Florida, formed a number of Latin health societies. They hired a prominent local physician, L. S. Oppenheimer, as their surgical director, with a generous contract. Oppenheimer had to resign from his county medical society in order to take the position, but his colleagues still put so much pressure on him that after two years he withdrew from his contract work.[104]

Physicians, insofar as they were organized, continued to work to exclude competitors, especially the quacks and remaining sectarians. The homeopaths and sometimes the eclectics were in many areas politically powerful enough to hold off the regulars temporarily, and the homeopaths and regulars would sometimes combine to try to exclude less well trained practitioners and frauds. Sometimes idealism triumphed, and sometimes MDs' economic motives showed through.

The Struggle to License Physicians

After the *Dent* decision in 1889, regular physicians intensified their campaign to license physicians and extricate medical care from the free-enterprise marketplace, which was causing so much inferior or even disastrous medical practice. As one physician commented in 1898, there were still in the profession "a vast number of incompetents, large numbers of moral degenerates, crowds of pure tradesmen." The political forces opposing licensure were powerful, however, including opinion leaders who distrusted physicians.[105] Eventually, however, another force intruded: members of the public recognized that regular physicians finally could show surprising effectiveness in, for example, surgery, vaccination, and public health. Therefore consumer movements such as that operating in the American Social Science Association appeared. They represented people who perceived the growing gap between well-trained MDs and ignorant and incompetent practitioners, and at a time when occupational licensing was finally becoming more commonplace, they provided an additional impetus to a movement to license physicians.[106]

State after state set up an examining board to license physicians, sometimes along with separate boards for homeopaths or, later, osteopaths.[107] Many initial efforts were weak. In Indiana, one county clerk issued a license to a man who presented a napkin from a Chinese restaurant and declared that it was a diploma from a Chinese medical school. States' circumstances varied, but one after another they began to require graduation from an "approved" medical school. The Illinois Board of Health, which began judging medical schools as early as 1883, soon set an implicit national standard. State boards of health also began to institute examinations for medical professionals, and in 1891 a number of them joined in founding the National Confederation of State Medical Examining and Licensing Boards. In a remarkably short period the confederation specified various realistic, national standards for states to adopt.[108]

Gradually the noose tightened around diploma mills and for-profit medical schools. But reform of medical education had a separate history that drew also on other streams in America's rapidly changing culture. The rise of hospitals made hospital experience more commonly a part of training. A survey in 1904 found that already half of the medical school graduates were taking hospital training, and many more were trying to do so.[109] Teachers of medicine in the late nineteenth century had found that didactic teaching of medical knowledge was no longer fully possible. There was simply too much to learn. Therefore, teachers began to teach students how to solve problems. The new laboratory sciences that were coming into medicine provided a model: the experiment. And so medical education increasingly became a hands-on series of trials and puzzles, just as in Cabot's case conferences. Students could use new instruments, especially the microscope, and they could examine actual patients and discuss what they were seeing. All of this required facilities and technologies. To a surprising extent, it was simply dedicated teaching that drove innumerable, aggregating small changes in classes and curriculum, including an expanded number of courses and faculty. These changes culminated in the revolution in medical education that began to become national in the 1890s and dominant in the early twentieth century (a story continued in the next chapter).[110]

Specialization as a Major Force

With the growth of cities and the development of medical science, specialism in medicine became much more common. Sometimes a journal, and later an organization, would signal the existence of a critical mass of specialists. The specialists often wanted to communicate with one another in order to keep abreast of the latest developments in their fields or to try out new ideas on the best-informed practitioners. Young Howard Kelly wrote from Germany in 1887, with evident surprise and admiration, "A German specialist is a true specialist, he treats no one whatever not in his own peculiar branch. . . . The German specialty is a brotherhood in which the individual members are all earnestly working for the good of the body at large."[111]

Local and then national specialty organizations were evidence of the trend in specialization. Ophthalmologists organized nationally as early as 1864, and otologists in 1868. In the last two decades of the nineteenth century, neurologists, dermatologists, gynecologists, surgeons, pediatricians, gastroenterologists, and proctologists, among others, organized formally—a movement that threatened the general practitioners, who were already dealing with overcrowding of the profession.

With the exception of pediatricians, the earliest specialist groups were organized around particular bodily organs. Hence it was easy for general practitioners to criticize them, especially if one considered the traditional role of physician as friend and counselor, as did an 1897 practitioner: "The specialist has to do more frequently with local diseases, which, important though they may be to the individual at the time, still as a rule are mere incidents in his life. In conditions which permanently affect health and happiness, which lengthen or shorten life, and which threaten death, it is the general practitioner whose advice, counsel, and support are sought."[112]

Yet one reason why specialization flourished was that Americans in the increasingly organized society operated more and more on the principle of finding experts for special functions. Specialization not only was essential to bureaucracy but also increased efficiency, a goal in industrialization.[113] As hospitals became more bureaucratic, they organized according to special services—in the United States now just as in Germany. At Massachusetts General Hospital the process was well under way by the 1870s. Even the American Medical Association acceded and began setting up specialty sections. In 1896 an editorial writer concluded that "the American Medical Association has become . . . a veritable confederation of medical bodies devoted to independent lines of thought and practice." A second reason why specialization flourished was the increasing respect for science. American middle-class people wanted disinterested, reliable information and hence experts. They supported civil service reform with the idea of impersonal, tested merit. Educated people resonated to the idea of an experimental, as opposed to a traditional, basis for medicine. And the most scientific physicians were those specialists who were closest to the labs. The clearest example was the pathologists, whose specialized work by the 1880s was making them leaders in introducing germ theory into medicine in general.[114]

For physicians, even before specialist residency became the major means of training, there had appeared an institution that could provide some specialist training, the polyclinic. The polyclinic provided a partial alternative to study in Germany. Beginning in the 1880s, in some large cities, specialists set up six-week advanced courses for physicians. The instruction, by leading specialists, was based in very practical experience in special dispensaries, the free or low-cost clinic model noted above. The polyclinic was designed to upgrade general practitioners with specialist instruction, for example, in the new technique of intubation for croup and diphtheria. The Mayo brothers started with polyclinic training as they launched careers as specialists. The polyclinics were so successful that some added hospitals or transformed into medical schools. After the turn of the century, however, improved medical teaching made the polyclinics unnecessary.[115]

Medical Research and the Growing Faith in Science

American physicians and scientists who returned from Europe with strong ideas about research institutions found virtually no support for research outside universities. Medical schools continued to focus on practice. Even hospitals, where clinical research could arise spontaneously out of the cases present, were not necessarily friendly to investigation. As a Cleveland physician declared in 1898, "No man should be a member of a hospital staff whose zeal for the advancement of science or the promotion of his own ambition will sacrifice the interests of his poorest patient."[116]

Around the turn of the century, some medical research funding began to appear. Agriculture schools, with federal money, often supported related research, such as that on animal diseases, nutrition, and physiology. But most American medical research grew at first out of university support for biology, a new, all-encompassing field that included botany and zoology. At the end of the nineteenth century, biology was moving out of the field and into the laboratory. Especially as a physiological viewpoint expanded, and it became clear that all life, human and otherwise, was based on cells and protoplasm, the subjects of study became universal life processes, such as respiration.[117]

Those who started out in medical chemistry often found themselves teaching physiological chemistry. And with teaching and even some research opportunities in service positions, both chemists and physiologists tended to focus on medicine, with the result, as Charles O. Whitman put it in 1896, that physiology in America "has limited its field to man and higher vertebrates"—a dilemma that tended to afflict all experimental biology, even as science provided a new central ideal for advocates of medicine.[118]

The issue was experiment, and in 1896 William H. Welch, at the new Johns Hopkins University and Medical School, founded the *Journal of Experimental Medicine*. This was almost exactly a century after the first medical journal in the United States, the *Medical Repository*, of 1797, and it is perhaps symbolic of the flood of changes that were coming in American medicine as the German-trained elite helped lead profound institutional and intellectual transformations.

From one point of view, the most spectacular outcome of the germ theory of disease was not therapy but a growing faith in scientific research, or more precisely, scientific technique.[119] As universities introduced graduate-level education, research in physiology and biology followed naturally. More and more frequently, younger physicians and scientists hoped to emulate the German research institutes and the endowed Pasteur Institute in France. But for medicine, progress was slow until 1897, when John D. Rockefeller's adviser on philanthropy, Frederick T. Gates, happened to read Osler's textbook of medicine. "I found . . . that a large number of the most common diseases . . . were simply infectious or contagious, were caused by infinitesimal germs . . . only a very few had been identified and isolated." With quick intelligence, Gates understood that despite great increases in knowledge, medicine had not yet found ways to counter most infectious diseases, each of which had a microbial cause. He began to work on Rockefeller to find a way to support research on these diseases.[120]

In 1901 Rockefeller's three-year-old grandson died of scarlet fever, illustrating that nothing was known about the causes of that disease or many other diseases. The Rockefeller Institute, devoted largely to bacteriology at first, was founded and began supporting research in 1903, with special independent laboratories in place by 1906. Advised by leading American medical scientists, the founders knew exactly what they were doing. Even a writer in the *Evening Post* understood the formal announcement: "The American Medical profession have been criticized for lack of original work. The new institute will provide for the release from cares . . . men of trained scientific intelligence, who will be enabled to devote themselves to the solution of definite problems."[121] A similar institute was founded in 1902 by the little boy's father, a member of the McCormick family, in Chicago, concentrating at first on scarlet fever and other "common, acute, infectious diseases."[122]

Change in American medicine therefore proceeded rapidly on different levels. Medicine in general was transformed significantly by antisepsis and surgery as well as by germ theory. At the same time, fundamental institutional changes were occurring in the United States. By 1914 most of the great figures in German medicine had visited the United States. In the 1880s such visitors had found backward conditions. By 1898 they were finding a better-developed profession and, here and there, instructive developments in medicine to take back to Europe.[123] The depth and speed of change in both science and organization was breathtaking, now on both sides of the Atlantic.

CHAPTER 6

Physiological Medicine, 1910s to 1930s

Even as antiseptic/aseptic surgery and germ theory continued to transform medicine and ideas about health in the first years of the twentieth century and later, another approach brought in a new era, the era of physiological medicine. On many levels and in many ways, physiological medicine began to transform health care workers' thinking in the years around 1910 by centering on the ways in which the body functions or can break down. At the same time, medicine and health became caught up and deeply involved in, and affected by, the general Progressive reform movement that, like the era of physiological medicine, continued into the 1930s.

The Rush of Science and Physiology into Medicine and Health

Physiological medicine represented the payoff from laboratory physiology, a field to which American investigators now made major contributions, particularly in the areas of vitamins and hormones. One later commentator tried to suggest the drama of the coming of physiological medicine: "Within a single decade, it was discovered that the course of three of the most complex and lethal of diseases—pellagra, pernicious anemia, and diabetes—could be turned around swiftly and precisely by the restoration of absent biochemical reactants. Thus a new era opened. . . . Human disease was curable by the use of scientific methods."[1]

In this era of physiological medicine the physician's role as the symbol of science was becoming increasingly clear in the United States. The signs were many and various. Almost imperceptibly, but surprisingly rapidly, for example, homeopathy had lost legislative and public support and recognition by 1917.[2] Now the story became not just the contours of ordinary medical practice but new views of the body that contributed to and derived from the interaction of science and medicine. Research not only increased but became organized, as did the hospital, medical education, and licensing. Once the major pillars of scientific medicine were in place, ideas of biological complexity shaped and animated medical thinking and practice. This took place in a setting in which in the language itself the word *scientific* took on an extraordinarily positive value. Even religion in the 1920s won praise from many opinion leaders for being "scientific."

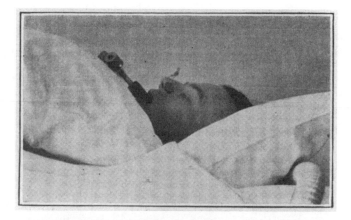

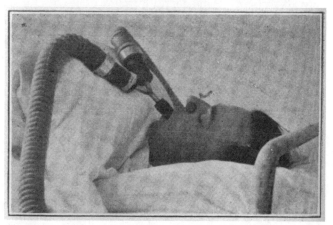

By the 1920s, investigators were deep into physiological medicine and had devised means to measure a variety of metabolic functions in the body. Here the subject has nose and mouth clips in place. The mouth clip attached to hoses carrying the products of respiration to a measuring device. Researchers hoped to identify patterns of physiological functioning in different kinds of patients so as to uncover the causes of a variety of symptoms and syndromes. B. S. Levine, "Technic of Basal Metabolic Rate Determination in Psychoneurotic Patients," *Journal of Laboratory and Clinical Medicine*, 8 (1922), 76.

Contours of Medical Practice

The pace at which medicine mutated was stunning. First of all, practice changed. A small-town Virginia doctor compared his practice in the summer of 1909 with what it was in the summer of 1922. In the summer of 1909 two-thirds of his patients had had malaria, diarrheal diseases, or typhoid. In 1922 he did not have a single patient with any of those diseases.[3] At the same time, practitioners could hardly believe the extent to which the profile of medical knowledge changed before they were even halfway into their careers. The new knowledge came from the lab and was first applied to practice by new kinds of specialists and subspecialists.[4]

Yet one could ask, what difference did personal medical care make to ordinary people? Individual medical practice was of course quite different from transformative public health programs—and even more distant from the factor of living standards as a determinant of health. Indeed, there is a much-repeated dictum that only after 1910–12 did an ordinary doctor, treating an ordinary patient, have more than a fifty-fifty chance of improving the patient's health.[5]

As late as 1923 a writer attacked the medical establishment because the intervention of individual physicians did not really improve one's health: "There are no certain cures and . . . death rates would have declined irrespective of anything doctors and health authorities would have done." In a response to that writer, Ray Lyman Wilbur, of Stanford University, mentioned some life-saving procedures and then went on to point out that the critic "thinks that doctors are concerned only with individual treatment of patients." Physicians, Wilbur argued, contributed

A group of children who as babies had suffered from and been blinded by a terrible eye infection served as reminders of the continuing inability of physicians to counter serious infections of all kinds, leading not only to death but to disability that families and society more broadly had to cope with. Enlightened leaders emphasized prevention and healthful living when there was no cure. "Ophthalmia Neonatorum," *Medical Review of Reviews*, 18 (1912), 370.

fundamentally to the prevention of disease through social measures, in that day generally still called "sanitation."[6] Both were correct: Physicians still could not cure most infectious or degenerative diseases, and yet death rates had fallen and people were living longer and more comfortably. Patients with chronic diseases, notably diabetes, underwent more complicated treatments and often did not die nearly as quickly as before. Indeed, diabetes was the model of an acute disease that new medical measures—insulin in this case—moved into the growing category of chronic diseases by keeping patients alive.[7]

Some basic patterns within health care structures crystalized dramatically between the second decade of the twentieth century and the 1930s. Institutions were created that lasted the rest of the twentieth century, not least large hospitals, a new standard pattern for medical education, and even single-disease pressure groups. For the first time, the bulk of Americans, regardless of critics and naysayers, came to believe that individual medical care was becoming materially effective, particularly through surgery and preventive injections based on germ theory. Typhoid vaccination, for example, which had begun tentatively in Europe in the 1890s, was introduced to the United States in a major way in 1908–11, when the U.S. Army began using typhoid vaccination with spectacular, well-publicized results, in contrast to the disasters of the Spanish-American War.[8]

Both access to care and the promise of biomedical science became major social issues. Infectious and constitutional diseases were still disabling and killing both adults and children, but, as will be described below, a "new public health" movement, based in part on fresh medical thinking, began to have effects as well, despite the many contradictions and distressing failures on both individual and social levels.

New Views of the Body: Dynamic Atomism

Over several decades, educated Americans' mechanistic and materialistic ideas about the human body—ideas largely embraced by biomedical scientists and advanced physicians—yielded to a new way of viewing the body and biological organisms in general: dynamic atomism. The shift was stimulated partly by the coming of quantum theory in physics and genetics in biology, both of which suggested a model of dynamic interactions of autonomous elementary particles. Following this more general model, physiologists now described within the body the interaction of a variety of other independent particles, including not only germs and the leucocytes that attacked the germs but any number of cells, chemicals, and chemical processes, all of which together maintained an internal balance that could interact with a person's environmental inputs.[9] In the 1920s the old, mechanical model of germs that spread disease by just entering the body gave way to a more complex

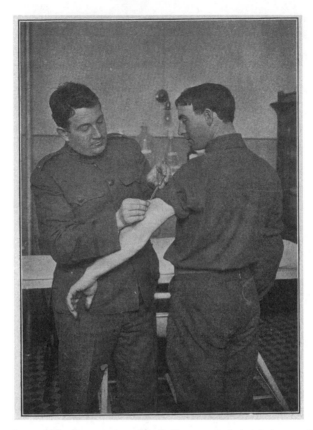

The widespread inoculation against typhoid after c. 1910 helped change disease patterns. In the process, inoculation became familiar to increasing segments of the population, including adults as well as children. Henry Smith Williams, *Miracles of Science* (New York: Harper & Brothers, 1913), 210.

understanding of the interplay and balance between general environmental factors, including microbes, on the one side, and individual immunities, on the other. The baffling flu pandemic of 1918–19 in particular left many thoughtful people, both within medicine and without, skeptical of simplistic models and explanations.[10]

Implicit in the atomism from physics and genetics was the way scientists more and more often described interaction: statistically, or quantitatively, rather than directly mechanically. An analysis of papers published in the *American Journal of Physiology*, for example, showed that in 1908 the authors of 27 percent of the articles were using quantitative or statistical evidence, but that proportion increased to 48 percent in 1918 and 58 percent in 1927. One important paper, for example, showed the correlation between two different types of blood cell counts in normal blood—illustrating incidentally how the statistical concept of correlation was coming into standard medical science.[11] Such a heavy emphasis on not only complex laboratory experiments but also sophisticated, quantitative interpretation signaled a changing configuration in the biomedical science underlying health care.

Dynamic atomism culminated in the late 1920s, when Walter Cannon, a Harvard Medical School physiologist, introduced the idea of *homeostasis*. The human or animal organism, he wrote, "lives in a fluid matrix, which is automatically kept in a constant condition. If changes threaten, indicators at once signal the danger, and corrective agencies promptly prevent the disturbance or restore the normal when it has been disturbed." The agencies in this regulatory system were the chemical and biological processes in a very complex body, acting in what Cannon, speaking for much of his generation, called "the self-regulation of physiological processes."[12]

At the same time, in medicine itself thinking changed. The new approach went beyond statistical interpretation. In 1928 Alfred E. Cohn, of the Rockefeller Institute, attempted to express his sense that there had been a shift in viewpoint: "A disease, instead of being merely a quantitative deviation from health, is a collection of new phenomena, a new complex, and is sufficiently different to be regarded as a qualitative change [in a person's body]."[13] When Richard Cabot classified heart disease by functional cause instead of by structural flaw as had been customary for two hundred years, "the revolution in point of view," the specialist Paul Dudley White commented later, "was amazing."[14] Beyond the ancient model of humors that were out of balance, in the new view physiological processes that proceeded too rapidly or too slowly could also produce illness.

Organizing Knowledge and Research in the Physiological Era

The major obvious sign that a physiological era in medicine was beginning came about 1910, when two new subjects became prominent in American medical journals: vitamins and hormones. The most spectacular therapeutic development of this physiological era was the discovery in the early 1920s of insulin, understood as a hormone, in a form that dramatically extended the lives of many of those previously struck down by diabetes. This era of physiological medicine continued into the 1930s.

Vitamins and hormones were not just fads but demonstrated clearly how thinking had shifted away from a localized anatomical emphasis to a strongly physiological emphasis in medicine. Along with a focus on physiological processes came two additional emphases. One was the idea of well-defined deficiency diseases, comparable to well-defined bacterial diseases. Many illnesses were caused by the *lack* of elements such as vitamins and hormones, in contrast with diseases that came from the *added* poisons of germs. Moreover, physiological illnesses were often chronic illnesses. Indeed, historians have pointed out that around 1920 the incidence of chronic diseases began to exceed that of acute, usually infectious diseases among Americans, although almost no one at that time noticed, except a few leaders who were already alert to intellectual shifts.[15] Cohn, for example, observed in 1924 that "the discovery of bacteria has for two generations been so absorbing as to dwarf the interest medicine has always displayed in conditions associated with derangements of the organs and with the ailments of advancing age."[16]

What was most obvious was a change in medical leadership, particularly in the medical schools. The Association of American Physicians, founded in 1886, continued to bring together physicians, typically younger physicians, who worked to apply science to clinical medicine. They did so primarily by invoking three disciplinary labels that embraced all the new medical understanding: "chemistry," "pathology," and of course "physiology." Moreover, in the opening decades of the twentieth century, medical schools often appointed research scientists with scientific degrees, not the MD, to teach physiology, physiological chemistry, and related subjects. Investigators' growing conclusion that the chemical structure of a substance correlated with the physiological effects of the chemical shifted the whole approach to scientific medicine, and a number of investigators with doctorates in chemistry and other subjects often contributed to clinical studies of patients.[17]

But there was still one more element. As in earlier decades, advocates of advanced scientific medicine continued to be markedly enthusiastic. By 1900 their numbers had increased, so that they needed additional venues for discussion and publication of what they found so exciting. In 1903 the Society for Experimental Biology and Medicine formed. Also in 1903, the Interurban Surgical Club was founded, followed in 1905 by the Interurban Clinical Club, each gathering together a small, elite group from urban centers on the East Coast. In 1907 some of the more impatient physicians formed what became the American Society for Clinical Investigation. The next year, constituent demand moved the American Medical Association to launch the *Archives of Internal Medicine*, explicitly dedicated to bringing scientific research into clinical medicine.[18]

These and other groupings and institutional expressions continued into the 1930s to reinforce the efforts of elite enthusiasts to change medical treatment through a variety of strategies: replacing or augmenting missing factors, as in insulin therapy; blocking normal processes (as happened later in birth control); and finding "magic bullets" that targeted specific disease factors. In the physiological age, biomedical investigators still looked for direct causes of disease, as in the germ era, but now they found themselves dealing with multiple factors operating simultaneously.[19] More significantly, the organizational activities of physicians not only symbolized the movement to upgrade medical knowledge but also showed how medicine took on great momentum just as other reforms were also reaching a peak.

University Ties and Research Support

Research support grew remarkably in these early decades of the twentieth century, especially as medical schools increasingly became attached to universities, where there were biological laboratories. Vitamin studies especially flourished in feder-ally supported agricultural research settings, alongside continuing veterinary med-icine investigations. In addition, the growing support for general scientific research, research that often included funding for what we would now call biomedicine, came from many sources. In 1902 the Carnegie Institution of Washington was endowed with the then enormous sum of $20 million. A pattern of similar foundation sup-port was in place by the 1920s, especially as, in addition to the Rockefeller Insti-tute (noted in chapter 5), other foundations funded with Rockefeller money were operating in medical areas. In science and medicine generally, the number of known research funds doubled between 1920 and 1934. Leaders in these areas understood that research foundations would encourage the spirit of research and stimulate "progress" with "independent" new knowledge, which good physicians, now more than ever agents of science, would eagerly incorporate into practice: "Every advance creates new problems, problems of increasing importance—exhaustion is impos-sible," declared a physiologist addressing a graduating medical school class in 1912.[20]

By 1900 medical schools were usually teaching pathology or pathology and bac-teriology. Pathologists became physiological not only by taking on bacteriology but also by emphasizing pathological (disease-causing) processes rather than just lo-calized evidence of disease. Americans pioneered much blood testing between 1910 and 1920, so that such factors as the acid-base balance had significant meaning by the 1920s. Then pathologists went on to ask what else happened in the body in dis-ease. Beyond a primary lesion, how did the body respond with other changes?[21] Pathologists and physiologists coexisted happily and profitably in the Section on Pathology of the American Medical Association from early in the century to the 1930s.[22] Meanwhile, medical students were learning biochemistry, "the chemical dynamics of living systems," consistent with the new dynamic atomism.[23]

The outcome of so many factors was an impressive network of laboratories and research facilities that eventually brought American physiology into a command-ing position in the world. As early as 1907, for example, Ross Harrison, at Johns Hopkins, introduced a practical technique for tissue culture, growing lines of in-dependent, living cells in an artificial medium in the laboratory, a technique that soon became fundamental to medical research. People were amazed and inspired

TABLE 4
National origins of significant contributions to the field
of physiology in representative years, 1880–1924

Years	Germany	France	England	USA
1880–1884	49	5	10	1
1900–1904	78	2	14	11
1920–1924	47	2	13	24

Source: Extracted from Gerald L. Geison, "International Relations and Domestic Elites in American Physiology, 1900–1940," in *Physiology in the American Context, 1850–1940*, ed. Geison (Bethesda, MD: American Physiological Society, 1987), 117.

by this demonstration that cells could function without the body. Alexis Carrel, of the new Rockefeller Institute, immediately took up this work and gained fame by keeping chicken cells alive in the lab for very long periods of time. His winning the Nobel Prize in 1912 was a landmark in the rise of American physiology to global prominence.[24] The leaders in physiology constituted an elite of only about sixty scientists, concentrated at first on the East Coast but by the 1920s spread throughout the country and especially in the Midwest.[25] Sometimes enterprises labeled "experimental medicine" and "experimental biology" dealt with physiological content.[26] Regardless, it was that kind of all-encompassing physiology that infiltrated and shaped changes in medical knowledge and practice, giving rise in the 1920s, for example, to such specialties as hematology. In the first three decades of the twentieth century, in standard medical indexes the topic "metabolism" expanded by a factor of five—and each decade the proportion of American authors increased even faster.

A European scholar at one point charted the number of major "discoveries" in physiology by country. Even with a bias that exaggerates German contributions, samples from three time periods suggest the direction in which research activity was moving, that is, toward the dominance of the United States (table 4).

Vitamins

The vitamins and hormones symbolic of the physiological era were initially discovered elsewhere but developed very importantly in North America. Both involved physiology in essential ways. Not least, as a 1922 popularizer wrote, "To what extent are we justified in comparing vitamines [*sic*] with hormones? . . . both are present in minute quantity, and a small quantity seems to go a long way."[27]

Because germ theory was so exciting, it took some time for physicians to accept that dietary patterns might cause common diseases in a way different from that posited by moralistic dietary reformers like Sylvester Graham. Proteins, carbohydrates, and fats seemed to constitute the elements of an animal diet. Only slowly, starting with beriberi, did some unknown trace element come to appear necessary to prevent one disease or another—although the background for the idea had been around for many years in the case of scurvy.

Before World War I, scientists and physicians began to pay close attention to the new literature that identified diet-related diseases.[28] In the first English-language book on beriberi, the author, an American doing pioneer experiments in the newly acquired U.S. territory of the Philippines, wrote,

> We are now in a position to prevent the disease in any community that can and will follow our advice just as surely as we can prevent smallpox and yellow fever. . . . Moreover, the facts that have been discovered with regard to beriberi have thrown a great light upon other diseases, such as scurvy, and have revolutionized our ideas with regard to the metabolism of the body, although the textbooks on physiology have not yet appeared to notice this.[29]

At the time, it was easy to notice that the terms *deficiency diseases* and *vitamines* were coming into use, even though study of the chemistry involved in vitamins was just beginning. Already, too, it was obvious to American physiologists that it would now be possible to pinpoint what was missing from the restricted diets of poor people.[30]

After American investigators, led by Elmer V. McCollum at the University of Wisconsin in 1913 showed that some fats turned out to lack an element essential to humans' diet, they were able to identify a fat-soluble vitamin (vitamin A). Investigators, often in the United States, soon showed the existence of other vitamins, three—A, B, and C—by 1921, with more to come.[31] Then chemists turned to purifying (1920s) and subsequently synthesizing (1930s) various vitamins and other trace elements needed in human nutrition. The process of figuring out how these chemicals contributed to health persisted and was still flourishing well after World War II. For pernicious anemia, in the mid-1920s a Harvard investigator identified a dietary cure in the form of heroic consumption of beef liver; this cure was fully explained only in 1948, when scientists in the United States and Britain, working with physiologists on blood chemistry, identified vitamin B-12.[32]

Pellagra

Another spectacular physiological contribution was solving the mystery of pellagra. This was primarily the work of a U.S. Public Health Service physician, Joseph Goldberger. Pellagra was a terrible disease, marked by serious dermatitis, severe mental symptoms, debilitating diarrhea, and other problems. The case fatality rate reported at one point in South Carolina was 40 percent, and between 1900 and 1940 at least 100,000 people in the South died of the disease, half of the victims African Americans and more than two-thirds women.[33]

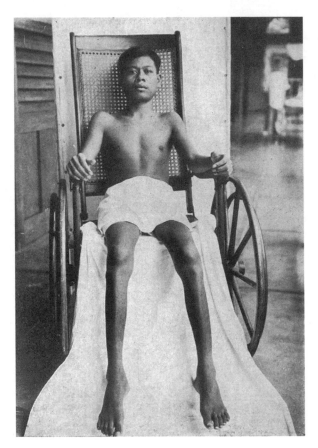

As the United States expanded to overseas colonies, American physician investigators followed the flag. In the Philippines, one American, Edward Vedder, who kept up with the most advanced research reports diagnosed this emaciated patient, who suffered degeneration of nerves and muscles, as a case of very serious beriberi. By feeding him a derivative of rice polishings, the physician was able to correct the symptoms. He reported the case in his pioneering contribution to vitamin studies, the first book published in English on beriberi. Edward B. Vedder, *Beriberi* (New York: William Wood, 1913), 304.

In 1908, journalists began to identify a new "scourge of the South" as reports came in that the disease was being diagnosed in large numbers, particularly in prisons, orphanages, and other institutions but also among poverty-stricken farmers of the rural South or among the millworkers there, who consumed mostly cornmeal, fatback, coffee, and molasses. Eventually it appeared that the disease had been common previously but had been diagnosed under a number of different headings.[34]

Goldberger ultimately concluded that a deficient diet was the key to this awful disease, contrary to the general opinion that it was an infectious illness or perhaps just a product of very bad living conditions, particularly under the sharecropping system, which, incidentally, discouraged the production of vegetables for home consumption so that the land could be used for cotton or tobacco. One could see that diet reflected poverty. To prove that pellagra was not a transmissible infectious disease on the germ model, Goldberger and his wife and volunteers in four locations in 1916 ingested or injected into themselves "matter" from "pellagrins" in various stages of the disease: blood, mucus, urine, feces, and scrapings from rashes. No subject in these "filth parties" became ill with pellagra. It was a dramatic answer to critics, who denied that the traditional southern regional foods could be implicated in a disease. In fact, the politics and criticism brought an unusually direct social intervention in a debate over the cause and cure of a disease.[35]

The pellagra story ended with an effective, targeted therapy using niacin (nicotinic acid, one of the B vitamins). Maxwell Wintrobe, the hematologist, recalled all his life seeing a patient in Baltimore as late as 1938:

> classic pellagra: brilliant red tongue; ugly lesions about her mouth, the backs of her hands, and over the whole perineum; suffering from diarrhea and severely demented. I had seen many such patients . . . , and I remembered too well what their fate had been. However, this time she was given nicotinic acid. The next morning she was oriented, the diarrhea had practically stopped, and she was comfortable, with the skin lesions beginning to heal. Her tongue was not at all sore, and the brilliant flow of the day before was gone.[36]

Obviously, vitamins were helping to prepare for the era of miracle drugs.

After World War I, Americans participated in expanding scientific understanding of deficiency diseases and also of the chemistry and physiology of trace elements, particularly amino acids—a whole area of physiological chemistry cultivated in U.S. labs.[37] Meanwhile, as the 1920s got under way, investigators' excitement spread to popularizers and commercializers. The American public in general learned about vitamins quite suddenly and perceived them not only as foods but as medicines. Advertisers and pharmaceutical manufacturers exploited that knowledge and helped spread it, often inaccurately. Even ethical firms took in huge amounts of money by marketing vitamins; in the 1930s accounting for 20–25 percent of Abbott Laboratories sales, for example.[38]

Hormones

Vitamins, as agents in physiological functioning, furnished a model for deficiency diseases. Hormones went further. Hormone malfunction often produced deficiency diseases, but hormones also had a coordinating and integrating effect that undergirded even more strongly the model of dynamic atomism. Moreover, beginning before World War I, Americans contributed conspicuously to the science of endocrinology and to attempts to use endocrine substances therapeutically.[39]

In 1889 Brown-Séquard, by then permanently resident in France, had reported administering to himself a testicular extract, presumably the product of a ductless gland, that dramatically countered the effects of old age. His work on glands had begun when he was in the United States, and his later report generated a lively interest among his American colleagues. At that point, in the 1890s, the slowly developing European science around various ductless glands and their secretions took on a serious clinical aspect. Glandular treatment for aging men generated failure, intense controversy, and sometimes, of course, profit. Beyond the possibility of rejuvenating patients, however, there was an immediate, dramatic payoff. Patients who suffered from thyroid failure shriveled, aged, and died very prematurely—a tragic, hopeless condition called myxedema. In the 1890s, however, for the first time these patients could often be saved by administering to them an extract made from animal thyroid glands obtained from local slaughterhouses. Even without standardized strengths or doses or any detailed knowledge, physicians had a rough-and-ready cure inspired directly by scientific investigation.[40]

Once the model was established, investigators explored both the scientific and the therapeutic potential of the hormones (secretions from endocrine gland tissue, first so named in 1905). Americans became increasingly conspicuous among investigators identifying and preparing active extracts of glands and isolating the active principles. One early star was the chemist Edward C. Kendall, who was working at the Mayo Clinic, where clinicians were treating hypothyroidism. He isolated and crystalized thyroxine in 1915.[41] Many similar findings by Kendall and a growing number of his colleagues active in endocrinology followed.

In 1922 several U.S. investigators were closing in on the problem of insulin when a Canadian team announced the practical isolation of the hormone that would save countless lives among those with diabetes mellitus, particularly children who, like the fourteen-year-old boy diagnosed by Cabot (see chapter 5), were otherwise doomed to waste away into inevitable, often speedy dissolution. Insulin therapy spread with astonishing rapidity, starting in the United States, and demonstrated beyond any words the practical impact of physiological medicine—and the need for more doctors to supervise the patients. Very soon after the first announcements, knowledgeable experts and journalists alike understood the epoch-making effects of "a cure for diabetes" (meaning that a person could usually have his or her life extended with medical assistance).[42]

Evidence of the powerful effects of gland products helped move physicians and scientists, and ultimately the public, into a new phase in health care. The obvious growth of experimental animals treated with endocrine products, as in this photo from a laboratory in 1917, increased awareness of a whole new area of physiology that had to be integrated into human biology and medicine. Carey Pratt McCord, "The Influence of the Pineal Gland upon Growth and Differentiation with Particular Reference to Its Influence upon Prenatal Development," *Surgery, Gynecology, and Obstetrics*, 25 (1917), 252.

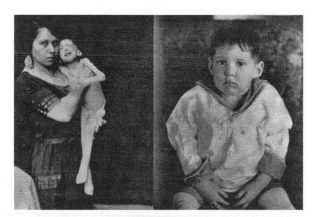

These before-and-after photos furnished convincing evidence that physicians could use endocrine products to perform a new kind of miracle, in this case saving the life of a child with diabetes. The picture on the left shows a boy who is losing weight and will soon die. He was one of the first people to receive insulin, and the results of the treatment are shown in the second photo, which shows the same boy not only surviving but plump and functioning well (his facial expression notwithstanding). Ralph H. Major, *An Account of the University of Kansas School of Medicine* (Kansas City: University of Kansas Medical Center, n.d), 68.

Even beyond clinical applications, popularizers of the glands, gland therapy, and commercialized extracts exaggerated not only hormone treatments but the idea that hormones, and the rates at which they were secreted, determined emotions and behavior. The areas of interest ranged from the amusingly named "male" and "female" hormonal chemicals to the emergency function of adrenalin, an idea of Cannon. This adrenalin work, which he began in 1911 and continued in the 1920s, demonstrated convincingly the connection between emotions, hormone secretion, and health.[43]

Investigators in general found that secretions came from various kinds of tissue. The multiple substances performing control and coordination functions inspired a New York physician, writing in the 1920s for the upper-middle-class public in terms they could understand, to compare the human body to a large, self-governing corporation. The body had no single "captain of industry" to run it, "but the glands of internal secretion are the directors."[44] By the end of the 1930s, investigators from fields as diverse as surgery and chemistry had learned a great deal about the endocrine glands and endocrine disorders, and the subject area was particularly well established in the United States. Yet as late as 1935, one pioneer recalled, "it was plainly evident . . . that great advances were about to be made in endocrinology. Scientific knowledge had developed just enough to stimulate speculation and not enough to control it."[45]

Complexities in Physiological Functioning and Malfunctioning

By the 1930s the mainstream of physicians engaged in physiological medicine had shifted their attention substantially to the functioning of the kidney and the circulatory and nervous systems.[46] Writing at the end of that decade, Robert W. Keeton, a Chicago internist, summed up "the physiological viewpoint in medicine." No longer limited by the direct, mechanical correlation of anatomical changes with symptoms, the physiological physician could add up a number of different symptoms in a patient and figure out what physiological processes were operating in concert to produce that combination of symptoms. One example he offered was a digestive dysfunction (steatorrhea) accompanied by a wild variety of symptoms: "The unabsorbed fatty acids produce the diarrhea. The diarrhea in turn produces nausea and vomiting and prevents further absorption of all types of food. Failure to absorb sufficient calories gives rise to emaciation and muscular weakness. Failure to absorb foods containing Vitamin B gives rise to further bowel irritability and the neuritic symptoms (numbness, tingling and alteration in reflexes)." Keeton then went on to describe still other symptoms in the patient and finally to indicate how cutting fat out of the diet could interrupt this lamentable cascade of interrelated physiological malfunctions.[47] More than ever, practicing medicine had become a matter of solving puzzles and working with complex interactions.

Gauges and Pharmaceuticals

Meanwhile, a second shift in physiology discernible in the 1930s had taken place among laboratory investigators. They were turning from studying the clinical actions of effective but previously unknown chemicals, especially vitamins and hormones, to characterizing the molecular structures and properties of those and other biological substances. The ultimate result would be not only isolating but synthesizing physiologically active substances—which in turn could lead to industrial production of those chemicals.[48]

Major pharmaceutical firms were setting up impressive independent research laboratories—Merck in 1933, Lilly in 1934, Squibb and Abbott Laboratories in 1938. Some of the laboratory scientists became distinguished figures. The number of pharmaceutical firm laboratories increased from eleven in 1920 to ninety-six by 1940. The pressure from medical reformers advocating "scientific medicine" to subject pharmaceuticals to controlled clinical trials profoundly affected the major manufacturers and stimulated work not only in their own laboratories but in collaboration with academic physicians and cooperating clinicians.[49]

Throughout the physiological era, both scientists and clinicians utilized more and more instrumentation to measure physiological states and changes. In cardiovascular disease, for example, the electrocardiogram, or ECG, which recorded heart impulses, had become useful in studying heartbeat arrhythmias. Then a Chicago physician, James Herrick, in 1918 used the instrument successfully to confirm his description of myocardial infarction (disease in a heart artery causing tissue in the heart to die). Over succeeding decades, the ECG would be widely used and applied to still other heart diseases.[50] To cite another example, the Van Slyke apparatus to measure the acidity of blood: "The simplicity of the technique and the few minutes required for the determination make it available not only for physiological experiments, but also for clinical routine." Again, the way was opened for commercial production and widespread clinical use.[51]

Physicians and hospitals came increasingly to depend upon commercial suppliers, not just the makers of x-ray and other machines but particularly the pharmaceutical firms, which were beginning to distinguish themselves. Just as the physiological era was dawning, a new type of product appeared: synthetic organic chemicals. The most notable was one discovered in Germany, Paul Ehrlich's specific for syphilis, Salvarsan, which came on the market in 1910. During World War I, Americans were freed from the constraints of German patents, and organic chemists could make their own copies of known drugs such as Novocain and Veronal that earlier had been imported from Germany. As they did so, the drug companies made great profits. After the war, the pharmaceutical firms continued their research. Because they were dealing with complex scientific problems, increasing governmental regulation (not least regarding narcotics and alcohol), and larger-scale marketing, in the 1920s many firms specialized and consolidated. The larger firms went further in serving less as general suppliers to drugstores, carrying the maximum number of products, and instead became expert in producing and marketing certain types of medical products. And they continued their collaboration with university and medical school scientists, most famously as Lilly did in developing and marketing insulin. In 1923 pharmaceutical firms paid for eight research fellowships at various universities; by 1934 the firms funded forty.[52]

Continuing Development in Surgery and Germ Theory

Physiological medicine overlay but did not stop the continuing momentum of surgery and the germ theory. Thinking and instrumentation in both germ theory and surgery looked different by the 1930s, and physiological thinking deeply affected leaders in both. Investigators faced with a disease still tended to turn first to bacteria to explain it, however. For decades, the number of "discoveries" of pathogens that ultimately did not prove out was very high. The most serious was perhaps the bacillus identified as the agent in the influenza pandemic of 1918–19, an identification that at the time caused many people to take an injection that they hoped would protect them against the disease—to no avail, as it turned out (it was not until 1933 that investigators showed that filterable viruses, not bacteria, caused influenzas). Those vaccine trials, however, did lead to standards for judging the efficacy of a vaccine, another sign of the shift to quantitative standards in scientific medicine.[53]

There was even a fad that began in 1912 and did not end until World War II: focal infection and systemic disease. Whenever there was a little-understood condition, such as arthritis or even mental disease, physicians might be able to trace it to systemic effects of a local infection, typically an infected tooth, tonsil, sinus, or appendix. One well-established model from which proponents extrapolated was a gonococcal infection that would cause arthritis. Before skeptics won out, untold numbers of teeth, tonsils, and other focal points had been removed surgically—and unnecessarily, as later critics held. Removing tonsils was particularly popular: patients favored it, and more than a few physicians may have found it profitable.[54] In some mental institutions almost all of the patients lost their teeth and tonsils in a vain attempt to extirpate the sources that were presumably poisoning their brains.[55] It was an excellent example of a combination of germ theory and surgery that physiological medicine would replace. And on the popular level, vendors of laxatives and gadgets and bran breakfast cereals to cleanse the intestines flourished on the basis of the focal infection stream in medical thinking.[56]

By the end of the 1920s, the germ theory model had produced two additional clarifying models, not only the viruses but the rickettsia.[57] The rickettsia, which at the time appeared to be intermediate between bacteria and viruses, were actually named for the American Howard T. Ricketts. He identified the pathogen of

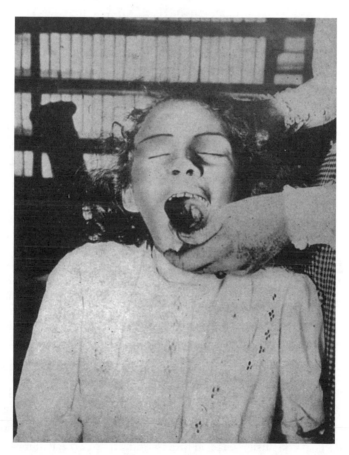

Physicians in the second and third decades of the twentieth century often used the common inflammation of the tonsils as a model of how a local infection could endanger the health of the whole body. This photo shows an extreme case of focal infection, an infected tonsil, such as would cause a clinician to remove the tonsils, which had no function in the body as far as experts at that time knew. Luther Halsey Gulick and Leonard P. Ayres, *Medical Inspection of Schools* (New York: Survey Associates, 1913), 128.

Rocky Mountain spotted fever and then was studying typhus when he contracted that disease and tragically died.

By the 1920s the center of virology was moving from Europe to the United States. The whole field, however, was in "a somewhat chaotic state," as one 1931 expert observer put it with substantial understatement, adding that "as a field for investigation the filterable viruses seem only in their infancy." More and more diseases appeared to be caused by viruses. Indeed, the expert detected "a tendency to place most diseases of unknown etiology in this group. It has been a very convenient waste-basket."[58] As early as 1912 it was possible to list thirty virus diseases, many in plants and animals but also such human diseases as not only yellow fever but typhus, polio, and rabies. In 1911–12 two Americans, Goldberger and his colleague John F. Anderson, had added measles to the list.[59] By the 1930s no one dared to make a list of virus diseases. There were too many.

The science of viruses was one thing. Applying the knowledge to the practice of medicine was another. Thomas Rivers, one of the major investigators of that period, later summed up events after recalling that in 1931 major figures in medicine had confused viral with bacterial pathologies. Some at one point identified a virus disease as a streptococcal disease. They simply found it difficult to escape the established germ model. "It took a long time," said Rivers, "to educate doctors to understand the nature of viruses."[60] Viruses did not fall into conventional categories, and unlike bacteria, they could not be cultivated in the laboratory. Were they alive or not alive? Moreover, many viruses actually operated in connection with bacteria, giving hope that they might serve to combat bacterial infections—a question many thought more important than the category into which the virus fell. Then later in the 1930s, biochemistry combined with pathophysiology to make a new era in virology (see chapter 8).

The Application of Physiology to Surgery

Physiological thinking also affected surgery. The direct application had begun at the turn of the century, most spectacularly in Cleveland. George Washington Crile, a surgeon there, began his own experiments to find out why he was losing patients from shock. He eventually concluded that in shock the body pulled blood out of circulation. That conclusion enabled him to undertake corrective measures. Up through and especially during World War I, Crile and others continued exploring the mechanisms involved in shock. But Crile saw further, as he wrote as early as 1904: "Borderland studies of physiology and surgery seem to teach us that the great principles established in the laboratory of physiology may be translated into the language of the surgical operating room. . . . Only the title of the great book of surgical physiology has been written, and the entire volume . . . remains for the

Health Care in America: A History

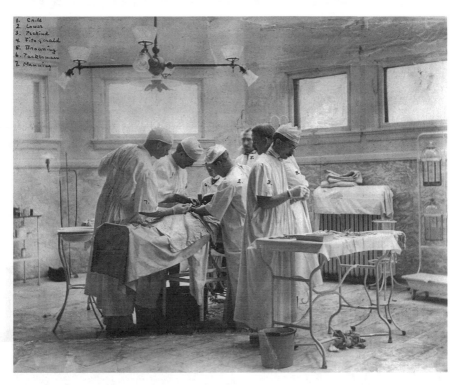

George W. Crile operating in St. Alexis Hospital in Cleveland in 1905. The patient's bed was tilted to take advantage of gravity in order to move the patient's blood in a direction that would decrease the amount of surgical shock. Dittrick Medical History Center, Case Western Reserve University.

younger generation to supply." Crile continued for years to evangelize his colleagues.[61] As physiological knowledge grew, surgeons brought into operating rooms ever more instruments to monitor blood pressure and other life processes as the patient's body responded to operative procedures.

By the 1920s and 1930s, surgeons were aware that ideas from physiology had shifted surgery in additional, more subtle ways. From one point of view, surgeons moved toward a much more conservative approach to interventions and repairs, as opposed to the sometimes dramatically heroic operations of an earlier day. From another point of view, surgeons thought more in terms of patient functioning. They were going beyond the earlier mechanical reconstruction and aiming much more consciously toward restoring the normal functioning of the patient. Instead of immediately cutting into a fractured skull, or draining fluids, they nursed the patient to determine what mild interventions might be appropriate.[62] And especially during and after World War I, they became more concerned with rehabilitation and prosthetics, again to bring the patient to normal functioning, following the model of enabling wounded soldiers to become productive workers again.[63]

Of course physiological knowledge facilitated radical surgical interventions as well. The various glands increasingly became the objects of surgery, and the gall bladder opened a whole field of work. Thoracic surgery in particular required physiological knowledge—most spectacularly how to keep the lung from collapsing during lung surgery.[64] In the field of breast cancer, which was dominated by radical mastectomy, a procedure inspired by an assumption that the cancer was a localized disease, a number of physicians began to explore a physiological approach, chiefly using radiation and radium implantation.[65]

In general, however, surgeons made innumerable small improvements in technique and tried to figure out how to improve the outcome statistics of their operations.[66] The leaders were obsessed with innovation, such as using a new kind of

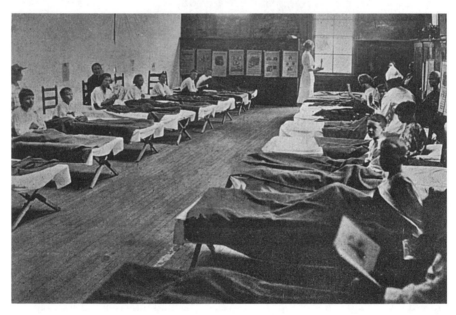

So dominant had surgery, and the ability of surgeons to remove tonsils—which often became infected in many, if not most, children—become that surgery became a preventive, rather than just a curative, measure. This photo shows a North Carolina elementary school classroom turned into a temporary hospital ward as the entire class had their tonsils and adenoids removed. Harry H. Moore, *American Medicine and the People's Health: An Outline with Statistical Data on the Organization of Medicine in the United States, With Special Reference to the Adjustment of Medical Service to Social and Economic Change* (New York: Appleton, 1927), 187.

clamp, and they traveled more than ever, but the general parameters of surgery did not change.[67]

The Revolution in Medical Education

The coming of scientific medicine was reflected dramatically in changes in medical education. Historians have traced the steps that led to the patterns that became established in medical schools in the 1910s and 1920s.[68] The changes constituted a veritable revolution in medical education. In 1902 there were 162 medical schools. By 1930 there were only 76. The rate of change accelerated when much critical foundation money came into the field of medical education after about 1910.[69]

The reformers had two different goals, both designed to bring more science into medical practice. The first goal, which was notably part of the agenda of the reformed American Medical Association (see chapter 7), was to raise the minimum level of knowledge and performance of physicians, so that any physician could be assumed to have a certain competence. The second goal was to tie the medical school to the university and to research, as in Germany, where the professor in a specialty was also the head of the institute doing research in that area. The aim of this second goal was to make the best medical care even better, not just to raise standards among the generality of physicians.

The preliminary steps toward a graded curriculum and university affiliation were well under way in the 1890s. Moreover, as noted above, medical teachers, even in the inferior schools, exerted relentless pressure to improve teaching. Then in 1893 an ideal medical school opened at Johns Hopkins. The medical school was able to open only because the school received an endowment that stipulated high standards and a guarantee that women would be admitted. The major requirements were (1) that it be an endowed, nonprofit educational institution; (2) that it require for admission a college degree with science courses and knowledge of French and German; (3) that it have a four-year curriculum, the first two years comprising scientific training, including courses in anatomy, physiology, physiological chemistry, pharmacology, pathology, and bacteriology, and the third and fourth years comprising clinical training, including training at the bedside in the hospital; and (4) that the faculty conduct research and contribute to medical knowledge.

By the 1920s and 1930s, the extent to which all surviving medical schools followed this pattern rigidly was remarkable.[70] Labs moved to the center of instruction, and hospital affiliations became mandatory. Obviously, to keep up with the competition, a for-profit proprietary school would have to furnish a substantial plant just for the laboratory instruction. Fewer than half of all schools managed to do this. The costs were too great. The successful ones generally merged into a university medical school. In that context, the medical school became a broader institution, tied to hospitals and dispensaries. The dispensaries, however, gradually faded away during the first decades of the twentieth century as both patients and physicians in training migrated to outpatient clinics operated by hospitals (typically connected to medical education). Whereas dispensary workers had come implicitly to perform social work as well as medical functions, now, in hospital outpatient facilities, medical and teaching concerns predominated, although here and there professional social workers began to work with hospital patients.[71] Meharry Medical College, in Nashville, Tennessee, for example, was not just a medical school but functioned as the center of a whole health care network for African Americans, who had few other resources in the area. But the new hospital clinics, it must be said, mostly failed to meet both medical and philanthropic goals. As earlier in the dispensaries, patients still went through the ceremony of "seeing a doctor" but very often did not benefit, because there was no follow-up or supervision to confirm the diagnosis or to find out whether the medication had had any effect.[72]

At the same time, the biomedical sciences became markedly more important in established universities and could tie into medical education. "The new university clinics," commented Alfred Cohn, "mean to take on new functions. Those on which they lay emphasis, indicate the adoption of a wider interest in the problems of concern to medicine. In addition to the traditional responsibility for teaching they avow the desire to contribute to an increase in knowledge."[73]

Meanwhile there were other pressures, and already by 1905 the revolution had come into sight. That was the year when the AMA Council on Medical Education began surveying and classifying medical schools, providing information that states could use to license graduates from different schools. Students were flocking, not to the least expensive schools, but to those with the best reputations. The dean at Western Reserve noted that "every school which has adopted increased standards of training . . . has ultimately been the gainer, not only in reputation but in attendance of students."[74]

And all this time, state licensing programs grew increasingly stringent, so that one had to have a minimum education and ultimately pass a test to be able to practice medicine. As the educational reformer Abraham Flexner observed in 1910, "The examination for licensure is indubitably the lever with which the entire field may be lifted; for the power to examine is the power to destroy." The state boards cooperated more and more until they united in 1912 in the Federation of State Medical Boards of the United States. In 1915 the National Board of Medical Examiners came into existence, offering a standard national medical examination for use by the state boards.[75]

Problems remained. As late as 1924, degrees from diploma mills could still show up. Now, however, they led to scandal and immediate corrective action.[76] Altogether, the extent of upgrading in standards and in medical education in a short time was remarkable. In 1930, none of the candidates in twenty-seven states failed the national medical licensing examination.[77]

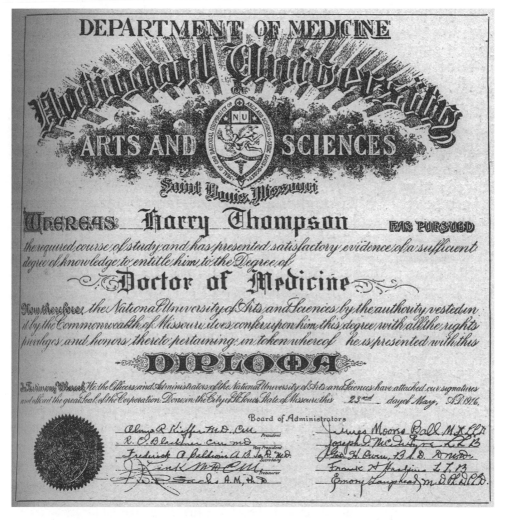

In 1923 a *St. Louis Star* reporter used the alias Harry Thompson to purchase an MD degree in Kansas City purportedly awarded by the National University of Arts and Sciences, Department of Medicine, in St. Louis, years after a university by that name had closed. N. P. Colwell, "Fake Diplomas and Unskilled Doctors," *Hygeia*, 2 (1924), 377. Reproduced with the generous permission of the American Medical Association. Copyright © 1924 American Medical Association. All rights reserved.

Where once there was a standardized curriculum thanks to Osler's textbook, now the national examinations and state requirements standardized medical education further. But, following the AMA goal, the official focus in different states was on how high to set the minimum passing level so as to guarantee that a licensed physician had a certain competence. There was no recognition for high rankings. The idea was just to make sure that all physicians had at least an acceptable level of competence.[78]

The Flexner Report

In 1910, in the midst of this upgrading process, came a famous exposé financed by the Carnegie Foundation for the Advancement of Teaching, in which the educator Abraham Flexner systematically compared each medical school with the Johns Hopkins model. Throughout the country, Flexner's report was front-page news as each local school was exposed.[79]

When Flexner visited the Chattanooga (Tennessee) Medical College in January 1909, he found 112 students and 25 faculty (presumably only part time). The entrance requirements were "nominal," and the school's total general budget was a miniscule forty-three hundred dollars per year.

> *Laboratory facilities*: The school occupies a small building, externally attractive; the interior, dirty and disorderly, is almost bare, except for a fair chemical laboratory in good condition. The dissecting-room contains two tables; the single room assigned to histology, pathology, and bacteriology contains a few old specimens, mostly unlabeled, and one oil-immersion microscope. The instructor explained that they "study only non-pathogenic microbes; students do not handle the pathogenic." There is nothing further in the way of laboratory outfit; no museum, books, charts, models, etc.
>
> *Clinical facilities*: Amphitheatre clinics are held at the Erlanger Hospital, which averages about 50 free patients. Students may not enter the wards. Perhaps ten obstetrical cases annually are obtainable, students being "summoned,"—just how is not clear. The students see no post-mortems, no contagious diseases, do no blood or urine work, and do not always own their own text-books. . . . There is no dispensary.

Flexner added that "this is typical of the schools that claim to exist for the sake of the poor boy and the back country."[80] As this excerpt shows, standards were rising, but as yet medical education could be very backward in any part of the country. The fact remains that from 1900 to 1930 the number of physicians decreased from 173 to 125 per 100,000 population and did not increase again until

1950. Most of the reduction occurred in rural areas, as better-educated graduates chose to practice in urban areas.[81]

Within a few years, medical teachers everywhere had enlisted the support of local citizens and philanthropists to transform the American medical school. Not only were there full-time teachers in the basic sciences for the first two years but a new group, the clinician scientists, a few of them also full-time staff members, were teaching in the third and fourth years, the clinical years. In particular they were combining research and patient care with the educational mission.[82]

Postgraduate Medical Education

Meanwhile, another kind of medical education was also undergoing profound change: postgraduate and continuing education. For ordinary education and licensing, an additional year of clinical training in a hospital, the internship, became standard. For specialists, there were the polyclinic and residency. By 1917–18 the average medical student, with wide variations, was taking one and one-half years of additional training, and states were beginning to require an internship after graduation.[83]

In 1905, the AMA Council on Education idealistically, and unrealistically, had advocated an additional year of training in a hospital for all medical students. But there had not been enough hospitals in which to place all the graduates. Indeed, it was 1923 before the hospitals could absorb the whole graduating class each year. By 1929 hospitals offered more openings than there were graduates. By that time, however, it was clear that the hospitals, not the medical schools, were controlling internships and specialist medical education. By the 1920s the medical schools did, however, recognize their obligation to find internships for their graduates.[84]

Moreover, since the larger hospitals were already running nurse training schools and occasionally midwife schools and had for years had some physicians-in-training on their staffs, a "teaching hospital" identity began to appear. Some teaching hospitals were connected to medical schools, but the American Medical Association was certifying others as "intern hospitals" as well. Among other intern hospital requirements in the 1920s was the provision that at least 10 percent of the deaths be autopsied, and by 1929, 15 percent. Those intern programs were separate from, but often accompanied, the hospital residency programs created to train physicians in various specialties. The first list of available specialty residencies appeared in 1927.[85]

Intern programs had ramifying, often subtle educational effects. The medical students came from schools where often their teachers had imparted great enthusiasm for the new "scientific medicine." The young MDs entered the hospitals with certain ideas and expectations. Despite their pupil status, they carried knowledge and attitudes from medical school to the hospital administrators and personnel—as well as to the general practitioners and specialists who brought their patients to the hospital.[86] The interns also learned hospital practices. All in all, the intern system was one more sign of how hospitals were becoming central to the functioning of American medicine.

The Hospital as an Established Institution

The hospital establishment was expanding rapidly. In 1909 there were already 4,354 nongovernment hospitals, with 421,000 beds. Most were small (the average number of beds was about 97), and, as earlier, many were the personal property of a single physician. In 1933 there were 6,437 hospitals, averaging 160 beds per hospital, and by 1943 the average hospital had 248 beds. In 1929, 23 percent of the U.S. medical costs went to hospitals (30 percent to individual physicians).[87]

In the 1920s, hospitals tended to provide services to private patients, chiefly those requiring procedures that were well defined, were quickly accommodated, and had a happy outcome, such as fracture repair, childbirths, tonsillectomies, and appendectomies. This image of hospitals tended to stick even when they functioned in very different circumstances. Most patient care at that time, however, still took place at home. One midwesterner recalled that when he broke his leg as a teenager in 1924, he was treated at home. "I never knew anybody from Lexington that was in a hospital. . . . Nobody ever went to the hospital. When people died, they died in their homes."[88]

The larger hospitals could function and attract patients as they did because they continued to draw on the two sources of free or almost free labor, interns and nursing students. In 1937 a survey of 534 hospitals revealed that the interns, who were on duty or at least on call twenty-four hours a day, served without cash pay, or for less than twenty-five dollars per month.[89] The student nurses continued to put in long, hard hours of work even as they paid educational fees. By the 1920s, however, again some standards issued by voluntary organizations came to be applied, and the nurses received more education and did less work. At the same time, fewer nurses went into private service in the home, and more cared for patients in a hospital setting, often at what then was considered a decent wage for a woman. There was even a budding movement to make registered nurses graduate with a four-year bachelor's degree from a recognized college or university, but that step in professionalization took two more generations to become a viable alternative.

Outside of government programs for veterans (federal) and chronic illnesses (state and local mental and tuberculosis hospitals), the new hospital system was based locally in charity, proprietary, and municipal institutions. In all hospitals, patients found at least some bureaucratic impersonality: the patient felt, as in one 1921 patient account, "surrounded by very busy, presumably very efficient, doctors, nurses, and employees, who are passing rapidly from one duty to another and have little time for *him*." The parallel to a factory was noticeable to observers even then. Yet the local philanthropic support and local control made hospital authorities sensitive to the opinions of those who would later be considered consumers or those acting on behalf of consumers.[90] Beginning in the late 1920s, for example, air conditioning appeared in a few places, but it remained too expensive for ordinary hospitals until many years later.

Hospital Inequities and Standards

The inequities of the local communities were replicated in hospitals and were particularly harsh where racial segregation and other ethnic and class biases prevailed. As middle-class people came to use hospitals routinely, the large wards suitable for charity patients were rapidly giving way to "semi-private" rooms. By 1930, 23 percent of all hospital rooms were semi-private. As one white-collar patient in Chicago pleaded: "Put me in a closet rather than in the ward!" Obstetric, pediatric, and emergency services, plus elective surgery, in particular expanded the middle-class clientele of hospitals.[91] At the same time, rural physicians believed that they were losing patients to urban institutions and technologies from which the country doctor was often excluded. So central had hospitals become that the institution became a symbol of all that was both good and bad in any community.

The hospitals set up to serve the indigent in large cities were chronically underfunded, and so while they contained the contraptions of modern medicine, they exhibited the neglect shown to the unfortunate. At Cook County Hospital in Chicago in 1936, Harry Dowling, who was there, reported:

> Patients and their families filled all the seats of the barnlike waiting room, and many were forced to stand for hours. All along a nearby corridor patients on stretchers endured the same interminable wait. . . . Examined at last by a hurried, overworked, abrupt intern . . . a patient ill enough to be admitted suffered through another long wait until an orderly or an intern was found to take him on a wobbly stretcher though endless passages to a drab, dismal ward.[92]

Quite independently, and again on a voluntary, nongovernmental basis, national standards for hospitals were put into place. The initiative came in 1918 from a private specialty organization, the American College of Surgeons, formed to upgrade standards in surgery. The extension to hospitals was natural, given the close relationship between hospital surgery and the growth of hospitals.

To be approved by the American College of Surgeons, a hospital had to meet some minimal standards: have an organized staff and good case records; laboratories suitable for physiological medicine; x-ray and fluoroscopic services; and regular clinical staff conferences. Above all, a pathologist had to be available, and in all cases of organ removal there had to be a pathological report confirming the existence of disease. The hospital pathologist requirement was the most difficult to meet, especially for small hospitals, and like marginal medical schools, weak hospitals tended to close. In 1918, when rating began, only 12.9 percent of the hospitals with one hundred beds or more met the minimum standard. When the board had finished determining which hospitals would be accredited, as one insider told the story, "the secretary was instructed to take the survey reports of the other 87 percent to the basement and burn them [in the furnace]. The board feared the consequences of letting the true conditions in hospitals be known." But change came quickly. Within a few years, personnel in most large hospital took the American College of Surgeons standards for granted.[93] Moreover, technology in the hospital such as x-rays and electrocardiograph instruments increased the authority of the physicians working there. In still another way, then, hospitals, originally philanthropic, came more under the control of physicians.[94]

Regularizing Competence and Specialization

Americans in the first decades of the twentieth century used the term *standardization* in two senses: to impose national uniformity, as in medical and nursing education curricula and hospital procedures; and to set minimum standards, as in the AMA approach to licensing physicians. Standardizing was in either case a reform tactic in medical care. In the mid-1930s academic and elite surgeons were still trying to upgrade surgery further by "standardizing" both hospitals and physicians through the American College of Surgeons. While they had made progress, noted a San Jose surgeon, more standardization was needed. His major concerns were inexperienced physicians and unnecessary operations.[95]

The other program of reform, cultivating excellence in medicine, took the form primarily of developing and recognizing specialized competence. As late as 1927, the following report appeared:

> In a small hospital in Chicago, a nonmedical practitioner [i.e., not an MD] attempted to perform the intricate and delicate mastoid operation on a boy supposed to be suffering from an abscess of the middle ear. Through his lack of skill, he punctured a large vein, the sigmoid sinus, thus doing with his chisel what the operation is intended to prevent the abscess from doing. The patient died and in the postmortem examination an abscess was not found.[96]

The trend toward specialization of physicians at the turn of the century continued. Many influential Americans believed that science and expertise in general could solve both social and personal problems. The specializing trend was reflected in many venues. In hospitals, for example, wards were more than ever before divided according to specialty.[97] More specialist organizations continued to appear. The most notable in the era of physiological medicine was the American College of Physicians, which recognized expertise in a new specialty, internal medicine. The College started on a small scale in 1915, and in the mid-1920s it began publishing what became the *Annals of Internal Medicine*. As one medical educator commented in 1928, the "rapid growth of medical science during the last few years has made it almost impossible for a single individual to master the entire field. In internal medicine, as [in] other branches of human knowledge, the age of specialism has of necessity arrived, and some of our ablest practitioners even devote themselves in great measure to one disease."[98]

World War I accelerated the trend toward specialization. The army gave formal recognition to two areas in which special competence was expected: surgery and neuropsychiatry. And the many physicians who served in the armed forces (about a quarter of all practitioners)[99] observed personally the advantages of specialized practice and the excitement of the new discoveries they learned about during training and service. After the war, vast numbers of general practitioners felt excluded as many of their younger colleagues, especially, continued to turn toward taking up a specialty.[100] By 1923 about fifteen thousand physicians identified themselves as full-time specialists. Many more acted as specialists part time, so that about 30 percent of all practitioners were specialists of some kind, a figure that reached 45 percent by the beginning of the 1930s. By 1934 there were almost twenty-seven thousand full-time specialists, or about 18 percent of all practitioners.[101] General practice continued to be the norm in medicine, but the 18 percent of practitioners who specialized furnished most of the leaders in the profession. They constituted the elite, and opinion makers in the United States recognized that elite status.[102]

Specialization was complicated not only by the confusion in identity between full-time and part-time specialists but also by the fact that any physician could list himself or herself as a specialist. And the definition of a specialty was unclear and fluid. Originally, as noted above, specialty groups were disciplinary interest groups. Specialists could organize around technologies (e.g., the x-ray) or organs (e.g., the nervous system) or patients (e.g., children). Moreover, by the 1920s subgroups were appearing in standard specialties, such as neurosurgeons in surgery.[103]

In addition, the nature of specialty groups was changing. Instead of just cultivating scientific and clinical advances, members began to see that they and their

Evacuation Hospital No. 2, with about 1,000 beds. During World War I the first stop for a sick or wounded soldier was the field hospital. The next step, if necessary, was the evacuation hospital, a few miles behind the front lines. The evacuation hospital was run by surgeons, specialists recognized as such by the U.S. Army, whose presence and practice influenced physicians and patients alike to accept specialization in medicine after the war. Charles H. Peck, "The Hospitals of the American Expeditionary Force," *Annals of Surgery*, 68 (1918), 463–466, photos between 464 and 465.

colleagues represented a public interest that might demand pressure group action for better medicine (famously the American Academy of Pediatrics, founded in 1930). But such scientific interest groups could also serve as occupational interest groups, advancing what benefitted the particular group of specialists and sometimes resembling a labor union. This transformation was particularly noticeable in the American Medical Association specialty group sections. Originally those groups were designed to educate general practitioners about standards and developments in different fields of medicine. By the second and third decades of the twentieth century, the AMA sections were focused on the trained specialists and their professional agendas.[104]

Confusion and Contentiousness around Specialization

A mix of motives drove specialist physicians and their private organizations. Often naked economic interests were involved. Indeed, one of the scandals of the medical profession in that day was fee splitting. In fee splitting, a general practitioner who referred a patient to a surgeon expected the surgeon to give him or her part of the surgeon's fee, which of course was set that much higher. The Mayo brothers and others disdained this practice (delicately and wittily referred to as "dichotomy"), but it survived in many areas into the 1940s.[105] Equally or more often there were professional and ethical concerns, such as trying to prevent incompetents from performing advanced procedures. And professionals who took pride in their work and scientific knowledge and what they believed they could do to help patients and improve the nation's health had a complementary set of motives.[106]

By the 1930s a significant development in American medicine in general had begun to center on the growing number of specialist physicians: the fragmentation of medicine. At some point doctors began to identify themselves as members of a specialty more than as just physicians.[107] By the end of the twentieth century, a practitioner commonly lived in a world of rheumatology or gynecology or some variety of microsurgery, not the world of medicine.

In the opening decades of the century, however, the most acutely articulated division was between the specialist and the general practitioner. The complaints of the general practitioners were traditional and straightforward: a physician should be broadly trained so as to be able to treat and take responsibility for the whole patient, not just a part or piece. Even some specialists, such as an Ohio surgeon in 1921 citing a "roentgenologist" [radiologist] who in one case had missed obvious signs and symptoms, deplored the trends they were seeing: "We are now almost in a frenzy of ultra-specialization. . . . Our medical colleges have all but ceased to train general practitioners and instead are practically training only specialists and research workers."[108]

Reformers in Medicine

In the early twentieth century it was the claims of the specialists who wanted to bring science into medicine that were most distinctive, and they could draw on laboratory-based physiological medicine. Within medicine, a group of "reformers" emerged, including many medical school instructors who were upgrading teaching and the enthusiasts in the interurban organizations. These reformers wanted to go beyond standardization in making medical treatments reflect the findings of laboratory and also clinical research—asking, for example, whether a given pharmaceutical would prove effective in an acceptable proportion of cases. Such clinical investigations flowered in the 1920s. The question of effectiveness also came up concerning preventive measures, as, notoriously, in the use of anti-influenza immunizations after World War I, which physicians employed widely (and futilely) without even basic controlled testing. Thus statistics in clinical outcome joined laboratory testing in advancing practice, including, the reformers hoped, the practice of the isolated country practitioner.[109]

The reformers seeking rational therapeutics targeted two forces. One was the conservatism and ignorance of ordinary physicians, who, for their part, could invoke the authority of clinical judgment and professional status. The other was the commercial bias of marketers of drugs. A Cornell University pharmacologist summed up this point of view in 1936: "Nearly all abuses arise because someone profits thereby." Already in 1905, just before federal law moved toward honest labeling of drugs, the American Medical Association set up the Council on Pharmacy and Chemistry. The Council, a typical Progressive nonprofit organization, reported on the effectiveness of new drugs on the market. The Council's guidance for practitioners appeared in the *Journal of the American Medical Association* and, beginning in 1907, in *New and Nonofficial Remedies*.[110]

In addition to pharmaceutical firm laboratories with their often questionable results, the therapeutic reformers had resources primarily in hospitals and espe-

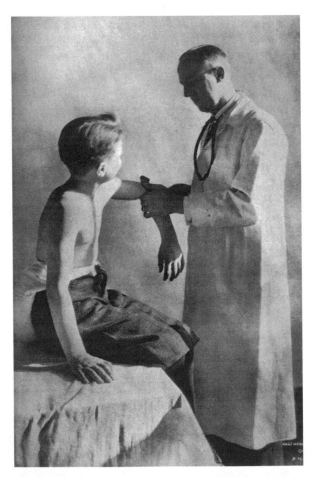

The stereotype of the physician wearing a white coat (and sometimes with a binaural stethoscope hanging around the neck) was well established by 1926, as demonstrated by this Illinois state doctor described as giving an annual medical examination to a patient in an institution in Chicago. *Welfare Magazine*, 17 (Jan. 1926), opp. 110.

cially, beginning in the 1920s, medical school laboratories and clinics. The results did not translate well to ordinary practice. The complicated technologies of big hospitals or the Rockefeller Institute did not fit into the doctor's traditional black bag and were not available to most physicians. At one point, just before World War I, most practitioners, for example, had not had experience in administering a course of drugs intravenously—one reason for the failure of a pneumonia serum treatment.[111]

Proponents of scientific medicine continued their struggle into the 1930s and increased the impact of physiology. They even established an identity with their "American style" of medical practice, which was carried abroad mainly by the Rockefeller Foundation's international programs. Americans notoriously emphasized laboratory tests rather than impressionistic clinical experience, even in very different cultural settings. They also favored a connection between university research and the teaching hospital, which was already an effective model in many centers outside of the United States.[112]

On another level, a new professional uniform for physicians had appeared and was spreading rapidly: "the doctor's white coat." Clinicians adopted from laboratory practice this symbol of science, which was different from, but functioning similarly to, the white gown worn by those who performed aseptic surgery.[113] During World War I, one visiting British surgeon returned from America and urged his colleagues to adopt the white coat to signal to the public that physicians were practicing scientific medicine.

Wide Variation in the Effectiveness of Medical Care

Outside of a few elite institutions, change in practice could be slow. Eminent practitioners continued to advocate idiosyncratic procedures. Alongside the best lab reports, many medical journals carried reports, written by physicians, of the therapeutic benefits of patent medicines or of bizarre procedures such as prostate massage.

The actual practice of medicine involved much care of the patient, for there was little else to do for victims of typhoid and tuberculosis. Large numbers of infants continued to perish from intestinal disorders. From the point of view of a practitioner writing in 1984, in the 1920s "we had no really effective medication" for infections or such conditions as dehydration. Blood transfusions remained extremely difficult, and the electrocardiograph was just coming into use. Another physician, who began her internship in 1934, recalled dealing with all of the "youngsters admitted with paralysis or with frank meningitis. We had no antibiotics. All that we had were intravenous fluids, blood for transfusion, bacterial cultures, and blood counts inexpertly performed by ourselves at 3:00 a.m."[114] Surgeons had more success with cures. Nevertheless, in spite of increased biomedical research, the problems of infection and loss of blood and body fluids continued to burden the profession, especially in obstetrics and even in surgery.[115]

Physicians were often impressed by what they were able to do, as opposed to a few years earlier. Many members of the public, however, remained not very trusting. A survey in Los Angeles in the early 1920s found that only 10 percent of the population employed a regular physician. The typical citizen in "Middletown," in Indiana, tried to avoid medical assistance, scientific or not, and did not think a person was sick unless he or she was too unwell to go to work. "I never go to a dentist until my tooth aches so badly I can't stand it any longer," said one middle-class woman. "If you go to a doctor or dentist he is sure to find something wrong with you."[116]

Yet the social prestige of the doctor continued to climb in the United States. Moreover, it turned out that being a physician in the first decades of the twentieth century could take one far beyond science, into a society deeply affected by Progressive reformism. The world around medicine was changing with overwhelming rapidity. As early as 1921 one official noted that a century earlier, what a doctor did when treating a patient was of concern only to the physician and the patient. But now, "Suppose a single case of bubonic plague were discovered tomorrow in New York City. The correct diagnosis in this single case would affect directly or indirectly every man, woman and child of the millions in New York and . . . in the eastern half of the nation."[117] The social significance of a physician's work was obviously greatly magnified by that time; moreover, health care in general had a growing impact on the whole country.

CHAPTER 7

Physicians, Public Health, and Progressivism

Before the twentieth century was more than a few years old, thinking Americans became acutely aware that a wave of reform, Progressivism, was transforming American society. Many physicians and other health care workers actively led or were deeply involved not only in reform in medicine but in many of the most important social changes that came with the social uplift movement, the momentum of which lasted, along with the physiological era in medicine, into the 1930s. The population was finally benefiting to some extent from the rising standard of living and the extension of water and sewerage infrastructure.[1]

Both reform and medicine at the time were inspired by the ideals of science to confront ill health as well as other kinds of suffering. The actual changes usually took place on the local level or involved nongovernmental activities. Historians who have focused on national politics have therefore largely missed the profound social shifts that affected and were affected by medical action and thinking. Physicians particularly began to contribute much more importantly to their society, not just as reformers but with public health efforts. On one level, then, medicine furnished specific public health campaigns, such as those against goiter and hookworm. Yet on another level, with the antituberculosis movement, medicine furnished one of the great models for delivering social improvement.

The record inevitably was not, from a later viewpoint, wholly positive or consistent. Physicians could educate many people, and yet they still could not cure polio or, pointedly, influenza. Meanwhile, after World War I a new set of leaders of organized medicine, representing the small-business thinking among practitioners, frustrated uplift colleagues by withdrawing from a health insurance reform effort. Medical care itself, however, with ever-expanding research and understanding of bodily processes, served as a model for encouraging hope and optimism among the Progressive reformers.

A Changing Population

Between 1900 and 1930 the number of people living in the United States increased from 76 million to 123 million. Life expectancy at birth was 47.3 years in 1900 but

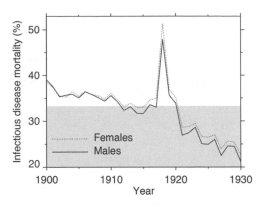

Figures from New England, which produced better statistical data than other parts of the country, show the declining percentage of deaths caused by infectious diseases in the first decades of the twentieth century, including the exceptional period of the great influenza pandemic. The shaded area indicates the one-third mark. U.S. Department of Health, Education, and Welfare, 1956, reprinted in "Causes of Death in Nineteenth-Century New England: The Dominance of Infectious Disease," by Andrew Noymer and Beth Jarosz, *Social History of Medicine*, 21 (2008), 577, and reproduced with the generous assistance of Andrew Noymer.

59.7 years by 1930. Since the birth rate declined by one-third in the same period, the population was becoming markedly older, which altered disease patterns. As noted earlier, the rate of deaths from infectious diseases dropped permanently just after the flu epidemic of 1918–19—a sudden decrease comparable to the one at the end of the 1880s and a third one after 1940.[2]

Before 1920 the population increased significantly through immigration. Sometimes a million people a year arrived, largely, as noted already, from southern and eastern Europe. The newcomers, part of a massive world labor market, typically came from the European countryside, with very few assets, and flooded into the cities. There they suffered from, and exacerbated, the usual urban problems. In addition, the immigrants generated many fears and prejudices based on health beliefs among people whose families had long been resident in North America. It is hard to explain to later generations how both informally and in official policies descendants of passengers on earlier ships often tried to protect local and national health by discriminating against recent arrivals from a variety of cultures.[3] Like migrants throughout the industrialized world, an unusually large percentage of newcomers suffered from disease and social problems that might have been pre-existing or exacerbated by stress and hardship. There was in fact a strong relationship between statistically "early" death and immigrant status in the United States.[4]

Moreover, ethnic and social class differences became ever more acute in a society that had very recently been (relatively) more culturally homogeneous. Social leadership at the turn of the twentieth century continued to embody evangelical and bourgeois traditions that showed up in the form of moralism and dedication to a work ethic. As was typical of people with individualistic, middle-class outlooks, they believed that they could control their own fate, a belief reinforced now by their everyday experience—electricity that turned night into day or even heavier-than-air flight. And, not least, surgery—even brain surgery—as well as diphtheria anti-toxin and typhoid immunization.

Inspired Progressive social leaders therefore had some reason to believe that with technology and education they could change the world. In some areas, particularly areas connected with medicine, they were often remarkably successful. Moreover, just as reform became a powerful cultural force in the first decade of the twentieth century, a number of reformers started a variety of reform efforts within health care itself. The same physicians who had formed the interurban clubs, for example, were behind the attempt to rein in commercial drug exploitation and upgrade treatment by issuing *New and Nonofficial Remedies*, as noted in chapter 6.[5]

Cultural Nationalism

Progressive leaders in general shared a strong cultural nationalism. Sometimes they turned this patriotic sentiment against fellow citizens, attempting by exclusion or labeling to keep distant those whose existence might threaten the dominant culture. Privileged sectors kept other population groups out of the medical schools, which many young Americans used to gain social leverage. Most notorious was the 1920s quota system for Jewish students and the various attempts to exclude women or keep female colleagues from becoming serious competition, to the point that the number and proportion of women physicians actually declined for decades after 1910.[6]

With the germ theory of disease and a great emphasis on contagion, it was even more likely than before for poor people and members of ethnic groups to find themselves the objects of aversion and segregation on hygienic grounds. In 1900, Chinese people in San Francisco were blamed for an outbreak of bubonic plague. In 1916, Italian immigrants in New York were falsely implicated in the polio epidemic.[7]

One side of cultural nationalism, however, appeared in an unexpected place: technical medicine. After 1910, Americans ever more frequently asserted that their country had claim to some status as a world leader in the field of medicine. Europeans would sometimes acknowledge some of these claims either very grudgingly or as a matter of diplomatic graciousness. But the continuing pilgrimage of Americans to German universities before World War I suggests that American medicine, if not wholly derivative and dependent, also did not have a starring role. How the situation was changing, however, is indicated by a statement as early as 1917 by a dermatologist who was not just bragging: "Europe is now understanding that America is to be regarded seriously. So far as Americans are concerned, Europe has ceased to be Mecca in dermatology, and such pilgrimages as they may deign to make in the future will not be with the idea of obeisance, but with the full knowledge that they will bring abroad at least as much as they receive."[8]

World War I did cut off ties to Germany and to some other Continental countries. Suddenly, Americans were acting as if they had developed a critical mass of institutions and talent that made them largely self-sufficient in the field of medicine. As a surgeon wrote in 1917 criticizing snobbish colleagues who cited only European publications, "Let our prophets in our own country come in for their just share of credit and place American surgery just where it belongs!"[9] Concrete evidence of this dramatic change appears in works cited in major medical journals. Before World War I, American medical authors cited German publications especially. Afterward, authors generally had recourse to medical literature published in the United States. The change was so rapid as to constitute a total reorientation

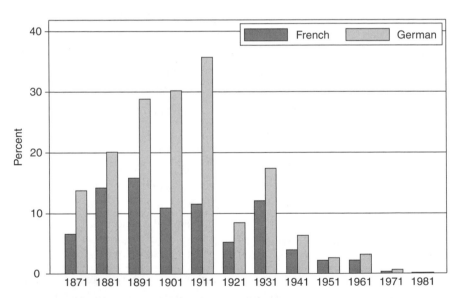

The proportion of citations from French and German medical publications in the *New England Journal of Medicine* in sample years from 1871 to 1981. The trends in this leading medical journal suggest that around the time of World War I, U.S. medical investigators and writers shifted dramatically to citing mostly U.S., rather than European, medical authorities.

of what people considered American medicine. By the mid-1920s a number of bright young British physicians, particularly those working in physiological medicine, were, as observers at the time noted, adding a new degree to their list of qualifications: "B.T.A.," or "Been to America."[10]

Changing Conditions and Professionalizing in Medicine

At the turn of the twentieth century, however, ordinary physicians in the United States had been finding existence difficult. In 1907 one survey showed that physicians' incomes had not increased at all in twenty-five years, and other surveys showed that about half of the 135,000 physicians in 1910 were barely scraping by. Many still blamed overcrowding of the profession, but others noted that while doctors' fees had remained flat, there had been a powerful inflation (about 24 percent in a quarter of a century). Complained one editorialist in 1910: "The wage earner has seen his wages steadily increase with the increased cost of living; the storekeeper makes the same margin of profit, or a larger one, than before, . . . but the medical man . . . sees his income steadily diminish . . . and his dollars shrink in size."[11]

Thus physicians sensed that something was wrong, even if they did not know what. Moreover, as suggested earlier, a new generation of leaders emerged in the medical profession, so that beginning just after 1900 the American Medical Association was essentially transformed. Those leaders, in harness with leaders from the Progressive movement, undertook a broad program that left the profession in a remarkably enhanced leadership position. One outstanding figure, Joseph N. McCormack, who had served successfully on the Kentucky state board of health, was the face of the reform program.

The first step was to reorganize organized medicine. The AMA leaders restructured the organization so that it depended upon the local county medical societies. And McCormack and others evangelized those societies, so that they became, on the one hand, educational, helping physicians keep up with the latest developments through postgraduate educational programs, and, on the other hand, potent political forces. Parallel to the dream of a world-class hospital in every community, AMA leaders promoted the idea that every locality could have highly trained, up-to-date private medical attendants.

Leaders also worked hard to persuade physicians to follow professional ideals of cooperation and service rather than continue the competitive and jealous demeanor previously so common among doctors. McCormack wanted each doctor to vow that "so long as God shall let me live, I will never say an unkind word of a fellow doctor."[12]

In a time when public relations was just getting started, McCormack spent years before 1911 crisscrossing the country to sell both physicians and the public on the idea that a well-educated, well-paid profession was in the interest of both the public and physicians. One of his standard talks was "The Danger to the Public from an Unorganized and Underpaid Medical Profession." A charismatic speaker, he also advocated public health and pointed out the costs of preventable diseases such as infant diarrhea. And he of course denounced quackery and patent medicines, still powerful competitors to physicians as well as hazardous to health.[13]

McCormack and other leaders of medical organizations used their political influence to promote legislation that leading physicians favored, such as stricter licensing laws in various states. They also joined with other national and community leaders to obtain governmental action in many areas of public regulation and public health—not least the federal Pure Food and Drugs Act of 1906.[14] In addition, as a private organization, the AMA for decades campaigned to supplement the weak federal legislation against dangerous and deceptive patent medicines. AMA investigators pursued quacks and the makers of nostrums relentlessly, publicizing activities that preyed on the continuing, amazing credulity of the public. Better Business Bureaus came to depend on AMA reports. In 1925, for example, AMA scientists discredited an alleged antiseptic with the trade name Listerine. As the AMA *Journal* noted, bacteria did not accept the flavor and color tests foisted on consumers: "Four hundred and ninety-five dollars' worth of Listerine has the antiseptic action of a cent's worth of corrosive sublimate; or fifteen dollars' worth of Listerine equals a cent's worth of carbolic [acid]."[15]

Practical Uplift

Many physicians acted in their private capacity as major figures in various Progressive reform campaigns. They also frequently acted out their uplift impulses using their professional identity—a remarkable sign of the growing deference to expertise and also, now, to members of a largely respected profession.

One of the most striking examples of the ways physicians worked within Progressivism was provided by Alice Hamilton, who in 1919 became the first female faculty member at Harvard University (with the proviso that she not try to join the all-male Harvard Club!), at a time when no women were admitted to the school. Hamilton was a midwesterner who attended the University of Michigan Medical School and in 1907 joined the famous informal uplift group at Jane Addams's Hull House in Chicago. There she became aware of the terrible labor conditions in American industry. She began to study, not the accidents that wreaked havoc in the factories, but industrial diseases, which were often slow and subtle in their effects but could be profoundly disabling and deadly—typically chronic diseases such as silicosis. Perhaps the most notorious was the radiation sickness of women who unwittingly worked in a watch factory painting luminescent radium on watch dials. Hamilton's contribution as a physician was to point out that hazards of the workplace could be hidden, not obvious to employers or employees. By 1908 she was a member of the Illinois Commission on Occupational Diseases. Few physicians anywhere knew the field of occupational diseases, and within a few years Hamilton was one of the best in the *world*. She personally toured industrial sites everywhere, observing the dusts and poisons and interviewing the workers and their families. From her 1910 notes, she recorded this account of a case of lead poisoning: "A Bohemian, an enameler of bathtubs, had worked eighteen months at his trade, without apparently becoming poisoned. . . . One day, while at the furnace, he fainted away and for four days he lay in coma, then passed into delirium during which it was found that both forearms and both ankles were palsied."[16] Social arrangements in those times were such that any workplace illness was the responsibility of the worker. As one otherwise good citizen exclaimed to Hamilton, "Why, that sounds as if you think that when a man gets lead poisoning in my plant I ought to be held responsible!"[17]

In combining the roles of physician and Progressive, Hamilton used the voluntaristic, persuasive direct method that was effective in American society when it was dominated substantially by upper- and middle-class elites. She faced down industrialists, engineers, politicians, and leaders at every level. Once, Hamilton confronted an executive in a white-lead factory near Chicago. "He was both indignant and incredulous when I told him I was sure men were being poisoned in those plants. He had never heard of such a thing; it could not be true; they were model plants." He summoned a clearly intimidated worker, who denied that anyone got sick. Hamilton insisted, and this woman physician impressed the executive to the point that he agreed to take appropriate measures if she furnished proof. Hamilton immediately produced twenty-two cases of severe lead poisoning. The executive, she reported, "was better than his word": he reformed all of the works, often using methods that had not been tried before.[18] Hamilton did not have to wait for a long regulatory process. Her efforts directly and immediately saved untold numbers of workers.

The Antituberculosis Campaign as a Fundamental Social Model

Beyond field work and educational campaigns, however, medicine furnished a general model for solving social problems. It was the great Progressive-era crusade against tuberculosis that provided basic patterns for American society in the twentieth century. The crusade began when it became clear that tuberculosis was an infectious disease. Since, as I have noted, the disease each year killed one out of five hundred Americans, over a period of time everyone came to know victims, usually people who took a very long time to waste away into invalidism and death.

Moreover, when people realized that the disease was communicable, anyone who showed symptoms could become the object of fear and discrimination. A person who coughed at work might be discharged. Someone who coughed at home might be identified as "a lunger" and turned out of a rented apartment.

The national antituberculosis organization, which grew out of earlier local and state efforts, came into existence in 1904. It immediately became a model for the single-disease pressure groups that continued to form throughout the twentieth century. The group contained two major components: physicians, and social workers and other lay reformers. The two groups joined together because poverty caused tuberculosis, and tuberculosis, in turn, by bringing lingering incapacity and death, caused poverty.[19]

By 1909, in addition to the National Tuberculosis Association, 350 other antituberculosis groups existed, mostly on the local level. They worked to educate physicians and the public. They sought legislation on the state and especially the local level to set up sanitariums and pass ordinances such as those forbidding spitting in public. They worked to educate physicians and the public, with the aim of fostering health in general and also identifying tuberculosis cases as early as pos-

In the first decades of the twentieth century, antituberculosis crusaders employed mass meetings publicized by brass bands and any other means to try to educate the public about the dangers of tuberculosis and to suggest ways to prevent and treat "the white plague." Edward A. Moree, "Public Health Publicity: The Art of Stimulating and Focusing Public Opinion," *American Journal of Public Health*, 6 (1916), 106.

sible. Someone figured that the press devoted a half mile of column space to "the white plague" each week. By 1911 the focus on education had broadened and intensified. People bought Christmas seals, which funded the campaign and simultaneously educated the public about tuberculosis.[20]

The crusaders developed a general strategy based on earlier attempts to contain or cure the infection. The afflicted person was to leave his or her ordinary environment and move to Colorado, New Mexico, or some other place with a high altitude and fresh air. Or the person was to move to a sanitarium, like the one Edward L. Trudeau famously founded at Saranac Lake, New York, in 1884, which became a symbol of hope for treatment. There healthful food, rest, and clean air would save those who could be saved. In a similar way, in sanitariums across the country tuberculosis patients ate well and slept out of doors on porches or in tents, even in very cold weather. The strategy of removal to an institution meant that the ill person would have a chance to get better and, at the same time, would not be able to infect anyone else. Treatment and prevention worked together. Especially after 1908, isolation as prevention expanded rapidly in importance and came to include poor people. By 1925 the United States had 536 tuberculosis sanitariums with 673,338 beds. This hospitalization had the serendipitous additional effect of shifting medical costs substantially away from individuals and families to private charities and then local and state governments.[21]

The model that spread from this strategy in the years around 1910 could apply not just to disease; it could apply to any social deviation or problem. If there was a problem in any area of society, a professional, often a doctor, would work with the person who embodied the problem. Eventually, the person could be institutionalized or simply restored to normal functioning. The one-on-one, one-professional-to-one-citizen system was costly, but it meant that general social arrangements did not have to be altered in order to bring improvement to many people even without general reform. At the same time, the individual person was saved from becoming an impersonal casualty of the social system.

Social Change to Combat STDs

Following the antituberculosis model, other physicians worked another, quite different major shift in American society by trying to fight sexually transmitted diseases, or STDs. These physicians had formidable obstacles to overcome. First, there was the conspiracy of silence, which meant that by consensus no subject anywhere near to sexuality could appear in print or be discussed publicly. Second, society in the 1890s was still operating under the double standard. Ordinary women were expected to stay pure and stay home. Men, however, had freedom to resort to prostitutes when they pleased, a custom that physician reformers now identified as the

The dining room in the Otis, New York, tuberculosis hospital, showing a diverse population who were separated from other people but maintained in the most favorable conditions possible. Some would recover, and others would not. Daniel W. Weaver and E. W. Weaver, *Medicine as a Profession* (New York: A. S. Barnes, 1917), opp. 102.

major source of "venereal diseases," as they were referred to at that time. In Europe, compulsory medical inspection of prostitutes was commonplace (albeit ineffective), but in the United States officially sanctioned prostitution was usually politically impossible. It made no difference. The double standard resulted not only in infected men but "syphilis of the innocent"—wives infected by their husbands, wives who then infected the babies produced in such marriages, babies who in turn infected wet nurses, who infected more babies. Even physicians hardened by practice were horrified when they had to confront a baby born deformed and diseased with syphilis.

Just at the end of the nineteenth century, the STD situation took an ominous turn. STDs were more common in the United States than most people had believed. Physicians were also discovering the profound damage that syphilis could bring, including, after many years without symptoms, syphilis of the brain, which left a citizen crazed and dying inevitably from degenerative brain disease. Furthermore, gonorrhea, which was generally considered no more serious than a cold, turned out to be a really menacing disease, particularly as it affected women.[22]

In the face of these discoveries, a dour, conservative New York specialist, Prince A. Morrow, led the founding of a single-disease organization that soon provided the leadership for what became the powerful social hygiene movement. Morrow proposed to end the conspiracy of silence by educating people about the dangers of venereal diseases. He further proposed to do what no religious movement had managed theretofore: abolish prostitution (at that time a fundamental source of STDs). And then in concert with the Progressives of his day, he proposed further to use education to try to get at the basis of prostitution by making monogamy work. Everyone, he believed, should be taught about lovemaking so that marriage would be fulfilling and prostitution would become superfluous. Amidst social changes under way at the time, many physicians and other kinds of reformers, such as educators and feminists, joined in the social hygiene movement or allied with it.[23]

What is most remarkable is the extent to which Morrow's program succeeded. The message did indeed change American society. Everywhere, middle-class people moved substantially away from tolerating prostitution—a movement accelerated by governmental action to protect army recruits from prostitutes during World War I. The idea of a sexually fulfilling marriage became more and more widely accepted. The conspiracy of silence was fractured forever. Sex education came into the schools. There was more sex education taught in Chicago schools in 1915 than in the 1950s. An educational drama sponsored by physician groups concerning the devastating effects of syphilis opened the way for sexual content in the theater and other places. The physician who succeeded Morrow as head of the movement admitted that "as the primitive centre for discussion of sex in this country, we must assume at least some of the blame for . . . the flood of 'Sexology' amidst which we struggle for breath today." But, he continued, the reformers stood for high moral standards, and besides, open discussion was a minor matter compared with the evils and suffering caused by venereal diseases.[24]

After World War I the reform element in the campaign against STDs was muted—a curious contrast to the noisy sexual liberals of the Jazz Age, who took advantage of the breaking of the conspiracy of silence. Only in the 1930s, just before World War II, did venereal diseases again become a major public issue. One reason for the declining attention to STDs was the new treatment for syphilis, alluded to above, which was imported from Germany beginning in 1910–11. In addition, in the 1920s Americans slowly adopted a malaria fever treatment, also imported, that seemed to arrest the development of deadly tertiary syphilis—giving patients malaria, which could be treated with quinine, and in the process causing the malaria symptoms to check the syphilis trepanoma.[25]

The new chemical treatment for syphilis, Salvarsan, was an early harbinger of the later miracle drugs: one chemical for one disease, a "magic bullet." Both medical and popular publications reflected enormous excitement at the thought of a simple cure for a dreadful disease. Of course, in addition to the dangers of

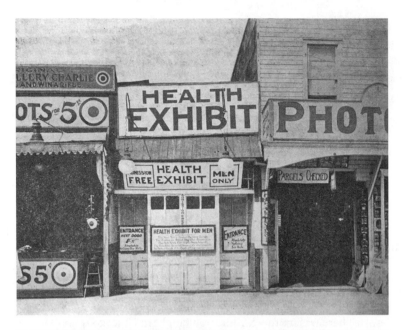

As part of the Progressive-era educational campaign against venereal diseases, the social hygiene activists set up a display "for men only" to avoid giving offense and also implicitly to attract men and convey to them in plain language, as in the bottom photo, showing the inside of the display, the dangers of contracting syphilis and gonorrhea. "Public Health Notes," *American Journal of Public Health*, 6 (1916), 1346–1347.

administering an arsenical substance like Salvarsan, the intravenous and intramuscular administration by needle caused endless trouble and patient resistance. Advocates noted that "it is logical that the greatest possible quantity of the medicament must be thrown into the organism at one time, directly into the blood stream, to reach the tissues." A clinician trying to teach proper Salvarsan procedure at the Mayo Clinic commented, "Even the poor can scarcely be expected to submit with good grace to repeated barbarities offered in the name of medicine."[26] But in the end Salvarsan became an emblem for advocates of scientific medicine.

Physicians in Other Reform Movements

Many other Progressive reforms involved physicians or medical content. Medical arguments were essential in the uplift movement that brought Prohibition, when federal laws prohibited the manufacture and sale of alcoholic beverages—a reform that was actually repealed in 1933. The Progressive idea that education could lead to a better world also received dramatic reinforcement from the rather sudden introduction of psychotherapy for troubled individuals around 1905. Psychotherapists found that they could use verbal and personal influence to change disturbed and dysfunctional behavior. The psychotherapy movement flourished in the 1920s and 1930s, first as positive "suggestion" and then in the form of well-defined techniques, including in the most advanced form psychoanalysis. The changes in intractable cases of invalided outpatients persuaded many reformers that environmental manipulation could alter the behavior of not only children but also adults. Those attempting to improve mental health therefore often had as an explicit goal creating a better society by manipulating everyone's psychological environment.[27]

After the development of genetics, taken up rapidly in the United States after 1900, many physicians assisted in campaigns to try to control the incidence of various clearly undesirable, probably inherited traits in the population. This "eugenics" campaign, mentioned earlier, was always disturbing to some religious zealots and later was misunderstood and misdirected when the world and science had changed. Of course many or most people, at least until World War II, confused social class / ethnic group cultural patterns with biological traits, and so social conflict around eugenics programs should not have been surprising.[28]

Public Health

The most important direct effects of Progressive reform on medicine came in the area of public health. In the wake of germ theory, public health workers gave special emphasis to finding individuals who spread disease. The most notorious case was that of "Typhoid Mary," Mary Mallon, a cook who worked for wealthy

"LOOK HERE, DOCTOR, YOUR BILL FOR SERVICES TO MY WIFE IS JUST DOUBLE WHAT YOU SAID IT WOULD BE."

Therapeutic-Electric-Psychological Specialist: I KNOW IT; BUT AFTER I STARTED IN TO TREAT HER I FOUND THAT SHE HAD A DOUBLE PERSONALITY.

This cartoon displays hostility to high-priced medical specialists, a sentiment common among many social leaders, even while referring to the psychiatric condition of double personality, which, along with other psychological diagnoses and treatments for mental illnesses, was already well known before World War I. *Life,* 20 Dec. 1917, 217. Courtesy of Rare Books and Manuscripts, Ohio State University Libraries.

New Yorkers. In 1906 there was an outbreak of typhoid in a home at Oyster Bay, where Mallon was a cook. It was found that for years she had left a trail of families infected with typhoid. She appeared to be an otherwise healthy "carrier" of typhoid, a phenomenon that had only recently been described in the infectious disease literature. She was isolated and then finally released. She broke her promise not to cook again and was found in 1915 at Sloane Maternity Hospital, in New York City, where twenty-five new cases of typhoid had appeared. She became the symbol of dangerous individuals with contagious infections in the community and, rightly or wrongly, spent the rest of her life in enforced isolation.[29]

As larger cities created public health departments, standard functions developed for such departments. They could, as already noted, furnish diagnostic laborato-

A "clean milk" distribution center in a baker's shop, Rochester, New York. One of the great public health achievements in the cities was furnishing children with milk to drink that was not contaminated or diluted with chalk water or any other fillers that private vendors were foisting off on urban residents. John Spargo, *The Bitter Cry of the Children* (New York: Macmillan, 1907), 235.

ries for use by both public health physicians and private physicians. They could conduct inoculation campaigns, particularly against smallpox and then later diphtheria, typhoid, whooping cough, and other diseases. Finally, they could isolate patients with infectious diseases, placarding homes, whose inhabitants were then legally quarantined. Public health officials could undertake other campaigns, such as drainage and screening campaigns to keep malaria-bearing mosquitoes away, and they worked with other citizens to try to provide clean foods, pure water, and sewage disposal.[30] Indeed, the pure milk and, after about 1912, milk pasteurization movements substantially improved the health of children who depended on milk for nutrition.

Many public health measures were very slow to reach smaller cities and towns, much less the countryside. Water and sewage systems were still faulty or absent in many places in the 1930s. Small municipalities did not have resources for food inspection or much else connected with public health. Moreover, at the local level, economic interests and politics in the most flagrant way stymied enforcement of even mild measures, such as requiring that milk and meat not be adulterated or that houses be connected to water and sewer systems. In one instance at the state level, California legislators failed to pass a general smallpox vaccination requirement.

By as early as 1912 there had been only about five hundred cases in the state each year. For 1921, however, after the failure of the vaccination requirement, the total was more than fifty-five hundred.[31]

Yet the smaller communities did often benefit from major public health efforts in the large cities. Public health laboratories were generally available either free or on a fee basis to physicians and public health officials everywhere, especially among populations tributary to the city. There was also a network of private medical laboratories centered in the cities but willing to support physicians anywhere. The most important function of this public health technology was diagnosis, which was essential for epidemiology and disease control. In a similar way, public health laboratories could furnish inoculation materials not available elsewhere, a service for which New York in particular was famous. When Robert Koch visited the lab founder, Hermann Biggs, in New York in 1908, Koch was impressed that

> a physician could leave a throat culture at a drug store in his neighborhood at 4 p.m. and be sure of receiving a report by telephone before 10 o'clock the next morning. . . . "You will agree, my dear Biggs, that most of these bacteriological and serological discoveries have come from Germany. For my part, I must admit with shame that we in Germany are years and years behind you in their practical application."[32]

Physiological medicine of course expanded the range of public health efforts. One of the most successful instances came in the 1920s, when authorities persuaded the marketers of salt voluntarily (in the Progressive style) to add an infinitesimal amount of iodine to their product, which became "iodized salt." This had a sensational effect in preventing goiter abnormalities in large parts of the country. In Michigan, the death rate from goiter diseases dropped dramatically in the last part of the decade, and the number of goiter operations declined by 60 percent just between 1927 and 1933.[33]

The New Public Health: Education

All these various types of public health activities continued into the second half of the twentieth century. Beginning around 1910, however, leaders in the "new public health" emphasized still another dimension: education, including direct intervention with individuals, for example, persuading people to obtain vaccinations.[34] Some of the education was national, and some was local, but the key units were local public health officials and the public schools.

In the 1920s a significant number of health experts turned their attention away from general environmental conditions such as poor housing and to the specific conditions of individuals. Their efforts included the health education of individuals carried out personally by physicians and nurses as well as through the new public health educational efforts. There was even a famous health campaign by the Metropolitan Life Insurance Company to send policyholders educational material and visiting nurses so as to reduce the payouts of death benefits. The campaign was clearly successful, because it lasted for many decades. More general public health reformers targeted mothers for education and care to lower the infant mortality rates.[35]

One of the best examples of how education worked in the first four decades of the twentieth century was the campaign to eradicate hookworm. A zoologist with the U.S. Bureau of Animal Industry, Charles Wardell Stiles, in 1902 identified a species of hookworm and subsequently found that it was endemic in the southern states, harbored by as much as 40 percent of the population there. The disease caused symptoms of anemia and malnutrition, leaving the victims who survived stunted, with sallow complexion, stooped posture, and a remarkable lack of energy and ambition. Journalists saw in this syndrome the stereotype of the unambitious, good-for-nothing southerner. When Stiles publicized his findings, the journalists dubbed hookworm infection "the germ of laziness" and suggested that there might be a medical remedy for the legendary shiftlessness of poor whites in the South, which had sometimes been ascribed to debilitating malaria. The hookworm publicity provided a powerful stimulus to many residents of the South to attend to a movement to Americanize, that is, modernize, the region in public health and, incidentally, other areas as well.[36]

Stiles, a typical Progressive, advocated practical measures to prevent hookworm. Eggs from the worms in the human intestinal tract passed in fecal discharges into the soil. There they hatched and waited to hook into the feet of people walking barefoot on the polluted soil. From the feet, the worms traveled through the body to the intestinal system, where they fattened on the food coming through. Therefore the main, surprisingly simple preventive measure was to educate the population, first, to use sanitary latrines so that the eggs and larvae could not get out, and second, to wear shoes, something many children, especially, were not accustomed to doing.[37]

In 1909 the Rockefeller philanthropies furnished a million dollars for a Sanitary Commission for the Eradication of Hookworm Disease to travel throughout the South and teach people how to prevent the spread of hookworm. Commission workers also offered diagnosis and treatment (cleansing the GI system with thymol), but 80 percent of their budget went into educational programs. Some railroads furnished trains that stopped in every small town and encouraged the local population to tour the educational cars. The Commission personnel made a point of always operating in cooperation with the local health department, which greatly enhanced the status of those public health officials. In some areas in which

One of the great victories in the new public health educational campaigns was controlling hookworm in populations in the southern states. The very presence of treatment centers like this one, where those already infected could undergo treatment, taught people that they could avoid as well as cure this debilitating regional curse. Robert Shaplen, *Toward the Well Being of Mankind: Fifty Years of the Rockefeller Foundation* (Garden City, NJ: Doubleday, 1964), 27, reproduced with the generous permission of the Rockefeller Archive Center.

there were no health departments, usually because of poverty or racism, the county governments were embarrassed into setting up public health departments. Seldom has so limited a sum of money brought such major health effects, mostly from education.[38]

By the post–World War I years, virtually everywhere an essential public health educational system had come into place, although of course better developed in some localities than others. Whenever a concern arose in public health departments, officials would issue press releases to local newspapers. In addition, appropriate information would appear in pamphlets and other media that were distributed, usually free of charge, to appropriate groups in the community—businesses, organizations, and particularly schools, where teachers acted as fundamental distributors of health warnings and information. Typically, large cities had a monthly publication featuring the current concerns of health officials. In 1923 the American

Medical Association started a popular health information magazine, *Hygeia*, aimed particularly at teachers.

For the public in general, an authoritative medical column in the newspaper became standard. The author was always a physician, and no personal advice was given, but letters that might set off general, standard explanations from medical knowledge or literature were welcomed. The first column in the *Chicago Tribune* appeared in 1911. Through syndication in the teens and twenties, the high point of newspapers in the United States, by one count about a third of the population over fourteen read these columns, which continue to this day.[39]

The Effects of Health Education

As public education developed rapidly in the early decades of the twentieth century, the schools became increasingly important in spreading ideas about health. "Health habits," such as eating right, brushing one's teeth, and exercising, became part of the drill. The "pageant," in which youngsters might dress up as carrots or toothbrushes or otherwise act out the health habits, was a public event that also educated parents who had grown up without some of the modern insights, such as knowledge about vitamins or taking care not to spit. Furthermore, the antituberculosis crusaders powerfully reinforced general health education so that children would be less susceptible to tuberculosis.[40]

The evidence is consistent in showing that educational programs improved health.[41] The death rates of the poor were consistently substantially higher than those of the rich. But the wealth-health correlations in the U.S. population became much more distinctive after education became a factor around 1900. Most suggestively, in the 1890s professionals' children died at an even higher rate than did children of those in other U.S. populations. By 1925, however, children of teachers and doctors survived at a rate far higher than the national averages.[42]

Later generations might find it hard to understand how, within the local understandings in which the system operated, official public health education moved so quickly and effectively to target groups. Literate, middle-class people especially did in that time pay attention to what experts and local officials publicized. In this way, medicine became ever more effectively the vehicle for modernization and "progress." The quest for modernity, carried by advertising and health education, included cleanliness and healthful behavior. Kotex and other commercial products for "feminine hygiene," for example, became standard in the 1920s. Of course some advertisers also added to the list of diseases to which Americans were subject—in the mid-1920s, "halitosis," offensive breath to be combated with Listerine; in 1930, "athlete's foot," to be cured with Absorbine Jr.—but those advertisers were selling cures rather than preventing diseases.[43]

The Influenza Pandemic of 1918–1919

The effectiveness of the combined nonprofit and official health dissemination system may explain why the flu pandemic of 1918–19 was a demographic disaster but did not leave any substantial institutional changes behind. The first cases in the world may have (or may not have) appeared on a military post in Kansas. From there the flu spread, possibly mutating, across the globe. It traveled especially with troops involved in the last stages of World War I. In the United States, probably one-fourth or more of the population suffered overtly from the influenza. Approximately 550,000 people, or one-half of 1 percent of the U.S. population, died. Moreover, those who died were not just the very young and very old, as is usual with influenzas; especially hard hit were people aged twenty-five to thirty-four.[44] Physicians were helpless at the time, and into the 1930s they were still puzzling over patients who suffered from an encephalitis that apparently was an aftereffect of the flu.[45]

In the flu pandemic, public health efforts were local. Local accounts show that authorities issued health advice, and many people wore gauze masks and tried to live healthfully. The authorities also forbade any activities that might cause people to gather in groups, a measure that seriously disrupted ordinary life. Both private and public efforts to decrease person-to-person contacts probably limited the extent of the pandemic. But in the United States, unlike in some other countries, the basic profile of health care and public health did not change. The one exception was that patients often went to the local hospitals, the beginning of a distinct new tendency to use hospitals for nonsurgical as well as surgical and similar conditions. In the end, however, the great pandemic disappeared from public memory until many decades later.[46]

Prevention and Health

Quite aside from the influenza incident, Progressive reformers both inside and outside medicine had good reason to be concerned about the nation's health. Illnesses were a source of economic inefficiency (another Progressive concern) as well as personal suffering. Sick people represented human resources that, like other natural resources, needed to be conserved. Life insurance companies supported the educational efforts of the descriptively named Life Extension Institute. "Even the family physician," noted a reformer in 1915, "is in some cases being asked by his patients to keep them well instead of curing them after they have fallen sick."[47]

People at the time identified two particularly telling national health scandals. The first was the rate of infant mortality. In the second half of the nineteenth century between 15 percent and 30 percent of all babies died in their first year, depending on location and social circumstances. As late as 1917 a national campaign to "Save the

Seventh Baby" grew out of the report that one in every seven babies died, a fact that was magnified by an increased social valuation of infants and children. The second scandal was that fully 26 percent of the army draftees in World War I were rejected as unfit, and 46 percent had some identifiable physical or mental defect.[48]

Patients came to have ever more respect for the doctor's office and the technology in it and to accept the idea that identifying and treating a disease early meant that there was a better chance to cure it or at least to avoid bad consequences. Both antituberculosis groups and insurance companies supported the idea of a yearly health examination as a preventive. Physicians therefore worked to make the annual physical examination a standard for personal medical care—and, as some cynics observed, to guarantee doctors an income even from healthy patients. In a world in which chronic diseases were becoming dominant, the idea that screening might be beneficial was not without foundation and found support among public health authorities who connected the annual checkup with prevention of chronic diseases. Following an earlier lead by life insurance company physicians, the formal campaign for an annual checkup began in 1923, but it did not achieve general success until midcentury.[49] Meanwhile, the increase in chronic disease incidence had another effect: whereas in the case of acute diseases, medicine and hospitals had relieved patients from needing home nursing, with more chronic disease the burden on women (who furnished most of the labor of the actual care) could increase as they had to look after family members who spent long periods, even many years, at home sick and disabled but still alive and in need of care.[50]

Physicians' Work Patterns

The patterns of physicians' work and their contacts with patients also changed with changing times. The automobile became closely identified with the profession. As noted above, doctors could see many more patients if they used an automobile to make house calls. At the same time, the pattern of office practice became much more prevalent, especially in the 1920s, when the explosive increase in cars and roads enabled patients to visit the doctor's office, where more advanced equipment was available than in homes. The office pattern applied not just to specialists but also to general practitioners and was reinforced by the growth of outpatient clinics associated with hospitals, which also helped make hospitals more central to health care. In 1921 one in thirty-five Americans visited such an outpatient facility; by 1934, one in thirteen.[51]

Ironically, in the context of the new pattern of office practice, practitioners did add one member to the private office: the office nurse, who was of course an extension of the doctor, not a competitor or a supervisor. "It has been found," wrote a physician in 1934, "that the physician who must work alone accomplishes the least.

If his work is supplemented by that of a nurse, the two together can accomplish three times as much as the doctor could do alone."[52]

One more pattern in "health care delivery" that appeared in the opening decades of the twentieth century became significant with World War I, when physicians in the army found that they could work effectively with their colleagues. As a Johns Hopkins surgeon recalled, "The one thing . . . that impressed me most and gave me more cause for thought later was the effective medical and surgical service given the troops. . . . There was no private practice; and high and low, the general and the humblest private, got the same treatment. It was superb, grand. . . . Nobody, so far as I could see, loafed when there was work to be done."[53]

The civilian equivalent was private group practice, where practitioners instead of working alone in an office worked in a situation in which they had backup. Sometimes specialists for different kinds of patients and diseases would constitute part of the group. Regardless, group practice spread particularly in the Midwest and West. In 1931 a study of fifty-five group practices found that almost all had one or more staff members who had worked in the group practice at the Mayo Clinic.[54]

The Distribution of Health Services

Changes in health care patterns revolved in one way or another around physicians' services. Yet, just as public health infrastructure and efforts were incompletely reaching all Americans, so, too, professional care was not reaching everyone. In 1932 a prestigious Committee on Costs of Medical Care, after a five-year study, concluded that adequate modern health care was available for the population. There were more than a million doctors, dentists, nurses, and other health workers for a population of 123 million. Yet those services were shockingly inefficiently, unevenly, and unfairly distributed and used. The result was that the bulk of the population, those in the lower income brackets, used only a fraction of the resources that those in the higher income brackets used. Even the middle classes were clearly underserved. Altogether, more than 38 percent of the population received no medical, dental, or eye care at all in 1928–31, and not because they did not need it.[55]

The Committee identified some major factors that helped explain why perhaps half of the population lacked substantial health care. The overwhelming factor was economic. Large numbers of people did not have money to pay for good health care—or any health care at all. As one observer commented in 1929, "I do not think I exaggerate when I say that the high cost of sickness, at least among the middle classes, is as potent a cause for social unrest as poverty among the poor."[56]

Moreover, many citizens stuck to their folk beliefs and practices (ignorance and superstition in the eyes of well-educated contemporaries). Patent medicines also persisted to a remarkable extent. Advertising of nostrums and quack doctors in local newspapers was perhaps more prevalent in the 1920s than in the 1890s. As one social service worker commented about typical health care, "People of the poorer class here like to keep taking things and going to see the doctor," and often the "doctor" they did see was not qualified.[57]

Often health resources were not available. South Carolina had just one doctor for every 1,431 people, while California had one for every 571 people. One hospital bed was available for every 749 people in South Carolina, compared with one bed for every 154 people in Wisconsin. Study after study commissioned by the Committee spelled out the disparities and generally unfulfilled promises of medicine. And while the Committee focused on economic factors, those factors in turn had political and social determinants.[58]

The Health of African Americans

The Committee on Costs of Medical Care omitted African Americans (10 percent of the population) from most of the study because the general health care and health status of a large proportion of them was so appallingly inadequate. Infant mortality and tuberculosis death rates among African Americans were far higher than in the rest of the population. In 1920 the Metropolitan Life Insurance Company statistician Louis Dublin showed that the company's nursing services, using education, had had remarkable success in reducing the death rate among African American policyholders. The evidence was clear, Dublin confirmed, to the consternation of many traditionalists, that adequate public health services would have dramatic effects on that part of the U.S. population, reinforcing the message that African Americans were not going to die out from tuberculosis—or anything else.[59]

Racism continued to pervade American society to an extent that later generations find hard to comprehend. Medical treatment typically varied not only by class but also by ethnicity. In 1927 the son of an African American physician was injured in an automobile accident near Athens, Alabama. The teenager was refused admittance to several hospitals because in that area "there were no hospital facilities for Colored patients, regardless of the severity of the disability." After several hours, an ambulance was located to take the boy to a hospital thirty miles away, in Huntsville, where he died, probably because he had had to wait so long to receive care. Similar stories were told by Mexican Americans in the Southwest.[60]

Most people, including many well-educated physicians, continued to be affected by traditional beliefs that there were diseases to which African Americans were peculiarly susceptible, just as there were "Jewish" diseases and distinctive susceptibilities of Mexicans in California and Texas. There were similar beliefs about differences in bodies, typically constitutional or hereditary differences. After a physiologist in 1910 detected a new painful, debilitating disease that became known

As part of the general 1920s program to improve national health, philanthropists provided these underprivileged urban children with a healthful—but segregated—camp experience, including good food and exposure to nature. "Negro children attending the camp are first introduced to the road to health through fresh air and sunshine, balanced diet, regular hours, and supervised play." Mrs. Cora DeForest Grant, "Two Weeks' Vacation on the Potomac," *Nation's Health*, 9 (1927), 38, reproduced with the permission of Sheridan Content Services on behalf of the American Public Health Association.

as sickle-cell anemia, for decades experts defined it as African American. But, as the historian Keith Wailoo asks, was that designation a sign of inherited difference that signified "Negro inferiority"? Or was it just another individual "susceptibility," such as those found in industrial poisoning or accident proneness, which industrial managers were dealing with? In the new social medicine that was coming in between the two world wars, physicians and other Americans began to view diseases more often in epidemiological rather than individual terms. Both bodies and their diseases could now be considered in the aggregate.[61]

Another major change appeared in the newly conceptualized social problem, African American health. In the first two decades of the twentieth century, destructive segregation and unfavorable health conditions were understood to be local. By the 1920s, however, health workers and African American leaders were treating the deplorable health status of this one group as a national scandal in the age of scientific medicine.[62] In addition to numerous local efforts, a whole infrastructure grew up around efforts of the private Rosenwald Fund and the National Negro Health Movement and Negro Health Week, founded in 1915 by Booker T. Washington. More and more African American citizens learned about personal health and sanitation. They received vaccinations and sometimes adequate medical treatment. Noted one reformer in 1934, "In the . . . years from 1910 to 1927 the death rate of Negroes decreased thirty per cent and their death rate from tuberculosis decreased fifty per cent. Since 1912 the life span of Negroes has increased five years." But the disparities remained massive, and by the late 1920s the Rosenwald Fund leaders had decided that the greatest priority for reaching the victims of a segregated society was to finance more black health professionals and institutions.[63]

Moreover, with the germ theory of disease, unhealthy slum dwellers came to represent a precise and material threat to the privileged classes. Older ideas of filth now combined with fears of microbes carried by people treated as "others."[64] As late as 1929, fundraisers reminded potential "white" donors to Provident Hospital in Chicago that "germs have no color line." One argument for fighting tuberculosis among impoverished ethnics of any variety who might contaminate the wealthy was based on the population of African Americans who acted as servants to the affluent and who suffered disproportionately from the great killer. As an Atlanta newspaper editor put it in 1914, "To purge the negro of disease is not so much a kindness to the negro himself as it is a matter of sheer self-preservation to the white man."[65] Part of the public support for turn-of-the-century public health in fact had come from monied people who feared that the clothes they bought in the new department stores might be carrying tuberculosis and other infectious diseases from immigrant sweatshop laborers.

The Compulsory Health Insurance Campaign

By the 1930s many medical leaders were speaking out about a growing social consciousness among their colleagues, particularly their concern that access to medical services was inequitable on many grounds. Yet from a later point of view, something that did not happen between 1910 and the 1930s is remarkable: unlike any other industrialized country, the United States did not set up a system of compulsory health insurance.

Despite the existence of many urban specialists, the majority of practicing physicians were spread throughout the country in small or medium-sized towns, often living a precarious existence from fee to fee collected personally from

This 1907 photo shows a mother with tuberculosis attempting to care for her children by contracting to assemble and sew clothing on a piecework basis in her living quarters, where she not only exposed her children to her tuberculosis but also, so many people feared, would infect potential purchasers of the clothing she produced. John Spargo, *The Bitter Cry of the Children* (New York: Macmillan, 1907), 173.

patients—often patients seen in their own homes. The sliding scale of fees continued to predominate. Even when a doctor worked under contract to a union or fraternal group, he or she had to set and often collect fees.

For their part, workers and other poor people had ways of dealing with illness. To them, illness or disability first meant lost wages. Until well into the twentieth century, medical care was a distant secondary concern. Therefore, through unions or otherwise, many American workers continued to participate in benefit funds, in which for a certain small amount each week the worker was insured against loss of wages through illness. Only sometimes did the ill worker receive a small cash payment to help with medical bills.[66]

Workers on their own or in groups (as in the contract system) had to seek and employ physicians. A few years into the twentieth century, a number of Americans came under arrangements whereby a third party, neither the doctor nor the patient, arranged to pay for medical services. Some special groups of workers began receiving medical care from company doctors and hospitals. As industrialization increased and factory work became a normal American experience, the incidence of occupational injuries became a major social problem. Some large firms set up private insurance schemes to compensate workers injured in workplace accidents

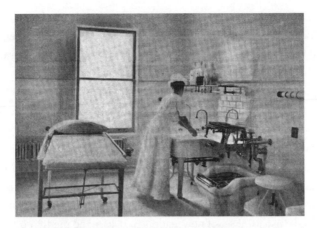

Operating room in the Southern Pacific Railroad Hospital in San Francisco before World War I. As part of the expansion of hospitals in the late nineteenth and early twentieth centuries, many large industrial companies maintained major hospitals like this one for their employees, many of whom in those days suffered accidents on the job. Daniel W. Weaver and E. W. Weaver, *Medicine as a Profession* (New York: A. S. Barnes, 1917), opp. 151.

(and incidentally to head off lawsuits). Since this made production more expensive, it became wise policy to try to get the state governments to require that all employers (including one's competitors) carry such insurance. Thus were born worker compensation laws.[67]

Finally, casualty insurance companies slowly developed some private insurance policies for individuals or groups (usually company contracts covering their employees). Into the 1920s these private, individual health, accident, and disability policies usually were not profitable; underwriters' risk calculations were upset especially by the flu pandemic of 1918–19, with claims exceeding premium payments. Most companies had to abandon health policies, although they began to return later in the 1920s. Eventually, in the 1930s, hospital insurance became more promising—but that led to a different situation, taken up below.[68]

In 1911 the British Parliament had enacted a compulsory health insurance system for factory workers. That act precipitated the compulsory health insurance struggle that convulsed the United States repeatedly over the next century, for the act left the United States as the only country in the developed world without such a measure. It was no coincidence that in 1912 U.S. reformers began a systematic campaign to bring compulsory health insurance—on the state level—to each of the states.

The same group that had spearheaded the state-by-state campaign to enact worker compensation, the American Association for Labor Legislation, led the movement. The original model state legislation that they used to start a national movement included not only medical care but also, for twenty-six weeks, cash payments equal to two-thirds of wages, plus a cash payment for funeral expenses. (Ironically, the life insurance companies campaigned against the measure because they had a profitable burial insurance business.) The proposal reflected both the poor reputation of medical care and the continuing strong connection between illness and poverty.[69]

By 1917 the leaders of the American Medical Association were clearly in favor of compulsory health insurance, which the AMA president referred to as the inevitable "next great step in social legislation." Then physician and business opposition appeared in New York and California, and the momentum for state enactment dissipated. Instead, a set of new leaders in organized medicine, working with unlikely allies such as Christian Scientists and drugstore owners, began an extreme campaign, based on fear, half-truth, and misrepresentation, against "socialized medicine," an inaccurate label that stuck. In 1919, for example, opponents claimed that compulsory health insurance was German and Bolshevik, "an Un-American, Unsafe, Uneconomic, Unscientific, Unfair and Unscrupulous type of Legislation [supported by] Paid Professional Philanthropists, busybody Social Workers, Misguided Clergymen and Hysterical Women."[70] In any event, in 1920 the AMA's House of Delegates voted overwhelmingly to oppose compulsory medical insurance.

Many observers at the time attributed physicians' opposition to their belief that they would suffer economically if they worked with compulsory health insurance. But there was another agenda as well. A number of advocates of compulsory health insurance had observed the efficiency and economic advantage of organizing physicians into groups and outpatient clinics.[71] Many practitioners, including important specialists, who were just beginning to be able really to work miracles in their offices, as with insulin, or in the hospital, with surgery, thought that they were performing a great service that brought an often good income. They did not want someone else to organize their practices, even aside from income issues. They feared losing their professional independence.

One more factor excited doctors' opposition to compulsory health insurance. Physicians had already had experience with third-party payers and found it bad. Under contract practice and private insurance schemes, or even employer health programs, the third party always wanted for the patient the cheapest treatment, regardless of professional expertise or scientific considerations. The widespread experience with the new worker compensation laws intensified what were already unfavorable encounters with third-party payers.[72] For many intelligent physicians, opposition to compulsory health insurance was therefore informed and responsible. Not even the amazed contempt of the rest of the world, the leadership of hospital and medical school physicians, or the scandalous lack of medical care for a majority of Americans revealed by the Committee on Costs of Medical Care could soften the stance of a core of traditional doctors and the organizations that they controlled for generations.

The negative, emotional anti-insurance campaign of the 1920s began to make organized medicine appear to be unprofessionally grasping and even unscrupulous. The hard fact was that away from reformers and philanthropists in the cities, the ordinary physicians in everyday life in the teens and twenties were still fighting for fees. Often of good character and with professional sense, they were extremely sensitive to any threat to their practices. To keep a patient, a physician might ignore the requirement to report a venereal disease to the public health authorities—all the while treating the patient with repeated visits. And the competing physicians in any community would quickly band together if a practitioner came from outside. In Muncie, Indiana, when the state sent in a venereal disease expert, the doctors drove him out of town. A local antituberculosis group reported in another instance: "We have no good chest man in town, and a man was

With the return of injured World War I veterans, physicians were challenged to save these heroes from being forced into a dependent class by integrating them back into the productive life of the nation using prostheses and industrial training, as in this rehabilitation workshop in a U.S. Army hospital. It immediately became obvious, incidentally, that such a program could also extend to disabled people who were not veterans. Garrard Harris, *The Redemption of the Disabled: A Study of Programmes of Rehabilitation for the Disabled of War and of Industry* (New York: D. Appleton, 1919), 234.

sent in from the state capital one day a week to hold a clinic. But the local medical men raised such a protest we had to stop, and now we use local [inexpert] men."[73] It was this fierce, day-to-day economic consciousness and self-interest that fueled many physicians' susceptibility to concerns about "socialized medicine."

New Factors and Strategies in the 1920s

The failure of the campaign for compulsory health insurance at the beginning of the 1920s led to a new implicit strategy: to introduce medical insurance step by step. Progressive reformers in 1921 got Congress to pass the Sheppard-Towner Act, which provided federal aid to the states for maternal and child health. The high infant mortality rate was too obvious to ignore or deny, and members of Congress were influenced decisively by the newly enacted women's suffrage amendment. Although conservatives and alarmed organized medicine leaders killed off appropriations for the program by 1929, Sheppard-Towner nonetheless set a precedent. Moreover, war veterans, who since 1919 had been treated for service-related conditions, in 1924 began receiving some medical and hospital services from the federal government. By 1936 the Veterans Administration operated hospital facilities with more than 64,500 beds in 80 localities in 43 states. About 90 percent of the admissions then were for nonservice disabilities. And so, almost imperceptibly, "the incremental strategy toward national health insurance" got under way.[74]

Another contingency appeared by the late 1920s, one implicitly highlighted by the Committee on Costs of Medical Care: the rising cost of health care. Thus, the need for insurance against illness became attributable less to poverty caused by work lost and more to even white collar workers' inability to pay for medical care, care that both doctors and patients believed was bringing ever better results.[75]

Distribution of health services thus remained mostly frozen in time even as health care in general otherwise adapted to cultural and technological changes. Both physicians and patients continued to find "scientific" a powerful ideological concept that made health care desirable. In the earlier Progressive years, science had been invoked to bring social improvement. By the 1920s, however, the science ideal applied to one's personal life. The technology and image of medicine thus represented a kind of idealism, exemplified in Sinclair Lewis' famous novel *Arrowsmith*, about a biomedical scientist who wanted to bring the benefits of scientific medicine to the world.[76] The science ideal of the 1920s was parallel to the art-for-art's-sake attitudes of avant-garde thinkers who in disillusionment tried to express their idealism at a distance from the dirty realm of politics and social problems. In the 1930s, however, science brought some surprises to medicine. At the same time, the sociopolitical world changed in unanticipated ways.

CHAPTER 8

The Era of Antibiotics, 1930s to 1950s

This chapter covers times of depression and war that brought new governmental and social arrangements and, at the same time, incredible changes wrought by antibiotics in patterns of medical care. The specialty boards came in, and in a countermovement, nonspecialist general practitioners organized. It was the time of the birth of the Blues—Blue Cross and Blue Shield. It was a time of great research achievements. It was, from one point of view, the golden age of American medicine.

By the beginning of the 1930s the basic institutions of biomedical science, hospitals, medical education, and even single-disease advocacy groups were set. During the 1930s and 1940s, further events reconfigured those institutions, so that a new, long-lasting profile of health care activities came into place in the United States by the middle of the twentieth century. And on a deeper structural level, health care consolidated around the hospital into a system that the historian Daniel Fox has named "hierarchical regionalism," a concept that explains much of the working of health care arrangements in the United States for half a century.

Nevertheless, two contingent developments brought revolutionary changes to medicine and health care. The first was the Great Depression, bringing a shift in public orientation, so that a new player became important, if not dominant, in the quest for health: the national government. The second development, which was perhaps even more radical, was the arrival of the antibiotic "miracle drugs." There had been major chemotherapies earlier—quinine, which implicitly targeted malarial parasites; Paul Ehrlich's Salvarsan, which targeted the trepanoma of syphilis; vitamins, which corrected scurvy, beriberi, and pellagra. But with the sulfas and penicillin came medications that acted powerfully against whole groups of bacteria. Seldom has a technological innovation had such immediate and lasting effects as did the antibiotics. Walsh McDermott, of Johns Hopkins, who was an important contributor as well as an eyewitness clinician, recalled the experience: "One day we could not save lives, or *hardly any* lives; on the very next day we could do so across a wide spectrum of diseases. This was an awesome acquisition of power."[1] Finally, the World War II experience reinforced the basic, familiar health care institutions. At the same time, the war accelerated and further intensified the role of the federal government and also, in parallel, the astonishing antibiotic revolution.

The Impact of the Great Depression

The Great Depression, which began in 1929, affected both patients and physicians in the United States. To begin with, there was a clear connection between health and economic circumstances: poor people suffered from sickness more, and more often, than did the rich. A health survey in 1935–36 showed that even among urbanized populations, crowded housing and the absence of private inside flush toilets both correlated with substantially more disabling diseases. Or maybe it was some other factor that caused more illness in poor people than in those with higher incomes.[2]

But the effects of the Depression cut across class lines and included doctors. The average net annual income of physicians at that time was only about one and one-half times that of a skilled laborer. It dropped, however, from $5,300 in 1929 to $4,500 in 1938. Even in 1938, after the Depression had moderated, one-third of the profession reported annual incomes of less than $2,000, and one-sixth reported less than $1,200.[3] It was not that physicians had no patients (their practices continued). It was just that patients did not pay their bills, even the patients who, the Committee on Costs of Medical Care had found, years earlier could afford medical care. Leaders of organized medicine often concluded that there were too many physicians and urged the medical schools to raise standards and cut back on enrollments. Arthur Dean Bevan, who had worked with Joseph N. McCormack, was still active and viewed conditions in the 1930s as the reverse of those in the Progressive era, when he and McCormack had made headway in building up a well-paid, competent profession.[4]

People during the Depression commented repeatedly that in addition to those who could not afford medical care, there were those who put off seeing a practitioner until their condition had become really serious. An Indiana dentist commented in the mid-1930s, "In the depression my working-class patients have tended to delay about two years longer in coming, and then the tooth has become so bad that there is nothing to do but to extract it. Relatively more of my practice in the depression has been this sort of emergency work, as compared with the preventive work I did formerly." As the Depression intensified, and as regular health professionals suffered, even the sales of patent medicines declined steeply![5]

The Depression had an even greater impact on health care. For example, in Detroit in 1932 all of the public school physicians—who had had great success in improving the children's health—were dismissed because of a public financial collapse. Thus a local health care initiative simply disappeared. The levels of children's health declined, and institutional physicians joined the unemployed.[6]

That failure in Detroit typified what occurred. Across the country, local and state governments and private institutions failed and ceased performing their social functions. In the realm of health care, for example, at the beginning of the Depression various public sponsorships supported eight thousand outpatient clinics, accounting for 50 million patient visits a year, in which physicians provided low-cost or free care. In the disaster of economic breakdown, local medical groups forced the closing of most of these "competitors," hoping to divert payments to physicians in private practice.[7]

The Federal Government Drawn In

There is a widespread but ill-informed belief that the federal government then and later embodied a drive to expand jurisdiction and power. Instead, it was a vacuum that pulled federal government action in where local government and community institutions simply stopped functioning at an acceptable level. The result was that when institutions everywhere ceased to work, as happened in the Depression, within a very short time the federal government began to support basic functioning in every community. In their desperation, leaders at all levels reoriented themselves and began to think in terms of national programs, both governmental and nongovernmental. Where federal agencies had once only advised state and local governments, now the federal government was not only propping the locals up with funding but monitoring how federal dollars were spent, in effect directing programs.[8]

Many of the federal programs affected medical services directly, albeit very unevenly. Those of the New Deal, beginning in 1933, are easy to trace. In that year, the Federal Emergency Relief Administration disbursed funds for states to use for medical services and supplies for people receiving direct government unemployment payments. The program also stimulated interest in health education and in health care in various venues, particularly, for example, rural areas without medical services. Beginning in 1937, the Farm Security Administration funded health services for poor rural families, including four hundred thousand people in six hundred rural counties. Elsewhere the agency facilitated or financed health care for many migrant families, such as the famous "Okies" uprooted from Oklahoma. These emergency relief measures were very limited, however, and totally inadequate for the needs. They highlighted the failure of local resources in many parts of the country.

Other government programs, such as the Civilian Conservation Corps, provided assistance to public health programs, for example, controlling mosquitoes to prevent malaria. The Public Works Administration built two-thirds of the

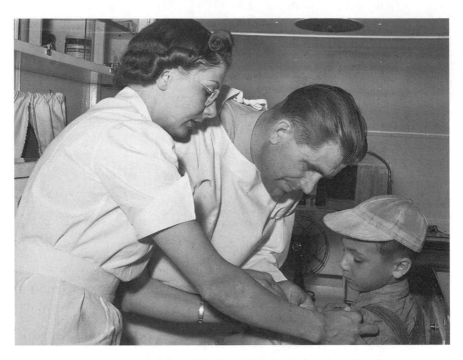

A doctor and a nurse at work in a U.S. Farm Security Administration migratory labor camp mobile unit in Wilder, Idaho. A government physician visited this neglected and exploited population twice a week during the later years of the Great Depression. FSA Collection, Library of Congress.

sewage treatment plants constructed in the Depression years. And by 1940 the Works Progress Administration, another agency, had built 100 hospitals and worked on 1,422 others, adding 120,000 hospital beds to the national inventory.[9]

The most important permanent federal legislation was included incidentally in the several parts of the Social Security Act of 1935. First, acting through the states, the federal government essentially reestablished the maternal and child services of the Sheppard-Towner Act. Second, the legislation set up programs of support for crippled children and disabled people. In practical terms, through this second Social Security program the federal government was making provision for medical care and maintenance of certain types of chronic disease patients, often now categorized as disabled. As an accidental consequence, that part of the act gave important impetus to the founding of private nursing homes, which began to take the place of the county poorhouse.[10] Finally, the act provided for assistance to the states for public health programs, funding that led to a major change in the role of public health in the United States.

Failure of the Campaign for Compulsory
Health Insurance

Still not included in Social Security was compulsory health insurance. With the Great Depression and the New Deal, reformers turned definitively to the federal, not the state, level to try to get health insurance provided by legislation. The leaders of organized medicine, who had discovered their political power in defeating compulsory health insurance at the state level in 1918–20, continued their campaign and fought vigorously to prevent the inclusion of health insurance in the Social Security Act. Indeed, anticipating a political brouhaha, the New Deal leadership did not even try to include it in 1935. In 1938 and again in 1943 the leaders of organized medicine frustrated major efforts by the U.S. Congress to enact a general health insurance similar to that in other industrialized, "civilized" countries.[11]

For decades, many of the best-informed and most sensible leaders in medicine and society in general in the United States favored a compulsory health insurance program. Historians have tried repeatedly to explain the emotional energy of opponents and why they were so effective politically—and the weakness of the reformers, whose ranks included many physicians, notably leaders from medical schools and other institutions.[12] A writer in 1939 noted how fragile the leverage of medical professional organizations was. Even aside from the drug companies, only one person in nine employed in medical care was a licensed physician, and physicians received just 30 percent of the money spent on medical care—and much of that went to overhead and other office expenses. As one physician noted in 1939,

> The traditional belief that the practicing physician is the central and controlling figure in our system of medical care is taken for granted by medical men and approved by the majority of the general public. Neither appear to realize . . . how tenuous is the arithmetical basis upon which the supposed hegemony of the regular medical profession is founded. . . . Physicians will require vastly more legal and regulatory authority than they now have if they are to continue to control medicine in fact as well as in theory.[13]

In 1946 President Harry Truman began one last campaign for compulsory health insurance, a campaign that lasted until Dwight Eisenhower won the presidential election of 1952, ending hope for a comprehensive government program. The campaign involved many factions, but basically the political stances had not changed since 1920. Truman had begun the campaign by declaring, "It is a crime that 33 ⅓ per cent of our young men are not physically fit for service. Let us see if we can't meet that situation." In actuality, the figure was probably higher than 40 percent—a true national scandal. The campaign ended when the president of the AMA observed after Eisenhower's victory in 1952, "As far as the medical profession is concerned, there is general agreement that we are in less danger of socialization than for a number of years."[14]

Extended Effects of the Defeat

The more profound effects of the campaign, however, reverberated for decades. First, as historians have noted, the constant drumbeat of physician groups against "socialized medicine" and "government medicine" had a significant role in eroding Americans' confidence in all government programs; they were an ideologically powerful force that many politicians and other interest groups used with great effectiveness, especially in the Cold War years and after. A medical journal editorial writer observed in 1950, "We are fighting not only socialized medicine but the socialization of everything else."[15] Leaders of organized medicine, in their quest for professional autonomy, that is, to have the ability to make decisions about their own work, did much to validate negativity and stymie social action against many societal ills, not just lack of health care. Indeed, the AMA in 1950 claimed credit for defeating prominent congressional leaders who had worked for many social uplift measures, not just health insurance.

The second effect of physician organizations' political success, however, was on at least one level to discredit organized medicine. AMA leaders raised from their own membership enormous amounts of money for political purposes and allied with racist southern politicians and extreme partisans on the right. By working so closely with anyone claiming to favor business and opposed to governmental controls of any kind, physicians were coming to look like just another selfish business or extreme interest group, not admirable professionals. Why else would they oppose governmental support for people disabled by illness, as they did in 1946? In the end, physicians appeared to be avaricious small businesspeople, no different from the garage owners whose interest in profits was known at that time to lead them to do unneeded work on customers' cars or to bill for work not actually done.[16]

In addition, the ruthless tactics of leaders of organized medicine went far beyond acceptable limits. In 1949 one AMA group begging for physicians' support against health legislation sent out a letter to doctors that was addressed, "Dear Christian American," in a blatant appeal to anti-Semitism (many leaders in the medical factions favoring health insurance were nominally Jewish, and everyone at the time understood the appeal). The AMA repudiated the mailing but that same year assessed its members and gave a notorious public relations firm a huge budget to persuade both doctors and the public to demand "independent" practitioners. The firm flooded the media and also used physicians' professional authority for political purposes, providing propaganda to be distributed in doctors' waiting

"Imagine! They're trying to bring in socialized medicine after I've spent twenty years building up a clientele of wealthy hypochondriacs!"

COLLIER'S   DAN KOERNER

This 1949 cartoon suggests how the AMA campaign against universal health insurance raised suspicion among members of the general public. The appearance of this cartoon in a major mainstream magazine suggests some of the public relations problems physicians were having at that time. *Collier's,* 7 May 1949, 73.

rooms: "Compulsory Health Insurance—Political Medicine—Is Bad Medicine for America!" The personal physician of every member of Congress was asked to write personally to "his" officeholding patient expressing opposition to the health insurance proposals.[17]

The AMA, which had once worked to upgrade standards and fight nostrums, had thus become overwhelmingly politicized and often partisan. In succeeding years in the mid-twentieth century, extremists took over the organization and even ousted the longtime spokesperson for private practice, Morris Fishbein. They hired nonmainstream propagandists and simultaneously closed two respected health care research programs, and the AMA stopped criticizing dubious pharmaceutical products.[18] In response, increasing numbers of physicians dropped their membership. Only in the late 1960s, after the then scandalous news broke that the AMA was supporting tobacco interests to gain political support and after Medicare and Medicaid had become law (see chapter 10), did conservative moderates begin to regain leadership. Decades later, however, the damage to the AMA image had not been repaired.[19]

There are powerful ironies in this story of the transformation of a professional organization into a discredited, albeit feared, negative political machine. The first irony is that the federal government's health programs of the Depression and war years were in fact well run and successful. Perhaps the best example of federally funded medical care was the Emergency Maternal and Infant Care (EMIC) program. In 1941, as men were being drafted into the army, they often brought families into the areas around military camps. The families could not find medical care there, and so

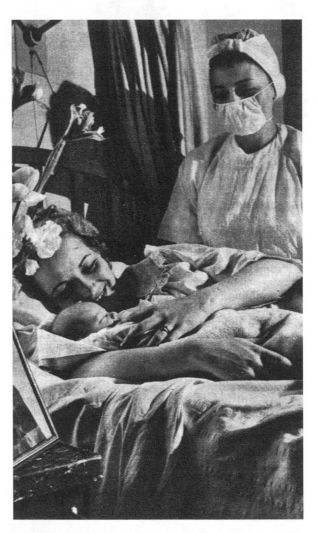

When a woman identified as Mrs. John Mulligan, whose husband was in the navy in World War II, had a baby near the end of the war, the Emergency Maternal and Infant Care program, or EMIC, took care of the bills. She and her husband were thrilled to have what they considered excellent, modern medical care from this government program. Amy Porter, "Babies for Free," *Collier's*, 4 Aug. 1945, 19.

the camp staff at Fort Lewis, Washington, and elsewhere asked the federal Children's Bureau for help in providing and financing medical care for servicemen's families. Pilot programs sprang up, and by 1943 Congress had formalized the work with official appropriations. The success of this government program made the AMA leaders frantic, and in 1946–47 they successfully lobbied Congress to phase out EMIC.[20]

A High Point for MD Prestige

A second irony was that even as the organized medicine extremists were alienating many of their most concerned fellow citizens and colleagues, public respect for *individual* physicians reached a high point. Opinion polls showed consistently that physicians were among the most admired people in society, sometimes ranking above Supreme Court justices. Particularly after the advent of the "miracle drugs" in the 1940s, the prestige of medicine was extraordinary, and criticism of individual doctors reached a low point. As a skeptical observer wrote in 1952, "Most patients are as completely under the supposedly scientific yoke of modern medicine as any primitive savage is under the superstitious serfdom of the tribal witch doctor." In fiction and on the radio and in other popular media, the physician (still almost invariably a male) was a standard hero figure.[21]

As the effectiveness and prestige of individual physicians rose dramatically before and after World War II, the public demand for personal health care escalated, a deep contrast to the distrust and indifference shown by much of the population at the beginning of the twentieth century. Indeed, many problems arose later in the twentieth century because the public in general had unrealistic expectations about the results of medical services.

A Growing Role for Technology and Hospitals

By the late 1940s, then, even if there was no action on national health insurance, circumstances had changed. Not least was the extent to which technology was reshaping medical practice. Already in 1937 a small-city general practitioner (GP) in California noted, "The greatest difficulty I have always found is for the patient to meet expenses when laboratory, X-ray and hospital work becomes necessary. These are the cases that have to be referred to county hospitals, clinics or some type of charity relief."[22]

Meanwhile, medical care that could not be obtained through federal legislation was extended piecemeal and partially through private initiatives. Often these initiatives originated with health care workers themselves. Such was the case with Blue Cross and Blue Shield, a story that starts with the hospitals.

In the interwar years, hospitals rapidly added technologies—everything from labs and diagnostic machines to contraptions hooked up to patients intravenously.

As early as 1929 the electroencephalogram began to be used for confirming cases of epilepsy. From the late 1930s through World War II, in the wake of the cyclotron at Berkeley, physicists began cooperating in procedures using radiation against cancers—an extension of applying radium to cancerous growths.[23]

But this obvious modernization hid an economic factor that was operating already before World War II: hospitals used charges for technology to balance their books. As one medical school faculty member observed in the mid-1930s, "Most of the hospitals in the country are essentially in the business of dispensing medical services for a profit. The money received from their dispensary admissions and for the ancillary services—basal metabolism tests, urine, blood, X-ray examinations—is being utilized to defray the red ink figures of their free ward services."[24]

Indeed, the equipment for treating many conditions existed only at the hospital, and admission rates continued to grow, tripling from the 1920s to the 1940s. Already by 1930 one-fourth of the deaths in the United States took place in a hospital; by 1940, it was one-third. At the time of World War II, one in ten Americans was admitted to a hospital each year.[25]

Hospital Insurance

A few hospitals began experimenting with pooled prepayment plans at the end of the 1920s. The best-known was Baylor University Hospital, in Dallas, which in 1929 agreed to a group prepayment plan covering fifteen hundred teachers. This successful venture was reported at the 1931 meetings of the American Hospital Association, where administrators of other hospitals, desperate to obtain income to keep their institutions economically viable, were very interested.[26] At first each hospital had a separate plan, but beginning in 1932 in Sacramento, a group of local hospitals collaborated in offering multihospital insurance. Eventually this became the Blue Cross plan. As a nonprofit enterprise, it was not taxed by governments, and usually states gave Blue Cross special status because the community benefitted.[27]

Blue Cross quickly became a social movement. Plans sprang up everywhere.[28] At first the hospitals contracted to serve groups, typically employee groups. After the enactment of Social Security in 1935, however, the idea of payroll deductions became commonplace, and employers became the major units included under the plan. In this way, accidentally, the American health care system became dependent upon an anomalous connection between health insurance and employment.[29]

By 1940 there were 4.5 million Blue Cross members in seventy-one plans. A dozen years later, about 40 million Americans had Blue Cross hospital coverage. During World War II, high employment made workers willing to pay for health insurance. In industry, both unions and managers became interested in health insurance coverage for employees.[30] Because wages were frozen by law during the war, employers could offer fringe benefits to attract workers, and health insurance became one of those benefits. And because neither the employer nor the employee was taxed for health insurance, Blue Cross coverage constituted a government subsidy of middle-class health insurance.

Prepay Medical Care Plans

Blue Cross covered only hospital costs. Doctors billed patients separately and were paid fees in the customary way—thus blunting AMA opposition to prepaid hospital insurance. But it was not many years before local physician organizations set up prepayment schemes for doctors' services. Throughout the 1930s, step by step, AMA leaders reduced their fervent opposition to voluntary medical insurance, as long as it was under the control of local physicians and physician groups, often pretending these were "experimental" efforts of members of the profession. In 1934 there were already 150 local plans; by 1939, there were 450.[31] In 1939 the California Medical Association set up the California Physicians Service, a nonprofit pooled prepayment plan to ensure that those having trouble paying their doctors could still afford medical attention and that the physicians in the plan would receive at least some compensation. Physician organizations in other states followed suit, and together they formed the Blue Shield plan. By 1948 the Blue Shield and Blue Cross administrators were cooperating informally.[32]

Meanwhile, during the 1930s some insurance companies began to wake up to the possibilities suggested by the Blue Cross movement and offer similar plans for hospital insurance and even private medical insurance. Blue Cross typically offered full coverage for specific services and a specific number of days in the hospital; private insurers typically offered dollar amounts. The directions of the two types of insurance would converge in a later period (see below), but for two generations Blue Cross and Blue Shield, covering middle-income Americans, were one of the symbols of American health care.

In 1943, in the middle of World War II, another symbol of medical care appeared. In his extremely popular book *Kaiser Wakes the Doctors*, Paul de Kruif, a successful popularizer of scientific, specialized medicine, described how the miracle-working wartime industrialist Henry J. Kaiser had founded a nonprofit medical system for his workers, first at Grand Coulee Dam in 1938 and later at his wartime West Coast shipyards. By getting each employee to pay a few cents a day for medical insurance, Kaiser would furnish modern medical care in advanced facilities without further cost. Furthermore, Kaiser employed specialist physicians, as one might expect in big-city practice. He explicitly created a "Mayo Clinic on the desert," proving to the satisfaction of many hardheaded people that access to medical care would pay off because early treatment prevented later, more serious, more expensive treatment— and incidentally kept employees at work and not off sick. Thus was born the health maintenance organization (HMO) movement. For decades this more efficient, more economical specialist medical care, limited to a few areas, stood as a constant reminder that there were alternatives to the unorganized patchwork system that most Americans had to work with. Medicine that was preventive, high quality, and democratic was also economically efficient and could be self-sufficient.[33]

Other versions of pooled prepayment medical cooperatives also appeared, particularly in the late 1930s. In each case, the promoters had to struggle against the hostility of local and national physician organizations. Leaders of organized medicine persecuted physicians who were involved in health cooperatives, denying them membership in county medical associations, for example. Such ugly behavior persisted on the local level into the 1950s even though it was clearly illegal.

The waiting room in the Kaiser Permanente medical facility near a wartime shipyard in Vancouver, Washington, in 1944. These patients received excellent medical care, for which they paid only a very small sum each week. Rachel Foster, "To Keep the Records Straight," *Modern Hospital*, 68 (1944), 71. Reprinted with permission, copyright © Crain Communications, Inc.

At the end of the 1930s the federal government instituted an antitrust case, and in 1943 the Supreme Court affirmed that the American Medical Association was "in restraint of trade." So far had the image of the organization moved from professional to commercial ("trade") that the AMA spokesperson at the time observed after the first decision in 1941 that the case had convicted "the AMA in the eyes of the people as being a predatory, antisocial monopoly." From that time, there began to be openings for physicians to practice in group cooperative arrangements offering prepaid medical care.[34] A significant accidental by-product of the case was that their being under indictment prevented the AMA leaders from getting control of the wartime biomedical research effort, which was thereby preserved for medical investigators more sensitive to science than to physician income and professional leverage.[35]

Hierarchical Regionalism

It was on a level deeper than the politics of organized medicine, and without depending directly on medical payment plans, that hierarchical regionalism, which was intrinsic to the basic structure of medical care, became dominant in the 1930s.[36] For generations, medical care had been widely dispersed and delivered by personal physicians, with some specialists in urban centers, as already described. Beginning in the 1920s the new pattern, based on hospitals, was coming into place. The first stop for the patient was the general practitioner. For a problem patient, the next stop was a specialist. But typically a specialist was attached to the local hospital. Patients with difficult conditions then tended to filter up to a biomedical research medical school, where the most expert specialists practiced, the highest unit in the hierarchy of medical care.[37]

Just before World War II, one eyewitness explained how the system worked and how the hospital system had become so central. In urban practice,

> each man has his own private office, though upon occasion certain of them band together and have communal offices. Some are grouped in a great so-called professional building. . . . There are also privately conducted clinical laboratories and X-ray offices. Most of the special work, though, is done actually in the hospitals, patients being sent there for special investigation.[38]

From one point of view, the hierarchical arrangements represented the way health care conformed to the bureaucratic or organizational society of the early twentieth century. And in health care, at least, the structure lasted until very late in the century.

The hierarchy also served to introduce medical advances, which presumably "trickled down" from the teaching hospital to ordinary practitioners, although this system was always imperfect.[39] In the report of the Committee on Costs of Medical Care, for example, it was clear that the aristocrats of specialized medicine held a low opinion of the ordinary practitioner of the 1920s. From the viewpoint of the elites, the usual doctor out in practice was resisting the exciting changes coming out of laboratory and clinical research. In a world in which the ideal of scientific progress galvanized medical leaders, those leaders deplored conservatively sticking to the old, customary routines in practice, not to mention failing to keep up with the latest science. As late as 1937 one North Carolina medical leader observed, "Seventy-five per cent of doctors are probably badly equipped to meet the demands of modern medical practice. Our laws have permitted the licensing of badly prepared physicians and quacks of every description."[40]

Experts in particular continued to target practitioners who claimed specialty status but in fact did not have either the training or the ability required of specialists. That the best medical care was that provided by the leading specialists was not only the view of medical leaders but a view reinforced for the public in novels, movies, and radio dramas. As late as 1939, however, the majority of medical students, by then required to serve a year as hospital interns, still went into general practice and did not, at least not immediately, seek residency training. Yet after the usual Johns Hopkins–model curriculum training and an often high-powered internship, the realities of general practice in the lowest part of the medical hierarchy often disappointed young practitioners.[41]

The element of regionalism in hierarchical regionalism operated through the hospitals, which continued increasingly to become a part of the experience of ordinary people. Within any region, hospitals, like practitioners, were arranged in a hierarchy. At the top were the great academic research centers, and at the bottom were the neighborhood and regional institutions to which patients were first admitted. In the United States, location also could imply a place in the social hierarchy. What could one expect from a hospital in a poor neighborhood or isolated farm community? Personnel in such institutions were expected to refer serious or challenging cases up the hierarchy, even if a patient had to travel a great distance to the state medical school.[42]

Hierarchical regionalism seemed to grow naturally because specialists more and more expanded their place in American medicine, particularly as advanced physiological and medical research continued to flourish. The connection between academic specialists and academic or elite hospitals was very close by midcentury. Medical students declined in importance in the operation of medical schools. No longer desperately recruited, they faced higher admission standards and had to compete for places in an entering class. By contrast, the academic medical specialists had more say in deciding the priorities of the growing medical centers. There

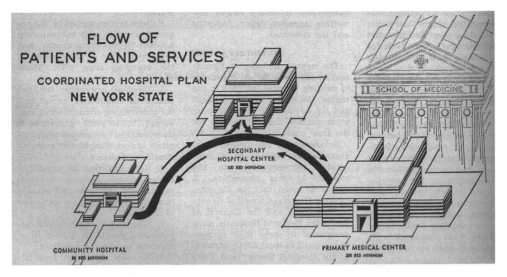

FLOW OF
PATIENTS AND SERVICES
COORDINATED HOSPITAL PLAN
NEW YORK STATE

SCHOOL OF MEDICINE

SECONDARY
HOSPITAL CENTER
100 BED MINIMUM

COMMUNITY HOSPITAL
50 BED MINIMUM

PRIMARY MEDICAL CENTER
200 BED MINIMUM

Hierarchical regionalism did not have a name until it was suggested in the literature of a later time, but administrative analysts even before midcentury were witnessing the pattern in the flow of patients up and down the hierarchy of hospitals in the United States, as shown in this diagram from a 1950 medical publication. Reprinted from John S. Bourke and Hildegard Wagner, "Regional Councils: Valuable Aids in Statewide Planning," *Hospitals*, Mar. 1950, 66, with the generous permission of Health Forum, Inc. Copyright 1950, by Health Forum, Inc.

research faculty could control hospital admissions and set up small empires focused on teaching and research in their own specialties. Moreover, by bringing in foundation and government grants and contracts, the academics had further leverage in hospitals and medical schools at the top of the hierarchy.[43]

Specialists and Specialty Boards

The fragmentation of medicine into specialties was therefore often driven from the top as well as by individual physicians who specialized to improve their prospects. Whereas in 1934 there were already 26,756 full-time specialists, by 1949 there were 62,688 (with only 110,441 general practitioners and part-time specialists). The bulk of the specialists remained in private practice, usually a practice more prestigious and lucrative than a general practice.[44]

Both practicing and academic specialists were, as their predecessors had been, deeply concerned that unqualified physicians were claiming specialist status. As noted above, state licensing laws were in place that restricted the practice of medicine to those who had graduated from a medical school and passed a national test. Yet there was no licensing of those claiming to be qualified specialists. In a major, sudden movement in the 1930s, not licensing but voluntary certification became available—through a still typically American solution to a social problem, the private, nonprofit organization.[45]

Specialists' first attempts to devise some badge of their superior qualifications consisted in making membership in specialist organizations a certification of competence. In a variation of this arrangement, in 1913 some leading surgeons had set up the American College of Surgeons, imitating the British royal colleges (organizations of colleagues, not educational institutions). In the American College of Surgeons, the professional leaders appointed one another to membership and set standards for others to be admitted. This did not work so well with internal medicine and other specialties, however. Meanwhile, in 1917 the ophthalmologists, who claimed special competence with special instruments, collaborated in setting up an examining board to certify specialists in ophthalmology (which was important for them because they felt acutely their competition with the non-MD optometrists). The professional organizations in otolaryngology, with a particular interest in tonsil and adenoid excision, followed in setting up a specialty certification board in 1924. Other specialists held back until in 1930 the American Board of Obstetrics and Gynecology was established, followed by a board in dermatology in 1932. By then the movement was well under way, to the point that existing boards as early as 1933 set up a coordinating advisory board. Within eight years, eleven more boards were approved, covering all of the major specialties in medicine.[46]

The pattern that quickly emerged was that to be board certified, the candidate had to complete an approved residency and pass an examination. The residency—specialty training in a hospital beyond the general internship—was already in place and controlled by the hospitals, but the number of residencies increased radically. In 1930 there were about 2,000 residencies in 338 hospitals; in 1940, 5,120 residencies in 587 hospitals. By 1948, under the stimulus of the GI Bill, there were 15,000 residencies. Like the interns, the residents were on duty twenty-four hours a day and received little more than board, room, and uniforms. For the hospitals, resident physicians added to the almost free labor provided by interns and nursing students. Few medical schools had offered more than occasional specialist training, and the board system bypassed the medical schools. As the academic specialists and the hospitals worked more closely together, it became possible to "standardize" specialist training, that is, to devise uniform requirements that became ever more stringent. A sign of the success and public nature of the voluntary board system was recognition by government agencies. From the beginning of World War II, for example, board-certified specialists in the armed services received higher rank and pay. Another sign of success was the decline in part-time specialists to an insignificant number by 1945.[47]

General Practitioners

Just at the end of World War II the success of specialization finally led to the beginnings of an uprising. As medical reformers tried to police the quality of medical care by restricting the ways private physicians used hospitals, the general practitioners, who were the main target of the reformers, rebelled. Perhaps it was inevitable that the upgrading of standards that came with specialist domination threatened the GPs. A major issue arose, for example, when a qualified physician who was not a board-certified specialist in surgery was not allowed to operate in a hospital: typically only ear-nose-throat specialists could remove tonsils or adenoids, and only obstetrical-gynecological specialists could perform some common procedures. The GPs, who were suddenly excluded from performing procedures that constituted a significant element in their practices, objected violently. To a large extent, they were losing the battle against specialization. But in one locality after

"It's all the swimming pool we can afford—Doctor doesn't specialize."

A comic, if bitter, depiction of how nonspecialists perceived the contrast between the financial rewards earned by a general practitioner and those earned by a specialist in the post–World War II world. The issues included envy and the pain caused by a changing world, but it was obvious even to outsiders how disproportionate the compensation levels in medicine were. *Medical Economics*, Nov. 1950, 107.

another, they organized, and a national organization, the American Academy of General Practice, came into existence in 1947, with important implications for the future.[48]

Before World War II, physicians who worked alone were on the whole conscientious, but they also were fee conscious and often relatively isolated in their everyday lives. Their social position could be a comfortable one, but many GPs were the objects of concern or contempt on the part of colleagues, especially colleagues higher in the professional hierarchy. One clinical pathologist reported, "Not long ago I was talking with a doctor who told me with pride how the best prescription for any kind of indigestion was one that his professors had given him thirty years before when he was an undergraduate. He is still using this shotgun remedy, making no attempt to recognize the particular nature of the patient's ailment and failing to keep up with progress in medicine."[49]

The issue that GPs could rally round was the doctor-patient relationship, for generations the traditional factor raised in denunciations of specialism. In 1946 the dean of a midwestern medical school underlined how the relationship persisted. "I am inclined to say that Dr. Smith is my doctor, emphasizing the possessive pronoun, but much less do we hear the expression 'Dr. Jones is my laryngologist or my surgeon.' We do say that 'Dr. Jones operated on me' or 'I go to Dr. Smith for my nose.'" Moreover, by the 1940s GPs could muster psychological and sociological evidence of the importance of the personal relationship with "my doctor."[50]

The Beginnings of the Antibiotics Revolution

It was into this system of general practitioners and specialists in the community, tied particularly through the hospitals into a hierarchical and regional system, that a technical transformation took place in medical care. Historians, reflecting the outlooks of later generations, have tended to play down the revolutionary impact of the coming of antibiotics. They forget that in the early to mid-1930s the major concern in everyday medicine was still uncontrolled infectious diseases. Some infections arose out of local injuries or surgery. Others, such as pneumonia, rheumatic fever, and syphilis, were systemic. In the 1930s one out of every hundred hospital admissions was to a special (usually state, and long-term) tuberculosis hospital.[51] Physicians often had no effective measures available to battle infections, which were too often fatal.

The first sign of dramatic change came in 1935, when a German pharmaceutical firm released news of a newly compounded substance, a dye derivative, that had startling success against streptococcal infections. This was the first of the sulfa drugs, with the trade name Prontosil. Reports reached the United States via medical journals and media reports in 1936, because, among other things, Prontosil worked in cases of childbed fever, which still haunted medicine. Between the 1930s and the 1950s just this one dreaded infection declined by 80 percent.[52]

It was another story, just at the end of 1936, that attracted medical and public attention in the United States, a story that takes on additional force when an apt contrast is introduced. In 1924 Calvin Coolidge was president of the United States. The younger of his two sons, Calvin Coolidge Jr., was a well-liked, handsome athlete and altogether talented and admired young man. One day Calvin Jr. developed a blister on his toe while playing tennis. The blister developed an infection—"blood poisoning," as it was commonly called then—that spread to his whole body. The best physicians could do nothing to save him, and he soon died. A dozen years later, at the end of 1936, another president, Franklin D. Roosevelt, also had a son, Franklin D. Roosevelt Jr. That son developed a deadly streptococcal infection in his throat. In his case physicians used the experimental drug sulfanilamide, a chemical modification of Prontosil. Franklin D. Roosevelt Jr. did not die but recovered nicely. The significance of the new drug was apparent to everyone, including newspaper reporters.[53]

In 1937 one pediatrician reported that in "cases of erysipelas the result was so prompt as to be positively shocking." Or, whereas earlier no one suffering from streptococcic meningitis had survived, now the cure rate was impressive.[54] As with diphtheria antitoxin and insulin, physicians had with sulfa drugs a new specific cure, one with which the physician actually produced immediate effects in the patient and astonishment in the doctor. It was 1939, however, before sulfas changed hospital routines, and that came with a sulfonamide that was effective against pneumonia.[55]

For some years, physicians considered the sulfa drugs to be just another type of antibacterial chemical agent. Many investigators had tried to find a specific chemical to target a particular pathogen, the "magic bullet" noted above. But either substances that targeted and killed bacteria in the lab killed the patient as well, or the patient's body rendered the substance ineffective. In one case, a chemical that successfully targeted pneumococci in humans also targeted the optic nerve and left many patients improved but blind. Altogether the record was discouraging. Therefore, the sulfa breakthrough, while encouraging, was at first very narrowly conceived by physicians. Only during World War II did the sulfas come to be part of the "miracle drug" phenomenon that dramatically and finally reversed basic therapeutic nihilism in medicine.[56]

It is true that Americans did not respond immediately to European developments. Not until 1937 did many of them begin to take up the sulfa innovations enthusiastically. In that year, a symposium in the *Journal of Pediatrics* began: "No therapeutic agent has appeared on the medical horizon for many years which has attracted so much attention and interest as sulfanilamide." At the same time, work in developing different sulfa drugs led U.S. pharmaceutical companies to significantly expand their research efforts, during World War II setting the pattern and laying the basis for big Pharma of the latter half of the twentieth century.[57] These efforts, commercial or academic, to find drugs were now increasingly often planned and rational rather than being based on a chance finding from screenings of many substances.

With the sulfas, investigators ultimately turned their attention not only to killing bacteria but also to substances that inhibited the ability of pathogens to grow in the body. Moreover, researchers and practitioners soon found that the sulfas were effective in diseases as diverse as gonorrhea and bacillary dysentery—two diseases that, along with pneumonia, happened to have a significant impact in keeping armies in the field. By the time of the Pearl Harbor disaster, in December 1941, physicians were routinely using sulfas in treating wounds and burns—as was just becoming standard practice in civilian settings as well.[58]

One of the great contributions of the sulfa drugs, therefore, was to reset the pattern of practice, so that when penicillin started to become available in 1943, physicians already knew how to use it. Indeed, by the middle of the 1941–45 war, strains of gonorrhea that did not respond to sulfa treatment alarmed military doctors, giving rise, along with other evidence, to specific recognition of drug-resistant pathogens. When small quantities of penicillin became available, physicians first used them to treat cases that did not respond to the sulfas.[59] A Harvard surgical consultant declared that "the sulfa drug era of wound surgery was . . . of relatively short duration" for two reasons: "(1) Bacteria rapidly develop their own 'immunity' to chemotherapeutic agents; (2) the discovery of penicillin and other antibiotics."[60]

Penicillin

The dramatic account of how British scientists developed penicillin, step by step, is one of the best-known stories in the history of science. From the beginning, American investigators assisted, and critical funding in the very first stage came from alert officials at the Rockefeller Foundation. After an experiment with just eight mice suggested that the substance had amazing anti-infective properties, the investigators attempted to treat a human patient in February 1941. Penicillin was a product of the growth of mold, and how to produce it in quantity was a major problem. Medical scientists tried to synthesize penicillin as they had vitamins, but they did not succeed until 1959. Under the pressure of war, however, scientists and U.S. Department of Agriculture experts developed practical production methods, so that immense facilities in the United States were soon devoted to growing incredible amounts of mold.[61]

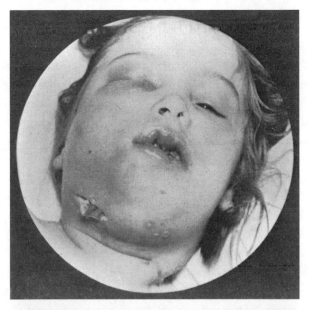

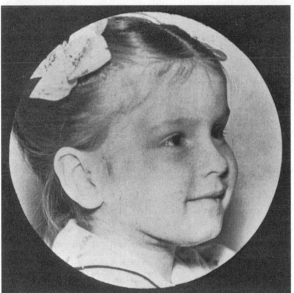

A young girl at the Mayo Clinic with a dangerous, uncontrolled infection that would likely spread to her brain. Her dire condition is apparent in the first photo. The second photo shows her remarkable and complete recovery after she became one of the first civilians in the country to be treated with penicillin. Such "miracles" powerfully impressed practitioners and everyone else and illustrate a basic change in medical practice in the 1940s. Reproduced from Lennard Bickel, *Rise up to Life: A Biography of Howard Walter Florey, Who Gave Penicillin to the World* (1972; reprint, New York: Charles Scribner's Sons, 1973), 226.

By the middle of 1943, enterprising American journalists had begun to flood the news media with stories titled "New Wonder Drug from Mold," "Magic Penicillin," and "Miracle from Mold." Armed forces medicine absorbed the entire output, however. In a handful of exceptional cases, an investigator would furnish a little penicillin for a dying baby or for an infection that was destroying a teenager. Such instances, along with tales about soldiers overseas, made irresistible human interest stories. The demand for the drug became acute and indeed overwhelming. Supplies eventually began to become available for civilian use, and in 1943 the government set up a rationing system. By 1945 production was becoming adequate. It had gone from 425 million units in 1943 to 646 billion in 1945 (and by 1951, more than 25 trillion units). The price of penicillin soon dropped precipitously. One administrator noted wryly at the time that it would cost less to cure a case of gonorrhea than it did to contract it.[62]

The world of medicine was never the same after the coming of penicillin, soon conflated first with the sulfas that it often replaced and then with later antibiotics (antibacterial agents "of microbial origin"). Between 1946 and 1958, worldwide from fifty to sixty new antibiotics were announced each year. Already by 1951, a chart for general practitioners provided a guide to which agent to use against which type of infectious organism, including intestinal worms, protozoa, bacteria, rickettsia, and the "largest viruses." The miracle agents now included not only penicillin and synthetic antimalarials like quinacrine but also the "mycins"—streptomycin, aureomycin, and terramycin—which had just been identified and brought into production, mostly in the United States. The abovenamed mycins were known by 1952 as "broad-spectrum drugs," because they were effective against such a wide variety of infections, a new phenomenon that made the news with chloromycetin in 1947. As one writer observed at the time, describing this "revolution" in medicine: "We have entered an 'antibiotic age.' "[63] The majority of the discoveries and development took place in the United States, even though it was hard to believe, as an observer at the time put it, that anything life saving, such as streptomycin, the first drug really effective against tuberculosis, came from the muds of New Jersey or that the best strain of penicillin came from a rotten melon in a store in Peoria. The fact remains that major diseases such as syphilis and tuberculosis, which had filled the pages of medical journals and had been the concern of specialists, now moved into the realm of routine treatment by ordinary general practitioners.[64]

Looking back, it is possible to see that by the 1920s and 1930s medical practice had come to include significant vaccines and serum treatments as a way of controlling major infectious diseases such as tuberculosis and particularly pneumonia. Some of this program was preventive and turned controversial as it came to include public health measures (which are almost never welcomed unanimously).

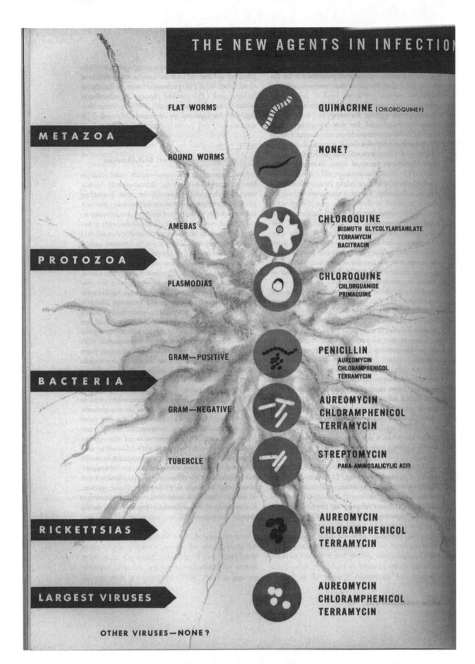

FLAT WORMS — QUINACRINE (CHLOROQUINE?)

METAZOA

ROUND WORMS — NONE?

AMEBAS — CHLOROQUINE
BISMUTH GLYCOLYLARSANILATE
TERRAMYCIN
BACITRACIN

PROTOZOA

PLASMODIAS — CHLOROQUINE
CHLORGUANIDE
PRIMAQUINE

GRAM—POSITIVE — PENICILLIN
AUREOMYCIN
CHLORAMPHENICOL
TERRAMYCIN

BACTERIA

GRAM—NEGATIVE — AUREOMYCIN
CHLORAMPHENICOL
TERRAMYCIN

TUBERCLE — STREPTOMYCIN
PARA-AMINOSALICYLIC ACID

RICKETTSIAS — AUREOMYCIN
CHLORAMPHENICOL
TERRAMYCIN

LARGEST VIRUSES — AUREOMYCIN
CHLORAMPHENICOL
TERRAMYCIN

OTHER VIRUSES—NONE?

This chart was an attempt to summarize the impact of antibiotics at midcentury. The infective agents are listed on the left, and the antibiotics, including particularly penicillin and the new "mycins," are listed on the right, next to the broad types of infection that each could counter. The chart suggests how widely effective antibiotics could be, including against some conditions only vaguely identified at the time. *GP*, Nov. 1951, 64.

Nevertheless, serum treatment, backed by pharmaceutical company advertising, was becoming standard in medicine. In the face of the miracle drugs, however, the whole serum strategy, based on a late version of germ theory and immunology, faded away.[65]

Indiscriminate use of wonder drugs and the development of resistant strains of bacteria, along with other problems, later tended to obscure the extraordinary change that antibiotics brought so suddenly to medicine. As early as 1950 a Philadelphia practitioner charted the amount of penicillin and streptomycin produced in just the few years from their availability to 1949 and found that "the phenomenal increase in the use of these drugs is out of all proportion to the known indicators for their use." He blamed media publicity for this indiscriminate use. Others have blamed physicians for overprescribing. One historian has labeled penicillin a tragedy as well as a triumph because while the model miracle drug led to extravagant expectations, the actual harvest was resistant strains of bacteria. In the backwash of discovery, however, it was still possible in pharmacology to observe that "potency is married to danger."[66]

The Decline of Infectious Diseases

Historians have found another approach that when taken out of context also unduly minimizes the impact of the wonder drugs. As noted earlier, scholars have shown that in the decades before 1940 there was already a great transition in the health of the overall population in the United States. It was easily summarized: "Infectious diseases declined precipitously as the major cause of mortality." Infectious diseases continued to claim enormous numbers of victims before World War II, but not at the rate that had been commonplace before about 1900. Infant mortality figures especially changed the overall statistics, decreasing from sometimes 40 percent to a small number, a decline caused chiefly by hygienic and preventive measures. But the rate of fatal infections went down for all age groups. Because of this shift and apparent success, later experts became more and more interested in the chronic diseases, which were constantly increasing as a proportion in physicians' practices even before the war.[67]

Years later, in 1980, McDermott, the Johns Hopkins antibiotic pioneer, realized that people were forgetting what medicine had really been like in the 1930s, before the antibiotic revolution. In spite of the falling death rate from infectious diseases at that time, in the crowded wards of a public hospital one could see a range of patients struck down by microbial diseases. Some beds contained patients with pneumonia. They had high temperatures and from moment to moment fought to be able to breathe. Treated with sera and oxygen tents, they often turned grey-blue as they expired. There were a number of young adults with congestive heart failure. There was the fourteen-year-old boy with rheumatic fever. Patients with pneumococcal meningitis were almost certainly doomed. In other beds were workers with wound infections and people of all ages with spontaneous abscesses and especially tuberculosis. In the average general hospital, microbial diseases furnished about half of the patients in the 1930s. Within a surprisingly few years, however, the hospital population had changed substantially, as the eyewitness McDermott recalled:

> Gone were the pinched-faced rheumatic children with their desolating gallantry; gone were the stoical men with bulging aneurysms pulsating like time-bombs right under the skin; gone were the vigorous young parents made bed-fast by . . . tuberculosis . . . ; gone were the young people with endocarditis living out doomed lives; and above all gone were those young or old with the hyperacute illness of pneumococcal pneumonia. . . . For some diseases such as rheumatic fever and the aortic aneurysm, the new technology was employed to prevent the actual occurrence of the disease; for others such as tuberculosis and some pneumonias, the technology changed treatment so that the patients were no longer admitted to the hospital. For still other diseases such as acute pneumococcal pneumonia early treatment stopped the disease in its tracks. The new technology thus drastically changed the pattern of disease and the illnesses as created by disease encountered in the hospital. Within its first decade of use, that part of the in-hospital disease pattern due to microbes . . . could be markedly suppressed and almost abolished.[68]

Or as Louis Weinstein, of Boston, recalled, in the infectious disease hospital the most common disease was scarlet fever—about twelve hundred cases a year. "The numbers of complications and the types of complications were fantastic": otitis media, various suppurative complications in the sinuses, the bone in the forehead, the nose, the skin, even vulvovaginitis in children. Children typically were in the hospital for almost six weeks. Then "penicillin changed all that . . . the streptococcus is gone in forty-eight to ninety-six hours, and you can send patients home." Soon many younger doctors had never seen a case of scarlet fever.[69]

An Age of Wonder Drugs

It was at that point that experts began to project the miracles into the future. In 1958, for example, an expert on tuberculosis, extending the trajectory of the disease, conjectured that "before the end of this century it may become so rare in the United States as to constitute a medical curiosity."[70] And seldom noticed was another impact: in such diseases as tuberculosis, miracle drugs dramatically but quietly at an early date reduced the huge inequality in treatment among "racial" and other social groups, where the cost of long-term care had been a factor.[71]

People who lived through those times and saw what the wonder drugs were doing had good reason to begin to expect miracles from medicine. Moreover, American culture had set the stage for those expectations. This was the period when science fiction became influential and popular, suggesting that futuristic technologies would transform everyday life and that applied science would solve all problems. The world's fair in New York in 1939 featured "The World of Tomorrow." By 1945 everyone knew about the actuality of what shortly before had seemed impossible: rockets and the atomic bomb—and penicillin.

The effects on individual physicians endowed with this new curative power were remarkable, just as happened when earlier practitioners, used to certain fatality, encountered diphtheria antitoxin and insulin. But perhaps more important were the changes in medical practice. Nowhere was the change more striking than in the practice of pediatric specialists. The statistics were startling. As early as 1945 one expert noted that since 1920, "deaths due to the communicable diseases of childhood have decreased 90 per cent, and infant mortality has been cut by nearly 60 per cent."[72] Both physicians and members of the public who implicitly witnessed

"It's the very latest wonder drug—has been for over forty-eight hours!"

A 1950 cartoon satirizing the drumbeat of new antibiotic and other medications that overwhelmed attempts to make sense of and track the discovery of "miracle drugs" at that time. *GP*, Oct. 1950, 89.

and sensed such changes, came, rightly or wrongly, to associate them with the miracle drugs. Older drug categories such as vitamins, along with new ones, got the label when investigators found an additional use for a substance, as happened, for example, with many endocrine products or the anticoagulant heparin. All became "miracle drugs."[73]

One other category of treatment that entered practice rapidly in the 1930s and 1940s was a technology that could at last affect some of the huge numbers of patients being cared for in mental hospitals (by 1936, mental patients occupied about half of the hospital beds in the United States). The new treatments were the "physiological therapies," shock treatments and psychosurgery. Many patients who suffered severe, disabling symptoms benefited remarkably from these drastic procedures. Like the antibiotics, physiological therapies were often overused and later got a bad reputation. The procedures were in fact effective in the cases of many patients who had been without hope of relief, especially deeply depressed patients. Electroshock in particular was widely used. These therapies coincided with the antibiotics in time and, like them, had remarkable effects in a number of individual cases and therefore added to the sense that additional wonder drugs had been discovered—even though people still first thought of magical injections and pills for infectious diseases as the wonder drugs.

Rheumatic Fever

Rheumatic fever, a major disease first fully described in the nineteenth century and now little remembered, illustrates how central the sulfas and penicillin became. The rheumatic fever patient, usually a child or young adult, would suffer from some combination of fever, atypical arthritis, involuntary movements, rash, and chest pain. Often the patient had a history of streptococcal tonsillitis. Death was frequent. On autopsy, damage to the heart was found, and in the 1930s, American investigators led the way in linking the heart damage to streptococcal infection. Rheumatic fever began to decrease dramatically, but in 1940 there were still forty thousand deaths per year in the United States. In addition, many victims were disabled permanently or for long periods of time because of damage to the heart. In 1930 it was estimated that 1 percent of the population had suffered some degree of damage from rheumatic fever. There was a resurgence of the disease in the crowded quarters of armed forces personnel during World War II, parallel to the high rates earlier found among poor people living in overcrowded conditions.

Investigators in the 1930s could not account for the cascade of symptoms that the streptococci set off until they finally decided that the original infection had triggered immune reactions. But since an infection was involved, physicians treated rheumatic fever first with sulfa drugs and then with penicillin. Penicillin would not cure the disease, but it was effective in preventing it from occurring or recurring. The effect of an antibiotic was important even if not in the usual role as a miracle curative agent. Physicians of course tried for the miracle effect before resigning themselves to a preventive role for a wonder drug. And even then, the fear that a sore throat would turn into a serious heart disease set the stage for what later appeared to be physicians' overuse of antibiotics. At the time, however, clinicians and particularly members of the public tended to attribute the decline of rheumatic fever to wonder drugs.[74]

Transformations in Medical Practice

Another curious effect of antibiotics appeared in pediatrics. There the dramatic caseload shift away from infections allowed physicians to turn their attention to other conditions that afflicted children. In a major movement, pediatricians turned to accidents, which, with the decline of infections, by 1946 had become the leading cause of death in children over the age of one year. Burns, falls, and poisonings, in addition to automobile accidents, caused enormous numbers of injuries and deaths. Pediatricians now joined with safety organizations to try to change behaviors in order to make children's lives safer.[75] Perhaps the greatest single direct accomplishment of the child specialists, once they were liberated from the extended care of infectious disease patients, however, was persuading pharmaceutical companies to use "child-proof" bottle caps on medicine containers—a voluntary innovation in the 1950s that immediately began to save untold thousands of lives of toddlers, who had been poisoning themselves with brightly colored or flavored medicines.[76]

Pediatricians were not the only specialists whose practices diminished or changed with antibiotics. "Entire categories of disease which used to keep doctors solvent year in and year out have been reduced to an easily managed and, from the M.D.'s angle, unprofitable estate," according to one report. Among the preantibiotic, steady-income items the authors of the report listed were venereal diseases and ear-nose-throat infections, with the previously ever-dependable tonsillectomies. Also on the list was another miracle drug category, the antihistamines, which were available without prescription for hay fever and other ailments and were "emptying many an allergist's waiting room to such an extent that he can now have what used to be an incredible experience for him: an extended summer or autumn vacation."[77]

The antibiotics left a major mark on other major streams in medicine as well, not least traditional surgery. Some surgeons found the wonder drugs a bit dizzying. "The goal of the control of all infections seems to come nearer with each six months," when a new antibiotic was introduced, reported a bemused Harry C.

Saltzstein, of Detroit, in 1950. Ordinary procedures in surgery did not change very much, except that antibiotics permitted more "open" surgery, in which the operator was on occasion freed from the earlier extreme concern with asepsis. "Methods for the technical performances of most operations seem to have become standardized," Saltzstein observed. "There do not seem to be very many more ways of doing the generally accepted routine operative procedures that are now in use," which would have included, as other witnesses ultimately confirmed, more endocrine organ and heart operations, along with increasing attention to cancer. The most remembered innovation came in 1945, when the Johns Hopkins University cardiologist Helen Taussig persuaded the new head of surgery, Alfred Blalock, to create a surgical shunt for infants born without an adequate circulatory connection from the heart to the lungs. This was the famous "blue baby" operation, which saved cyanotic children and incidentally called attention to the fact that a woman, Taussig, was having more than usual success in medicine.[78]

But now there was a new emphasis, Saltzstein continued, because "methods of 'making the patient safe for surgery,'" that is, in pre- and postoperative care, "are improving with kaleidoscopic rapidity." In other words, earlier concern about the physiological processes in the surgical patient intensified greatly in the 1930s and 1940s. Means of keeping the GI and urinary systems functioning, including simple diet and intravenous feeding, rose in priority. The major change, initiated particularly in treating the wounded in a military setting in World War II, was early ambulation. Instead of being allowed to rest and recover, patients were forced out of bed and bullied (or so it often appeared to the patients) into activity so that healing would proceed much faster. This new approach, Saltzstein pointed out, "has doubled the surgical bed capacity of our hospitals, because beds are turned over twice as fast as they were." As in pediatrics, then, in surgery too the antibiotics reduced physicians' preoccupation with infection, freeing them to pay attention to other matters, such as potassium balance in the surgical patient.[79]

Continuing Germ Theory, Physiology, and Science

In many such ways, the wonder drugs ramified into health care. In an area beyond immediate care, biomedical science took on characteristics and dimensions based directly on the experience with the sulfas and then penicillin. Team research and international cooperation became much more common. The antibiotics especially, as one witness reported, brought unprecedented teamwork: "Disciplines as far removed from each other as mycology and physical chemistry, clinical medicine and chemical engineering, have joined hands to attain the desired goal." Moreover, academic scientists worked in close collaboration with staff members of large pharmaceutical companies. And those great companies became major players not only in medicine but in American public life in general, particularly because of their economic power and advertising.[80]

Just when physicians were bringing bacterial diseases under control, the other kind of "germ," viruses, brought further excitement to biomedical scientists. Biomedical investigators in the 1920s and especially the 1930s had begun shifting to chemical approaches to understanding and fighting diseases. They brought new views of the immune system, with which human bodies fight diseases, moving beyond the phagocytes to serum antibodies in the body, such as those that triggered rheumatic fever.[81] But in the mid-1930s the shift to chemistry in studying viruses

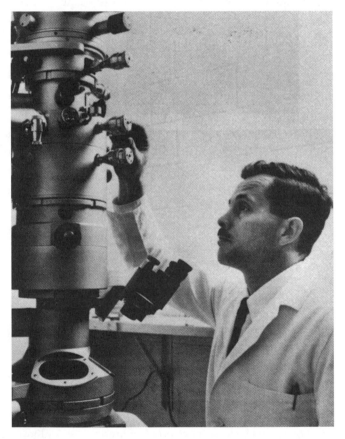

The electron microscope, an early version of which is shown in use here, became fundamental in biomedical research, permitting investigators to visualize even subcellular processes and structures and thus expediting the development of molecular biology and other fields of research. *Closing in on Cancer: Solving a 5000-Year-Old Mystery*, NIH Publication 87-2955 (Washington, DC: National Cancer Institute, 1987), 30.

became spectacular when in 1935 an American chemist, Wendell M. Stanley, succeeded in crystalizing a virus, finding "a giant protein molecule," in the words of a contemporary journalist. The finding made it clear that a virus needed a living cell in order to reproduce. The next step, taken within four years, was to use the new electron microscope to visualize a virus. By 1950, virology had become a separate discipline, and a new age of biochemistry had arrived in medical research.[82]

Virology started out as an international endeavor, with contributions from Britain, Germany, and now the United States. But by the late 1930s the configuration of American biomedicine was changing with the arrival in the United States of some of the most creative scientists in the world, investigators whom the Nazis had hounded out of Germany and other countries. Out of just the 7,622 professionals who came to the United States from Germany and Austria, 2,352 were medical professionals. Many of the refugees ended up in, at best, marginal positions in a relatively unwelcoming Depression-era United States. A significant group, however, many of whom had already been functioning under various handicaps in Europe, blossomed when they encountered the wealth of laboratory facilities and opportunities in the United States and were able to make great contributions themselves and to inspire American students and colleagues to do likewise. Elite American science, including medicine, was never the same after being enriched by knowledge and technique transferred from Europe with the refugees, who typically worked with more respect for theory and different styles of doing research than did many well-placed Americans.[83]

The gap between university- and laboratory-based biomedicine, on one side, and the practice of ordinary physicians, on the other, widened during the 1930s. The reformers trying to bring science into medicine continued their efforts to make recommended therapies dependent upon valid clinical testing rather than on just the judgment and experience of a physician, even an eminent specialist. Attempts to set up cooperative clinical tests of a defined medical treatment, for example, serum therapy for pneumonia, were, however, still only partially successful.[84]

In 1938 Congress passed a new regulatory law in the wake of the Elixir Sulfanilimide scandal, in which 107 people died after taking a commercial preparation of sulfanilimide, the effective but also dangerous sulfa mentioned above. The company chemist (who afterward committed suicide) had found a substance in which to suspend the new drug and so make it marketable in liquid form. Unfortunately, in ignorance he had used ethylene glycol (antifreeze), a terrible poison, and the manufacturer did not test the medicine (the manufacturer denied responsibility). The resulting publicity from numerous patient deaths, however, permitted Congress to enact a law stipulating that new drugs had to be tested for safety and labeled for effectiveness. The intent of the law was to curb abuses by patent medicine vendors, but the reformers suddenly had unexpected backing to encourage the testing of therapeutic effectiveness, because in many cases the Elixir Sulfanilimide had been prescribed by a physician. Both the public and health care workers needed protection from the economic calculations of commercial interests, and for once the government had acted. In addition, a second disaster occurred in 1941, when a superior sulfa, sulfathiazole, became contaminated in the production process. The deadly results motivated the rapidly expanding pharmaceutical manufacturers on their own to institute systematic safeguards, much improving manufacturing practices.[85]

Concern about the safety of medications was one thing; concern about the effectiveness of pharmaceutical preparations was another. The biomedical scientists' quest for systematic clinical testing of the efficacy of medications remained unfulfilled. In the World War II era, large studies for the armed services on penicillin and then streptomycin, using human populations from government groups, showed that organized research could produce better results using control groups under constant conditions and following other scientific standards. Specialist physicians had to curb their individual "clinical" judgments and variations in practice and defer to controlled clinical trials with many patients. And the reformers may have learned that one had to be especially careful when using human subjects not to vary from the prescribed protocols, which might impair fairness and violate the conditions the subjects had agreed to. Altogether the practical details of the randomized clinical trial were becoming clear to American researchers. The quantitative standards of judgment already present in physiological experiments now started to become more effectively a model for clinical investigations and drug testing.[86] Reformers of course had continuously to contend with commercial firms' resistance to most regulations.

World War II and Research

Not all of the effects of World War II were as favorable to biomedical science as were controlled clinical trials. The diversion of scientists to military service and research disrupted many lines of investigation. Some were never continued. Then after the war, investigators had to start anew and generate momentum again. They used what they knew of prewar and wartime research, but in a new context (see chapter 9).

Perhaps the area the war interrupted most dramatically was cancer research. For decades, cancer had been a growing health problem. Few ambitious researchers would touch the subject, because no one knew whether malignant growths constituted one disease or many different diseases. Moreover, existing research had to a surprising extent led to dead ends. In the United States, perhaps only a half-dozen investigators were working full time on cancer before World War II. Just as the war came on, however, some important work was beginning to appear. Most centrally, the consensus developed that any external cause of neoplasms (abnormal growths), such as poisonous tars, operated not through physiological systems, and not through endocrine changes, but through direct action of the agent on individual cells. As the Public Health Service surgeon general noted in his 1941–42 report, new research "demonstrates conclusively that the cancer process is started in individual cells and is not dependent upon constitutional conditions of the whole body."[87] Because investigators were called to work on critical wartime problems, however, follow-up to this line of thinking simply did not take place. When the scientists returned to their laboratories in the late 1940s, they did start to focus on the cellular level, but it was in a different world.

What did happen in World War II was that, as in the case of cancer, investigators put aside many problems and turned to applied biomedical research. Certain lines of investigation that had been developed here and there in the United States became coordinated and financed in unprecedented and unanticipated ways. Or wholly new lines, like tropical diseases, came in. Any illness or health concern in the military, from nutrition to surgical techniques and medication dosage levels, became an emergency. Hence it was possible to list among the projects of the federal Committee on Medical Research (who controlled wartime medical research) not only antibiotics and wound management but also nutrition, crash and decompression problems in airplanes, blood plasma management, protection against poison gas, and parasite control.[88]

Many of these military problems had much in common with civilian practice, and physicians who treated the ill and wounded in the armed services carried that experience back home. In 1943 almost half of the private practitioners under sixty-five were in the armed services. Fifty-eight percent of them reported in a poll that they preferred working in a group. They also carried back to civilian practice the experience of having their work evaluated by other members of the group in which they were working and of being organized by specialty, even if only in a tent in the desert.[89] Both doctors and the tens of millions of patients who served in the military came to expect hospitals and clinics. Basic institutions—elements in hierarchical regionalism—persisted after the war, but the world of health care was never the same. Physicians especially were aware of how military authorities could frustrate professional judgment. It was not just the outrageous segregation of donated blood so that "white" troops injured in battle and needing blood replacement were, for political reasons only, protected from being helped with "Negro blood." On the everyday working level, service physicians complained about "inefficient administrative control" of practice and talked about how good it would be to be back in civilian life. But they had also seen how necessary organization was to maintaining health and standards of medical care. Since almost all of the younger physicians had some service experience, the potential for the future was ambiguous, as they observed the advantages and disadvantages of organization.[90]

Wartime research did not bring great scientific breakthroughs, but usually just further development of earlier lines of work. Looking back on small improvements in surgical technology, one investigator noted, "As is often the case in any field of medical investigation, similar projects had already been undertaken independently. The emphasis had been on some particular phase of interest to the individual or group in the laboratory or clinic." Coordinating such independent research efforts for patriotic purposes brought a sense of immediate accomplishment.[91]

Multiple small improvements could add up, however, and people at the time were aware of a number of substantial advances. In addition to early ambulation of surgical patients, using blood plasma and blood substitutes (already known before the war) advanced treatment dramatically in many cases. Experts did introduce some major changes in handling specific diseases. Malaria control became more effective with chloroquine and other synthetics. An effective influenza vaccine was in use as early as 1943. And a new antilouse insecticide powder helped check typhus. Then DDT was introduced. In the hands of the American military, it famously cut short a typhus epidemic in occupied Naples in 1943–44 and went on to have profound public health consequences, not least, for example, controlling leishmaniasis, a parasitical tropical disease with which Americans had not previously been well acquainted.[92]

Many research efforts did not pay off in terms of treatment, for example, filariasis (infection by a largely tropical parasitical worm, causing elephantiasis and other syndromes). In one ironic case in 1942, civilian, not military, medicine made a contribution. A terrible fire caused 492 deaths at a crowded Boston night club, the Cocoanut Grove. The number of burn cases allowed local surgeons there to do a clinical comparison of different ways of treating burns. The results from this civilian tragedy were passed on to the military, greatly upgrading standard treatment for burns.[93]

One area of medicine in which clinical experience in the field carried over with major consequences to civilian health care was psychiatric treatment of battle fatigue (earlier usually called "shell shock"). Especially in the South Pacific, physicians trying to deal with neurotic disability among service personnel found that the dynamic psychological explanations of their psychoanalytic colleagues made the best sense and provided the basis for the most effective treatment. Psychoanalysis thus gained sudden, very strong acceptance in medicine and more

An army psychiatrist in the South Pacific Theater in World War II conducting group psychotherapy with psychiatrically disabled soldiers, at a time when psychological treatments were becoming widely used in medicine. Medical Department, United States Army, *Neuropsychiatry in World War II*, 2 vols. (Washington, DC: Office of the Surgeon General, Department of the Army, 1973), 2:670.

generally in the 1940s, changing the profile of psychiatric treatment and medical authority substantially.

Public Health

The war also continued the federal impact on public health. After the new funding to state and local public health bodies under the Social Security Act (1935), previous public health efforts had increased greatly, leading to an acute shortage of qualified personnel. In Missouri, for example, the number of people served by full-time public health workers had doubled already by 1937. In addition to local expansion, as defense facilities sprang up everywhere, the U.S. Public Health Service moved in to provide safe environments for troops and workers. For example, the Public Health Service cooperated with local agencies in antimalarial campaigns, largely eliminating that disease from the continental United States. In the name of defense, the Public Health Service even conducted inspections of food processing and other activities to which sanitary standards could apply. Altogether, the public health apparatus of the country was enormously increased in just the ten years 1935–45.[94]

The most spectacular changes in public health after 1935, transparently understood then as modernization, took place in geographical areas, particularly in the

South, where poverty and tradition had prevented the institution of modern sanitary measures, measures that included education, infrastructure, and vaccination programs. The surgeon general of the Public Health Service reported as early as 1938 that "a greater advance has been made in public health in the United States in the past 2 years than ever before within a comparable period." But the emphasis nationally continued to shift. Not only were earlier venereal disease control programs resurrected but, in a symbolic recognition of chronic diseases besides tuberculosis, the National Cancer Institute was founded in 1937 to conduct research into "the cause, the prevention, the diagnosis, and the treatment of cancer."[95] The appearance of the antibiotics, however, shifted the grounds for disease prevention, and after 1949 a whole new era opened.

Not least among the shifted circumstances were the growing health efforts of voluntary organizations such as the American Heart Association (founded 1924), which increasingly added research programs to their educational and patient support agendas. The national anticancer society had become very active in cancer education, which health leaders believed could get victims to see their doctors for early diagnosis, the only hope against the disease at that time. The anticancer propaganda, however, gave a major assist to legislators who created the federal National Cancer Institute.[96] Joining the single-disease groups was the March of Dimes, an antipolio organization that enjoyed spectacular success in the wake of media scares about polio and the general knowledge that a popular president, Franklin D. Roosevelt, was a polio survivor.

Still another kind of organization that began to have major social input in the 1930s was the consumer advocacy group. These groups, using laboratory testing (some carried out by the AMA), revealed the cupidity of commercial drug and cosmetic vendors. One of the founding documents was a book that characterized the hapless U.S. population as "100,000,000 guinea pigs," describing products that merchandisers were still foisting on an ignorant public, such as cadmium-plated milk cans or hair preparations that, when they were not ineffective, were poisonous. The black hair dye Inecto, for example, contained a chemical that in one documented case caused a woman to suffer serious symptoms for two years, including decreased sensation, dizziness, headaches, and impaired mobility.[97] This sort of exposé from consumer groups greatly assisted in the passage of the 1938 federal Food, Drug, and Cosmetics Act, which was designed to curb some of the most offensive labeling and safety violations.[98]

Changing Expectations from Medicine

Besides the work and influence of organizations, health care workers between the 1920s and the 1940s had to deal with a changing "public." The mass media became ever more important in shaping what patients expected from physicians and nurses and medicine in general. So well had the idea of scientific medicine and then the miracle drugs penetrated into media and popular discourse that, as noted above, health care became extremely desirable by the 1930s and 1940s, even though typical Americans still spent more on advertised health aids and self-dosing than on physicians. But now when physicians were called in, their pay tended to be in money, not barter goods.[99]

The prestige of physicians translated into two things. First, the patient's trust, which the anti–socialized medicine campaigns heavily reinforced, was mobilized by the fact that many people knew from personal observation that antibiotics and other technological devices, starting with the x-ray, had saved lives and prevented invalidism. The second result of physicians' personal prestige was freedom for doctors to function as they wished, that is, to exercise professional autonomy.[100]

Physicians in the 1930s and 1940s found, however, that increasingly often the social circumstances in which they worked actually restricted their authority. A significant number worked in industrial or business settings, where they had to be careful not to agitate for too many costly protective measures, which would impair their employers' profits.[101] Physicians practicing in clinics or hospitals encountered similar constraints. Repeatedly, hospital administrators had to discipline or coerce the medical staff while appearing to defer to professional judgment. In one case, a hospital switched from buying each surgeon his or her preferred kind of rubber glove to making economical mass purchases of one standard kind. This simple, small adjustment required much diplomatic maneuvering.[102]

More than ever, the popularizers of medical practice were selling a picture of a kindly physician who could work miracles with science. Particularly influential were the popular fiction authors of the 1930s and 1940s who wrote about doctors and nurses. Their stereotypes of kindly care took on unusual potency not only through the development of cheap, paperback books, but also through soap operas on the new popular medium, the radio, and the powerful impact of sound motion pictures. Hollywood's romance with doctors and nurses is well known. The romance was sharpened particularly by the remarkably successful Dr. Kildare movies, in which a brash, talented young Dr. Kildare played against the wise old Dr. Gillespie. Mary Ellen Avery, who later held a chair at Harvard, reminisced about her call to medicine: "I was about twelve when the Kildare movies appeared. . . . The medicine excited me. Kildare could cure almost any disease in the course of a two-hour movie." These fictional physician stereotypes, along with those of smart, caring nurses, reappeared far past the point of cliché in those years and after. Yet "the public" could not seem to get enough. Between 1937 and 1947, Metro-Goldwyn-Mayer produced fifteen Kildare movies.[103]

Even before television, the direct effects of popular media in raising expectations of results from medical care had a major impact on the health care system. In a book written for children in 1953, the young reader meets Dr. Cramer, a pediatrician. Dr. Cramer goes to the hospital, where one of his patients, Betty, has had her broken arm set in a cast, "and Betty feels much better." He goes to see Tommy "to be sure he gets the right care." Then he visits Joan in her home (house calls still were possible then) and examines her.

> "I'm going to give Joan some medicine, Mrs. Daniels," he says. "It will make her more comfortable. Here's the prescription."
>
> . . . Joan's mother is holding the prescription in her hand.
>
> "Are you sure Joan will be all right, doctor?" she asks.
>
> "Quite sure, Mrs. Daniels," says Dr. Cramer. . . . He knows that Joan's mother will take good care of her. . . . The doctor has other patients waiting for him. . . . Like Joan, they need the doctor. He will help them get well, too.[104]

Such an image of necessary and successful medical care was not limited to care for children in the years before and after World War II.

Nurses in Depression and War

Beyond family, the people who provided most of the day-to-day care of the ill were the nurses. In the 1920s and especially the 1930s and 1940s, as patients moved into the hospitals, so did nurses, as was noted earlier. At first, registered nurses did private duty in the hospital as private employees of the patients. Then they became employees of the hospital and a standard part of medical care, gradually working with ever more technology, including lab samples, x-rays, and injections. The circumstances and timing of the responsibilities varied with local conditions. It was mostly into the 1930s before a nurse was trusted to take a blood pressure reading, for example. Nurses also moved into industrial employment, and with the increasing technology of medicine, still more physicians employed nurses in their offices.[105]

By the time of the Great Depression, there was an oversupply of nurses, in large part because of the large number of two-year hospital nursing schools that continued to supply low-cost labor to the hospitals. Nursing leaders criticized the caliber of recruits, many of whom did not have a high school diploma—a factor that was becoming more important as nurses handled more of the technology of care, such as various intravenous diagnostic procedures. One leader in nursing complained about the high school dropouts, who "stayed out late at night and were slightly incorrigible. . . . These undereducated, unprepared women make trouble within the profession." Hospital administrators and physicians generally opposed efforts to upgrade training, which they argued was a mere economic tactic to decrease the supply of nurses and student nurses. Yet the movement to close schools in small hospitals, to increase academic content in training, and ultimately to make nursing a baccalaureate-degree program was at least launched in the 1930s, though it was not destined to set the standard for another half century.[106]

Then came World War II. Immediately the surplus of nurses was replaced by an acute shortage. Many programs were set up to increase the training of nurses, capped by the creation of the United States Cadet Nurse Corps, founded in 1943 to train both military and civilian nurses. At the same time, nurses' aides became common. At first these were patriotic volunteers. The position soon became a paid (badly paid) entry-level position for women moving into the job market. The aides moved the nurses up in the hospital hierarchy, for registered nurses' duties now tended to include select procedures and supervision and less emptying of bedpans. One aide recalled that with all her experience, she still did not get respect from doctors, from nurses, or even from nursing students: "I would very diplomatically have to direct them, although they resented the hell that I was both black and a nurse's aide. But I had to do it in such a way that they didn't feel that I was claiming to know more than they did."[107]

After the war, the shortage of nurses continued. Fewer graduated. Many followed the prescribed gender roles in society at the time: they married and left the job market. Moreover, in the postwar inflation, nurses' salaries did not keep up. The persistent sexism of the workplace continued. As one nurse observed, "I feel a little more understanding and a little more kindness and consideration from doctors, hospital executives, and also from private individuals would have prevented the present shortage of nurses."[108]

Chronic Disease versus Acute Disease

It is ironic that in the face of the antibiotic revolution, the American health care system remained structured for at least half a century more to care for acute, rather than chronic, diseases. Two major categories of infectious chronic diseases, tuberculosis and sexually transmitted diseases (STDs), did yield to antibiotic treatment, at least for the time being. The other major chronic disablers of the mid-1930s were accidents, cardiovascular-renal diseases, rheumatism and allied diseases, nervous and mental diseases, and cancers—plus, of course, polio, the aftereffects of which were worse than commonly realized. Late in the 1930s, a few experts did try to point

Nurses Are Needed Now!

FOR SERVICE IN THE

ARMY NURSE CORPS

IF YOU ARE A REGISTERED NURSE AND NOT YET 45 YEARS OF AGE
APPLY TO THE SURGEON GENERAL, UNITED STATES ARMY,
WASHINGTON 25, D. C., OR TO ANY RED CROSS PROCUREMENT OFFICE

This iconic World War II recruiting poster shows that the status of nurses in American medicine and society was substantial and might even be changing as society changed. It also flagged the shortage of nurses during wartime. Courtesy of the National Library of Medicine.

to the high rate of chronic diseases—affecting perhaps a sixth of the population—and the social consequences. One expert wrote in 1940:

> Rehabilitation of the chronic sick, and the prevention or postponement of disablement is to the public interest, for it prevents dependency of their families and the consequent tax on the public treasury. . . . Methods of diagnosis and treatment have made such progress in the past twenty-five years that many persons with chronic diseases may now be restored to comparative health, to an extent not thought possible in the past.[109]

World War II and the antibiotic revolution refocused attention on acute conditions. The decline in the threat of infectious diseases, however, meant that many chronic conditions remained major social and personal problems, particularly in a population in which, in the second half of the twentieth century, many more Americans were surviving and expecting to survive into old age and very old age.[110] A Gallup poll in 1949 showed that 88 percent of Americans expected to see a cure for cancer within fifty years. (Only 15 percent thought that a human would land on the moon in that period.)[111]

What Americans had developed by the 1930s was the hospital as the symbol and carrier of medical technology—the lab, the x-ray, the beginnings of air conditioning, even the familiar over-the-bed table—alongside legacies from tradition and the continuing streams of surgery, germ theory, and physiology. By the end of World War II, however, the antibiotic pill had become a symbolic element in medical technology. Regardless, in succeeding decades it was technology as such that dominated health care.

PART III

Medicine and Health
in an Age of Technology

Somewhere around 1950 the story of health care in the United States became much less about individuals and much more about groups, statistical units, systems, and general configurations in society. Even though change was taking place on a larger scale, it continued to be dramatic. Indeed, the large scale underlines the drama.

Moreover, the process of modernization intensified. The impact of technology, along with further organization and bureaucratization, dominated health care after World War II. For a while after the war, people in health services did not much notice the extent to which the world was in fact a different world. Familiar roles and institutions persisted. Yet health care was increasingly a mass phenomenon carried out by teams. Events became statistical changes, and individuals' roles became indistinct in the records of historical shifts. By the 1960s the United States, already flooded with material goods, had become a consumer, not a producer, society. Then slowly that society fragmented into many different elements, a process that tended to hide the fact that in the continuingly affluent, consumerist society were layers of deeply disconnected people who complicated generalizations.

Nevertheless, physicians and their institutions and paramedical personnel rose steadily in income and prestige. After World War II, increasing numbers of Americans made health care an important part of their lives and demanded more and more of it, especially as health insurance became commonplace after the war. As one historian has remarked, the benefits of medicine became so obvious that the major question became how to pay for it.[1] Meanwhile, medicine moved from a transforming, hopeful golden age in the 1950s to become a major political, economic, and social issue by the 1970s.

Formal health also became more a part of the culture. It was not just that on average 72 percent of the population visited a physician once a year in the 1950s, compared with 48 percent in 1931.[2] Mass media intruded markedly into life, often featuring the new "disease or cure of the month." As middle-class women became even more engaged in activities outside the home, including the workforce, hospitals and then nursing homes more often provided care for family members. What had been free care in the home (or care implicitly financed by earnings forgone, typically by women) now was a service for which one paid. As traditional religious institutions retreated, physicians and paramedicals furnished support during birth, death, and troubling times. Some commentators even suggested that the expanding domain of scientific medicine was a gauge of the extent to which Americans were transferring their concerns and preoccupations from their souls to their bodies. Inevitably, as society changed profoundly, so did health care.

Underlying what was happening in medicine, therefore, was a new outlook in American society as it moved from an individual and entrepreneurial to a technological mode of functioning. Earlier in the twentieth century, in society as well as in science, the model had been the lone professional or lone innovator/discoverer. Around midcentury, however, opinion leaders began to idealize technology and technological systems. Even research was conducted by teams. Indeed, it became Big Science, typically in a bureaucratic framework. In American society, if a problem of any kind arose, investigators and policymakers increasingly sought a technological solution even more single-mindedly than they had before. The term *technological fix* entered the language officially in 1966.[3] Even on the level of personality, in American culture the association between technology and medicine was marked. In the standard midcentury vocational interest test, the profile of doctors' personal interests was remarkably similar to that of engineers.

A pattern of seeking technological solutions for any problem was not new, but the extent to which it dominated and shaped society, and especially medicine, became overwhelming. It was this preponderance that marked a new era in medicine and in culture. Both industrial and academic research grew amazingly. Physicians in clinical settings very greatly increased their use of technology in the diagnosis and treatment of disease. They also used technology to further professional agendas.[4] In epidemiology, disease itself became, in important ways, a statistical or social entity rather than a condition in a particular body. Particularly with the growth of health insurance, more and more Americans fell into the insurance model of society, in which individuals at risk were treated as just parts of statistical aggregates—risk groups (a transformation explored further in part IV).[5]

Technology affected health care, therefore, as both a general reference point and an organizing principle and also in the form of innovations, innovations with which physicians and paramedical personnel had to work every day. People at the time were well aware that biomedical scientists, with their technologies, were moving beyond the wonder drugs. Accompanying the technologies was an aggressive style of medicine described by a shrewd observer as "science-based, disease-focused, technological, and interventionist."[6]

In addition, the amount of information available about all aspects of medicine and health increased by at least one order of magnitude. Whereas statistics and analysis concerning medicine and health had increased substantially before World War II, in the 1950s and 1960s whole categories of study appeared and expanded, most notably those in the areas of medical economics, medical sociology, and medical anthropology. Collection of statistics, often financed by the federal government, multiplied remarkably, and close study of those statistics produced much knowledge about how health care was functioning. This information was a sign that increasing governmental intervention in health care furthered organization. And in modernization, of course, organization could manifest itself as any private or public bureaucracies, standardizations, "systems," or regularized coordinations.[7]

Health policy was remarkably complex by the postwar period because so many private businesses and organizations were interacting singly and in groups with all levels of government. As in the model of hierarchical regionalism, various kinds of local government were furnishing or supervising health care activities, as did state governments and the federal government. Just to run a medical school, for example, involved fundamentally not only local, state, and federal governments but also myriad private groups—all operating simultaneously on many levels and in an often unstable network of interactions.[8]

Changes of an even more subtle nature were perhaps more important than merely reactive attempts at policy. Historians of medicine have underestimated the impact of the technology that became part of health care after World War II. It usually arrived incrementally, in small steps, so that the cumulative impact of machines and methods has been misread as politics and "discovery." And then in the 1970s and 1980s there was a new context for this technological enterprise. The culture produced a major shift whereby American opinion leaders started viewing humans, including their illnesses, as part of the environment and in dynamic interaction with the environment. Indeed, the idea of environment came to include not only natural surroundings but also social relations.

For decades, therefore, the informed public, including health workers, tended to reconceptualize disease as a complicated process with many interactions. And to top it all off, the immediate social context of health care was deeply affected by a cultural upheaval that began in the late 1960s and reverberated for generations afterward. If adjusting to technological innovations was unsettling, the generational upheaval of the environmental era was often distressing. With germ theory and antibiotics, many health policy leaders had hoped to eradicate diseases in the same way that smallpox had been eradicated in the United States by 1949. Preventive vaccines in particular seemed to constitute a realistic tool. But with environmental consciousness, the competing approach, using education and social means to prevent diseases, came to appear equally or more effective. The times were clearly changing, and health care with them. As chronic diseases became ever more dominant, medicine focused less on saving the lives of ill persons and more on preventing illness. "If this fails," wrote an observer as early as 1974, "the main concern becomes enabling the patient to live happily, productively, and comfortably in spite of somatic and/or psychic disorders. . . . Our challenge is not so much to prevent death *from* disease or disability as it is to preserve the quality of life *with* them."[9]

At the beginning of the 1990s, another scholar summed up the experience of a generation:

> The upheavals brought about by the embrace of the scientific enterprise and the proliferation of innovative technological and pharmacological tools radically altered both physicians' and the public's notions of responsibility and possibility in medicine—from care to cure to recreation—and that transformation happened in one lifetime. The *process of change* itself and, more important, the *ramifications of change* for the tasks of medicine and for American society were unplanned and, with few exceptions, unexamined. . . . Both the rapidity with which that transformation occurred and the fact that it was unpredictable, unanticipated, and not subject to scrutiny or evaluation have contributed to contemporary medicine's moral quandaries.[10]

Landmark Dates

1949	Cortisone publicized
	First molecular disease announced: sickle-cell anemia
1950	First statistics suggesting link between smoking and lung cancer
1951	Wide acceptance of fluoride as preventive of tooth cavities
1952	American College of Surgeons turns accrediting of hospitals over to Joint Commission on Accreditation of Hospitals
1954–55	First psychoactive tranquilizers available: chlorpromazine and Miltown
1955	Salk vaccine for polio succeeds
1959	Major public criticism of health care begins
1961–62	Kefauver drug hearings followed by legislation for screening drugs
1965	Medicare and Medicaid enacted
1968	Health care "crisis" perceived and named
1968–71	Beginning of substantial increase in proportion of female and minority medical students
1969	American Board of Family Practice approved
	Federal funding of health research reaches a high point
1970	L Dopa treatment for Parkinson's disease approved

	Legislation facilitates foreign medical graduates' immigration
	First Earth Day celebrated
1971	Smallpox vaccination ended because disease eradicated in the United States
	National Cancer Act (war on cancer) enacted
1972	Disability insurance federalized and made more medical
1974	Economic emergency shifts concern to health care cost containment
	Last attempt to enact compulsory health insurance
1976	Legionnaires disease stimulates new fear of infectious diseases
1981	Beginning of Reagan-era defunding and destruction of health efforts and programs
	First reports of AIDS epidemic appear
1983	Diagnostic related groups (DRGs) made basis of payment for medical care
1986	Federal law requires emergency treatment regardless of ability to pay
Late 1980s	Problem of cancer reconceptualized on the basis of basic science

CHAPTER 9

The Age of Technological Medicine, 1940s to 1960s

As waves of technological innovation and the technology itself swept into health care in the 1950s and 1960s, the roles of physicians and other health personnel began to change. Providing health care now demanded attention in the worlds of government and business. Above all, the scale of medical care and medical research made the whole activity look different. It was not just vaccines and statistics and molecular biology. Technology intensified earlier ideas and enabled a new understanding of body functioning in a transformed society.

The massive redistribution of income in the United States meant that large numbers of individual working people for the first time had a significant income. They learned as never before to buy on credit, and in general what had once been a producer society driven by the work ethic was transformed decisively, and on a massive scale, into a consumer society. Americans came to be identified primarily as consumers of particular goods and services and only secondarily as workers. In the 1950s, the makeup of the industrial economy itself shifted from manufacturing to more than half service industries.

A Changing Profile of Health Care

Among the new service industries was health care, which became more and more conspicuous as an economic as well as a consumer and lifestyle force. Between 1947 and 1967 the dollars that Americans spent on health care increased an average of 8.2 percent *each year* (as the total expenditures rose only 6.2 percent annually).[1] In 1948 the average American spent 4.3 percent of his or her income on medical care. In 1969 the figure was 7.2 percent. Health expenditures went from 4.6 percent of the national expenditures in 1949–50 to 6.7 percent in 1968–69. This was, as an observer at the time put it, a mass shifting of funds and people into the health sector.[2] By 1970 health care was the third largest "industry" in the United States.

In the mid-1960s one major national survey had produced some astonishing total figures about health and medicine in the United States. As usual, poor people suffered more illness and disability than did the more affluent. By the 1960s, urban populations were no longer more unhealthy than rural ones—in contrast to conditions a century earlier. In a single year,

> Americans suffered from more than 380 million acute illnesses serious enough to require medical attention or at least cause limitation of the patient's normal activities. . . . Collectively these acute conditions accounted for well over half of the bed-disability days that year and of the time lost from work. . . . 80 million [people] suffered from one chronic condition or more. Of these 80 million[,] 58 million experienced no resulting limitation of activities; four million were sufficiently disabled so that they could not work, keep house, or go to school; 12 million could engage in these "major activities" only to a restricted extent[,] and six million, although able to do their basic work, were limited in other activities.

These figures translated into ten days of "disability for each man, woman and child" and "844 million visits to physicians" each year.[3]

At the same time, in the postwar decades a number of talented and well-informed science writers were conveying to the educated and even to the general public some sense of what was coming out of biomedical sciences. By the 1950s the public acceptance of biomedical science and the popularization of "wonder" technologies had reached the point that mothers in a number of communities organized marches demanding gamma globulin for their children, which scientists at the time believed

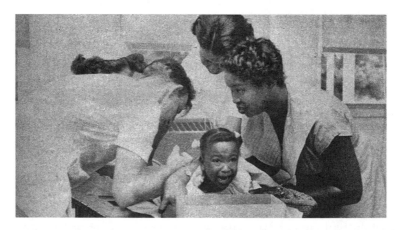

With no therapy effective against polio, parents turned to substances that might have protective properties and save their children from "infantile paralysis." Here a child is receiving a gamma globulin injection in the early 1950s, which presumably offered hope that the child would escape the disease or that the shot would "lessen the crippling effect of polio." "What Parents Should Know about Polio This Summer," *Look*, 2 June 1953, 49.

would protect youngsters from polio.[4] The contrast of these demonstrators, demanding scientifically justified medicine, with the urban riots against smallpox vaccinations only six decades earlier, also often led by women, could not be more stark or eloquent. A 1966 poll found that 84 percent of the population believed that the likelihood of their enjoying good health was higher than it would have been a century earlier—evidence confirming that personal experience, as well as propaganda, was causing most Americans to idealize innovation, not tradition, in medicine.[5]

Chronic Disease and Death

During the 1950s and 1960s, even as technology transformed many of the specifics of health care, the problem of chronic diseases and an aging population persisted. Between 1949 and 1969, the population increased by more than a third, from 150 million to 201 million. The numbers of old people and, for once, children increased at rates substantially higher than those for the total population. Life expectancy at the beginning of the century had been 49 years. By 1940 it was 63 years, and by 1970, 71 years. From 1935 to 1969 the infant mortality rate dropped from a scandalous 56 per 1,000 births to 21 per 1,000.[6] The annual total death rate per 1,000 people fell from 17.2 in 1900 to 9.9 in 1948 and 9.3 in 1954, but then it remained steady into the 1970s, largely owing to the number of people who were living into old age.[7] Overall, from 1950 to the 1980s the chief killers continued to be the familiar ones: heart disease, cancer, and stroke, plus accidents and pneumonia/influenza—not, except for pneumonia/influenza, the common acute infectious diseases of a former time. "What ever became of those quarantine signs?" asked a journalist in 1967, referring to public health measures to contain infectious diseases, signs rendered obsolete by "medical advances."[8]

One medical leader in 1952 noted how far medicine had come. In surgery he could mention "improved anesthesia, new technical skills and instruments, antibiotics and anticoagulants, improved knowledge of electrolyte and water balance," along with blood and fluids to control hemorrhage and shock. Surgeons now operated "upon the heart, blood vessels and lungs." But then he noted that "we still have not solved the problem of neoplastic [cancer] diseases and leukemia, influenza, the common cold, peptic ulcer, functional [mental] disease, poliomyelitis and virus infections, hypertension and the ever-increasing degenerative diseases."[9]

Subsequent to that 1952 summary, some major technical developments changed health care substantially. Despite the technology in medical care, for the rest of the twentieth century, health policy experts continued to be concerned primarily about chronic diseases: besides heart disease, cancer, and stroke, such crippling ailments as arthritis and asthma. Already from 1949 to 1956, chronic disease was the subject

> # QUARANTINE
> # DIPHTHERIA
> All persons are forbidden to enter or leave these
> premises without the permission of the HEALTH OFFICER
> under PENALTY OF THE LAW.
>
> This notice is posted in compliance with the SANITARY
> CODE OF CONNECTICUT and must not be removed without
> permission of the HEALTH OFFICER.
>
> Form D-1-D   _____Health Officer.

During the 1920s and 1930s, public health officials used signs such as this typical quarantine placard from Connecticut to mark homes and thereby reduce the person-to-person spread of many infectious diseases. Courtesy of the National Library of Medicine.

of a special commission. The Commission on Chronic Illness reviewed the major known chronic diseases, such as tuberculosis and epilepsy, and discussed the social issues and patient options. The commission ended by overwhelmingly recommending prevention, because cures were not in sight for most of the diseases, although antibiotics modified the makeup of the category, especially by diminishing tuberculosis, and the boundary with welfare disability continued to evaporate. A decade and a half later, another survey found little change. About half of the population "had one chronic condition or more." Twenty-two million people had to limit their activities, and 6.3 million had limited mobility. The sufferers included not just older people but young and middle-aged adults as well.[10]

In another category of concern were diseases that did not answer to miracle drugs, chiefly the virus diseases, most notably in the late 1940s and early 1950s the terrifying epidemics of polio. But now the list of concerns included the common cold as well as influenza. Moreover, the medical journals were filled with numerous articles about one baffling but very common serious illness: stomach ulcers.

At the beginning of the 1950s the press heralded another miracle drug, cortisone, along with the more general pituitary extract corticotropin, or ACTH. Such substances, according to promoters, would finally bring relief from the pain and crippling of arthritic diseases and other conditions—which, significantly, were conspicuous among the chronic diseases. Indeed, after the first announcements of cortisone, the demand was so great that a black market in the drug developed. Unfortunately, clinical trials ultimately showed that although the drugs were useful, they did not come near to fulfilling the promise of enthusiastic journalists.[11] Pharmaceutical firms also marketed the antihistamines, beginning particularly in the late 1940s. While antihistamines did not help most asthma patients, they did relieve the symptoms of another very common malady, hay fever.

Polio Vaccine

One new miracle drug of the 1950s, however, had spectacular effects: polio vaccine. In ten years, 1955–65, the incidence of polio in the United States dropped an incredible 97 percent. This saga of biomedicine is well known, and it thrilled Americans of the 1950s, as they not only learned about it but experienced receiving the vaccine and suddenly feeling securely protected against a terrifying disease that had been increasing in incidence and severity. As one prevaccine observer explained, "Of all the experiences the physician must undergo, none can be more distressing than to watch respiratory paralysis in a child with poliomyelitis—to watch him become more and more dyspneic [breathless], using with increasing vigor every accessory muscle of neck, shoulder and chin—silent, wasting no breath for speech, wide-eyed, and frightened, conscious almost to the last breath."[12]

The polio miracle occurred because, to begin with, there was an impressive virus research community among biomedical scientists.[13] And even at this late date there were special heroes. First, as noted in chapter 8, was Franklin D. Roosevelt, who was crippled by polio but rose to become president of the United States. The single-disease private organization that his admirers initiated, the National Foundation for Infantile Paralysis, started out in the late 1930s by buying iron lungs to keep polio victims alive and later financed research and clinical trials. The banks of huge iron lung machines providing breathing for paralyzed victims were an unforgettable symbol of polio before vaccines.[14]

Another hero was John Enders, of Boston Children's Hospital, who had already pioneered the antivirus vaccine used against influenza during World War II. Jonas Salk, of the University of Pittsburgh, and Albert Sabin, of the University of Cincinnati—both driven researchers from humble beginnings—became international figures. They made and promoted the practical vaccines that saved untold numbers of lives. There was even a heroine in the story, Elizabeth Kenny, a nurse who arrived from Australia in 1940 and courageously advocated active treatment rather than the bed rest that had been conventional for polio victims in the United States. For those with a taste for such things, all of these events also generated both scientific and personal controversy.[15]

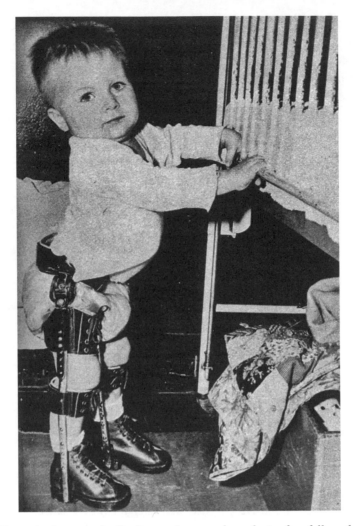

If a child survived an attack of polio, weakness and paralysis often followed, not only during childhood but for the rest of his or her life. Children wearing braces like this were a common sight before 1955. "What Parents Should Know about Polio This Summer," *Look*, 2 June 1953, 51.

The story became particularly dramatic in 1954–55. Leaders of the National Foundation for Infantile Paralysis bypassed the usual scientific and governmental steps and launched a controlled test of the Salk vaccine in 1,022,684 schoolchildren in forty-four states—all volunteered by their parents. During the next polio season the vaccine was very effective in two of the three strains included and had a marked effect on the third strain. When the results were announced in 1955, typi-

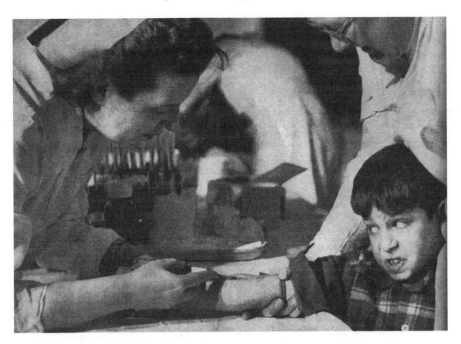

The big breakthrough, and a model for protection against all virus diseases, came with the Salk vaccine to prevent polio. In this possibly staged photo a child is none too happily receiving the miracle of protection against a much-feared disease.
Roland H. Berg, "The Beginning of the End of Polio," *Look*, 11 Jan. 1955, 30.

cal newspaper headlines were "Polio Conquered" and "Triumph Over Polio." Scientists learned much from the test, and members of the public, including children, learned the difference between a controlled and an uncontrolled study. In a notorious incident, the vaccine supplier Cutter Laboratories shipped vaccine that caused some fatal cases. Private firms and the government thereafter greatly increased their monitoring of how well manufacturers followed protocol in their production processes. In succeeding years, however, manufacturers began to find that vaccines of any kind, even promising vaccines, carried too much potential liability to be tested or produced, and so legal and financial considerations retarded work on disease prevention.[16]

Learning about Preventing Other Diseases

In the end, the polio vaccine altered most people's viewpoints, especially after the oral (Sabin) version became available in 1963. For scientists, the virus hunters, along with the epidemiology and mass clinical trials they employed, helped raise standards for biomedical research. For the public, the "conquest of polio" as a miracle drug story raised expectations from medical science ever higher. For medical and policy leaders, the success demonstrated that the way to attack virus diseases was to devise effective vaccines. Mass personal prevention by activating individual immune resistances offered hope where cures were elusive. As the journalist author of one of the early reports noted, "Techniques developed in production of the Salk polio vaccine now can be applied to help surmount previous stumbling blocks. . . . Man one day may be armed with vaccine shields against every infectious ill that besets him."[17]

In the 1960s, scientists in fact did devise vaccines to prevent three more major viral diseases. These were commonplace childhood acute contagious diseases, but they could have very serious consequences, even death or permanent disability for a percentage of victims. Measles had continued to be a dangerous illness (vaccine in 1963; from 1965 to 1968, a decrease in cases from 262,000 to 22,000); mumps, a source of danger and, for young men, notorious sterility (vaccine 1967–68); and rubella, or German measles, a terrible danger to women and their fetuses (vaccine 1969).[18]

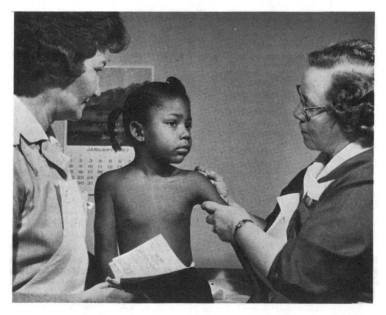

By the 1960s most children were accustomed to immunization as a normal part of life, and one could assume that not only individuals but the general population enjoyed immunity to many major infectious diseases when the rate of individual immunizations reached a high enough point. *Pennsylvania Department of Health Annual Report, 1967, 60.*

As if to confirm the strategy of using drugs to prevent disease, a major public health measure was introduced in 1945 and spread rapidly in the 1950s and 1960s: adding fluoride to water to prevent dental caries. One of the most remarkable demonstrations of professionalism was dentists' campaigning for fluoridation of public water supplies, knowing that their reparative work would diminish dramatically. Antifluoridationists appeared, however, and played on paranoid fears of mysterious effects of artificial fluoridation, pitting possible effects on old people against certainly protecting children's teeth. The antifluoridationists also invoked the prevailing Cold War anti-Communism against what they believed was insidious poisoning by "the government." Antigovernment dentists and physicians found themselves torn by ideology, science, and professionalism. By 1967, however, more than three thousand communities were fluoridating water for 60 million inhabitants. This ugly controversy continued for decades, but dentists turned more and more to orthodontics and periodontics rather than to cavity filling. They joined toothpaste makers (who added fluoride to their product) in equating healthy teeth with beautiful teeth.[19]

The other great disease that dominated biomedical journals and popular publications through the 1960s was cancer. Cancer was providing a different kind of model of a disease, specifically a chronic disease with multiple possible causes, in comparison with the classic germ theory model of one cause for one disease. As will be noted below, however, a new form of technology, statistics, would reconfigure ways of understanding this common menace. And still another technological approach revealed a new disease, one in which there were initially no symptoms but rather a laboratory-tested chemical imbalance, what became known as type II diabetes (see chapter 13).[20]

Meanwhile, the decline in infectious diseases and the growth in laboratory and technological medicine enabled experts to amass an enormous amount of knowledge about most diseases. Particularly important were tissue culturing and special media for isolating and studying agents, including viruses, which caused the infectious diseases, even Lassa fever in faraway Africa.

Signs of Transformation

Shifts in general patterns in both social strategy and scientific thinking affected the particulars of how health care came to function in society. Signs of transformation were everywhere. Many of the advertisements in medical journals around 1949–50 were still for baby foods and formulae and especially for laxatives, along with essentially proprietary preparations for discomfort such as aches and itches. Absorbine Jr. was prominent. Cigarette ads were featured alongside ads for medical equipment. But occasionally a medication such as the wonder drug terramycin was advertised with a specification in fine print of the chemical formula, along with references to relevant research in the medical literature and lists of possible side effects, a model that large ethical pharmaceutical companies—the now huge "Pharma" industry—later followed consistently (unlike proprietary medicine makers).

By contrast, in the late 1960s the medical journal ads were overwhelmingly technical and dominated by the Pharma model. Cigarette and patent medicine ads had disappeared, and clinical trial outcomes were everywhere in the advertising. Coca-Cola still ran innocuous announcements, and in *Texas Medicine*, for example, as a sign of physicians' changing socioeconomic status, there were ads for securities brokers and even an airplane charter service. But the tone of the journals, even in the ads, was remarkably professional and technological, not aimed at the friendly general practitioner with his favorite bowel and itch recommendations. This was, incidentally, one more indication that prescription drugs had become more profitable for Pharma than over-the-counter preparations.

By the 1950s both physicians and patients were aware of the transformative changes a cascade of specific chemicals, instruments, and machines was bringing to medicine. In 1941 a new standard textbook, *The Pharmacological Basis of Therapeutics*, had appeared. A second edition did not come out until 1954, and it took eight years to produce. "The task would have been much easier," commented one of the authors, "if medicine hadn't crammed a century of discovery and change into one narrow decade."[21]

The line was constantly shifting between innovators' dreams and the reality of practice, both of which were reported in the media. A patient never knew what new medication, machine technology, or classification would appear in the clinic or hospital. Many of the new techniques in fact did not work out, either because they were ineffective or because they had unacceptable side effects—notorious in high-pressure oxygen treatments or, for gastric ulcer, freezing the stomach.[22] Usually a new gadget or a pill would go through a cycle of enthusiasm, disenchantment, and then partial guarded final acceptance or rejection. In the 1950s, for example, from Europe came interferon, touted as a penicillin for viruses because it would interfere with virus growth. It did not work. Later, interferon was tried as an anticancer drug, attracting enormous foundation, scientific, and media support in the United States. Again, the numbers did not confirm significant effect.[23]

One of the most tragic instances of medical innovation was the simple procedure of using concentrated oxygen, pumped into an incubator, to increase a prematurely born infant's chances of survival. In 1942 investigators noticed that alarming numbers of premature infants—ultimately many thousands—were developing permanent blindness in the weeks after birth, but it took more than a dozen years to establish that the well-meaning technological innovation of oxygen enrichment was the cause.[24]

Physicians Trying to Cope with Change

Despite mechanical and chemical innovations, physicians in the 1950s were still talking about "the art of medicine," by which they meant wisdom and intuition that came from experience and tradition. A medical journal in 1949 offered an example of such common sense: if a female patient came in complaining of vomiting, the doctor, instead of immediately ordering an expensive gastrointestinal workup, should first ask when her last menstrual period had occurred.[25] Physicians generally held "science" to be a foundation for medical practice. Knowledge and technology aided clinicians, who used "clinical judgment" to apply the available science and devices to individual patients. By the 1960s and into the 1970s, however, the presence of science and technology had increased to the point that "scientific medicine" took on a new meaning: ideally the doctor would apply the scientific method to his or her practice, to make each patient diagnosed and treated the subject of a scientific experiment guided by standard rules, standard procedures, and available technologies.[26]

Yet as they worked with new technologies, physicians to an ever-increasing extent found themselves acting as coordinators rather than one-on-one healers. As described by a sociologist in 1960, the doctor

> is suffering from a drastic loss of function, . . . reduced to a kind of ceremonious middle man whose chief work is to mediate between the patient and an array of technical and specialized resources for both diagnosis and treatment. Thus, for instance, he may hospitalize his patient on the basis of findings announced by a laboratory technician and treat him by means of the brilliant pharmaceutical armamentarium. In all this, the physician's own role has become simply that of "gatekeeper" or communications switchboard, bringing therapeutic resources and patient together but having no immediate involvement with either.[27]

Repeatedly patients complained about the impersonality of technologically dominated medicine, yet physicians and paramedicals were also critical of impersonal procedures. A California physician as early as 1952 complained, "What with all our increased scientific knowledge, technical skills, and laboratory tests, we are in danger of becoming diagnostic and therapeutic mechanics." Indeed, there were claims that the number of laboratory tests per patient was doubling every five years. And that rate increased into the 1970s, even though the quality of perhaps a quarter of the tests was dubious.[28]

Walter Alvarez, a leading consultant and also popularizer, recounted, with his own biases, an illustrative incident:

> One day, I was called in consultation by a young doctor, a graduate of a good school, who said his patient, an old brewer with nephritis, was not responding well to diuretics. As we drove out to the patient's palatial home, I asked my young friend how he had made the diagnosis, and he said a laboratory girl had reported albumin in the urine as well as high blood urea. On walking into the room, I was shocked to see an old, fat, cyanotic, alcoholic man with pneumonia, stertorously breathing his last. I had to take my young friend out into the hall and tell him he had better start breaking the news to the family that the man would probably not last out the day. This may seem an extreme case, but I could tell of scores more like it in which an almost inexcusable mistake in diagnosis was due to the tendency to accept a laboratory diagnosis and not look at the patient as the old-time masters of clinical diagnosis used to have to do.[29]

The fact remains that physicians employed technology because technology worked, often spectacularly, not least in the form of the miracle drugs. Scholars are still sorting out the histories of the many technical innovations in diagnosis and treatment that were introduced in the two decades after World War II and were still spreading, from one medical site or practitioner to another, years later, often along with the paramedicals who operated or dispensed the technologies. The obvious technologies that patients observed involved treatment, either pharmaceutical or surgical, and diagnosis. Beyond those were other types of technology, notably at that time statistics and psychotherapy.

Pharmaceuticals

For pharmaceutical technology, the firms constituting big Pharma combined research, innovation, and, increasingly, marketing, using branded rather than generic products. Mostly well established by World War II, these firms continued what they had started in the antibiotic era but now pursued on a much larger scale. Between 1951 and 1961, 4,562 new "prescription products" came onto the market.[30] Sales figures increased enormously. Earlier, pharmaceutical manufacturers had sold drugs based on natural products, such as opium and quinine, or based on dyes (mostly of German origin), a category that included even Salvarsan and, later, sulfas. Especially in the physiological medicine era, pharmaceutical firms made great sums from vitamins and endocrine products. After World War II the industry was solidly established with the production of antibiotics, but the firms also introduced other substances, such as acetaminophen (Tylenol, 1951), that were useful, effective, and profitable.

TABLE 5
Leading drugs used in U.S. medicine, mid-1940s and mid-1960s

Mid-1940s	Mid-1960s
Penicillin, sulfonamides, other antibiotics	Anti-infective agents
Whole blood, plasma, blood derivatives	Tranquilizing agents
Quinine, quinacrine	Cardiovascular, diuretic agents
Ether, other anesthetics, morphine, cocaine, barbiturates	Steroids
Digitalis	Antidiabetic agents
Arsphenamines	Analgesics, anesthetics
Immunizing agents, specific antitoxins, vaccines	Antihistamines
Insulin, liver extract	Antianemics
Other hormones	Hormones
Vitamins	Vitamins
	Biological products

Source: Based on Chester F. Keefer, "The Contributions of the Pharmaceutical Industry," in *The Medicated Society*, ed. Samuel Proger (New York: Macmillan, 1968), 213–214.

Already in 1945, medicines were no longer conceptualized as individual substances the doctor might use but were classified into general categories. Within two decades even the categories broadened. One leading clinician listed the categories of drugs most frequently used in the mid-1940s and the mid-1960s (see table 5). The disappearance of specific drugs from the list and the appearance of whole categories of ever more powerfully intrusive and effective pharmaceuticals in just twenty years is striking.

Drug Regulation

Beginning in 1938, federal law, by means of labeling, had begun to distinguish between medications that could be sold over the counter (directly to consumers) and those that should be dispensed only by prescription by a licensed medical professional, typically a pharmacist (chapter 8). The regulations were clarified in 1952, specifying that pharmaceutical manufacturers must identify "by prescription only" substances by including instructions for use by physicians (package inserts), not the public, for whom the druggist would provide instructions.[31] Of course patent medicines and quack medical devices continued to sell as before. One critic estimated the proprietary device and dose market at a billion dollars a year and suggested that exploiters, by diverting people from proper treatment, accounted for more human lives lost than did all violent crimes.[32]

In 1961 the story broke that thalidomide, a drug widely used by pregnant women in Europe, had caused a large number of babies to be born (or stillborn) without hands or feet. In the United States, a Food and Drug Administration examiner, Frances Kelsey, in 1960 had not been satisfied that the drug was safe and had repeatedly delayed its approval. Kelsey of course became a heroine, symbolizing federal protection of the public. Subsequent federal legislation in 1962 upgraded and increased surveillance and standards for new drug testing, standards that pharmaceutical firms often resisted but finally ended up following.[33]

The legislation, which followed some famous congressional hearings led by Senator Estes Kefauver, was also the result of work by reformers in academic medicine, who were notably alarmed by the now commonly noted overuse of antibiotics. These reformers were mobilized particularly by brand-name, broad-spectrum antibiotics that were very widely used in the 1950s. Particularly pernicious was the prophylactic administration of antibiotics. As one sarcastic commentator put it, could a person dying of heart failure or accident "be denied the antibiotics that may prevent a secondary pneumonia, and thus fail to assure him of an adequate antibiotic concentration in his tissues as he is laid to rest?" In later decades, the quest for rational prescribing ultimately led not only to proof-of-efficacy requirements and evidence-based medicine but to a continuous battle with Pharma over marketing strategies.[34]

Drug Discovery and Marketing

When after World War II Pharma executives saw that prescription drugs were more profitable than the old proprietary formulas, some of the firms that had been producing proprietary medications switched to new lines—most notoriously the makers of a very successful laxative, Carter's Little Liver Pills, who in 1955 reluctantly introduced the tranquilizer, Miltown, which to their astonishment made them enormous profits. Altogether, it became clear that newly discovered drugs could be very profitable indeed. The postwar industry thus became based, not on the prewar techniques of screening vast numbers of chemicals for therapeutic effects, but overwhelmingly on tinkering with a successful chemical to make a slight but patentable variation. Already in the 1950s, twice as many such "congener" drugs as really new substances were introduced. In 1950 Pharma took in $1 billion from sales of prescription drugs; in 1969, more than $4 billion.[35]

In addition to ever more numerous antibiotics, new drugs available to physicians for their patients included preparations to relieve or control chronic diseases. The substances included not only cortisone and antihistamines but also, in the 1960s, beta-blockers for circulatory diseases.[36] And as just noted, in the mid-1950s the mood-changing drugs, or tranquilizers, began to come on the market. All of these drugs for chronic illnesses had repeat customers, often over many years—unlike one-occasion antibiotic users—and so could be particularly profitable.

Two additional lines of drugs came on the market in the 1950s, with great profit for manufacturers and profound social effects. The first were the antipsychotic drugs, which in many patients could bring the symptoms of severe mental illness substantially under control. The changes were so dramatic that by the 1960s many enthusiasts and politicians, including President John F. Kennedy, mistakenly concluded that hospitals for the mentally ill no longer needed financial support.

The second type of drug was the birth control pill. "The Pill" made a major contribution to the lives of untold millions of women. Developed by American hormone experts, it became available by prescription from a physician, which dramatically brought family planning into the realm of medicine, although, of course, that process had already begun when many physicians began installing contraceptive diaphragms in the decades before 1960. Many people believed that the Pill was a factor in social changes symbolized by *Playboy*, but that magazine had been founded in 1953 and had already reached a circulation of over a million before the Food and Drug Administration approved the birth control pill in 1960.[37] Step by step, major pharmaceutical products, with physician intervention, were affecting the lifestyles of the physically well in the United States as even patent medicines never had. For many women, the personal control the Pill brought to their lives went far beyond mere lifestyle.

Commentators at the time and also later historians criticized the commercial and marketing strategies of pharmaceutical firms: copycat drugs, exploitive pricing, and particularly the marketing strategies, with advertising, marketing disguised as education, and not always entirely scrupulous "detail men," who went from doctor to doctor showing the advantages of a particular drug.[38] A 1956 British physician explained that, as he had observed in America,

> considerable enterprise is used by the drug manufacturers in bringing their products to the notice of physicians. The average doctor receives 2,500 pieces of advertising and samples each year. . . . The large drug companies send detail men, usually qualified pharmacists, to visit doctors' offices and encourage their prescribing. One firm has gone so far as sponsoring nation-wide close-circuit [*sic*] television programmes. . . . A national conference is shown simultaneously on screens . . . at selected halls and hotels, and all members of the profession are invited to attend.[39]

Marketing was implicit in all of the favorable publicity and journalism about antibiotics and other miracle drugs, a context that marketers encouraged.[40] Even a ho-hum new substance could be announced as a "miracle drug." By the 1960s another strategy of big Pharma had become commonplace, a strategy that expanded in succeeding decades: creating a perception of a disease not as yet defined but for which a new product had been designed.[41] Among the new commercially created diseases of the 1950s and 1960s that historians have identified was anxiety, a spiritual state that became a disease treatable with tranquilizers. Other syndromes were hypertension, created by Merck's Diuril at the end of the 1950s, and asymptomatic type II diabetes, created by Upjohn's Orinase in the same time period.[42]

By the 1960s, then, a familiar pattern of conscious technical innovation was in place. First one laboratory and then another announced a new drug or a synthetic version of a natural drug or a chemical variation of an older drug. It did not matter whether the lab was commercial, typically in a big Pharma firm, or a nonprofit university, government, or foundation lab. Most likely, research from all such sources underlay any innovation. But now a new pattern was emerging. Investigators were learning more not only about the physiology of disease processes but also about how each chemical affected those basic processes. Drug discovery became better targeted and less accidental.[43] One of the best examples was L-dopa, approved for Parkinson's disease in 1970. Unlike many other pharmaceuticals, this substance was based on fresh research and provided an unequivocally useful therapy.

Surgeons and Engineering the Body

The pills and injections from big Pharma were one thing. But the engineering dimension of medicine—devices for diagnosis and devices for surgical repair of the body—was another. When, as noted above, surgery had largely reached the limit of small improvements of standard surgeries, by the 1950s, technology permitted surgeons to expand what they could do to extend the functioning of a patient's body. This was the period described by one historian as "the heyday of aggressive surgery in the United States. Surgeons who had served in World War II, performing risky and novel operations, had returned with a great confidence about what they could accomplish. . . . Surgeons removed the ribs and limbs of many patients with metastatic cancer, with the hope of eliminating all remaining cancer cells." This aggressive style, reminiscent of heroic therapy of another era, even showed up in the operating room, where prima donna surgeons "screamed at residents and nurses and flung instruments against the walls." And they mostly got away with it, for they were the only ones who had the skills that would provide hope, particularly in cases of cancer.[44]

Technology also made it possible to do more than simply restore functioning. By the early 1950s, writes another historian, "medical procedures now enabled individuals to transform their own bodies to be compatible with a private identity that had been hidden from public view. . . . People could change what they looked like and the stories their bodies told about their identities." The extreme, of course, was the publicity accorded in 1952 to an ex-soldier's "sex change operation."[45] It was difficult to draw a line between, on the one hand, restoring function and, on the

"There's nothing to it. I just had a little operation."

By the 1960s, the creator of this cartoon could expect a general popular audience to know about the many ways in which surgeons could alter a body to suit the owner of the body—and to laugh at this all-too-human desire to possess a socially ideal body. *Medical Economics*, 4 Nov. 1963, 108.

other hand, enhancing function and beautifying and gratifying personal ideals. Physicians could do it all.[46]

The engineering dimension opened up almost unlimited possibilities that further inspired both professionals and the public with ideas of more medical miracles. In 1955 an industrial designer working with the Veterans Administration devised an extremely successful stainless steel, split-hook prosthetic hand. Eventually there seemed to be no end to the prostheses and transplants that could keep parts of a person's body functioning.[47] Beginning in the early 1950s, for example, enormous numbers of people, particularly old people, who were losing their sight from cataracts benefited from implanted artificial lenses made of plastic, an innovation imported from Britain.[48]

One could often predict a new clinical machine ahead of time. What was the need for it? And how could it be engineered? Engineering innovations came in incremental steps. It was common to refer to generations of an instrument, as each few years a new, improved model appeared, whether it was an electron microscope, a blood chemistry analyzer, or a prosthesis.[49]

Technological Innovation

But now not just surgeons and other practitioners but also teams that included engineers and even inventors were competing to devise the next practical stage of development. Typically, a machine or technical improvement would be very large and very expensive, immediately reinforcing the move of the practice of medicine to the hospital, a process that intensified in the 1950s and 1960s. This migration was transparent, especially with big, clumsy devices such as a heart-lung machine, used to keep a patient alive during heart surgery and the heart immobile and accessible. In this case, after the first use on a human in 1953, significant improvements in keeping the blood balanced and oxygenated followed.[50] And when any such large new machine appeared, innovators all over the world competed to devise each small improvement and export it across national borders.

As early as 1954–55 one group of innovators had already founded the American Society for Artificial Internal Organs. The initial forty-seven members were mostly MDs doing pioneering work on artificial kidneys and hearts, although one strange man, who appeared at the first meeting and then disappeared, reported memorably on his idea for an artificial uterus! As a founder from Philadelphia recalled, at that meeting

> we were, after all, a motley group held together only by common curiosity about machines performing the function of various organs. Few of us were highly regarded in the most prestigious of research circles and some of us had to defend what we were doing among our colleagues back home. I think we realized for the first time in 1955 that we did have soul mates in an important field. I suspect none of the work reported then will win a Nobel Prize. Taken together, however, the papers suggested that a new body of unknowns would emerge which could not only withstand rigorous challenges, but might undergird therapeutic advances the like of which had not before been seen.[51]

So it was that clinicians, scientists, engineers, inventors, and entrepreneurs all rushed to find a new material or instrument or to improve one already in use. The fact that at about the same time that the first commercial jet airplane service began, in 1952, new materials were making hip and knee replacement more and more feasible suggests the excitement and hope of technology at the time. Heart surgeons and technicians introduced first one and then another mitral valve prosthesis or implant, along with artificial tubing to repair blood vessels. Then came the pacemakers, a whole story in itself as they evolved from external to implantable devices. By the late 1950s the most spectacular surgeries, exploring, clearing, and bypassing the blood flow to the heart, were well on the way to becoming clinically use-

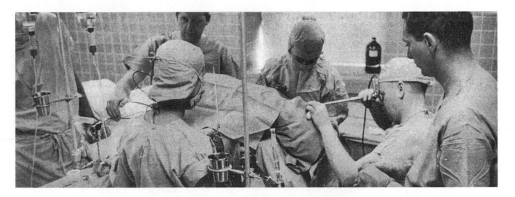

A patient at the National Heart Institute in Bethesda, Maryland, undergoing heart catheterization, a new procedure that typified the extraordinary ability and willingness of surgeons by the end of the 1950s to undertake extreme measures even with the living human heart in place. "Your Heart: Your Doctors Are Learning More about It All the Time," *Look*, 19 Feb. 1957, 64.

ful. Cochlear implants began to be used for loss of hearing. Meanwhile, the use of radiation to scan internal processes became more frequent, and not only radioactive tracers but also biochemical and electrical means of diagnosis increased. Altogether the technological innovations brought what one historian has labeled "The second revolution of surgical practice."[52]

The era of actual organ transplants began with first the kidney and then, in 1958, the heart. The idea of transplants originated early in the twentieth century with two procedures. The first was grafting skin from one part of a person's body to another. The other was blood transfusion, which was really a replacement of a body part. With new techniques for reducing immunological reactions or rejections, the road to transplantation and implantation was clear in the late 1950s. By 1965 a science writer could look backward, particularly at artificial heart parts, and forward: "Within approximately the past decade have come a remarkable series of developments. . . . Even more ambitious aids, or substitutes, for vital human organs and systems are being perfected in laboratories."[53]

More and More Technology in the 1960s

The pace of technological development quickened in the 1960s. Many of the innovations were highly technical, developed by complex collaborations between nonprofit and for-profit interdisciplinary teams. Diagnostic and imaging tools based on very advanced physics, such as the radioactive tracers, were being introduced into clinical practice. Lasers, which first appeared at the end of the

1950s, affected the practice of surgery with *photocoagulation* in ophthalmology and soon with other procedures. The usual incremental improvements continued, as in pacemakers, and, dramatically at the end of the 1960s, heart transplants. In particular, cytoscopes and gastroscopes enabled surgeons to look into almost any part of the body.[54] While it was not until the 1970s that the sensational scanning devices based on x-rays and sound combined with computers, as in CAT scans and magnetic resonance imaging (MRIs) (see chapter 11), in the 1960s the revolu-

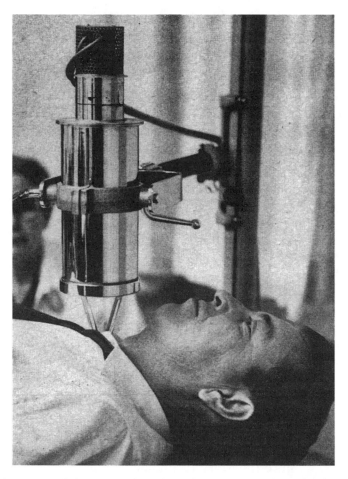

Using radioactive iodide to trace physiological processes was one of the amazing technological innovations of the 1950s. Physicians thus gained great understanding of what was happening in an individual's body to produce a pathological condition. This early shortcut to an exact diagnosis was transformative for clinicians. "In One out of Three Cases, Childless Couples Can Have Babies," *Look,* 17 Sept. 1957, 42.

tion in diagnostic imaging was already coming into sight as experts adapted the industrial use of ultrasound for medical uses.[55]

In 1965 the Association for the Advancement of Medical Instrumentation was founded to set standards for safety and manufacturing as more and more people encountered technology. Within two years, physicians gained control of the organization and began to prepare everyone for the governmental regulation they all believed was inevitable. That was a sure sign that medical technology was established, and indeed in 1967–68 the courts ruled that medical devices should be regulated as drugs. In 1976 Congress provided special legislation for regulation of medical devices. Some of the impetus came from a surgeon's claim in 1969 that twelve hundred patients died each year in American hospitals from accidental electrocution by medical instruments, a figure the reformer Ralph Nader later increased to five thousand. Neither figure proved to be accurate, but the alarm did have a basis in the widespread, unregulated use of technology in hospitals.[56]

Altogether, clinical and engineering innovation was intensive before 1970. After that, however, economic factors became dominant in determining how and when medical technologies reshaped health care.[57] But that change came with the beginning of another era.

One exceptional development illustrates both the problems and the possibilities of medical technology in the 1950s and 1960s: kidney dialysis machines. Even as kidney transplants for failed kidneys started to become practical in 1954, innovators were devising machines that would take blood from the body and filter it as a kidney would. In 1959 a University of Washington team devised a permanent shunt made from Teflon, a neutral plastic newly available, that opened into a patient's circulatory system, so that the patient could hook up to the machine easily and frequently without surgery. Thus those with end-state renal disease (inevitably fatal kidney failure) could be kept alive for an extended period of time—even years.[58] The problem was that the procedure was very expensive, and in the early days doctors had to decide which patients to save. Eventually Congress intervened and in 1972 placed all of the eligible patients under Medicare so that the federal government would pay to keep them alive. Experts figured that the total cost for the program could run to a billion dollars a year, a gigantic sum at that time and clearly unacceptable. Nevertheless, the program started out with a few thousand patients. But no politician moved either to impose rationing or to cut off funding, and fifteen years later the number of patients in the program was 147,000. The actual costs soon ran into a staggering several billions, more often than not unexpectedly covering dialysis for patients with other diseases, such as diabetes.[59]

In the postwar decades two other technologies, of totally different kinds, affected the nature of medicine. Neither was new, but in the 1950s and 1960s they moved from the margins, affected thinking about disease, and even modified the practice of medicine. The first was psychotherapy; the second, epidemiology based on statistics.

Psychotherapy and Psychosomatics

Those not trained in psychotherapy often have trouble understanding that high-caliber psychotherapy was, and is, a technology. Practitioners follow set procedures and rules. As with other medical procedures, the outcome is not always measurable or positive, but the input is still a carefully controlled technology. A highly technical psychoanalysis became the model for psychotherapy in the 1950s and 1960s among both physicians and, it turned out, the public. Like the x-ray, psychological probing could reveal hidden thought processes inside the patient that were preventing normal functioning.

The wartime impetus to psychoanalytic thinking in medicine, noted in chapter 8, led to familiarity on the popular as well as the professional level. Educated people knew that in the case of internal psychological conflicts, thinking could cause physical as well as behavioral symptoms. As early as 1947 a Chicago clinician commented on "the recent tremendous upswing of interest in the psychosomatic approach to an ever-widening range of medical problems." By 1955 a New Jersey internist was writing about the problem of some physicians' "resistance to the psychosomatic approach," a resistance often driven, even in the case of medical students, by preoccupation with exciting material medical technologies. In fact psychosomatic disturbances constituted a frequent topic in American medical journals, particularly pediatric journals.[60] Awareness of this kind made many physicians who were not interested in practicing psychotherapy of any kind smarter doctors, because they listened to patients better.[61]

But psychological medicine also changed how experts viewed disease and the patient. Hardheaded clinicians believed that a set of mental mechanisms, whether psychoanalytic or behavioristic, could cause physical disease symptoms. Very sophisticated neuroendocrinological researchers, building on the earlier work of Walter Cannon showing the physiological effects of emotion, connected psychological functioning with significant bodily changes. Psychosomatic (or in the words of one witty surgeon, "somapsychotic") illnesses therefore attracted both psychological and physiological investigators. Meanwhile, beginning in the 1950s and 1960s advocates of a new concept, stress, largely appropriated the popular idea of anxiety and tied mental events specifically to mechanisms of neuroendocrinology.[62]

"You're giving me an ulcer!"

A cartoon using a phrase common in the 1960s in a comic way. The phrase came from a widespread belief that many illnesses were psychosomatic in character, caused by stress and pressure. So familiar was this belief at that time that people employed it manipulatively as well as to help explain malfunction and pain with an uncertain cause. *Medical Economics*, 1 Apr. 1968, 117.

Epidemiology and Lung Cancer

The change that epidemiology, the other transforming technology, wrought was based on statistical technology. This technology was accompanied by considerable drama. Beginning in the 1930s, clinicians began to notice more cases of a rare condition, lung cancer. The most plausible explanation for this sudden increase was that diagnosis had become better and physicians were therefore noticing the disease more often. Not until 1949, when some Danish investigators reported their statistics at an international conference, did experts realize that they had an epidemic on their hands. The most probable cause of lung cancer was breathing fumes from road tar, a known carcinogen (cause of cancer) that had increased greatly with the automobile highway movement just as the epidemic materialized.

In 1950, however, several studies appeared virtually simultaneously in which the authors independently reported that deceased lung cancer victims had generally been smokers. This evidence was nevertheless dubious. It was based only on statistics, and those statistics were gathered after the fact. No laboratory tests could show a clear causative connection between tobacco and actual cancer, and so hard-headed scientists were skeptical. They wanted a demonstrable causal sequence such as one could find in the case of a germ using by now classic methodological postulates, or an industrial poison with clear chemical effects that could be confirmed on autopsy. Experienced physicians knew that because so many Americans were smokers, much more proof would be needed than a mere numerical association between smoking and lung cancer. Common sense too made a connection unlikely. Moralistic health advisers had been teaching for generations that smoking was harmful, but since social leaders, including physicians, overwhelmingly smoked and showed no obvious ill effects, who would believe the uplift hygienists? The moralists were, after all, the same people who had warned of the harmful effects of masturbation, alarmist warnings that turned out to be without foundation. Either mechanical laboratory demonstrations or personal experience, not statistics, would convince professionals or even the educated public that smoking caused cancer.[63]

TABLE 2.[1]—*Expected and observed deaths for smokers of cigarettes only and mortality ratios in seven prospective studies*

Underlying cause of death	Expected deaths	Observed deaths	Mortality ratio
Cancer of lung (162-3) [2]	170. 3	1, 833	10. 8
Bronchitis and emphysema (502, 521.1)	89. 5	546	6. 1
Cancer of larynx (161)	14. 0	75	5. 4
Oral cancer (140-8)	37. 0	152	4. 1
Cancer of esophagus (150)	33. 7	113	3. 4
Stomach and duodenal ulcers (540, 541)	105. 1	294	2. 8
Other circulatory diseases (451-68)	254. 0	649	2. 6
Cirrhosis of liver (581)	169.2	379	2. 2
Cancer of bladder (181)	111. 6	216	1. 9
Coronary artery disease (420)	6, 430. 7	11, 177	1. 7
Other heart diseases (421-2, 430-4)	526. 0	868	1. 7
Hypertensive heart (440-3)	409. 2	631	1. 5
General arteriosclerosis (450)	210. 7	310	1. 5
Cancer of kidney (180)	79. 0	120	1. 5
All causes [3]	15, 653. 9	23, 223	1. 68

[1] Abridged from Table 26, Chapter 8, Mortality.
[2] International Statistical Classification numbers in parentheses.
[3] Includes all other causes of death as well as those listed above.

In the summary of the famous 1964 Surgeon General's report, this table was critical. Not only was there an obvious statistical association of smoking with lung cancer but also, unexpectedly, an association with coronary heart disease. *Smoking and Health: Report of the Advisory Committee to the Surgeon General of the Public Health Service,* Public Health Service Publication No. 1103 (Washington, DC: U.S. Government Printing Office, 1964), 29.

Experts, however, wanted to confirm or, more likely, explode the idea of a connection between smoking and lung cancer. During the 1950s investigators followed very large samples of populations (prospective rather than retrospective studies). Researchers even identified and followed control populations, such as Seventh-Day Adventists, most of whom had never smoked. Despite methodological disagreements, the emerging statistics showed an impressive connection between smoking and lung cancer. Public health and often surprised medical leaders became alarmed. Eventually skepticism and enormous political and economic pressure from tobacco marketers and advertisers could no longer mute the warnings, and in 1964 the Surgeon General issued a landmark report confirming expert opinion that there was a probable connection between smoking and the incidence of lung cancer. Indeed, the experts accidentally and unexpectedly found figures that suggested that smoking might also cause other dangerous conditions, including heart disease.[64]

The Struggle over Statistics

Statistical epidemiology originated when a cause was sought for aggregated cases of parasitic diseases such as malaria. Yet by the 1950s infectious diseases were generally understood to have mechanical processes, and by then they could be controlled relatively well with antibiotics. Also under way was the epidemiological "transition" to chronic diseases, diseases that had complex causes acting over a period of time, not a single identifiable agent. Lung cancer caused experts and other well-informed people to think in terms of "risk groups," general population groups at risk of developing a disease.[65] Even decades later large parts of the population still could not convert to thinking in terms of mere probability rather than mechanical certainty, and they could not grasp the idea that there were group variations in health rather than just mechanisms operating on one's own personal health.

Biomedical scientists, however, had to think in terms of complex patterns of factors that caused illnesses. More and more frequently through the rest of the twentieth century, physicians worried less about single-factor cures and more about risk factors that might make their patients ill. Biomedical scientists were also finding that they had to have statistical backing to justify any therapy they recommended. Therapeutic effectiveness was no longer confirmed by the AMA Council on Pharmacy and other institutional authorities but by test methods, in the 1950s and 1960s by the randomized controlled trial (RCT), which was based on statistical techniques.[66]

The transfer of authority was not an easy one. In the therapeutic realm, the new statistical standards appeared clearly in the medical literature after World War II in tests for the effectiveness of streptomycin in treating tuberculosis. Even in evaluations of preventive medications, it took time for statistics to be employed. As one of the pioneers of pertussis (whooping cough) vaccine recalled, "In the 1940s statistical procedures were not in general use in the evaluation of biological products." Statistics came to play a bigger part, however, with testing of polio vaccines in the mid-1950s.[67]

For years after World War II, clinicians fought a "shadow war" against the statisticians, many of whom came from outside of medicine or who had learned statistical techniques in agriculture, biology, and psychology. But in 1962 Congress mandated "well controlled studies" to demonstrate the effectiveness of pharmaceutical products, and in 1970 the legal standard for any therapeutic procedure became explicitly the randomized controlled trial, in which experimenters included random subjects and asked about negative as well as positive outcomes and data, all in order to obtain "objective" results.[68] Between the construction of disease categories and determining therapeutic effectiveness, statistics, statisticians, and statistical thinking gained enormous authority within a short time. Statistics could not match the certainty of laboratory tests, but the two together superseded individual physicians' clinical judgment.[69]

Meanwhile, mathematical thinking was gaining importance through another technology, computers. Although, as noted above, the stunning instant analysis of computers did not come into its own in diagnostic devices like CT scanning and magnetic resonance imaging until the 1970s, already in the 1960s a growing, often dedicated group of medical professionals wanted to apply the power of computers to patient care and medical records to enhance the work of physicians. Many investigators also tried to figure out how to harness computers to analyze a patient's symptoms and come up with a diagnosis. By the late 1960s, computer enthusiasts were focusing more on improving the information and reasoning that went into medical decision making. They wanted, as one leader put it, "to quantify . . . the complex phenomena emerging out of the changeable nature and variability of disease processes and resulting from the interactions of man, disease agents, and his environment." As late as 1978 one analyst believed that "the computer's use can lead to revolutionary changes in the health-care system. . . . It may alter the role performance of physicians and nurses, perhaps even change the occupational hierarchy within the health professions. The computer can help with medical education and . . . change the nature of what physicians should learn."[70] The results, unfortunately, were not what innovators had hoped for. Instead, computer technology affected medicine in other ways.

As research accounted for a bigger share of the budgets and personnel of the institutions at the top of hierarchical regionalism, another technical procedure, peer review, very rapidly became a factor in medical communication and funding. For a long time, both private granting agencies and scientific journal editors had occasionally turned to experts in a field for advice about funding a grant or publishing an article. Peer reviewing to decide how to award research grants was actually stipulated in the law establishing the National Cancer Institute in 1937. But within a very few years after World War II, by an implicit consensus, peer review procedures became absolutely standard in medical publishing and research funding.

Peer reviewing to determine which articles should be published in medical journals unintentionally served as a transformative educational device for medical investigators. Not long after World War II, an author who submitted a paper that was not supported by controlled experiments and/or statistical tests soon learned from the reports of peer reviewers that regardless of his or her previous education, reputation, and practices, a new standard had to be met.[71] The language and general culture of biomedical science changed in less than a generation. Practitioners in the 1950s and 1960s learned that they were working in a world in which not only medicine but society, and science in that society, was producing new ways of viewing humans and their illnesses. In a survey of medical research at midcentury, before federal input became spectacular, the message was still clear: medicine depended upon "the broadening and deepening of fundamental research directed not only toward diagnosis and treatment of specific 'diseases' but also toward understanding of the total living organism."[72] What is surprising is that for two decades and more after 1950 even public policymakers understood that pure science, as evaluated by other investigators, was basic to medical innovation.

Federal Funding and Support for Research

More than ever after World War II, the policies of the federal government dominated life and institutions in the United States. There was no return to a simpler, more localized way of life. In particular, large parts of the health care workforce came to expect some sort of direct federal support. Leaders in all sciences conspicuously were determined to obtain federal support for applied and especially pure science such as the government had furnished during the 1941–45 war. The biomedical sciences got a head start with legislation in 1944, when Congress authorized the National Institute of Health of the Public Health Service (NIH) to continue and expand the wartime research projects of the Committee on Medical Research. The NIH was now also authorized to go beyond its own walls to fund research. Following the example of the National Cancer Institute (1937) and the Committee on Medical Research, the NIH began to fund research by qualified institutional investigators across the country. By 1951 the National Institutes of Health (now plural) comprised seven institutes and had a budget of $60 million, of which $16 million was for outside, or "extramural," grants. By 1960 the NIH budget was an astonishing $430 million, and in 1968 the figure reached $1.6 billion, a high point.[73]

Looking back, the Nobel Prize winner Robert Gallo recalled, "The incredible opportunity available to an eager, interested young man at the National Institutes of Health in the mid and late 1960s cannot be overestimated. Every chance for success was provided: stimulation, space, support, encouragement, equipment, and whatever else was needed. Anyone who was lucky enough to be part of this great place at that period was indeed fortunate."[74]

What happened was that members of Congress discovered that medicine and health was an extremely popular issue. The legislators had four obvious ways to fund health efforts: enacting compulsory health insurance; providing hospital and health facilities; funding education for health workers; and supporting biomedical research. Compulsory health insurance was, in the end, politically impossible. AMA leaders were also able to hold off assistance to medical schools, claiming that there was no shortage of doctors even after the drafting of physicians during the Korean War (1950–53) made it obvious that there was indeed a shortage of doctors. That left the two routes for funding that were politically feasible: hospitals and medical research.[75]

Hospital support came through the Hill-Burton Act of 1946. The act furnished funding for facility construction through each state, which had to conduct a hospital survey, submit a comprehensive plan, and furnish some matching funding. The goal was to provide facilities for underserved populations. In fact, the funding served further to establish the hospital as the central organizing unit in American health care, particularly in reinforcing regionalism by encouraging local hospitals. A similar federal program for health care facilities for veterans had the same effect. Within twenty years Hill-Burton funding alone totaled $2.6 billion spent on 8,400 projects.[76] It was not politically feasible at the time to use funding to bring racial integration to the medical care system (that began in 1964), and there were other imperfections and exclusions. In the end, however, the impact of Hill-Burton was enormous in confirming not just the hospital but the technology of the hospital as the center of the health care system.[77]

The Hill-Burton Act was designed to furnish health care indirectly. Support of biomedical research was even more indirect. There the model was "trickle-down," and it of course operated through the regional hierarchies. That is, the research in academic medical centers at the top of the hierarchies would show physicians how to improve medical care. The research would reach the most advanced specialists and hospitals directly. From there, the new knowledge would flow down the hierarchy as far and as fast as was appropriate. If the research was effective somewhere in American health care, that was justification enough. As Congressman Melvin

In addition to funding hospitals, the Hill-Burton Act of 1946 contributed to the construction of other health care facilities across the country, such as this building for the Baltimore League for Crippled Children and Adults. In many such indirect ways Hill-Burton provided federal funds to subsidize health care. U.S. Public Health Service, Division of Hospital and Medical Facilities, *2 Decades of Partnership: Hill-Burton*, Public Health Service Publication No. 930-F-9 (Washington, DC: Superintendent of Documents, 1966), 11.

Laird remarked in 1960, for the American people, "medical research is the best kind of health insurance." The trickle-down strategy also did not disturb other hierarchical arrangements then current in medicine and society in the United States.[78]

From at least 1950 until the late 1960s there was in place a consistent, remarkably generously financed U.S. health policy based on the trickle-down theory. As early as 1951 the federal government suddenly became the largest provider of direct support for medical research, accounting for 42 percent of all support, followed by industry, 33 percent; philanthropy, 25 percent; and hospitals and medical schools, 11 percent (the total is more than 100 because of joint sponsorships). Unlike most government policies, this policy was not defined and led by the executive branch; it was a congressional policy. Regardless of party, and despite opposition from the administration, in particular the Bureau of the Budget, Congress year after year increased the funds appropriated for biomedical research. Moreover, public opinion polls justified this congressional policy.[79] Congress funneled most of the funding through the NIH but sent significant amounts for research through the Department of Defense, the Veterans Administration, and the National Science Foundation, one official goal of which when Congress established it was to improve health.[80] The result of all of these initiatives was a biomedical research establishment that overawed even those who worked in it.

The Politics of Research

The congressional leaders who kept increasing the research appropriations commonly started out with two models in mind, both from World War II. One was the Manhattan Project, which, given unlimited funding, turned theory and experiment into the atomic bombs. The other was the penicillin story, in which, again, generous funding enabled the practical development of a lifesaving technology that reached millions of people in a remarkably short time. Beginning in the 1950s, year after year congressional leaders called in the top biomedical investigators to testify about what increased funding could do. Over time, many congressional leaders, particularly Senator Lister Hill (of Hill-Burton) and Congressman John Fogarty, became remarkably expert. One of the ideas they and others finally absorbed was that one starts out with pure research, without a specific application in mind. The best-informed scientists knew that much of the most important scientific innovation is unforeseen or an accidental by-product of other kinds of research. The result was that generous federal funding turned many creative scientists loose on general, rather than specific, problems. Remarkable investigations took place, far removed from any idea of medical advances, in the name, for example, of "cancer research." (What had the theory of genetic mechanisms to do with tumors of the pancreas?) A generation later, some spectacular medical applications came only because Congress earlier had provided money that might have appeared to be wasted on studying, for no particular reason, abstract biological processes.[81]

The emphasis on basic research was well suited to the quest to contain, if not conquer, the chronic diseases that leaders both within and outside medicine viewed as the chief health problems of the population. But in pursuing basic science the quest for medical understanding took investigators far from the clinic and pushed doctors to familiarize themselves not only with information but also with approaches that often seemed to them utterly irrelevant and foreign to the practice of medicine. There was evidence everywhere of a significant tilt toward research that was not about humans, diseases, or therapy but about fundamental life processes.[82] The new term *biomedical research*, instead of just *medical research*, was already in general use by 1960.

In addition to the congressional leaders, two pressure groups supported research. One was the growing number of scientists who were attracted to the research effort. They, together with the administrators of the universities, medical schools, and laboratories where they worked, formed an increasingly demanding and effective pressure group.[83]

The other group was a self-appointed health lobby. Although many able people operated on behalf of the health lobby, there was one charismatic leader, Mary Lasker. She held no office, but for a generation she managed the funding and diplomacy and leadership of the movement that created the great American biomedical research establishment. The wife of a leading advertising agency executive, Lasker had some background working with the American Heart Association. Others likewise came out of or worked through nonprofit health advocacy groups. Lasker and other leaders of the health lobby had friends and supporters in journalism and politics, and they knew how to direct funding to support their allies in

Mary Lasker at dinner with the elder stateswoman Eleanor Roosevelt and the well-known Hollywood movie director and producer Dore Shary, illustrating Lasker's comprehensive role as a major source of power and influence in public affairs and as a leader of health policy in the United States after World War II. This undated photo is from the 1950s. Mary Lasker Papers, courtesy of the Rare Book and Manuscript Library, Columbia University. Reprinted with the further permission of Darlene Studios, Inc., New York.

Congress—again without regard to party. Like Mary Lasker, movers in the health community had no commercial or personal motives and indeed were themselves philanthropic donors. Altogether, the health advocates were potent and surprisingly effective. As a Center (later Centers) for Disease Control official recalled, "If you wanted anything done, you touched four bases: the surgeon general, Senator Hill, Congressman Fogarty, and Mary Lasker."[84]

So vital to the health research effort were the lobbying groups that when they split up in the late 1960s amidst disagreements over strategy and the federal budget crises resulting from the Vietnam War, the appropriations for medical research stopped increasing, beginning in 1969–70.[85] The major divisive issue was practical results. The health lobby began to demand that there be some payoff from all of the funds spent, and members of Congress joined in a movement to divert funding from pure science to efforts to develop applications of all of the research that had been financed for two decades. Less sophisticated administrators and legislators, along with simple-minded journalists, wanted lists of dramatic scientific "breakthroughs" in the campaigns against disease. One camera crew even, at one point, entered a lab demanding that they be permitted to photograph the staff making a scientific breakthrough!

Clinicians and Scientists Facing Physiological Complexities

Practitioners in the postwar decades had to adapt to an enormous output from laboratories and workshops. The extent of that adaptation is suggested by an observation in the 1970s by the pediatrician Thomas E. Cone Jr., who was trying to describe how much medical science had changed in a quarter century: "In 1950, the thymus remained an enigmatic organ that was still being irradiated by some physicians for respiratory stridor [harsh breathing], gamma globulin determination was still in the realm of research, organ transplantation was science fiction, the function of the lymphocyte remained a mystery, and immunodeficiency diseases were yet to be discovered."[86]

Cone was particularly impressed by changes that had taken place in immunology, but he might have mentioned almost any other physiologically related area of medicine, because all changed in the 1950s and 1960s. It is easy to see a continuation of the growing complexity of physiology from the prewar decades, and change did have many aspects, but the most conspicuous was the pervasive role of technology, some of which was already present, though in rudimentary forms, prior to World War II. Investigators into cardiophysiology, for example, used technological means of visualizing processes in the circulatory system—but by the 1950s they also were on the road to devising defibrillators.

A very different impact of the continuation of postwar physiological medicine in the era of technology was an altered view of what went on in both the healthy and the diseased body. Particularly noticeable in biomedicine was a huge jump in the levels of complexity within which biomedical investigators worked. Continuing themes from 1930s physiology, the internist Stewart Wolf wrote in 1963:

> Traditionally, physicians have been inclined to consider diseases more localized than they really are, often on the spurious assumption that they involve only a single organ or organ system. Thus, pneumonia is thought of as a disease of the lungs although many of its manifestations—including fever, leucocytosis, increased antibody titers, tachycardia, etc.—are brought about by changes in organs distant from the lungs, such as the brain, the bone marrow, the lymph nodes, and the heart.[87]

With this complexity, it was possible finally to speak of a "revolution in pharmacology," in which investigators concentrated not so much on the cause of a disease as on the steps by which the disease process proceeded. The strategy was to find a chemical necessary to the disease process and to try to inactivate or disrupt the working of that chemical, often by substituting a closely related synthetic version that did not contribute to pathology. All of this involved minute, often intracellular events.[88]

New and improved instruments permitted biomedical scientists to investigate and measure ever smaller, more complex processes. For example, chemicals from both inside and outside the body were effective because they operated through specific receptor structures in the cells. The research on receptors had proceeded by slow increments until the 1950s. At that point, investigators, chiefly American but with much international interchange, employed a rush of new instruments that stimulated a whole new generation of conceptualizations. Within two decades this research profoundly affected both laboratory biomedicine and practical development of pharmaceutical products such as receptor subtype-specific drugs.[89]

Molecular Biology and Cybernetics

The conceptual changes that came from biomedical research after World War II stretched observers' imaginations. As a science popularizer wrote in 1963, "Great advances in medicine must come from detailed knowledge of myriads of molecular structures and functions. No revolution in man's history could be more portentous—penetration to the very core of life. We are in the seething midst of just such a revolution. It engages scientists of virtually all disciplines. . . . It has a name, 'molecular biology'—comprehension of life at the level of molecules and atoms."[90]

Molecular biology was more than the study of subcellular structure and function. It did indeed rest upon understanding chemical processes, but molecular biologists were not just "practicing biochemistry without a license," as some of their colleagues suggested at the time. Especially in the 1950s, the study of life processes went to such a basic level that the study of viruses and genetics at the molecular level joined with physiology and the newly emerging science of immunology. With the help of their instruments, starting with ultracentrifuges and electron microscopes, scientists were reducing life processes to chemistry and physics in often new ways. In particular, they identified nucleic acids as agents in the synthesis of proteins. On an almost unimaginably minute scale, the dynamic atomism in balance in homeostasis in the body, as understood earlier in the century, now appeared to work through units of information.[91] Nature operated as a feedback, or in the language of that day, cybernetic, system, even in the human body. When James Watson and

Francis Crick identified the structure of DNA in 1953, it meant that they had found a controlling, that is, informational, template for each live organism.

But in an age of molecular biology, or, more broadly, an age of cybernetics, there were still pathological processes to investigate, though now on the same infinitesimal level as other life processes. The lines between abstruse biology and medical application were hard to draw in laboratories defined by types and techniques of inquiry. In 1949 the colorful physical chemist Linus Pauling and his associates at the California Institute of Technology announced that they had identified a molecular disease, sickle-cell anemia. With newly available technical equipment to conduct electrophoresis, the Pauling group found that normal hemoglobin carried a distinctive electrical charge slightly different from that of sickle-cell patients. And indeed, further work showed that a gene chemically caused the pathological variation and the disease, which had confounded investigators since its discovery thirty years earlier. By 1956 Pauling was proclaiming, "I believe that chemistry can be applied effectively to medical problems, and that through this application we may look forward to significant progress in the field of medicine, as it is transformed from its present empirical form into the science of molecular medicine."[92] Pauling's statement expressed the optimism of many experts at that time.[93]

The signals that one complex chemical sent to another, with cascades of further stimulations, became the basis for a new level of understanding of the body that investigators were exploring in laboratories across the United States and elsewhere in the world in the 1960s. Standard pathological anatomy no longer offered satisfactory explanations of disease. Pathways of chemical effects, signals from enzymes, provided better explanations. Bacteriologists shifted from the study of conventional epidemiological evidence to phage typing to identify strains of bacteria.[94] Immunology and virology became independent disciplines even as their subject matter tended to converge with that of genetics and ultimately pharmacology. Studies of human cases of disease gave way substantially to animal experiments as scientists focused on basic biological processes common to all organisms. Pathologists, hoping that early detection might lead to treatment, tried to identify disease processes at the subcellular level before the organism/human showed any subjective or clinical symptoms.

The Biomedical Image of the Body and the Technological Future

The scientists in biomedicine were constructing a new view of human beings. The subcellular human was a collection of structures of interacting molecules, not just chemical reactions. A parallel structural model could be found in psychological functioning in the deep unconscious (even the public knew about Freudian psychological "mechanisms" such as conversion, rationalization, and projection). Reductionists often thought of the human as a holistic organism operating on the basis of extremely complex, interacting internal systems.

A second strain of thinking that came out of the 1950s and 1960s took the form of a vision of an extreme technological utopia. In 1956 a New York investigator concluded that the "logical experimental approach to the challenge of disease assures us that the Golden Age of Medicine which we now enjoy will extend far into the future." By the late 1960s major figures in medicine occasionally blurted out their

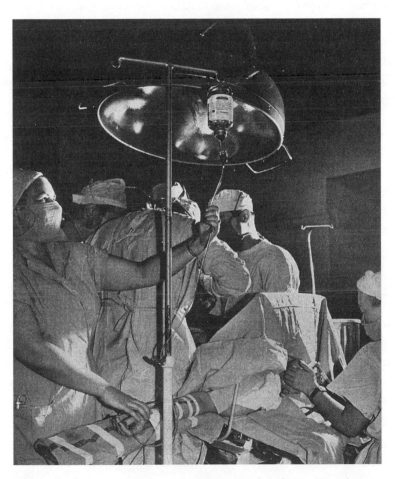

Giving plasma intravenously in the early 1950s, as in this photo, was only one way health care personnel could extend a patient's life by adding essential elements, including food and water, directly into the bloodstream. The bottle and tube to which the patient was attached became ubiquitous in hospitals. William H. White, "New Blood 'Substitutes' and Risks of Plasma," *Look*, 21 Apr. 1953, 105.

conclusion that at least ordinary infectious diseases were a thing of the past. Writing in 1968, one eminent physician cautiously observed, "The scourges of mankind—certain acute infectious and communicable diseases—that once ran rampant through entire populations had been slowed to a mere walk, if not a crawl, by last year."[95]

In fact, the antibiotics and other miracle drugs *were* working miracles anyone could see. People did not usually die of ordinary infections. Not only the tuberculosis hospitals but even the mental hospitals were being emptied and abandoned. Disability resulting from external physical injury (accidents) was being repaired or overcome. By the mid-1960s, patients in hospitals everywhere were aware that intravenous feeding and other life support via a tube could keep a person alive indefinitely.[96]

The accomplishments of biomedical scientists were impressive, and the promise for the future was even more impressive. A writer in *Newsweek* in 1959 could foresee within ten years transplants and replacements of body parts "fabricated from plastics and transistors." Others could predict, on the basis of what was already known, immunization, genetic engineering to control inherited diseases, and better detection of diseases.[97] Clearly some sort of medical utopia was in sight that went beyond the ordinary "progress of science and medicine" idea.

Flaws still existed in medicine. Health care workers of all kinds were aware of the threat from bacteria newly resistant to even the most recent antibiotics. Very often, patients did not take medications as directed after leaving the hospital.[98] Even when they did, the side effects of pharmaceutical and surgical technologies complicated everyday medical care. As one biomedical expert commented in 1953, "Serious reactions occur with all therapies—the safe as well as the hazardous, the useful as well as the useless, the old as well as the new, the folk remedy as well as the modern miracle drug," and he cited a report that in one large hospital, one of every twenty patients "was there because of adverse reaction to treatment."[99] In general, those who had glimpsed what the technological utopia in medicine could be were also leaders in identifying flaws in the actual performance of the health care system.[100] But the story goes much further. Health care and health care technology functioned in an often dysfunctional as well as rapidly changing society.

Doctors, Patients, Medical Institutions, and Society in the Age of Technological Medicine

The world of medicine and health in the 1950s and 1960s was difficult and challenging for the people who experienced it. The past, if only in attitude and approach, continued to have a presence in the streams of surgery, germ theory (not least in the form of virology), and certainly physiology. This chapter focuses on the social forms and circumstances of health care, which increasingly influenced what happened to medical staff, policymakers, and patients—the social structures and institutions within which the whole population operated, including health professionals, hospitals, public health, and economics. In this period, the economics of health care indeed became decisively a part of mainstream U.S. history.

The now familiar health care institutions persisted, and hierarchical regionalism helped tie them together. These continuing elements in medicine and health gave scholars a sense that health care institutions were permanent phenomena that could be studied and understood and implicitly conserved. And always, fee-for-service compensation continued to complicate every level of care.[1] For each aspect of health care, continuity from the 1930s and 1940s was a reality in the 1950s and 1960s. At the same time, technology, including medical technology, constantly affected social institutions and processes.

Multiple Realities among Americans

Ill and injured people continued to live in a United States filled with social problems and contradictions. Discrimination and racism pervaded the society, including medicine and health, despite the shift to consumerism and the first signs of desegregation. And then health experts began to talk in more specific terms than earlier about an overriding factor that was rarely discussed openly: social class. In a widely cited early study, the sociologist Earl Koos reported on a community study of health practices in the pseudonymous town of Regionville in the early 1950s. Koos was clearly shocked by the extent to which social class, as generally understood there, determined major differences in the ways people understood and used the health care system. He divided people roughly into three classes and stated that "in almost every example of opinions, attitudes, and behavior in health and illness, there appeared a significant difference . . . among the three social strata."[2] The groups' attitudes toward health care affected how they sought and used health resources, to the point that classes embodied distinct subcultures. When was a symptom severe enough to warrant seeking medical help? For fifteen symptoms that 51 percent to 90-plus percent of upper- or middle-class people would have considered serious, such as loss of appetite or persistent headaches or pain in the chest, only 20–30 percent on average of lower-class people would have considered medical care appropriate. As a local visiting nurse commented when shown the figures, "I would expect something like this. Poor people in our part of the state don't know much about sickness. . . . As for backache, tiredness, and stiffness, why should they worry about those? Heavens, they're second nature to poor people with bad diets, poor postnatal care, too much work, and all the rest."[3]

As in Regionville, throughout the nation people who worked in the health care system had to adapt to the range of people who sought help. For these diverse patients, in 1950 there were in the United States about 232,000 physicians and 375,000 nurses. By 1969 the figures were 325,000 and 700,000, respectively. While the number of physicians increased in those two decades by 40 percent, the number of physicians per 1,000 persons increased only from 1.49 to 1.60.[4] In the late 1960s the number of workers in the entire health care field was at least 3.7 million. Dentistry included about 237,000 (only 90,000 of them dentists), and there were numerous paramedicals, including 100,000 in radiological services, 130,000 in physical therapy, and 1.5 million in nursing services above and beyond those who were registered nurses. An office force of 250,000–275,000 provided direct support. And there were more than 19,000 nursing homes with 837,000 beds. One telling sign of the impact of technology and bureaucracy on this health care apparatus was that just between the years 1955 and 1965 the increase in the number of nurses and technical support personnel was almost four times the increase in the number of physicians.[5]

The Demand for Health Care Services

Between 1950 and 1970, while the consumer price index rose by 61.5 percent, the increase in health care costs was 124.7 percent, a figure that included physician fees, up 119.7 percent; dentist fees, up 86.7 percent; and hospital costs, up a staggering 398.1 percent.[6] High-level health professionals thus made substantial economic gains in twenty years, but behind the figures, and particularly behind the rise in hospital costs, were some other kinds of changes. The most obvious variable was the impact of technology on health care. And in fact, the major acute price increases were for "laboratory tests, x-ray studies, use of operating and delivery rooms and anesthetists' services."[7]

"Mr. Jones, you have tapeworms."

Even in 1950, physicians' use of technology could appear comically inappropriate. This cartoon reflects the artist's limited imagination when it came to machines to hook a patient up to; what was to come in the next few years might have astonished him. *GP*, Aug. 1950, 94.

Two economists in 1972 were baffled when they discovered that the actual use of medical services, which had increased after 1956, had declined in the years immediately before 1956. Then they considered the technological innovations suddenly apparent beginning in the late 1940s. The wonder drugs, particularly, had greatly increased physicians' efficiency, as they could clear up many serious conditions rapidly, and so the need for physician services had decreased into the 1950s. But medical advances from the late 1950s to the late 1960s included no such quick, powerful innovations. "Those advances that have occurred, such as renal dialysis, cancer chemotherapy, and open heart surgery," asserted the economists, "have typically been of a kind that make for only marginal improvements in general health indexes, despite occasionally dramatic effects in particular cases. . . . Since 1956, most medical advances have required substantial inputs of physician time for their implementation and have not had such pronounced effects on health." And so the success of some technology had led patients to demand additional technology, driving prices up and requiring more effort from physicians to achieve good results.[8] Although "the ratio of doctors to population has remained constant," complained one expert in 1966, "the rate at which people seek medical service has tripled in one generation."[9]

This simultaneous demand for technology and for more physician services set a pattern that would mark the rest of the century, as economists led others in showing that the popular belief that price would maintain a balance between supply and demand did not hold in medicine. Rather, the greater the supply of technological and personal health care that became available for people to use, the greater the demand. If more hospital beds were provided, the demand for hospital beds would increase. The price of health care services therefore went up, not down, when the supply increased, which was contrary to what one would have expected on the basis of popular, "commonsense" belief.[10] A similar lesson was learned when Britain instituted a universal health care system after World War II for its undersupplied population. Clearly, beginning in the late 1940s, if not earlier, more technology begat a demand for more technology.

Technology and Individual Physicians

The great increase in ever more expensive and complex medical technology had a subtle effect. Physicians could no longer carry their instruments with them in the

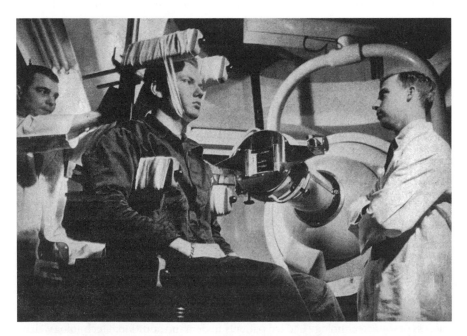

A patient at the Walter Reed Army Medical Center in Washington, DC, undergoing treatment from a million-volt x-ray therapy machine, as it was labeled in the mid-1950s. "Air Force Doctor," *Look*, 17 May 1955, 66.

emblematic little black bag. A doctor's private office might have an x-ray machine and a small lab, but the larger pieces of technology, such as many of those described in chapter 9, were located at, and belonged to, the hospital (or sometimes the clinic). Many physicians were aware, on some level, of this significant shift. Moreover, the work of physicians, not to mention paramedical personnel, increasingly became organized around, and dependent upon, one or more of the traditional medical institutions, typically a hospital and/or medical school.

One of the classic social distinctions of workers who do not own their own tools is that they are proletarians, not independent artisans. From one point of view, then, the professional physician, even one identified as a solo practitioner, was functioning often as a proletarian worker in a large bureaucratic setting. As the years passed, many adjusted. Yet doctors still aspired to have professional independence, or autonomy. In addition, as some historians have observed, large numbers of physicians, even in the age of technology, continued to think of themselves as independent entrepreneurs, in the old fee-for-service mode. Particularly in smaller urban centers, physicians identified heavily with local businesspeople. Doctors practicing in the 1950s and 1960s came overwhelmingly from white middle- or professional-class families, and they drifted easily into the entrepreneurial model. In one sociological study, a local leader was asked about important figures in the town: "I put Doc X on that list because . . . he is one of the best-educated men in town, and makes good money—drives a good car, belongs to Rotary, and so forth. . . . Of course, some doctors aren't as important as others, here or anywhere else, but unless they're drunks or drug addicts, they're just automatically pretty top rank in town." It was because of this social dynamic and "top-rank" self-image as well as social image that many physicians were sympathetic to doctrinaire campaigns against public nonvoluntary programs to promote the health of the mass of workers.[11]

Thus in the two decades after World War II, American physicians in general had to make enormous adjustments in their work. One way of adjusting to an ever more organized society was to practice in groups. In 1946 just over 3,000 physicians, or about 2.6 percent of all practicing physicians, were in private group practices. By 1969 there were 38,834 physicians, or 12.8 percent of all practicing physicians, in 6,152 groups, fierce AMA opposition to groups of any kind having become less effective. For some time, most of the groups were either single-specialty or multi-specialty groups. But physicians beyond the 12.8 percent also were involved increasingly in groups in clinics run by HMOs and nonprofit and governmental organizations of many kinds, including those concerned with veterans and public health agencies.[12]

Specialization

The other major way in which physicians adapted to the challenges of knowledge, technology, and society was to specialize. By 1965 only 37 percent of physicians classified themselves as general practitioners. Half of the GPs were over sixty-five. Only 15 percent of medical students planned to go into general practice, which was not surprising. Medical students were seeing much less of faculty members, who were too busy, becoming ever more obsessed with research projects. Instead, the students received most of their clinical training from residents who were narrowly focused on becoming specialists—or subspecialists, as was happening in many programs, such as pediatric endocrinology. In one medical school, a former student reported, "faculty openly told us 'no one from a decent school ever becomes a GP any more.'"[13]

In 1956 a small-town practitioner explained in a popular magazine why he was leaving general practice and taking up a specialty, radiology: He was seeing too many patients, sometimes forty to sixty a day, which meant that he was just giving first aid, and indeed he was in effect providing the services of a twenty-four-hour clinic, not available locally. He did not have time to keep up with "new medical discoveries." He also found himself neglecting family and community responsibilities. Although he had a good income, he had to struggle with charity cases and cases in which insurance covered the hospital and surgeon but not the GP. He believed that as a specialist he would be able "to be the good, steady, cheerful, competent and cautious doctor I want to be to all my patients."[14] In this one case, at least, what had been an ideal life for a physician fifty to one hundred years earlier obviously was not ideal in the 1950s.

This ex-GP clearly embraced the doctor's ideal of mastering and excelling in a narrow field as he had seen it exemplified in his mentors in medical school. Unsaid was the fact that specialists had almost always practiced in an urban setting. But in the 1950s and 1960s a specialist could also practice and live in the rapidly growing suburbs.[15] Specializing thus contributed doubly to the maldistribution of medical resources that afflicted medical care even more after World War II than before. For the rest of the twentieth century, both rural and inner-city areas were short of medical services.

The Board System and Additional Dimensions of Specialization

One of the continuing institutions that undergirded specialization was the board system. Although the boards offered only certification, not a license, specialty certification became de facto a license. Since the boards were the most recent of the distinctive, presumably permanent health care institutions, they were still expanding and developing after World War II. In 1947 there were fifteen boards, which certified 2,424 physicians as specialists that year. In 1969 there were twenty boards, certifying 6,296 physicians.[16]

Underlying all of these figures were a number of major developments. To begin with, the existing boards resisted the creation of any further specializations after thoracic surgery and preventive medicine in 1948. Instead, most boards ultimately created subspecialties with special qualifications, such as allergy, gastroenterology, urological surgery, and child psychiatry. There were already forty-five subspecialty certificates available from various boards in 1970.

The last major board to be added for a long time was the American Board of Family Practice, in 1969. Politically, this board accommodated the rapidly disappearing basic unit in American medicine, the GP. By 1950 the American Academy of General Practice already had ten thousand members and a journal, *GP*. In 1970 there were thirty-one thousand members, many of whom, as noted above, had had to fight board-recognized specialists in order to obtain hospital privileges in their own communities.[17]

But the new board was an astounding comment on medicine and on society. In a world of specialization, there had to be a new type of specialist, one who could specialize in generalization! Moreover, the generalist was reconceptualized as a functionary in the world of specialization, the *primary care specialist*—a term that first appeared in the literature about 1970—whom one visited first, before proceeding on to a recommended specialist (a secondary or tertiary care specialist). "Primary care" was the last building block in the hierarchical structure of medicine. Indeed, internal medicine specialists recognized the generalists' important social and economic role and tried to appropriate it (but as specialists, avoiding the idea of "general practice").[18]

The creation of the American Board of Family Practice and GPs' adoption of the name primary care physician had an additional significance. In the 1950s and 1960s the family GP could function at a high level. He or she could perform routine surgery and practice in the areas of obstetrics, gynecology, cardiology, pediatrics, and psychiatry, along with the rest of medicine. Supplementary training and literature allowed motivated family doctors to keep up with new developments and new technologies, while still practicing on a personal and family level in a community.[19]

But the new board "by the very act of rendering family practice a specialty was to assign it a domain in a self-limiting manner," one historian notes. "Once family practice became a specialty like other specialties, it necessarily fell back on a range of practices that defined the specialty and formed the basis of residency training." The primary care physician therefore depended upon surgeons and other specialists in the medical establishment and was no longer defined by the patient-doctor relationship. One simply could not keep up with everything new in the field of medicine, and so one's functions were limited and defined.[20] This major shift in American medicine occurred unevenly, but in a remarkably few years.

Residents Training for Specialization

By the end of the 1960s, hospital residencies were absorbing the general internship as the boards added an extra year to the number of years of training required for certification. Because so many teaching hospitals were closely tied to the increasingly research-oriented medical schools, physician trainees, as noted above, found that specialization set the pattern for all they knew in medicine. As another historian has concluded, "Medicine became a variety of skills, tied to definitions of 'specialties' which were developed in the 1920s and 1930s"[21]—and, she might have added, increasingly shaped by a variety of technologies.

In the decades after World War II, hospital residents in the various specialties, for a while along with interns, became even more essential to the delivery of medical services. They continued to provide fundamental, cheap labor ("quasi hospital servants," as one observer put it) to hospitals as hospitals became ever more central to the health care system, although residents began to be paid a living wage (their compensation tripled during the 1960s). In 1940, 6.6 percent of all physicians in the United States were in training. In 1949 the figure was 11.4 percent, and in 1969, a stunning 17 percent. The demand for interns and residents was so great that typically one-fourth to one-half of the vacancies went unfilled each year, especially as fewer and fewer trainees had support from the GI Bill.[22]

Soon the hospitals were recruiting "foreign medical graduates" (FMGs), graduates of foreign medical schools, particularly for vacancies in small hospitals or public mental hospitals, positions that were not considered desirable. A change in the immigration law in 1965 brought many physicians from countries such as India, the Philippines, and South Korea. In 1967 in 31 percent of U.S. hospitals FMGs made up three-quarters or more of the house staff. When a high percentage of these often valuable clinicians failed their board examinations (it was believed usually because of cultural and language differences, if not overwork in the hospitals where they were residents), major steps had to be taken to keep the labor force in the country. In 1966 the number of FMGs who entered the United States almost equaled the number of MDs produced by U.S. medical schools. In just the seven years from 1963 to 1970 the number of FMGs increased from 31,000 to 57,000—an addition of 26,000 physicians to meet the doctor shortage. Moreover, instead of returning to their home countries after training, a large percentage stayed in the United States to practice, creating an embarrassing, cruel "brain drain" away from developing countries.[23] Coming from all over the world, the FMGs often created and confused racist discriminations, as many provincial Americans had trouble comprehending that a person with dark skin and a heavy accent could possess technical and clinical expertise.

"Integrating" Medical Institutions

Preexisting and changing "racial" discriminations were particularly difficult for hospitals and for the health care professions in the two decades after World War II. People classified as African American constituted about 10 percent of the U.S. population, but only 2.2 percent of physicians were African American in 1950, a percentage that did not change much even as the number of doctors increased along with the population. In 1937–38 there had been only 350 African American medical students in training. By 1947–48 there were 588, and by 1955–56, 761.[24] Clearly the numbers were increasing significantly even if the proportion was not.

Any change was extremely local. Did a particular "white" medical school admit "Negroes"? Twenty-eight out of eighty did as early as 1956, and the number increased each year, but the schools had trouble getting students to apply. Did hospitals take African Americans as interns and residents or grant them hospital privileges? These were, again, local questions, often decided in practical terms on the basis of membership in local medical societies, which often excluded nonwhites. In some cases, segregated black hospital physicians did not want a change in their comfortable status quo, and integration threatened the very existence of some such color-line institutions, which were important elements in local communities. Although integration was progressing rapidly in schools and hospitals, for years the proportion of physicians and paramedicals who were African American continued to remain relatively low.

Local conditions varied greatly even in northern cities and rural southern communities. Each institution had its own particular set of steps required to bring African Americans aboard in any capacity. A Philadelphia hospital superintendent, for example, told of employing African Americans only as ordinary workers. Whereas before World War II all of the employees were "white," during the war the superintendent began to hire "Negro employees." "The white employees . . . threatened to strike if I did this," he reported. "I told them to go ahead and strike if they wanted to and said 'If you strike or quit, what I will probably do is put a Negro in your job.' There was absolutely no trouble." This same scenario was repeated in the 1950s and 1960s in many hospitals and with many levels of employees, including students, nurses, and physicians. After World War II, Veterans Administration hospitals desegregated, and in the 1960s court decisions gradually forced integration of both staff and patients even in southern hospitals. Since new technologies were being introduced at the same time as desegregation, people often confused the two. Which was more important, the adjustable hospital beds or who was in the bed next to one?[25]

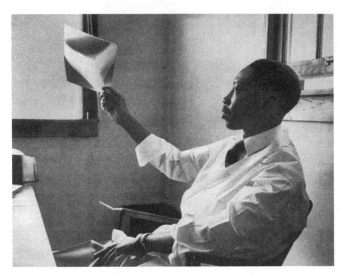

A program that operated without challenging racial segregation in 1953 nevertheless brought the general practitioner Dr. Rupert Searcy to an underserved rural area in the Mississippi delta, where he posed in his office looking at a patient x-ray. This program continued the earlier strategy of upgrading health care for African Americans by trying to provide more African American physicians. J. C. Furnas, "Mississippi Trains Its Own Country Doctors," *Look*, 17 Nov. 1953, 96.

From a High Point to Criticism of the Medical System

Even as vast amounts of money were pouring into health care in the 1950s and 1960s, the quality and experience of doctors still varied to an extreme, and that worried leaders in the profession. As a Florida pediatrician asserted in 1962, physicians should have the knowledge to act with "confidence and competence" and "use therapeutic tools with judgment, rather than to be a 'shot doctor' or a 'penicillin and prayer' doctor, whose public image undermines that of the entire medical profession."[26]

Overall in the 1940s and 1950s, however, the prestige of medicine and of physicians reached a kind of high point. In public opinion polls, physicians still ranked at the top of all professions, and it was not unusual to find that a physician's mere presence—charisma—was an important part of healing. Then sometime around 1959, beginning with complaints about the ways in which personnel and managers treated patients in many hospitals, the media of the day began to carry much criticism of physicians, along with criticism of hospitals and health care delivery in general. Such criticism of course reflected the fact that people both outside medicine and within had extremely high, no doubt unrealistic expectations of the health care system.[27]

Stories about the wonders of medical technology continued to raise patient and public expectations. As early as 1960 both rheumatic fever and polio seemed to be under control. Simultaneously, the incessant propaganda of the antigovernment medical establishment praising the individual, personal physician further raised expectations of the personal care every doctor should be able to offer—an unrealistic portrayal of consistent perfection among the 200,000 to 300,000 practicing physicians. And within medicine itself, reformers started calling attention to seriously defective patterns of organization and incompetent, ignorant, careless, impersonal individual actions that led to bad outcomes and unhappy patients. Each of the critics cited particular horror stories of malpractice and neglect, stories that investigative journalists began to detail for the general public.[28] As one journalist in 1964 summed up the complaints, "Three things worry Americans about their medical care: the poor quality of care too many of us receive, impersonal doctors and high, rapidly rising costs."[29]

Already in 1966, for example, Martin L. Gross, in a well-informed exposé, wrote about, as he titled his chapters, "Modern Medical Practice versus the Patient: Arrogance, Availability, and Assembly Line Medicine"; "The Doctor as a Scientist: The Fallacies of Doctor Erudition"; and "Life and Death in His Hospitals: Surgery Anesthesia, Obstetrics and Negligence," along with many other topics, including iatrogenic diseases (diseases caused by medical treatment) and the refusal of physicians to discipline incompetent or negligent colleagues. Gross's best informants about the worst problems in the health care system were conscientious physicians who found themselves unable to do much to change their colleagues or the system. In the 1950s and 1960s, journalists also wrote about physicians who had economic interests in drug stores, rest homes, medical laboratories, and physical therapy facilities, to all of which they might refer patients. Idealism and idealization had progressed far enough, however, that this profiteering was now considered scandalous.[30]

A continuing major problem was that so many physicians retained the mind-set of the independent, fee-collecting small businessperson. The mind-set was reinforced by a constant barrage of "pamphlets circulated by reactionary paranoids," as one practitioner put it. As late as 1961 a sociologist studying the experiences of patients in group practices characterized independent physicians as "entrepreneurial practitioners."[31]

Physicians' Attitudes as a Problem

Even without profiteering, most doctors in this postwar world of insurance and prosperity, as noted above, moved up economically, which in the United States usually meant a change in social role. Gross quotes a pathologist: "When I go past the hospital cafeteria at breakfast time, I see all the doctors with their papers open to the stock page. Perhaps they should be busily reading medical journals." One physician, noted the pathologist, knew the price-earnings ratio of hundreds of firms but had not mastered the technology of a machine he was using on babies: "The wrong pressure would blow the child's lungs out."[32] Clearly the prosperity of so many practitioners began to raise envy both inside and outside medicine. In the 1952 presidential campaign, a physician who was trying to organize his fellow doctors for an Eisenhower-Nixon, that is, Republican, parade notoriously advised his colleagues to use "a small car," rather than their usual big, expensive car, "to make the best impression on the general public."[33]

By the 1960s, patients with increasing frequency were actually voicing complaints about doctors, and also nurses and other paramedical personnel in hospitals and elsewhere, whose actions did not meet either realistic or unrealistic expectations. Physicians especially, even those who were not surgeons, could be unbelievably arrogant as well as rude and impersonal. In the earlier excitement about wonder drugs and the assertions in organized medicine propaganda that American medical care was perfect, the doctor-patient relationship somehow became less of a concern. When obvious trouble finally arose, the reaction of the leaders of organized medicine was direct but unwittingly ominous: they used a business model and tried to educate all physicians in the techniques of public relations.[34]

Any number of issues complicated the business model. Patients wanted personal attention and continuity, having his or her "own" physician. In the United States, however, the country "doc" of golden memory, who had had lots of time, had disappeared. Or was becoming extinct. A 1950 survey of twenty rural counties in Missouri found only one physician for every 1,760 people. The average physician in that group had received his MD degree forty or more years earlier, and one-third were already over sixty-five.[35] Clearly, even in rural areas the local family physician was fading out rapidly, and this in the face of nostalgic media images such as that carried in a 1961–65 television series featuring a kindly physician in private practice, Dr. Kildare.

One sign of change was the disappearance of the house call in most practices everywhere after World War II. Physicians had too many patients waiting in the office and the hospital to take time out for a home visit. As early as 1957–58 only 10 percent of physician encounters were house calls, and that proportion was dropping rapidly. Moreover, doctors believed they needed technology that was in the office, the hospital, and the clinic. As one physician recalled, "The house call became the hostage and ultimately the victim of high tech. You'd say to the patient, who would describe a high fever, coughing, headache, and something severe, 'Go to the emergency room.' There you had the equipment for a proper exam. That was the beginning of subverting the emergency room into an alternative, very costly, primary care setting." And in fact physicians tended to cluster around hospitals and to use the facilities there.[36]

By the 1960s, so pressing was the demand for medical attention that, as alluded to above, there was a doctor-shortage "crisis," long disavowed by leaders of organized medicine, who just ignored the evidence of the economists. In the 1950s, leaders of organized medicine cited as evidence that there were enough physicians the fact that people who could not pay were not crowding doctors' offices. That is, these physician spokespersons confused economic "demand" with what reformers believed was social neglect. Some leaders in medicine actually called for reducing the number of medical students to avoid an oversupply of physicians, a strategy quickly dubbed "professional birth control."[37]

But patients were aware of the critical shortage, particularly in the 1960s, because their own medical encounters, obtained only after a long wait, would be hurried and short—a horrible blow to the self-indulgent narcissism usually so transparent in a patient consulting a physician. "He's a good doctor, I suppose," said a woman in 1963. "Medically speaking, I mean. But he acts as if he couldn't care less about me." People were impressed by the new technologies, but as one 1963 journalist observed, for the patient who longed for the old-fashioned doctor with lots of time, "it is somewhat unrewarding to love a scientific instrument."[38]

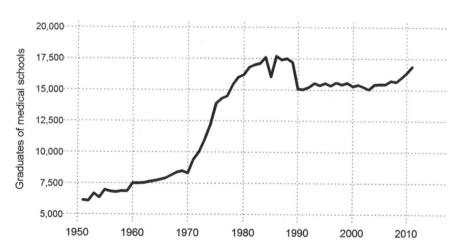

Graduates of U.S. medical schools, 1950–2010. The number of medical school graduates increased dramatically within a single decade, the 1970s, in part as an attempt to meet the perceived crisis in American health care and the maldistribution of physicians. Then the number of graduates remained relatively constant for a quarter century.

The Patient as a Problem

In 1951 the California Medical Association hired an industrial psychologist, Ernest Dichter, to suggest how individual physicians could improve the image of the profession, or "medical public relations." Dichter suggested that physician groups use manipulative slogans like "a personal physician for every person." Moreover, he urged doctors to recognize that patients wanted to feel like participants in their treatment. The patient "wants to be loved, not in a condescending, paternalistic fashion but as an equal partner in the fight against the difficulties of modern life." Dichter made the business model explicit: "Your personal physician is your modern medical general manager; your investment in his fee is always a wise investment in good management of your total medical problem," advice for the patient that was perhaps more meaningful for the dispenser of care. So it was that the art of medicine had come to be framed in business managerial terms. One of the books from the California Medical Association effort was actually titled *The Management of the Doctor-Patient Relationship*.[39] In the 1950s and 1960s the physician public relations movement spread throughout the country.

Both patients and doctors had to deal with organizational forces over which neither had control; in addition, patients and doctors were coming from two different worlds. Patients did want professional competence. Yet even fairly sophisticated patients also wanted personal attention.[40] As bureaucratic, impersonal medical institutions developed, the worlds of doctors and patients moved further apart. Moreover, the character of those who went into medicine was changing. Fewer students entered the competition for entrance into medical school with the goal of becoming wise personal counselors. As an informed psychologist observed, "Naturally as people with more technical and less conversational interests enter medicine, they will shape their practice further along the direction of those interests to the exclusion of the role of the comforter."[41]

Shaping the World of Doctors and Medical Education

Probably the best evidence of the separateness of the world of medicine was the way medical school socialized medical students into the profession. As the competition for entrance became intense, in 1948, to aid admission committees, the nationally administered Medical College Admissions Test (MCAT) became available, although the validity of the test as a screening device was unclear. By 1956 only half of the applicants gained admission to any medical school. At one midwestern university the premed students were lined up and told to look at the two students on either side of them: "Only one of the three of you will actually get into medical school." It is true that while "racial" and religious quotas and discriminations began to soften or disappear as general social attitudes changed, particularly in the 1950s and after, still in 1970, 90 percent of those admitted were male. Meanwhile, the National Board of Medical Examiners expanded and standardized testing of medical students and interns for licensing.[42]

The medical schools continued to maintain essentially the same curriculum as set up at Johns Hopkins in the 1890s: two years of basic science and two of clinical training. More and more information was pressed upon the students, however. As one shrewd student put it, "I think I have figured out their philosophy. They are giving us more than we can possibly do, but they want us to work up to our own capacity."[43]

Eventually, beginning in the 1950s, Western Reserve and other medical schools started loosening the requirements, so that students gained more electives and flexibility in what they were doing. But as sociologists showed, often marching in lockstep gave the students the sense that they were part of a special group. Medical school was a rite of passage that bound them together. More subtly, their shared intense concern for how their learning would be useful in clinic and laboratory afterward distinguished them as members of a single-minded group who would

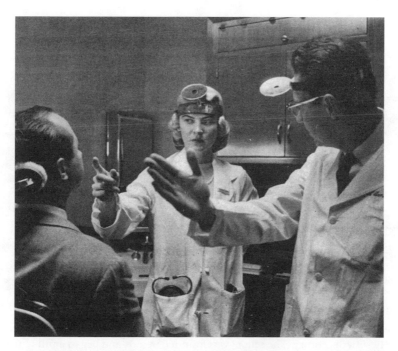

In 1960, Susan Cook, a medical student, received clinical training in her fourth year of medical school. Here she is working with a practitioner-researcher in ophthalmology, learning how in appearance, bearing, and technological mastery to fit into the physician community. Jack Star, "Our New Hospital Crisis," *Look*, 29 Mar. 1960, 29.

fit into and continue the separate world of medicine. Even in the clinical years and the internship, the struggle for access to more and more medical "experience" separated medical students from the rest of the population.[44]

Regular physicians began to notice that in response to the physician shortage that entered public discussion in the late 1950s, for two generations osteopathic physicians' education was upgraded to be equivalent to that of regular medical schools. Slowly, following the actions of leaders in California, the MD and DO became equivalent and followed the same standards.[45] In addition, new medical schools finally began to appear. As noted above, the number had shrunk drastically after the reforms and the Flexner Report at the beginning of the century. There were six new schools in the 1950s and fifteen in the 1960s, with twelve more under development—all aided by federal assistance for construction beginning in 1963, when Congress passed the Health Professions Educational Assistance Act to help expand not only medical schools but also training facilities for nurses and other paramedical personnel. Meanwhile, existing medical schools expanded in many ways. In 1951 the average school had just over fifty faculty members, but by 1970 the number was 257.[46]

The Biomedical Sciences Building at the University of Hawaii's John A. Burns School of Medicine was representative of the new medical schools of the second half of the twentieth century. The school was founded in 1965 and began a full four-year program in 1973, during a period when policymakers detected a shortage of physicians that would grow in the future. John C. Bowers and Elizabeth F. Purcell, eds., *New Medical Schools at Home and Abroad* (New York: Josiah Macy, Jr. Foundation, 1978), 44, reproduced with the generous permission of the Josiah Macy, Jr. Foundation.

Historians have described in detail how in each decade medical schools became more closely coordinated with the teaching hospitals at the top of the health care hierarchy. Nevertheless, in the 1950s and 1960s, research and patient care, not education, dominated and configured the medical complexes. In 1947, medical schools had an income of almost $70 million; in 1970, $1.9 billion. Research and training grants and contracts constituted a very large percentage of the growth. Tuition income, by contrast, had shrunk to well below 10 percent of medical school income by 1970. No one knew, for example, whether the clinic patients were more important for training students, interns, and residents or for serving as a research pool. And as more patients in the hospitals and clinics had insurance, how could they be subjected to serving as "material" on which medical students could practice or as subjects for investigators' research? The academic reward system, however, showed that research was driving the medical schools and academic medical centers.[47] Critics, usually at first (as noted above) within medicine, condemned the educational system for neglecting the human side of medical practice and for graduating badly educated physicians "trained in institutional, 'crisis' medicine, while the typical patient is ambulatory and apparently well, or chronically ill."[48]

Hospitals

For the public, by the 1950s the hospital had become unmistakably the center of the health care system. Three-quarters of the population had stayed in a hospital at least once, and one-half of the population had had two or more hospital experiences. Some people had insurance, and some did not. Nevertheless, as physician house calls virtually disappeared, anyone might end up where the technology increasingly was concentrated, that is, in the hospital. Moreover, it was an arrangement that suited a population who could now mostly reach medical facilities by automobile. By 1960 virtually all births took place in hospitals, except in the case of "nonwhite" mothers, for whom 85 percent did.

In the 1950s, nursing shortages drove most hospitals to establish intensive care units (ICUs), and the first neonatal intensive care unit opened in 1965.[49] By that time, for adult patients there were often elaborate special care areas with correspondingly elaborate technologies for coronary, respiratory, and other conditions. Both doctors and patients emphasized the importance of a hospital's being "modern" and up-to-date. And as chronic diseases became the major medical problem, hospitals responded, not with preventive or controlling programs, but with more and more technology.[50]

Meanwhile, almost without anyone's noticing, another kind of institution, the nursing home, began to exercise a hospital function. As noted above, nursing homes were first stimulated in the late 1930s by the money available to Social Security beneficiaries, who no longer had to live in county poorhouses. The number of proprietary nursing homes increased greatly in the 1950s, and such institutions took care of many chronic patients, freeing hospitals to focus on acute care. Indeed, the occupancy rates of nursing homes were consistently far higher than those of hospitals. Especially after the coming of Medicare and Medicaid in 1965, expenditures for nursing homes increased dramatically. In 1970 the *California Financial Journal* reported that "the future of nursing home and extended care facilities is wide open and virtually gold-plated."[51]

In the late 1940s and 1950s, hospitals served the local community even more than earlier. Hill-Burton funds awarded to underserved areas reinforced the idea of a public community hospital, which of course also had local social and political bases. At first, half the Hill-Burton projects supported hospitals of fifty beds or fewer; many of these projects were part of the effort to bring equalizing but still segregated health care to African Americans in the South.[52]

Then in the 1960s, hospital public relations officers began emphasizing the accessibility of the institution, how the hospital brought "mainstream medicine" to everyone. As in fact it did. Indeed, the rapidly growing use of emergency room, or ER, services at hospitals (a 400 percent increase from 1945 to 1962) in part substituted for the neighborhood clinics that failed to materialize in the United States, in contrast to what occurred in some European countries.[53] Until better organization came some years after World War II, typically what had happened in a real emergency was that untrained ambulance drivers took the patient to the hospital emergency room, where inexperienced interns under the direction of a nurse treated the patient. Whether for emergency or other patients, hospital outpatient services increased greatly. By the late 1960s there were five times more outpatient visits to hospital clinics than inpatient admissions. The outpatient visits, however, were substantially undercompensated and constituted a heavy drain on hospital finances.[54]

Hospital Economics

In the 1950s a trend became obvious. As noted above, total expenditures for medicine and health rose. Yet in 1958 hospitals received twice the income they had just ten years earlier. Some people believed that increased insurance coverage made the difference, others that health care was more affordable in the general prosperity of the time.[55] But when the average daily cost of a hospital bed rose from $15.62 in 1950 to $44.48 in 1965 (before Medicare) to $69.93 in 1969, other factors were clearly involved.[56]

During the 1950s and 1960s it became apparent that hospitals were charging too little, and as administrators introduced business methods, they asked users of the

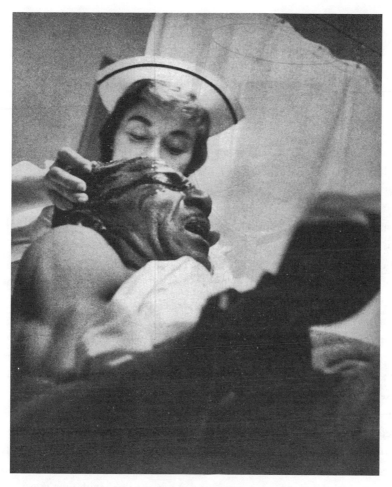

Emergency room personnel in Los Angeles sewing up head wounds of a patient thrown through a car windshield in a crash. The ER was a well-established institution by the 1950s. "Emergency Angel," *Look*, 2 Apr. 1957, 107.

hospitals to meet the costs. Business efficiencies could help, but there was much inertia because of the philanthropic traditions under which hospitals had always operated. Over the years hospitals had been able to charge "below-cost rates" by paying low wages, and they had received charitable contributions and sometimes government subventions. In the decades of intense, persistent inflation after World War II, local and state government subventions consistently failed to meet the higher costs. And charity represented an increasingly infinitesimal fraction of the yearly budget, down to about 10 percent in 1949. As one analyst concluded, the full-pay patient made up any difference, so that hospitals and indigent patients were supported by an implicit tax on the sick![57]

The fact remains that hospitals had survived largely by paying the least amount possible to unskilled laborers, particularly women, who were nurses' aides, cleaners, and kitchen help—or newly arrived immigrants, African Americans, and other exploited groups who had little choice of employment.[58] With changing times, the almost unbelievable exploitation of often helpless workers began to come to an end. During one 1958 labor confrontation, a *New York Times* editorial writer put the issue in terms of fairness: "Isn't it unfair and inhuman to ask hospital workers to help meet hospital deficits by accepting substandard wages?"[59]

As noted above, even the hospital resident physicians received increasing amounts of compensation for their work. For nonprofessional workers and some nurses, effective unions occasionally secured better wages, with an effect that over a number of years spread from New York and other centers to hospitals across the country. By the mid-1950s, Social Security finally covered hospital workers, and gradually state governments made minimum-wage laws apply to them. As hospitals increased in size and became more bureaucratic, managers had less recourse to personal loyalty and other informal means of persuading workers to accept inadequate compensation. All these changes, along with the mushrooming costs of new technology, made hospital rates rise rapidly. Members of the public, including taxpayers, no longer received as much indirect subsidy as they had from ill-paid workers, much less deliberate philanthropists.[60]

In the face of escalating hospital costs, health system planning, long advocated by reformers, became a force across the country. Funding officers in government and insurance deplored duplication of effort and instruments and strongly endorsed planning. Yet there was little other than moral and rational argument to enforce it, and the planning movement did not survive much past 1980.[61] Rational planning also encountered policymakers interested in rehabilitating and building up the great urban medical centers, often as part of efforts to save American cities. In was in this context that hospital programs after the 1950s turned from the underserved rural community to bringing modern medicine to everyone. As two experts commented, "We are shifting from the disease orientation . . . to the questions of health care delivery."[62]

In the 1950s and 1960s, observers noticed that huge amounts of income were transforming hospitals from charitable institutions into industrial institutions, particularly as trained and professionalized hospital administrators gained traction against physicians' influence. While Hill-Burton funding saved many small hospitals from consolidation, the most familiar model was that of the large hospital, which took on the characteristics of a huge bureaucracy with many specialized functions, as sociologists pointed out at the time. A hospital nurse said of her work,

In the 1970s, hospital workers from three local unions in Alameda County, California, staged demonstrations to protest their low wages. Worker organizations had some success over the years in obtaining better conditions for lower-level hospital employees, who were among the lowest paid and most exploited in the economy. Walter P. Reuther Library, Archives of Labor and Urban Affairs, Wayne State University.

"Well, . . . I don't know. There isn't anything that I find unpleasant. I have done it so long, I just automatically do it." The hospital was never a perfect bureaucracy, however, because, as noted above, it contained so many centers of real power. At the same time physicians and nurses were struggling for professional autonomy, and so they did not always or easily accede to the essential elements of bureaucracy: order, standardization, and discipline.[63]

Standardization, which in one sense still meant national pressure to upgrade, came to hospitals largely through the accrediting inspectors of the American College of Surgeons. In 1952, however, the College gave up and, along with other national groups, formed an independent agency, the Joint Commission on Accreditation of Hospitals. In 1969 the Joint Commission completely recast the standards in both detail and language. Included, for example, was the requirement that hospital medical staffs be suitably organized to maintain medical standards—a requirement that became momentous after the courts in the 1960s determined that a hospital was responsible not only for housing and feeding patients but for the medical procedures carried out within its walls.[64]

It was in the area of industrial-style discipline that medical personnel found their autonomy most constrained in the hospital setting. The personnel whose situation was most difficult, however, were the nurses, who were often caught between the competing interests of patient care, physicians, and administrators.

Changes in Nurses' Roles

The 1930s migration of nurses from private duty to hospital positions intensified after World War II. Virtually all nurses were women, and so gender roles also played a part in the occupational conflicts in which they found themselves as they tried to assert professional judgment in the decades after the war. Nurses became supervisors of other personnel on the ward, and they had much more administrative paperwork to do. Thus they became part of the administration and often substantially removed from direct patient care. Moreover, in the words of one historian, "once incorporated into the administration of the hospital, many nurses quickly gained a stake in its smooth functioning."[65] In 1965 a sociologist wrote about the nurse:

> The nurse is the one functionary of the hospital who is at the patient-care unit continuously. All others, including the physician, come and go. The nurse is the coordinator, the mediator and observer for all the patient services. . . . The nurse must understand the principles of organization and administration because, in reality, whether she likes it or not, she has become *de facto* administrator in the complexity of patient care.[66]

And all the time, beginning as early as the 1950s, the strict discipline of the factory model or even of the old-fashioned authoritarian head nurse was changing. Relationships between staff members altered as the "social distance" between physicians and other staff members decreased. A Minnesota physician wrote in 1966 that "there is no longer a gulf of unique expertness separating the physician from all other workers in the hospital. More and more, the doctor's status is that of one of the associated professional persons there. Although the physician remains the leading professional, he is not unaware that some hospital authorities would like to change that also."[67]

One consequence of the changing social relationships was the development in a number of hospitals of a "team" for patient care, parallel to the research teams mentioned in chapter 9.[68] In some hospitals, the old lines of command, obviously based in part on gender, continued for decades. In others, staff had a sense of constant change.

"Here's your treatment team, doctor," the administrator beams. "Of course, you'll be entitled to an administrative salary for directing them." Such teamwork helps high-volume doctors see—and admit—even more patients than they could before.

The concept of a team of health care workers was well established in the 1970s. The team was a human counterpart to complex technology and suggested a very high level of care. As the detail in this cartoon illustration suggests, however, the team could give the impression of thorough treatment by rushing patients through efficiently, a strategy not all physicians and patients were comfortable with. Mark Holoweiko, "Why Hospitals May Be Giving Your Colleagues an Edge," *Medical Economics*, 21 Jan. 1980, 87.

Steps in Expanding Health Insurance Coverage

A key factor driving change was the startling increase in voluntary health insurance coverage that continued after World War II. The figures are eloquent, including the appearance of a new category, major medical expenses, in the mid-1950s. The numbers of people covered in three different years are shown in table 6.

These figures bring the story down to the implementation of Medicare and Medicaid in 1965–66. They reflect the fact that only at the end of the 1940s and the beginning of the 1950s was there a major switch to having the employers administer health care benefits, for which employees and employers paid—a development noted above as a source of major problems. And once Medicare and Medicaid were in place at the end of the 1960s, it was possible to see that, at least in part, the expansion of health insurance had intensified the problems in the health care system, insofar as there was a system in the United States. The Health Insurance Institute in 1969

TABLE 6
Americans covered by health insurance, 1940, 1950, and 1965

Year	Hospital Expense	Surgical Expense	Medical Expense	Major Medical
1940	12,312,000	5,350,000	3,000,000	—
1950	76,639,000	54,156,000	21,589,000	—
1965	153,133,000	140,462,000	111,696,000	51, 946,000

Source: Based on *Source Book of Health Insurance Data*, 1968, 19.

identified three major challenges: "shortages of physicians and other medical man-power, poorly distributed manpower and facilities, and duplicate health care facili-ties," making "adequate health care inaccessible to many and expensive for all."[69]

Medicare and Medicaid constituted a major incremental step in providing health coverage for all Americans. After the defeat of a national program of compulsory health insurance at the beginning of the 1950s, the reformers did not give up. They continued to advocate social justice and to cite the embarrassingly poor health of Americans compared with citizens in other developed countries, even as private health insurance expanded dramatically. And occasionally there were "incre-ments." In the Cold War era, in 1956, without many people's noticing, Congress provided for medical care for dependents of those serving in the armed forces—about two million people! During the Vietnam War, the number of those partici-pants reached six million. The program was significant not just because it essen-tially reinstated the Emergency Maternal and Infant Care program of World War II (see chapter 8) but also because it established a precedent: where there were no military medical facilities available, private physician and hospital care could be contracted directly by the federal government.[70]

By 1965, before definitive Medicare, about three-quarters of the population was already covered by hospital and surgical insurance, as shown in table 6. Much of the coverage did not include regular physician care at all.[71] And attempts at ma-jor reform continued to be frustrated by the combination of conservative Repub-licans and conservative (mostly southern) Democrats. The American Medical As-sociation politicians provided those combined factions in Congress with lucrative and effective backing. Nevertheless, in 1956 Congress amended the Social Security Act to provide benefits for the permanently disabled. This measure was aimed at those who had suffered injuries, but it also opened the door for those disabled by chronic diseases.[72]

Reformers proposed other schemes, but none was politically viable. By the 1960s, suggested reforms were coming more frequently and more powerfully, in what both advocates and opponents believed was the incremental path to compulsory health insurance. In 1960, Congress, in the Kerr-Mills Act, provided for funding, through

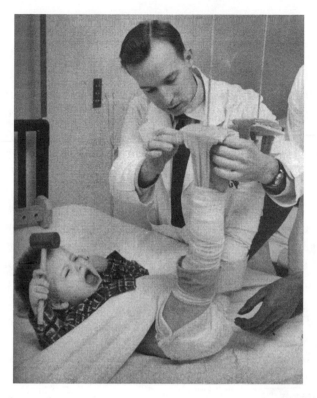

U.S. military hospitals treated many civilian patients, such as the child with broken legs in this photo. The writer of the accompanying article noted that in Air Force hospitals around the world "a baby is born every nine minutes." Government doctors providing health coverage for this segment of the U.S. population seem to have presented no political problem. "Air Force Doctor," *Look,* 17 May 1955, 69.

the states, medical care for the elderly who were indigent.[73] Most states, however, did not provide the necessary matching funds to receive federal assistance for welfare medical care. Only after the stunning Democratic victories of 1964 reduced the influence of conservatives of both parties could a major measure pass.[74]

In 1965 Congress created Medicare and Medicaid. Medicare, providing for old people, had two provisions. First, as part of Social Security, everyone who worked, along with their employers, began to pay for compulsory hospital insurance that would become effective at age sixty-five. It covered a limited number of days in the hospital for different kinds of services. Second, retirees who elected voluntarily to pay for medical coverage could do so for a small fee each month. Since well over 90 percent did elect to pay for this coverage, it, along with the hospital insurance, effectively constituted universal health insurance for the old folks. And their "gray lobby," mobilized in the 1950s and 1960s, continued to be a powerful political player in succeeding years.[75]

Medicaid for the Poor

Although Medicaid and Medicare were part of the same law, Medicaid got a different label. Medicaid was support for state programs to furnish medical care for the poor, those who passed a state means test showing that they had inadequate income (often defined in terms of the new federal poverty-level income standards). Thus Medicaid was a welfare program, only incidentally providing health care. In 1950 Congress had provided funds to the states for direct payments specifically for medical care for the poor, including the elderly poor. This program of direct funding was expanded, and, importantly, it further subsidized the nursing home explosion. Beginning in 1965, then, Medicaid continued programs to provide medical services for children of impoverished families, begun in the 1920s with the Sheppard-Towner Act, revived in the 1930s, and expanded thereafter. But it also provided a safety net for the elderly, including middle-class elderly, who might spend their money on nursing home care and then not have any funds left for health protection. Within a few years, about a third of the Medicaid funds were going for indigent elderly in nursing homes. It was under this provision, and even beginning under the Kerr-Mills bill, that the number of nursing homes increased so steeply.[76]

Historians have described Medicaid as a continuation of the Western tradition of furnishing medical care to poor people. Sometimes the community provided more, and sometimes less. Sometimes the support was direct, and sometimes, as in the sliding scale of fees, indirect. The religious hospitals were one expression of community care. So were tax-supported hospitals and programs. But always the poor had a separate, usually inferior stream of health care—most vividly the crowded wards that informed middle-class people wanted to avoid.[77]

Medicaid maintained that separate, usually cheaper stream of care for the poor. Sometimes either the federal government or state governments would reduce the money available. Then the qualifications for Medicaid could be tightened, or coverage restricted. In either case, very sick poor people ended up in hospitals, where administrators had to raise rates to cover patients who could not pay and who could not usually be turned away, in practice still levying a tax on the insured sick to pay for caring for the poor, as noted above. Legislative cost-cutters, however, would never admit their shortsightedness or their implicit social Darwinism, which denied the poor access to medical care except as parasites on a grudging health care system. This continuing confusion of welfare with medical care was a distinguishing feature of U.S. health care.[78]

Medicare-Medicaid was designed as one of the pillars of President Lyndon Johnson's 1960s Great Society program, and it constituted a major battle won in the War on Poverty. In fact, the law erased one disadvantage of the poor who were enrolled. Before, lower-income people had averaged 3.9 physician visits per year (compared with 5.2 for the more affluent). Afterward, the poor averaged 5.6 visits per year, more than the 4.4 of the highest income group but of course partly proportional to the health problems suffered by the poor. When the president signed the law, he articulated the aspirations and beliefs of the bulk of the population: "No longer will older Americans be denied the healing miracle of modern medicine."[79] From one point of view, Medicare-Medicaid was another step toward the goal of universal health insurance. From another, by taking care of the populations who most needed medical services—poor children and the elderly—Medicare-Medicaid functioned politically and economically to undercut further steps toward universal coverage. The government also in effect relieved the health insurance companies of many of their most expensive claimants.[80]

The story of course was not so easy or simple. The biggest problem was that at that time about one-fifth of American hospitals were not "racially" integrated, and it took some time for them to arrange to accept all patients on an equal basis—or lose accreditation (and payment) from the U.S. Public Health Service. Another problem, and one not seriously anticipated at the time, was the impact of Medicare and Medicaid on health care charges. The managers of the legislation simply did not consider costs. And yet within a few years, as I have already indicated, the rate at which health care costs rose became a major national problem. Medicare did increase demand for health care. As early as 1966, immediately after the Medicare and Medicaid legislation, in just one year physician fees rose by 8 percent and hospital costs rose by 16.5 percent. Within a few months an official government report was signaling trouble, because the federal budget was now heavily involved.[81]

Public Health

Even as so much effort and money went into curing and caring, preventive health programs in the post–World War II decades proceeded in surprisingly independent ways. Many public health efforts in the United States changed remarkably little even as ever larger sums flowed into community, state, and federal public health efforts. Between 1950 and 1970 the demand for public health personnel expanded so much that the number of schools of public health doubled. But with growth came specialization within public health, and with specialization, fragmentation. Therefore, a number of quite different efforts worked in parallel, largely in agencies at all levels of government. The whole effort suffered from "overlap and gap."[82]

Yet the impact of public health was substantial, especially as public health measures affected larger and larger parts of the population. Beginning in the 1920s and through the rest of the century, the effects of public health work, along with rising standards of living, caused socioeconomic status to have a markedly decreasing effect on death rates (presumably a measure of national health). That is, the gap between death rates of people of higher socioeconomic status and those of lower status narrowed consistently at least through 2000.[83]

The old epidemiology, with attention to outbreaks of contagious diseases and the specific sources of infection, continued and expanded. During the 1950s a wartime federal agency for malaria control evolved into the Center (later Centers) for Disease Control, with outstanding workers in the laboratory and in the field, ready to jump on a train and later a plane to anywhere in the world where a disease outbreak occurred.[84]

Authorities busy with such traditional public health work were slow to respond to changes beyond technical improvements in their labs. The new emphases on chronic diseases and diseases of old age only gradually generated new alliances with welfare and other agencies.[85] Meanwhile, in the wake of the pressure-group model of governing established during the New Deal, the old single-disease groups, and many new ones, that influenced policy changed. In addition to charitable work with victims and then, increasingly, biomedical research, the single-disease groups became more openly pressure groups and at the same time provided vehicles for the formation of local and national communities of victims and their families, as in deafness and diabetes.[86]

The new biostatistical epidemiology and the shift to prevention eventually engulfed many public health officials. They now often looked at disease prevention from a long-range perspective, taking up such issues as family planning and poverty, which they presumed would have health effects. Therefore educational programs, which for generations had been part of public health, in the 1960s became much more conspicuous, with a momentum that grew into the new era of environmental medicine. But already in the 1960s, in the rising antitobacco crusade, epidemiology and education came together.

Another fundamental example of the new epidemiology was the Framingham study. In that study, initiated in 1949, a large portion of the inhabitants of one unexceptional town, Framingham, Massachusetts, was examined over a long period of time to discover whether there were any obvious correlates of factors that might cause heart disease. When the first results appeared in 1957, they caused a sensation, because they confirmed a possibly lunatic suggestion, or at best a clinical impression, that there was a connection between eating fats (and probably being fat) and cardiovascular illnesses. The study made the life insurance industry idea of a "risk factor," for an individual, common in medicine and greatly expanded traditional, often unpopular governmental guidance as to what one should and should not eat (a subject that soon became more important).[87]

Preventive Medicine

Preventive medicine was always a contested area. Traditionally, medical authorities held that early diagnosis, with subsequent treatment, would prevent more serious disease. This was the pattern that seemed to prove out in the Kaiser Permanente HMO model in the World War II period and after. It was also the pattern that caused groups of physicians to recommend the annual physical examination, and employers, including the armed services, to require periodic checkups, with special emphasis on detecting tuberculosis in the early stages. A short-lived attempt to screen whole populations for illnesses proved to be impractical (see chapter 12).

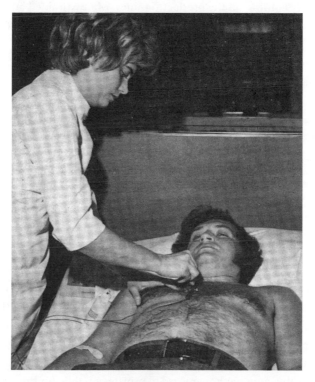

A patient having cardiovascular testing as part of a whole-population comprehensive health screening program in Pennsylvania in 1972. Programs of universal screening were terminated in favor of individual annual physical examinations, which produced more useful results for less money. Or perhaps just less total screening. *Pennsylvania Department of Health Annual Report, 1972, 26.*

Instead, the ceremony of the individual annual examination became even more commonplace and sometimes did turn up pathologies.[88]

Another pathway to prevention was well launched in the 1960s but grew rapidly in the 1970s: promoting a "healthy lifestyle," a phrase and concept that entered health and popular discourse in a major way in those decades. Again, this strategy could be directed toward particular risk groups.[89] Such educational efforts became increasingly important in the 1970s and 1980s in new technological and social contexts (see chapter 11).

Outlook for the Future

By the late 1960s the future of medical care in the United States looked clear. For most people, infectious diseases were no longer the threat they had once been. One could project that incremental improvements in technology would extend physicians' and technicians' abilities to diagnose all kinds of diseases. New treatments were constantly coming off the production lines. Many disabled bodies could be repaired with new parts. Medicaid and Medicare appeared to be just the first steps in extending health care to the entire population, so that the United States would eventually have the kind of prevention and care that other developed countries did.

Meanwhile, health workers in the 1950s and 1960s witnessed the continuing decline of mortality and morbidity. In just the twenty years 1949–68 the decline in *death rates* for a number of serious diseases was startling: polio, whooping cough, and dysentery, all nearly 100 percent; syphilis, 95 percent; tuberculosis, 88 percent; hypertensive heart disease, 78 percent; kidney diseases, 76 percent; maternal death in birthing, 73 percent; appendicitis, 72 percent.[90]

Health care personnel did have immediate concerns and problems, and they worked with institutions that had faults and flaws. However, despite complaints and concerns on the part of the public, it seemed likely that the institutions, with correction, would serve well into the future. When the head of preventive medicine at Harvard, David Rutstein, called for a "revolution" in medical care in 1967, it turned out that he really wanted just to improve the system already in place or in sight.[91] Of course neither Rutstein nor anyone else could foresee "the medicine of the future." And they did not anticipate the extent to which contingencies and forces outside medicine would intrude to shape health care from the era of the Vietnam War to the 1980s and after. Nevertheless, looking back, historians could see that already by the 1960s, without regard to private or public funding, there existed a huge, organized, heavily technological health care "industry."[92]

CHAPTER 11

Medicine in the Environmental Era,
1960s to 1980s

In the new era that began around the end of the 1960s there were several interacting currents. Tangible improvements in medical treatment and new understandings in the science of the human body and disease constituted one current. At the
same time, new diseases challenged physicians and scientists, who meanwhile were
trying to adjust to the sociocultural impacts of environmentalism and the rebelliousness of the 1960s, not to mention the continuing activities of traditional and
new health care workers and, now, patient support groups. All of these developments contributed also to the story, taken up in chapter 12, of powerful political
and institutional forces that affected medicine and health in this same period.

A Record of Gains in Doctoring and Living

It seems implausible that in the 1970s and 1980s the mind-boggling technical innovations of the 1950s and 1960s became routine and even, as much as was possible, assumed. Yet Americans who were working in direct patient care, in Pharma,
or in industry all routinized keeping up with the technical services they could offer patients. Within a generation or so after World War II, physicians were able to
suggest efficacious treatments (usually pharmaceuticals) for perhaps half the diseases they commonly identified (compared with fewer than a tenth around 1930).[1]
And more therapies were on the way.

In 1985 a team of journalists summarized what they perceived as "new triumphs
of medicine" of the previous few years: "Today, doctors replace organs with plastic parts, vaporize tumors with laser rays, make babies in test tubes, isolate memory cells and draw maps of people's genes." The writers featured in their account microsurgery, sophisticated drugs to control disease, "artificial skin,"
and other wonders, alongside technology like the incubator, which they described as essentially a "glass-and-steel womb for 2-pound premature babies."
It was, in fact, easy to make such a record of constantly advancing medical care
and "new discoveries."[2]

In this saga of progress, physicians continued to be the center of attention.
In the 1970s, pollsters found that the public still ranked doctors first among

professionals. A few years later, in another poll, they ranked fourth, behind dentists, but at that time there were doubts and questions about all professions. By the 1980s even many physicians found fault with the practice of medicine and said they would not recommend it as a career to young people. Yet at the beginning of the 1990s the overwhelming majority of doctors in still another poll said they enjoyed their work.[3] All of these contradictory results reflected the remarkable, often upsetting and confusing alterations in the circumstances in which not just doctors but all people in health care were functioning.

Moreover, despite the sense of dramatic change, more and more often, in a paradox that will be noted below, scholars were talking about an "American health care system," an articulation of their continuing sense that there were some relatively permanent working arrangements. Most of the usual statistics also suggest linear change and therefore relative stasis rather than something new. The population increased from 203 million in 1970 to 248 million in 1989. The proportion who were elderly and would anticipate suffering and expiring from chronic diseases continued to increase. Life expectancy even rose at an accelerated pace, from 70.9 years in 1970 to 75.3 years in 1988, compared with the rise from 68.2 years in 1950 to 70.9 years in 1970. People who reached sixty-five could expect 15.2 more years of life in 1970 but 17.2 more in 1989. The infant mortality rate dropped from 20 per 1,000 live births in 1970 to 10 per 1,000 in 1989.[4]

From 1970 to 1990 the number of physicians active in the United States increased from 310,845 to 547,310. The total number of people working in the health care field doubled between 1970 and 1989, to about 9 million. That figure included about 750,000 registered nurses in 1970 and 1,666,000 in 1989, with comparable increases in other ancillary and support personnel.[5] The number of physicians per 100,000 population, which had increased to 155 in 1970, had reached 233 by 1988, with similar increases in the proportion of nurses (from 366 per 100,000 population to 670 per 100,000, but still in short supply) and other health personnel. Despite problems in distribution, more Americans were using more health services. Already by 1978 one government report ascribed the increase to "higher incomes, expanded insurance coverage, higher levels of education and increased awareness of health, as well as the overall quality and effectiveness of the health care delivery system."[6]

Altogether, from the late 1960s to the late 1980s Americans put more and more effort and money into their already huge health apparatus, which, arguably, may have contributed to Americans' greater longevity. Moreover, discussions around medicine and health included another goal: improving the quality of life. And in that area there was an important change.

In all of the concern about diseases and the actual work and distribution of health services, few people noticed that the configuration of the life course of a typi-

An image of a harried nurse caught in action in a health facility was used in a 1978 plea to recruit more nurses into the health care system. Henry A. Foley, "Assuring the Nation's Health Resources," *Public Health Reports*, 93 (1978), 632.

cal American had shifted by the 1980s. One has to look at life expectancy, which was increasing rapidly. In the first half of the twentieth century, infants and children made by far the most gains as infectious diseases declined in importance. But three-quarters of the way through the century it was adults whose life expectancy increased. Between 1970 and 1980 the life expectancy of a forty-five-year-old person increased by 6.6 percent, while the life expectancy of a newborn child increased by only 4.0 percent. The change was in chronic diseases. In that decade the death rate for stroke declined by 37.4 percent, and that for heart disease, by 19 percent (reflecting in part, as we shall see, more healthful lifestyles as well as the containment of rheumatic fever). At the same time, the age of onset of chronic diseases that caused death did not remain the same. Had it done so, for most people that would have meant an extended old age beset by any number of chronic diseases. What happened instead was that the age of onset of chronic diseases was postponed in a way parallel to the postponement of death. Therefore the terminal diseases and terminal arrangements that previously had afflicted people of a certain age now afflicted people at a later age. The longer life that the average person enjoyed at the beginning of the 1990s was not, as might be expected, necessarily burdened with more illness simply because one lived longer. One set of authors named it "the Age of Delayed Degenerative Diseases."[7] These bonus years of relatively good health provide a background for the profound shift in medicine and health under way near the end of the twentieth century.

The continuities and triumphs of medicine always were shadowed by reserva-
tions. At the end of the 1980s two Johns Hopkins epidemiologists summarized the
cautions: "While we have reduced morbidity and mortality from infectious diseases
such as pneumonia and influenza and even completely eliminated the risk of smallpox, we are continually being challenged by the appearance of new diseases such
as AIDS and by chronic disease and accidents."[8]

Some of the projections from common statistics could confirm that in addition
to growth, a number of major changes were taking place in medicine itself. In 1971,
13.7 percent of all the students starting regular medical school were women, and
the figure had risen to 37.1 percent by 1988.[9] Similarly, the percentage of entering
students identified as "minority" rose from 4 percent in 1968 to 26 percent in 1988,
with the most significant increases in the years 1968–71. Those classified as Asian
rose from 0.2 percent of all entering students in 1968 to 12 percent in 1988. Afri-

This image of aging adults apparently happily exercising was typical of the health
promotion literature in the decades around the turn of the twenty-first century. It
carried the implication that those who did not seek health were to blame for their
need for health care. It also inadvertently illustrated the disconcerting fact that
older people were in fact enjoying a longer, healthier existence than in previous
generations. National Aeronautics and Space Administration and the National Institute on
Aging, *Exercise: A Guide from the National Institute on Aging*, Publication No. NIH 98-4258
(Washington, DC: National Institutes of Health, [1998]), 27.

can Americans, who had constituted fewer than 3 percent of the entering students in 1968, made up more than 7 percent by 1972 but then continued at 7 percent for the next sixteen years.[10]

In the midst of changes that could be projected into the future, one set of figures was alarming. In 1970, as noted above, health expenditures already consumed an astonishing 7 percent of the U.S. gross domestic product. By 1989, however, that proportion had risen, incredibly, to almost 12 percent, with no end to the increases in sight.[11] In 1969–70 intelligent participant observers of life in the United States began to identify a crisis in medicine and health care. By the late 1980s, however, those same observers were talking about being in a transition period. One writer surveying "modern medicine" in 1987 was even more basic: "Why," he asked, "is it changing so fast?"[12] Perhaps from a distance one can ask what happened in health care to move it from crisis to transition.

In the 1970s and 1980s two different but related currents intervened to redirect the flow of events in health efforts. The first fundamental circumstance was that Americans dealing with health concerns were profoundly affected by new ways of perceiving how humans lived in their environment. Second, a whole series of shifts in American society and culture took place. These sociocultural changes, labeled in historical memory as "The Sixties," intruded into the everyday world of medicine and health care, particularly affecting the many familiar, persisting institutions of medicine.

The Sixties

It is easy to look back and trace the origins of The Sixties to many years earlier. But it was not until the end of that decade and particularly into the 1970s that the real impact hit. Signs of fundamental social change had appeared when the new criticism of medicine began in the late 1950s. At that time the media portrayals of cities and cars, index images for the American dream, had begun to turn negative as opinion leaders associated them with pollution, poverty, and other undesirable social products. At the same time, the consumerism that had begun earlier expanded, particularly as television became a dominating, universal—and commercial—force. In addition, Americans in the 1970s and 1980s overall achieved unprecedented material well-being even as significant elements in society questioned or attacked many basic social arrangements.

Historians are still sorting out what happened in the late 1960s and after. As a thoughtful journalist wrote in 1997, "So powerful were—are—the energies let loose in the sixties there cannot now be, and may never be, anything like a final summing up."[13] The most important negative elements of the upheavals to affect health care were two: a powerful revulsion away from the social institutions that had flour-

"They can take out my appendix. They can have my kidneys, too. But they can't transplant my loving heart, 'cause that belongs to you."

A cartoon satirically combining the popularity of a musical group symbolic of rebellious young people of the late 1960s and 1970s with the casual public acceptance of the invasive surgery that represented some of the most advanced technology in medicine. Both transplants and new types of popular music were characteristic of changing times. *Saturday Review*, 5 Oct. 1968, 64.

ished in the United States for a long time—not least hospitals and the medical profession—and broad, consistent challenges to authority of all kinds, of which the authority of physicians was a convenient and close-to-home target.[14] Yet simultaneously The Sixties included movements toward social equality in which the long-established caste ("race") system began to give way with surprising rapidity, and not just in medical schools and hospitals. Also, in the 1970s a powerful women's movement began to undermine many other social arrangements, so that within a very few decades it was no longer clear that medicine was a male preserve and that nurses were a specially exploited group. In addition, idealism motivated many people throughout the society. As one enthusiast declared in 1972, "The fundamental shift in emphasis from order and technical knowledge to equality and freedom, from efficiency of the expert to participation of the citizen, will affect the roles of doctor and patient. As much as loss of revenue, this is what frightens the physician."[15]

A significant number of younger rebels and reformers began to question all the social injustice they could see—including the elements in health care that critics had already identified. At the same time, in medicine many physicians had become remarkably wealthy by working hard and doing obviously beneficial things that patients and other people appreciated. It was hard for those doctors, given the good they were doing day by day with individual patients, to understand attacks on their privileges and authority. Leaders in medicine expressed great concern about the image and morale of their colleagues.

Environmentalism

What was most fundamental in the time from the late 1960s to the late 1980s, however, was the change in how opinion leaders, especially in health fields, viewed the relationship between a person and his or her environment. In medicine itself, the shift to environmental perspectives was profound and was reflected in many ways, great and small. In 1969, for example, a textbook that had been issued for years as *Preventive Medicine and Public Health* became in new editions *Human Ecology and Public Health*.[16] Clearly, health care was passing through an era of distinctly environmental medicine and health. Indeed, those who lived through it would later find it startling how frequently Americans working in health care at that time explicitly invoked environmental thinking.

In the 1950s and 1960s there had begun to be concerns about toxic effects of radiation and also about pesticides and agricultural food additives such as endocrine growth stimulants. All these potential toxins were not observable but existed everywhere. It also turned out that the toxins persisted in the environment and collected in animal and human tissue. And above all, there was a growing conviction that these toxins could and did affect a person's health. Over time, at least some evidence linked such substances in the environment to "cancer, neurological disorders, reproductive defects, lowered resistance to bacterial infections, genetic change, and premature aging," according to one list.[17]

Evidence of the dangers had accumulated relentlessly from media reports and scientific studies after World War II. As antibiotics reduced people's concern about the power of infectious diseases, Americans reconceptualized their fears and aspirations for their health. The commonplace idea of communicable, infectious diseases diminished significantly in popular discourse (as reflected, for example, in the *Reader's Guide*, the index to general periodical literature). Meanwhile, one circumstance and then another made Americans aware of other factors in their surroundings that could contribute to disease and death.

The tragedy at Donora, Pennsylvania, in 1948, when "smog" (then a new word just imported from England) killed twenty people and sickened six hundred in a town of fourteen thousand, made leaders pay much more attention to the air as well as to asthma and other diseases believed to be allergic in character.[18] In addition, environmentalists, benefiting from the work of two generations of personnel in occupational health, were able to convey how a special environment, such as that in a factory or a mine, could cause disease.[19]

When the media reported on the killing smog in Donora, Pennsylvania, Americans began to understand the potential dangers the environment posed to their health. It took many years, however, for most people, inside medicine or out, to move "the environment" fully into their thinking about the causes of ill health. Courtesy of the National Library of Medicine.

In the 1950s and 1960s a grass-roots antipollution movement fed into the more general environmental concern. The movement had begun among residents in the rapidly expanding postwar suburbs. Their drinking water was frothing with dishwater detergent that had made its way into the water supply, and the air they breathed was stinging, dirty, and dangerous. Even farm workers joined in. Explained one: "I was picking tomatoes near Oxnard. My fingernails became infected as a result of poison that was on the tomato plants. Some of my fingernails fell off. It was very painful to work." People reasoned that if a lake or river was "dying," they might be next. Moreover, at the end of the 1960s the antipollution movement took on a new edge and gained unexpected power with help from countercultural "eco-freaks" (as critics named them). One young leader recalled that his contemporaries had imagined "that by the time we were our parents' age we would be sardine-packed and tethered to our gas masks in a skyless cloud of smog." The amazing popular reaction to Earth Day in 1970 was a signal that something remarkable was happening. Pervading and fueling the movement was the concern that the environment could damage a person's health or was already doing so.[20]

The Growing Focus on Health Consequences in the 1970s and 1980s

By the end of the 1960s, it was reasonable to ask more urgently whether the general environment was a major factor in health, an inquiry reminiscent of colonial and early national times when people believed that one's constitution could be affected by one's natural surroundings.[21] For generations after the late nineteenth century, people believed that the human organism had to adapt to a constant, usually normal environment or perish. Physicians and other health professionals helped people adapt. In the new environmental perspective symbolized by the publication of Rachel Carson's famous *Silent Spring* (1962) the organism was normal and natural, and it was the environment that was unusual and toxic. In 1965 René Dubos, in another frequently cited book, *Man Adapting*, laid out in detail all the environmental challenges to human bodies. At one point, for example, Dubos concluded that "excluding dust, pollen, and fog, about 100 air pollutants potentially dangerous to human health have been identified so far."[22] Not until the 1970s and 1980s, however, did these two books have their full impact.

To a remarkable extent, many opinion leaders reacted sympathetically to the environmental movement.[23] As early as 1970 the then newly reformed American Medical Association assembled a congress on the subject of environmental dangers. Carson in her 1962 book had described how pesticides and other contaminants disrupted and destroyed the natural world. What most people fixed on, however, was the section in which she described how chemicals made people, as well as birds and plants, sick. She quoted one health official: "We all live under the haunting fear that something may corrupt the environment to the point where man joins the dinosaurs as an obsolete form of life. . . . Our fate could perhaps be sealed twenty or more years before the development of symptoms."[24]

Some successful initiatives sought governmental action to salvage a natural, less threatening environment for all Americans. Not only did nuclear testing come under control but the air and water began to get cleaned up. A series of federal laws passed between 1970 and 1976 reflected one culmination of these concerns, acts with titles such as "Clean Air," "Pesticide Control," "Water Pollution," "Safe Drinking Water," "Resource Conservation," and "Toxic Substances." Americans were in fact less exposed to particles and chemicals than before the regulations. Perhaps the most striking directly medical effect was a program culminating in regulations removing lead from gasoline burned in motor vehicles and so reducing the amount

In the 1960s, sewage flowing into streams, as in this 1967 photo, was shocking and symbolic of unacceptable environmental pollution. Only two or three decades earlier, by contrast, it had been much more common. Expensive sewage treatment spread only incompletely in various parts of the country, as people assumed that flowing water would purify the waste rather than spread contamination quietly through the countryside. *Pennsylvania Department of Health Annual Report*, 1967, 44.

of lead in the air. The regulations became effective in the mid-1970s, and as the amount of lead in gasoline declined, the average load of lead in American children's blood in an amazingly parallel movement dropped by 40 percent in just four years.[25]

Policymakers also tried to change Americans' general environment and prevent ill health through another series of social technological measures. One initiative was aimed at the alarming rate of death and injury by accident. By the 1980s the leading cause of death in children and young adults was automotive injuries, and falls caused elderly persons to die at a rate eight times that of the rest of the population. Over a long period of time, engineers quietly redesigned the technological environment in the United States to remove common causes of accidents. In factories, safety devices and layouts prevented injuries. Automobiles acquired seat belts and airbags and were redesigned so that even drunk people would have difficulty killing themselves. Engineers also removed hazardous installations from the right-of-ways of highways or replaced them with breakaway poles and guardrails. In home environments, designers added safety features to appliances, and in public spaces handrails and other devices helped reduce injuries.[26]

Expansion of the Concept of the Environment

In this same time period, the idea of what constituted "the environment" expanded greatly. In 1979 a presidential commission concluded that "the major causes of mortality and morbidity are now recognized as having environmental factors in their etiology [causation]. The effects of various environmental factors on mental and physical health, and ways to control or circumvent hazardous factors are a significant focus of prevention research."[27] Commission members had in mind such things as chronic diseases; personal habits, such as the use of alcohol or tobacco; stress; accidents; and workplace conditions. Other leaders in health and medicine focused on still wider varieties of environmental factors. Beyond obvious surroundings such as housing and sanitation, now many other factors constituted a direct personal threat to individuals and populations. The long-understood connection between poverty and illness could be reinterpreted in terms of environmental stressors. Statistical epidemiologists were studying "the social etiology of disease."[28] Accompanying this extended awareness of environmental dangers were two closely connected ideas: lifestyles produced disease, and ordinary personal medicine, despite antibiotics, was not providing adequate protection for a person's health.[29]

The first stage of public consciousness of environmental dangers therefore was associated with chemicals such as insecticides or industrial poisons that caused clear-cut, symptomatic cases of impaired health. Then local disasters such as the notorious 1976–78 Love Canal tragedy led to the 1980s being labeled "the toxic decade."[30] Meanwhile a second stage supervened, with dire threats from agents that were ordinarily undetectable in one's surroundings. The original model, as I have noted, was radiation, which became well known by way of politics, if not science, in the years when atomic bomb testing flourished.[31] The detection of radiation in U.S. milk supplies after the bomb tests became notorious. But added to dangerous invisible radiation were now the innumerable chemical poisons and carcinogens in air, water, and dust. There was no escaping them. They were all around, even right in one's house. Where once the environment had been the out of doors, now it was everywhere, and it included even social relations with one's intimates.[32]

Experts, in attempting to explain how undetected environmental factors could impair health, often used the medical concept of a "subclinical" disease, that is, a disease present in one's body but undetectable by ordinary clinical means. The many new diagnostic technologies in fact frequently suggested that a person could have a disease and be unaware of any illness. The idea of subclinical or asymptomatic pathologies became ever more conspicuous in the 1970s. As a New York University toxicologist wrote in 1976, "There is and has been in the recent past a remarkable proliferation of new procedures in toxicology designed to give predictive information on the effects of environmental chemicals on humans. Instead of studies on humans themselves (clinical studies) a variety of biologic as well as biochemical systems have been exposed to environmental chemicals under many different conditions."[33] So environmental agents could indeed sicken a person without the person's being aware of it, perhaps even over an extended period.

Abandoned hazardous waste sites such as this one could harbor unexpected health dangers to people living nearby. Environmentalists sounded an alarm about such unknown sources of poisons, which could be present in anyone's surroundings. *National Water Quality Inventory*, 1988, xviii.

By the 1980s, experts had assembled an astonishing record of the ways a degraded environment could affect health. Ever more advanced and sophisticated scientists rapidly added to the list of chemicals that Americans were dumping into their surroundings and that over time could poison physiological processes or cause genetic changes. Eating, drinking, breathing, or just existing could introduce possible health hazards into the body. Or one could add up the ways each organ or physiological process was encountering possible hazards. According to one expert, "Chemicals may enter the body as deliberate additives to food in fertilizers, pesticides, herbicides, preservatives, oxidation inhibitors, etc. However, much more common chemicals, particularly hydrocarbons, enter in air or water, contaminated by industrial products and effluents." Even noise from one's surroundings could damage the ears and cause stress. Or radiation and chemicals everywhere could affect DNA. Incorporating statistics, laboratory and clinical reports, and details of events at the molecular level, the total record was intimidating and alarming, especially as chronic diseases became ever more prominent.[34] In addition, general social factors also contributed to ill health. One infectious disease expert pointed out to his colleagues that, in addition to microbes, the extent of crowding in housing could affect infectious disease rates.[35]

To top it all off, in environmental-era medicine even the work of the doctor could change, altering aspects of the standard doctor-patient relationship. Traditionally the doctor matched the patient's complaint and symptoms to a pattern, a process termed *pattern recognition*. By the 1980s physicians were expected to analyze the *functional relations* of the patient. What were all the factors, internal and environmental, that affected the functioning and malfunctioning of the patient? The complexity in the doctor's job made it enormously challenging.[36]

Confronting Health Hazards in the Environment

The extent to which proponents of general reform programs to improve the environment justified their programs on the basis of protecting human health is striking. In 1970 (the year when both Earth Day and the Environmental Protection Agency were created) a Senate committee confoundingly reported that "the health of people is more important than the question of whether the early achievement of ambient air quality standards protective of human health is technically feasible." In practice, scare tactics succeeded in producing many rational attempts to clean up the environment. Especially where even the possibility of birth defects or cancer was concerned, skeptics did not stand a chance. In 1985 a witty British scientist observed that Americans had become "thoroughly persuaded of the teachings of the Book of Genesis—carcinogenesis, teratogenesis and mutagenesis." Politicians had simply discovered that people longed for absolute flawlessness and purity so that, as they imagined, they and their children could lead lives free of disease and disability, without exposure to undetectable hazards in the environment.[37]

When those hazards became detectable, it was natural that opinion leaders would infer that more existed. Therefore, people with asthma and other allergies were the canaries in the mine for environmental medicine and environmental legislation.[38] Although allergy was identified in the nineteenth century, such a reaction was treated as an individual illness, commonly found among educated and monied people. After World War II, however, allergies, as typified by breathing problems, took on new social implications. Epidemics of serious asthma were identified in urban slums. By the 1960s and 1970s it was clear that asthma was particu-

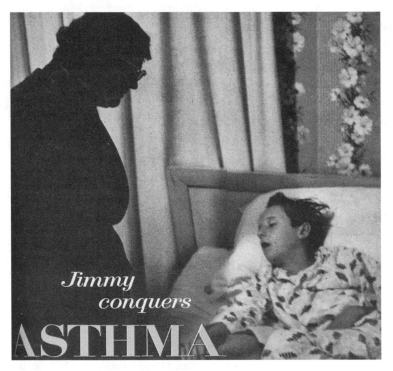

Even before the full onset of environmentalism, asthma, like that suffered by this clearly distressed boy, caused great concern and gained medical attention. In this case, midcentury journalists were attempting to show how individual medical care could offer at least a little relief to people who suffered from this crippling affliction, which could become fatal when the air became heavily polluted. A few years later, what had been an individual concern became a social concern as people perceived poisons in the environment that could affect everyone. "Jimmy Conquers Asthma," *Look*, 18 Sept. 1956, 79.

larly common among the people living in degraded environments the world over. In the United States the incidence of asthma reports increased by 29 percent just in the 1980s. As a leading specialist declared already in 1969, allergists "are intimately concerned with the total environments of our patients, and this includes, in addition to physical and chemical factors, social, economic, psychological, and educational factors as well."[39]

Three factors became prominent in the 1970s and 1980s. First, the line between poisons and allergenic (allergy-producing) substances was not clear, as in anaphylactic shock from peanuts or bee stings. Second, many of the allergenic substances were products of industrialization, not just polluted air but also, for example, useful things like latex, insecticides, and, for some people, penicillin. And, third, investigators from immunology joined allergy specialists to show how the process of allergic reaction triggered a cascade of somatic effects much more complicated than simple individual hypersensitivity. At the subcellular level, biochemical change was always possible. Pharma firms joined in and reinforced awareness by offering not only antihistamines but also, especially in the 1970s and 1980s, bronchodilators and other medical alleviators.[40]

Studies of the environment kept uncovering new items that sickened some people—and potentially everyone. In the 1960s and 1970s it was dust mite feces, found in every household environment. In the 1980s it was secondhand tobacco smoke, which made cigarettes a threat to everyone, not just to those who chose that particular personal indulgence. An article in *Good Housekeeping*, for example, posed a troubling possibility for female readers: "Can your husband's cigarette give you cancer?"[41] Indeed, a series of new diseases in the 1970s and 1980s, which showed that the germ theory was still operating, also underlined the endless potential of the environment to sicken people. Finally, by the 1980s some people believed that they and others were sensitive to many things in their built environments, the interior spaces in which they lived, especially their homes—generally termed *multiple chemical sensitivity*.[42]

Cancer as an Environmental Disease

The afflictions known collectively as cancer provided the model "environmental disease." During the 1970s especially, experts' thinking about cancer shifted radically: they began to think that, as one put it, cancer was "a disease primarily related to the environment," particularly to chemicals.[43] One expert in 1976 suggested that cancer could come from "what we breathe," "what we eat and drink," "skin exposure [including radiation]," "exposure to infection," and "social and psychological factors."[44] From the time investigators traced malignant growth to factors operating directly on the cell, most experts believed that factors that were not genetic had to come from the external environment. The question was, how, exactly? The quest to find a virus that caused cancer and to devise a vaccine on the polio model reached a peak in the 1950s and 1960s and then receded rapidly in the face of environmental factors. Meanwhile, animal testing for carcinogens (especially after a test for mutagens was devised in 1973) uncovered many possibilities among chemicals that could alarm the public. In 1979, officials at the National Cancer Institute finally had to issue a pamphlet, *Everything Doesn't Cause Cancer.*[45]

As investigators announced that first one and then another element in the environment could initiate the growth of cancer cells, the U.S. Public Health Service issued this pamphlet in 1979 to try to counter fear and hysteria set off by evidence that large numbers of environmental factors were making everyone vulnerable to developing malignant growths. *Everything Doesn't Cause Cancer* (Washington, DC: U.S. Department of Health, Education, and Welfare, Public Health Service, 1979), cover. Courtesy of Kent State University Library.

So it was all very well for statistics to show that tobacco smoke or an industrial chemical such as, in the 1970s, asbestos or vinyl chloride, or too much fat in the diet, or some other environmental constituent was a probable factor in setting off malignant growth. But the way in which any factor operated in a cell remained obscure. That question, as noted above, drove billions of dollars of research into subcellular and molecular biological processes, but the investigators' findings, no matter how scientifically interesting, did not appear to contribute to curing anyone of cancer.

At the beginning of the 1970s, then, cancer had several faces in the United States. First, and most conspicuous, was the longstanding, incredible popular fear, exploited and driven by fundraising campaigns beginning in the 1940s.[46] The public image was that the disease affected primarily "white," middle-class women, who suffered most from breast and cervical cancer, real problems that many women talked and learned about. The more popular image of cancer was loaded with stereotypes from "the feminine mystique," reminiscent of delicate early nineteenth-century heroines dying of tuberculosis. Then in 1972, physicians discovered a great increase in cancer in the "black population." Probably this increase was an artifact of reporting. Racist thinking had long contaminated the statistics, and with better diagnosis of more people, more cancer was turning up among minority populations, including those economically disadvantaged. It was an ironic aspect of equality in the civil rights era that "cancer crossed the color line."[47]

Cancer often meant a horrible, lingering death, a fact that was all the more significant for members of a population who, unlike their ancestors, were increasingly unfamiliar with death and dying. Just the mention of an unobvious malignancy would typically cause an emotional and perhaps irrational reaction. A dentist who had been (probably incorrectly) informed in 1962 that he had cancer of the liver and pancreas recalled: "I freaked out. . . . All your friends and relatives are ready to put you in a box when they hear the word 'cancer.'"[48] Scholars have studied the representations and rhetoric with which people expressed their horror, which was often so extreme that it would be comic if it were not so tragic. The historian James Patterson recounts the experience of a Cleveland surgeon:

> A seventy-five-year-old woman . . . suddenly could not speak. Her family doctor was baffled, but thought she might be suffering from a thyroid cancer that had spread to the brain. There followed a battery of tests. . . . [Then the surgeon] assembled the family and told them the news. It was far worse than cancer, which he might have been able to treat. "There is nothing that can be done," he said. "Your mother has suffered a stroke from a broken blood vessel; the brain is irreparably damaged. There is no operation or treatment that can

help." . . . The oldest daughter leaned forward, tense, and with a quaver in her voice, asked "Did you find cancer?" "There was no cancer," [he] replied. "Thank God!" the family exclaimed.[49]

Before and after midcentury, surgeons, with new technology, were operating more and more boldly to extend cancer patients' lives. But there were no other satisfactory ways of treating malignancies. In the 1940s and 1950s, radiation treatment (utilized with occasional success since the 1920s, alongside radium treatments) and chemotherapy, both with heavy side effects, became popular as additional ways of retarding abnormal growth. Like surgery, they had only limited effectiveness. Nonetheless, physicians and patients had recourse to them on a grand scale, inspired in part at first by utopian stories about benefits of atomic power. By the 1970s about one-third of the patients diagnosed with cancer could expect to survive five years or longer, which people at the time considered a notable improvement over the survival rates in preceding decades.

Trying to Prevent Cancer

In the presence of such a persistent disease, for which there was limited or no therapy, the best strategy was of course prevention. In general, clinicians and leaders of the American Cancer Society, with the help of producers of potentially carcinogenic products, portrayed prevention, not as social efforts to remove poisons from the environment, but instead as medical measures applied to individuals.[50]

In practice, then, the best "prevention" was early detection of a malignancy, which might lead to effective surgery. The Pap smear, particularly, had been around since the 1940s, and beginning in the 1950s it had been widely used to screen for cervical cancer.[51] In similar ways, prophylactic surgery for breast cancer became remarkably common for both "precancerous" and healthy breasts. The early detection campaign did bring patients into physicians' offices for examinations, but by the 1970s, when mammography screening changed the detection of breast cancer, physicians increasingly thought of patients in terms of risk groups (see part IV).[52] In the 1970s and 1980s another kind of prevention began to flourish. Researchers reported that first one and then another substance inhibited the development or growth of a malignancy. Immediately people thought that increasing their intake of the new substances might ward off cancer, and the media quoted experts advising people to eat one kind of food or another.

The War on Cancer

It was in this atmosphere of fear, plus "new hope for a cure for cancer" propagated by media reports of even the most tenuous and tentative laboratory studies, that

Mary Lasker and her allies in 1970 set out to make cancer their special cause. As the newspaper columnist Ann Landers famously wrote, "If this great country of ours can put a man on the moon, why can't we find a cure for cancer?" They persuaded President Richard Nixon to declare a "war" on cancer in 1971 (comparable to Lyndon Johnson's War on Poverty) and Congress to make the National Cancer Institute semiautonomous and answerable directly to the president, with enormous sums for research. The war on cancer was first financed in part by funds taken away from other National Institutes of Health units, it is true, but within six years the cancer institute budget had risen from $233 million to $815 million.[53]

One reason for separating the National Cancer Institute from the National Institutes of Health was to be able to use the results of all the expensive research that had already been carried out to devise cancer treatments and make them available to physicians. Two things happened. First, aside from fine tuning of existing measures, no "cure for cancer" emerged. Second, scientists continued to use funding targeted for medicine to do purely biological research. In the labs much was going on. In the 1970s investigators began to develop means for transforming DNA. And in the 1980s they identified regulatory molecules acting on the immune system that might produce therapies to contain abnormal growth.[54]

All in all, the huge governmental investment in research paid off. By late in the 1980s investigators using new laboratory resources, beyond electron microscopy, were learning so much about subcellular processes that they were reconceptualizing the problem of "cancer."[55] Scientists eventually concluded that the question was not just why abnormal growth occurred but why normal cell "death" processes became disabled, so that unwanted cells proliferated. The key factors could still be environmental, interacting with DNA or other factors in the body. But already by the end of the 1980s, "cancer" began to look very different from what it had seemed a couple of decades earlier. In 1991, two Ohio cancer experts wrote that "conventional thought" was being challenged. Four years later, two Massachusetts investigators looked back and declared, "It is mind-boggling that earlier pathologists (ourselves included) paid so little attention to the mechanism of organ shrinkage during atrophy," which would have led them to look into cell damage, atrophy, and death.[56] Once again, there was a dramatic payoff for funding pure, rather than targeted, or "applied," research. There was still no immediate cure, but now investigators were usefully reframing the question of how the environment (and genes) generated cancer.

Public Awareness of the Persistence of Disease and Death

Despite the attention to heart disease, stroke, and cancer, physicians, as well as patients, in the 1970s and 1980s found that in fact there were still infectious diseases all around. Most were treatable or self-limited. Beyond colds and the like, the most common of the traditional infectious diseases by the 1970s was gonorrhea, a fact that was probably also a comment on a changing social environment—which would soon have even more serious consequences.[57] Reflecting another social factor, just as the amount of alcohol consumed in the United States reached a new high point in the early 1970s, some startling laboratory findings suggested that alcohol consumed by expectant mothers was poisoning the physical environment of fetuses, setting off a concern about fetal alcohol syndrome. This concern stimulated a troubling controversy that continued for the rest of the century.[58] But the other health effects of alcohol, probably greater than those of tobacco, including deaths, gained little attention in society in general except for a self-contained drunk-driving campaign that did cut the death rate from accidents.[59]

From 1970 to 1986 the age-adjusted mortality rate in the United States dropped by 24 percent, continuing the trend noted above. Yet according to one poll, three-

Public health authorities used posters such as this one from South Carolina to warn women and their families about the danger of consuming alcohol when pregnant, based on studies showing that a fetus could be damaged even before birth when exposed to alcohol that was transported through the mother's body. South Carolina Commission on Alcohol and Drug Abuse, from the collections of the National Library of Medicine.

quarters of the population believed that life was riskier than it had been twenty years earlier. The facts were that people aged 34 and younger died most frequently from accidents; those from 35 to 64, from cancer; and those 65 and older, from heart disease. Nothing in this pattern was new except perhaps a decline in strokes and an increase in diabetes.[60] The paradox of public perceptions lay in the long tradition of fads not only in quests for cures but also in a continuing series of medical, public, and political concerns about diseases. Moreover, a number of public figures talked openly to the media about their afflictions. Television, radio, newspaper, and tabloid stories of sick celebrities "democratized the subject of illness," because accounts of diseases reached all social classes and even ethnic groups. In the 1970s and 1980s those accounts were amplified by environmental perspectives on infectious diseases.[61]

In the mid-1970s a new pattern of disease discovery and alarm appeared. Americans suddenly learned through the media about two infectious diseases that appeared to be new. These intruders into the health landscape showed up in the wake of the "Hong Kong flu," a serious outbreak in 1968–69 that was followed by a scare about "swine flu," when many responsible government scientists wanted to inoculate the whole population against a new, dangerous-looking influenza strain isolated early in 1976. Such a program happened to suit those with political agendas but little expertise. The scare, however, fizzled out, and among those vaccinated a statistically miniscule number developed Guillain-Barré syndrome. The vaccination program was suspended, and many opinion leaders lost a significant measure of trust in government science in general and vaccination in particular. Ominously, vaccination rates already were declining, leaving one-third of the children unprotected.[62]

"Discovering" Startling Environmental Diseases

The two new model disease "discoveries" were newsworthy because they revealed dangers that still existed to challenge "modern medicine." Fearsome, "unconquered" infectious diseases were in fact still about. The more deaths they caused, the more media attention they received.[63] Moreover, these new diseases, it became starkly clear, came from the environment.

The first of the new model diseases appeared in 1976, when physicians reported a serious lung infection, with fever, in attendees at the American Legion state convention in Philadelphia in July. Under great pressure, biomedical experts, using complex technological testing and standard epidemiological reasoning, soon eliminated all obvious toxins and known pathogens as possible causes. In the wake of the swine flu fiasco, news media and some medical authorities criticized the public health authorities and especially the Center for Disease Control for not immediately diagnosing the fairly distinctive disease that affected 221 people, of whom 34 died. The press named the mysterious disease Legionnaires disease, a name that ultimately stuck.[64]

Clearly, the malady was not passed from person to person, and suspicion rested on the headquarters hotel of the convention. Finally, five months later, on 18 January, the Center for Disease Control announced the isolation from lung tissue of dead Legionnaires a bacterium that investigators soon showed had to be the agent in the disease. Further investigation suggested that the bacterium, which was not unusual in fresh water, had been distributed by the hotel air conditioning system. Moreover, a similar chain of causes could explain several previously described but unsolved minor local epidemics—each one set off by a local environmental distribution of the bacterium, such as by modern ventilation systems.[65]

The alarm of the media spread to medical personnel. As two of the experts involved in the investigation concluded in 1981, "The most startling fact to emerge from the Philadelphia epidemic is that a major bacterial disease existed without detection until 1976, despite documented occurrence of earlier outbreaks. This serves as a clear and permanent warning to health professionals that nature may have similar unpleasant novelties in store for us." The surprise was one thing; the interpretation of the disease as a product of the air-conditioned environment was still another.[66]

While the mysterious Legionnaires disease was in the media, another baffling disease appeared. It too would get media attention, not just because it was a new disease but for two additional reasons as well. First, the initial cluster of cases appeared in a very upscale community, Lyme, Connecticut. If a disease could affect that kind of community, no one was exempt. And second, it was discovered by two ordinary citizens, women who were concerned about their families and their neighborhoods. They became exemplars of women activists who took on the Establishment just as doing so was becoming trendy.

Cases of Lyme disease clustered in wooded areas with expensive homes. One of the activists was Polly Murray, an otherwise inconspicuous wife and mother who saw her whole family, starting with her athletic, high school senior son, affected by terrible symptoms in 1974: "We had our water tested. We had everything we could think of tested. We wondered if we should move. And still the doctors just shook their heads. We were living in a kind of medical limbo." When she found that other families in the area were having similar problems, Murray, and, independently, another wife and mother, Judith Mensch, contacted the state department of health and ended up with a superb team of investigators who, because of the clustering of cases and their inability to find any obvious cause, quickly concluded that they had encountered a new disease. They learned only much later that it had been described under another name in Europe.[67]

In Lyme, Connecticut, Polly Murray poses with her husband and two sons, the victims of Lyme disease. She led a successful effort to get medical investigators to identify the disease and figure out how it was spreading. Elisabeth Keiffer, "Mrs. Murray's Mystery Disease," *Good Housekeeping*, Mar. 1977, 80.

By 1977 the investigators had concluded that the disease was spread by ticks, and eventually they figured out that deer, which had started returning to wooded areas in large numbers (there were 30,000 in Connecticut in 1976), along with the white-footed mouse, were harboring what at first they thought was a new kind of tick. Later, birds also appeared to be vectors. The pathogenic agent was not a virus but a spirochete, confirmed by additional advanced technology available only at the beginning of the 1980s. For biomedical scientists, the twists and turns of knowledge evolved into a decades-long controversy about the pathogen and the carrier tick, a controversy complicated by political factors introduced by patient support groups.[68] For the public, particularly outdoors people and suburbanites, who were noisy elements in national opinion leadership, Lyme disease was a new hazard in the supposedly healthful and desirable natural environment.

Still another disease alarm in the new model came in 1979–80, when a number of physicians discovered in American patients a syndrome named by a British investigator in 1978: toxic shock syndrome. These cases had an epidemiological link to women, and publicity about toxic shock syndrome spread especially through media aimed at women. Most of the cases first described were young women who happened to be menstruating. And it turned out that all of them had been using tampons, which had recently become very popular, especially among girls and young women. The symptoms were extensive, fast-acting, and alarming, because they came from a toxin affecting many body systems, with a drop in blood pressure. It was "like being hit from the rear by a Sherman tank," reported one victim who survived. There was a significant death rate, perhaps 5–10 percent.[69]

The public health and biomedical authorities worked quickly, so that by 1980 the connection was made between tampon use and a strain of staphylococcus. The exact strain was not known until well into the 1980s. Meanwhile, the alarm spread. The alarm did not reach everyone, however, and it was deeply affecting when a healthy, active sixteen-year-old would become violently ill and then die—from using a standard commercial product advertised with words like "confidence" and "security"— part of the comfortable technological environment in which Americans then lived. Makers of one brand of tampon, for example, boasted in a 1978 ad that this "internal protection" was "biodegradable and environmentally sound." Moreover, in succeeding years other fast-killing strains of streptococcus and other bacteria common in the environment appeared, leading a science writer in 1990 to observe that "bacterial diseases are still far from being understood at the molecular level."[70]

The Shock of AIDS

The ultimate disease alarm came with the recognition of AIDS (acquired immune deficiency syndrome) in 1981–85. As one chronicler of AIDS wrote, "There was life after the epidemic. And there were recollections of the times before. Before and after. The epidemic would cleave lives in two . . . [as would] a war or depression."[71] Nor was it just victims and people who knew the victims who realized that the world had changed. One physician recalled his intern and resident days in a big, busy hospital at the opening of the 1980s, that is, "before":

> My fellow interns and I thought of ourselves as the *vaqueros* of the fluorescent corridors, riding the high of sleep deprivation, dressed day or night in surgical scrubs, banks of beepers in our belts. . . . We strutted around with floppy tourniquets threaded through the buttonholes of our coats, our pockets cluttered with penlights, EEG calipers, stethoscopes, . . . hemostats. . . . Seven-inch needles with twelve-inch trails of tubing. . . . Only cancer was truly feared, and even that was often curable. When the outcome of treatment was not good, it was because the host was aged, the protoplasm frail, or the patient had presented too late—never because medical science was impotent.[72]

And then the spell was broken. In 1981 clinicians and public health workers identified the first victims of AIDS as sexually active men who were homosexual. As the months passed, other groups showed up with the infection: hemophiliacs, drug injectors, and, ultimately, women and children, indeed anyone. The numbers of African Americans and Hispanics in urban slums were disproportionately high, although it was not long before small towns also produced victims. AIDS brought fear, prejudice, concern—and relentless death.[73]

Historians have traced the first reports of AIDS in early 1981 to physicians in Los Angeles and New York who were encountering inexplicable pneumonias and skin cancers in otherwise healthy young people. These patients had in common a startling destruction of their immune systems, so that they had little resistance to infections. Their T cell counts could reach an impossible level of zero. (T cells, critical units in activating adaptive immune reactions, had been characterized in just the few years preceding.) And it appeared that the case mortality rate for this syndrome was actually 100 percent. All patients would die. Already in August 1981 U.S. epidemiologists had counted 108 cases of a disease that was "unexpected and undefined," as one investigator recalled.[74] Within two years there were 1,972 known cases, with 759 deaths. By 1987, 12,000 cases had been identified in a process that horrified medical experts.

Biomedical scientists in France and the United States with amazing speed identified as the infectious agent a retrovirus, a type of agent known to exist in humans only since the end of the 1970s. At first the disease, which elements of the religious Right sometimes called "the gay plague," became well known only among communities of male homosexuals, but by 1983 some general media carried alarming stories, including evidence that heterosexuals had also been infected. As early as 1982 NBC News had reported that "an unknown and mysterious disease is spreading. It has killed more people than toxic shock syndrome and legionnaires' disease combined." In 1983 a Gallup poll showed that 77 percent of the population had heard of AIDS. *Newsweek* in April of that year quoted a series of experts: "In my professional career, I have never encountered a more frustrating and depressing situation"; "People who you know are likely to die ask what they can do to help themselves, and you are forced to say . . . 'I have no idea.'"[75]

By 1984 the magnitude of the danger was becoming apparent. When 60–70 percent of asymptomatic homosexual men in a new test tested positive for the HIV retrovirus, the NIH investigator Robert Yarchoan received a huge shock when he did "a rough mental calculation of the number of gays in the country and the percentage who were likely to be HIV-infected, . . . estimating that there were half a million to a million people infected with this lethal virus who did not know it."[76] The more general public became more aware, if not educated, when the media in

1985 revealed that a famous all-American leading-man movie star, Rock Hudson, was dying of AIDS and when journalists reported that panicked local school officials in Indiana, to protect their students, had refused to allow the teenage hemophiliac Ryan White into his middle school classroom after he was infected (fatally, it turned out) by a blood transfusion. Two policy analysts later pointed out that television news programs were especially effective in publicizing AIDS. Just as Vietnam had been the first "living-room war," so AIDS was the first "living-room epidemic."[77]

The Impact of AIDS

Even aside from the wildly escalating costs of supporting dying AIDS patients, the social impact of the AIDS epidemic was enormous. At first scientists thought that the cause might be a chemical, but soon it became clear that the source was a pathogen that existed in the social environment. In a society that had been rushing headlong toward a kind of anything-goes, "pornotopian" social system, AIDS was the payoff for sex, drugs, and rock-and-roll, according to some. In fact, new attention to sexually transmitted diseases made physicians more conscious of histoplasmosis and chlamydia. Unlike earlier generations, young people in the time of AIDS discovery could not remember any such terrifying epidemic, not even polio.[78] Eventually, between 1987 and 1995, investigators devised a costly drug regimen to control cases. But public discourse and, surprisingly, public policy and public behavior had changed. AIDS had become "normalized," a presumed part of the environment.[79]

Nevertheless, for the next thirty years the specter of a new infectious disease plague like AIDS emerging from the natural or social surroundings haunted those concerned with health. Indeed, that fear was realized in 1987, when a veterinarian in the United Kingdom announced that he had found a disease in cows that would ultimately alarm populations everywhere and lead to an enlarged idea of infectious diseases. But that story belongs in another chapter. Meanwhile, victims and clinicians in the 1970s and 1980s who wanted to obtain recognition for specific disease syndromes as worthy of fear and study found that they had to manipulate the media and politicians. Such was particularly the experience of those concerned with the tragic syndrome known for many years as "crib death," in which, inexplicably, an apparently healthy baby died suddenly. As infant mortality rates declined, the steady incidence of crib death became more noticeable. Experts estimated that by 1969, when an international conference devised the term *sudden infant death syndrome*, or *SIDS*, there were at least ten thousand such devastating deaths each year in the United States—and yet there was hardly any research support. Parents had formed support groups and with some physicians obtained publicity and political leverage, so that even before they learned about AIDS, a majority of Americans had heard of SIDS.[80] But for another generation the process that brought death to the babies remained a mystery.

A New Factor: The Immune System

It is a profound irony that clinicians saw the destruction of the immune system as the central feature of AIDS. In fact, in the 1970s the immune system was just becoming established as a dominating component in health and in a new model of one's body, a model in which the environment functioned as part of normal human body processes rather than just impinging on disease states. Immunology since the late nineteenth century had involved identifying agents of infection and finding ways to counteract them, as with the long-established smallpox vaccine or the diphtheria antitoxin of the 1890s. For decades, further advances were few and hard won. The eradication from the earth of a second disease, after smallpox, did not occur until 2011, and it was a disease not of humans but of cattle: rinderpest. Meanwhile, investigators used less biology and more chemistry to try to figure out how antibodies actually worked to "fight" disease. Then after World War II, with many new technical developments, immunologists shifted from a chemical back to a biological explanation of immunity as they uncovered the complexity of pathological and immunological events. They now had the advantages of working on the molecular level and of having experience with the extreme form of allergy, immune reactions to skin and organ implantations.[81]

Beginning in 1967 and into the 1970s, immunologists combined protein biochemistry, genetics, and cellular/subcellular biology into one model and developed a major new view of events. The earlier, traditional idea, using a fortress model of the body, had been that a person should try to keep pathogens out and away by care and cleansing. Now, with the coming of environmental medicine, immunologists brought to both popular and medical views of the body the idea that huge numbers of antibodies in the body interacted with the environment to activate defense reactions.[82]

Thus was born "the immune system." The system idea involved both extreme complexity and built-in automatic mechanisms that kept the system in balance, akin to the cybernetic models used in military and business systems. By 1979 it was possible for biomedical theorists to use a systems model and view "health" as "the ability of a system (for example, cell, organism, family, society) to respond adaptively to a wide variety of environmental challenges (for example, physical, chemical, infectious, psychological, social)," while "disease" could "be seen as a failure to respond adaptively to environmental challenges resulting in a disruption of the overall equilibrium of the system." And within another decade, investigators found that immune reactions to microbes originated with local cells everywhere in the body.[83]

The Immune System and the Environment

All such thinking involved complex reasoning about interactions among many types of factors in a person's operating in a *biopsychosocial* world, a term introduced by the psychiatrist George L. Engle in 1977. This concept embodied an attempt to move beyond simpler models of psychosomatic medicine and narrow, reductionistic laboratory medicine. What rendered the new model particularly plausible was the continued development of knowledge about molecular and subcellular processes in basic biomedical science. Indeed, by the late 1980s, investigators' growing ability to show a connection between genetic factors and environmental factors in the production of disease processes had suggested that a major shift in medicine was under way. Medical students were admonished to look at the whole world of the patient. Yet medicine had gone so far from the bedside to the laboratory that even in the *Journal of Clinical Investigation* few articles concerned clinical material. The rest came out of molecular biology laboratories, in which factors in immunity were just part of general pathophysiology.[84]

Then the concept shifted, to the complex immune system interacting with complex systems in the environment to produce a dynamic working arrangement. The immune system thus became part of very intricate, everyday physiological processes. In the words of a medical student who found himself having to abandon the military metaphor of defeating an invader, "You are interacting with the environment on a kind of more interface level . . . you are part of an environment and this is what's happening to you."[85] From one point of view, the earlier cybernetics now extended to the external environment.

The new model fitted the experience of people at the time. Most had various immunities induced routinely as they grew up, such as through the childhood DPT and polio shots. They did not experience the traditional infectious diseases as special, specific threats. What they did know about was interaction with the environment, as in Legionnaires disease or Lyme disease, or a systemic breakdown that could become total, as in AIDS. By the beginning of the 1990s both writers in the popular media and biomedical investigators were treating the immune system as a major component of human health.[86] Breast feeding babies was good because of what it did for the immune system. Exercise was similarly interpreted.

Stress, the suddenly popular new disease arising presumably from environmental social factors in the patient's biopsychosocial world, could affect one physically, usually, it was now believed, by inducing changes in the immune system, along with other body systems. Stress had continued to be a more frequent topic in both techilical and popular media since the 1950s. Despite competing with "anxiety," which tranquilizer merchandisers publicized to sell Miltown and Valium, stress by the 1970s and 1980s had been incorporated into environmentalism. It did not seem to matter whether the stress factors affecting a person were physiological or psychological. They came from the biopychosocial environment, including housing and working conditions, and changed not only emotions and the immune system but fundamental operations of the body, that is, one's health.[87]

New Versions of Previous Historical Stages

In the environmental age of medicine, all the eras described thus far continued to evolve on the basis of momentum from the past and adaptations to changing times. In the 1970s and 1980s, characteristics of earlier eras often had new technological faces, but there were the same problems and aspirations from the eras of germ theory, surgery, physiology, and antibiotics. Antimicrobial technologies not only continued but enabled tremendous changes in medical practice, such as ever more daring surgeries or commonplace knee and hip replacements or control of many streptococcal problems, thus also changing patient expectations.[88] In the 1970s and 1980s, however, the legacy of technologies also included not only specific products and techniques but ideas about how technology and the environment interacted, processes often identified by statistical epidemiology.

Technology seemed unstoppable, as if it were a force of nature. Indeed, investigators and inventors still could see what the next step in a pharmaceutical or device should be, and then when they reached that presumably inevitable next step, they had the illusion that they were seeing and making "progress." Medicine was thus cast as a major component of a better life and the American dream. In practice, U.S. physicians and hospital administrators felt pressure to "keep up" with whatever was the most up-to-date medicine, which, with the help of commercial marketing, they typically read in terms of new gadgets and chemicals. And this pressure affected physicians and scientists everywhere. All health care workers, however, knew that the United States was critical, for it was there that one found by far the most demand, and the most money, for ever more modern medical technology.[89]

The most spectacular development was physicians' gaining the ability to see what was going on in a patient's body beyond the x-ray. So great was the rush that hospital executives who ordered the latest technology might find that it was out of date by the time it was delivered. In the 1970s computers, which already had promise for communication and analysis, combined with radiology and variations on radiology to produce imaging. The official dates are: 1972–73, CAT scans; 1975, positron emission tomography (PET); and 1977, magnetic resonance imaging (MRI).

Meanwhile, ultrasound joined these resources that permitted noninvasive exam-ination of tissue in the living body. In 1976 a medical journal editor complained that "CAT fever has reached epidemic proportions and continues to spread among physicians, manufacturers, entrepreneurs and regulatory agencies. . . . The costs of this epidemic are staggering."[90]

The devices permitted physicians to visualize structures inside a patient's body so completely that they could make diagnoses that previously had been possi-ble only on autopsy, such as rare diseases of the meninges. Indeed, the interior

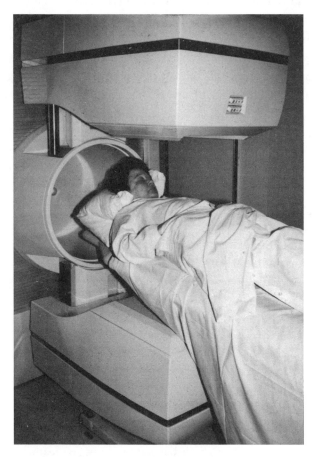

When one of the earliest positron emission tomography, or PET, machines was first employed at UCLA, experts realized that the occasion was worth photographing. This machine was a harbinger of the widespread use, in the 1970s and after, of computer-assisted readers that revealed in extraordinary detail events inside a patient's body without harm to the patient. William G. Myers Papers, Medical Heritage Center, Ohio State University. Reproduced by permission.

environment of bodily processes provided another dimension for environmental medicine.

Beginning in the 1960s, radioimmunoassay, a technique pioneered by Rosalyn Yalow and Solomon A. Berson, at the Veterans Administration Hospital in the Bronx, to identify insulin in the bloodstream, became a revolutionary force in many aspects of medicine and medical research and led to a Nobel Prize. By the 1970s and 1980s the technique would be used to identify not only chemicals but even viruses.[91]

While the most spectacular technological developments came in mechanical aids to diagnosis and in endless numbers of new laboratory procedures, physicians more and more often used technologies in treatment. New pharmaceuticals were developed, not least an antidepressant SSRI, Prozac (approved in 1987), and omeprazole, or Prilosec, antacid (1989). The physiological monitoring equipment used in surgery was repackaged for use at every hospital bed, incidentally giving the nurse monitoring the monitors a major new role in patient care.[92] For patients in danger, the mechanical respirator, a descendant of the iron lung, was constantly improved and joined dialysis in keeping the body functioning when otherwise the patient would have died a definitive death. The respirator became familiar particularly from press coverage of an unfortunate young woman, Karen Ann Quinlan, who in 1975 entered into a coma from which she never recovered. In the media, her case became emblematic of a machine that complicated questions of when one was alive or dead and whether the machine could be turned off so that the patient could return to nature.[93]

In the 1970s and 1980s, then, numerous new pharmaceuticals and devices continued to appear. Not least were new materials and gadgets for replacement parts, popularized as "bionic" spare parts out of which a human being could be constructed.[94] Where once there had been unmodified pieces fitted in, like a wooden peg leg, now there were artificial limbs that could bend and grasp, and prosthesticists, continuing ideas from the two world wars, emphasized restoring function—as was particularly appropriate in an age of environmental medicine. Indeed, it was hard to discern the line between cosmetic surgery and restoring function. At what point did imagined ugliness or shortcoming become socially so extreme as to become a medical condition?[95] Technology was raising increasingly difficult questions.

Sensitive physicians too wondered about the patients on whom they were using technology. The physician Abraham Verghese recalled a young man, unconscious and collapsed, whom he and his team had successfully resuscitated, pumping oxygen into his lungs. Several hours later the patient awoke in the ICU, hooked up to innumerable technological devices, besides the respirator tube delivering

Two nurses monitoring a patient at Doctors Hospital in Lake Worth, Florida, in 1973. The nurses could read vital signs and other information at a distance from the patient on television screens that would, among other things, permit one person to keep watch over instruments recording the conditions of several patients simultaneously. "Modular Monitor: Fix It Now, Make the Repairs Later," *Modern Hospital*, Nov. 1973, 83. Reprinted with permission, copyright © Crain Communications, Inc.

oxygen through the trachea, which he gagged on and tried to tear out. "One can only imagine his terror at this awakening: naked, blazing light shining in his eyes, tubes in his mouth, tubes up his nose, tubes in his penis, transfixed by needles and probes stuck into his arms. He must have wondered if this was hell."[96]

People in the health field were also pointing out that the technology in itself could reconfigure what was happening in a medical situation. Monitoring a fetus when a woman was giving birth implicitly created an adversarial relationship, possibly in opposition to the interests of the mother, which many people considered a horrible perversion of birthing. One MD remarked, "Call to mind an ICU with monitors blinking and beeping and remember how all eyes (even family members') go to the monitors—and away from the patient." Clinicians continued to be alarmed as machines and laboratory tests substituted for clinical examination of the sick person.[97] Moreover, the expanding array of drugs and treatments in the 1970s made physicians suddenly aware of new dimensions of an old problem: patient noncompliance. The shocking figures on how seldom patients followed instructions, especially for taking complex and potent pharmaceuticals, was met with "awestruck disbelief" on the part of physicians, who had not seriously calculated this possibility into their work.[98]

New technology, moreover, still produced diminishing returns. In this later stage of technological medicine, physicians again had to exert greater effort to produce less total health benefit. In 1974 the economist Herbert E. Klarman pointed out what should have been obvious: "Renal dialysis, cancer chemotherapy, and open heart surgery may achieve dramatic effects in particular cases, but bring about only marginal improvement in general indexes of health."[99]

Surgery

Surgery was more than ever the conspicuous face of medicine, and not only in the media. Per capita, surgery increased 12–20 percent in just the years 1963–65 to 1976–78.[100] Surgery continued to be dominated by technology, both old and new, and was increasingly sensitive to the environment of the operating theater.[101] Organ transplantation, well in place in the early 1970s, was further extended and improved every year, to the point that there was distressing competition for parts of donated bodies. People noticed that enthusiastic surgeons built hope without perhaps weighing adequately their failures and the "quality of life" of their temporary successes with kidneys, livers, lungs, and hearts. By the 1980s, coronary artery bypass surgery and balloon angioplasty were undergoing explosive growth, transforming the treatment of heart patients.[102] Yet another trend of the time was using technology, including lasers, to carry out surgery that was "less invasive," a phrase found with increasing frequency in the technical literature. Fiber optics, color television, other technical elements, and cost saving helped transform many surgical procedures.[103]

But many older techniques were subjected to technology as well. As one senior surgeon complained, "Three-quarters of what I do today I never learned in residency." Indeed, in the mid-1970s there was a howl of journalistic protest when people found out about the suddenly widely accepted custom of having surgical equipment salesmen "scrub up" and join the surgeon in the operating room during an operation to instruct the physician in how to use a new or improved instrument.[104] Moreover, the record suggests that many U.S. surgeons were both conservative and not as talented as they might be, a conclusion inferred when they were very slow to adopt new techniques successful in Europe, such as those for repairing fractures.[105]

Technology and changing economic conditions set the stage for a different rhythm for surgical patients. Instead of being kept in the hospital for days to recover, patients were discharged rapidly. Indeed, more and more operations became one-day procedures, done on an outpatient basis. Furthermore, although ambulatory surgery expanded dramatically in hospitals in the 1980s, freestanding, privately owned outpatient surgical facilities sprang up everywhere in competition with the hospitals. By the late 1970s even Blue Cross organizations were privately encouraging physicians to do as much surgery as possible on an outpatient basis to save on hospital bills. This was the template for a major change determined by both business models and new technology. Health care units now reported in terms of the actual cost of surgical procedures, the volume of patients, and the price and service competition.[106]

Biomedical Research

As noted in chapter 10, governmental research expenditures leveled off at the end of the 1960s. Private foundation money likewise did not increase. Corporate expenditures, however, expanded in the 1970s and 1980s, confirming the observation that economics became the dominating driving force in medical technology after the 1960s. In 1970 the great Pharma firms' spending on research was much less than $1 billion; in 1990 it was more than $9 billion (both figures in constant dollars).[107]

Biomedical research in the United States was not the same by the 1970s and 1980s. As one eyewitness explained, something had been lost, "and it is best described as the loss of a boundless optimism about the future of academic medicine. In the middle 1960s, all may not have been right with the world, but it was about as all right in the microcosm of American medicine as it had ever been before, or would ever be again. . . . It was a simple matter to identify the 1960s as the decade during which experimental medicine peaked." There was money, and there was intellectual excitement.[108] In succeeding decades, despite the exciting work, the frontiers of knowledge appeared to be more limited, as all of science and medicine came under attack and as money became more the measure of science even among opinion leaders.

There was another notable change in biomedical research, commented upon at the time: more and more investigators held PhD degrees. Even in medical schools, research units became ever more populated by PhD doctoral and postdoctoral students. The MDs who did research, not only laboratory research but clinical research, were increasingly outnumbered by either MD-PhD investigators or those holding only a PhD.[109] It was another sign of the fragmentation of medicine as even researchers developed independent identities.

Further Shifts in Physician and Specialist Functioning

In the 1970s and 1980s the activities of the half-million or so recognized physicians became more and more public. They had to bargain with third-party payers. They had to deal with multiple bureaucracies, governmental regulations, and ethical injunctions. The profile of the body of practitioners continued to change. As in the general population, older and younger cohorts in medicine grew proportionately larger as middle cohorts shrank. Increasing numbers of doctors were in some kind of group practice; already by 1975, sixty thousand, or 17 percent of practicing physicians, were in private group practice, compared with 38 percent in private office practice. Other kinds of group practice included a strong movement into HMOs and physician organizations (see chapter 12). Waves of superficial trends that affected physicians were caught in a statement overheard in a doctors' room in a hospital in 1972: "I was into general practice, then into holistic medicine, then into community medicine, then I was a generalist, then a primary physician, and now I am into health care delivery as a physician of the first referral. But my patients still call me doc."[110]

Although more and more MDs worked for a salary in a bureaucratic organization, in the 1980s physician compensation held steady or increased significantly in constant dollars. Moreover, physicians worked hard—in 1971 an average of fifty-six hours a week, when forty hours a week was standard among U.S. workers.[111]

And physicians continued to specialize. The number of doctors in general practice actually declined in the 1970s, even as the total number of physicians in practice increased greatly. Within that increasing total, in 1970, 80 percent were in specialty practice, and by 1989 the figure was 89 percent, including now, of course, those certified in family practice.[112] In the 1970s four new specialty boards were approved. None was approved in the 1980s, but between 1972 and 1989 within the existing boards a total of thirty-five more subspecialty certificates were approved, in areas ranging from pediatric nephrology to surgical critical care.[113]

Physicians in general practice, now usually called primary care physicians (which included many internal medicine specialists, not to mention pediatricians), still were an essential part of the system. A disproportionate number of the residents who were training in primary care programs, 25.7 percent in 1990, were graduates of foreign medical schools.[114] Unlike in most of the rest of the world, in the United States patients often went directly to specialists, without waiting for a referral from a primary care doctor. Because there were so many specialists in the United States, specialists tended to expand their services to include primary care. Primary care physicians continued to perform specialized services, and there was an implicit, disorganized competition. Yet it was the specialist who fitted the doctor image best at the end of the 1980s, as one major commentator observed: "The heroic family doctor, lovingly portrayed in Norman Rockwell paintings, is no longer the model popular physician. He has been replaced by the white-coated specialist describing the latest life-saving advance in diagnosis or treatment on a news broadcast."[115]

Beyond the virtual communities of specialists, in actual geographical space, physicians in urban areas tended to locate in medical office buildings, which could be very large, easily accommodating the groups into which MDs were tending to

The storefront clinic in a strip mall represented a new pattern in delivering medical care, following a retail, commercialized model that pharmacists, dentists, and other health care providers had pioneered. Storefront clinics presented a challenge to both hospital medicine and the long-term, personal care many Americans imagined physicians had once provided. David Charles Sloane and Beverlie Conant Sloane, "Los Angeles, 1990s," *Medicine Moves to the Mall*, 124. © 2003 David Charles Sloane and Beverlie Conant Sloane. Reprinted with permission of Johns Hopkins University Press.

cluster. These buildings generally were not in the old central cities. Instead, unless they were located near a hospital, medical office buildings were in the suburbs, near the new shopping malls. Eventually, especially beginning in the 1980s, as hospitals established small outpatient or specialty units, clinics could even spring up in strip malls, areas where podiatrists, acupuncturists, massage specialists, dentists, optometrists, and other health care retailers had already appeared. More and more, medicine was just another consumer item that location and parking made "convenient." The freestanding "emergency care centers" became known as "doc-in-a-box" (or "quack-in-a-shack"). In a "drive-through" culture, the continuity of care that earlier had been so important became less so, as in the case of "one-day surgery." Moreover, in the postindustrial mall, clinics could come and go. By the mid-1980s in Houston there were more urgent-care clinics than hospitals, in part because the cost at a clinic was half that of a hospital emergency room but also because of the niche marketing that was another mark of a postindustrial consumer society.[116]

Paramedicals as Competitors to MDs

Each group of specialty doctors had a corps of paramedicals and technicians who often worked in teams with physicians and nurses.[117] In the 1970s, when patients sometimes felt neglected and many people detected a doctor shortage, a new kind of paramedical appeared, in this case one who offered possible competition as well as opportunities for physicians to attain wider coverage or a more advanced practice. This was the nurse-practitioner, or the physician extender, that is, someone knowledgeable in medicine who could act as a primary care specialist and relieve physicians of routine patient care. One of the original models was the former medical corpsman, who was familiar to those who had served in the military and who, it was hoped, might ease the doctor shortage. An independent physician assistant training program began in 1965, and by 1972 there were twenty programs, drawing mainly male students. Also included in the category of recognized practitioners were specialists such as longstanding nurse anesthetists and a new small corps of educated nurse midwives, most of whom practiced in rural areas and urban slum neighborhoods. By 1984 there were twenty-eight nurse-midwifery training programs, and the number was growing. All of these types of physician-aide groups, each with a unique history, became well established in the 1970s, when primary care physicians were in particularly short supply. By 1990 special training programs were graduating more than three thousand paramedical practitioners a year, and the number increased in the following decades.[118]

Even though the physician shortage eased in the 1980s, nurse practitioners especially continued to find increasing acceptance in primary practice medicine, typically but not always in outpatient settings, not least because they offered another path to lowering the cost of medical care. Pediatricians in particular were friendly to nurse practitioners. But both medical organizations and nursing organizations were ambivalent about accepting and certifying primary care paraprofessionals and often opposed legislation for licensing or certifying advanced-practice paraprofessionals even as university training programs sprang up and federal funds supported paraprofessional training. The result was that nurse practitioners and other medical assistants became part of health care teams or physician practices in very different circumstances and local conditions.[119]

The Profession of Physician

In the 1980s sociologists detected that physicians, for good reason, were concerned about their status as professionals. Some experts wrote of the "deprofessionalization" of physicians. What was happening to physicians' monopoly over their knowledge, which had given them an advantage over other members of society? Not only were patients much better informed than in earlier times but the dawning age of computers was giving lay people access to medical knowledge. As one expert pictured it,

> No longer is the individual doctor expected to continue to dispense knowledge and skills learned during his/her socialization in medical school, hospital training, or even practice. Now a well-defined decision/support system can dictate the flow of events, prescribe optimal information, and provide standards for proper clinical decisions (or standards to judge clinical performance as quality assurance) that the practitioner must meet. While these decisions/support systems may appear to exist for meeting the diagnostic and treatment needs of the patient via the doctor/patient relationship, they may be more concerned with meeting corporate needs of accountability, profitability, insurability, and bureaucratic standards. Such use of information technology has an impact on the doctor's job.[120]

Many physicians noted that the image of their work was changing in the 1970s and 1980s. As Marc Berg put it, medical leaders began to think in terms of the new fad of cognitive psychology, and "the scientific status of medical practice was redefined as *a feature of the physician's mind.*" That is, the individual physician was supposed to think and act like a computer so as most effectively and efficiently to apply standard "scientific knowledge." That meant using statistical information to understand the patient. It also meant that the language and thinking of economics would become part of medicine as the doctor calculated what treatment would be "cost-effective." In 1981 a group of medical leaders actually organized a Society for Medical Decision Making. Both the institutions and the language of medicine revealed how practice was reacting to the cultural context of those times.[121]

The actual content of medical practice was in fact generally unexciting. The most common health insurance claims were for procedures connected with childbirth, including uncomplicated deliveries, cesareans, circumcision, and episiotomy. Back problems and gastrointestinal problems were also common complaints, along with cardiac and circulatory afflictions. The only acute infection that was common was pneumonia, mostly in older people.[122]

As always, local practices and specialty communities embodied traditions and continuities in health care. But in the era of environmental medicine social and political changes brought jarring new viewpoints and circumstances to Americans' attempts to understand and implement ways for humans to maintain their health. It was a time of rapid and, now, as the new diseases featured in the media suggested, unexpected change. Finally, and ironically, by the late 1980s, with the dawning of genetic medicine, clinicians and scientists would focus once again on the ill person and his or her particular illness.

CHAPTER 12

Environmental-Era Health Care in a Hostile Social Climate

In the course of the 1970s, especially after the oil crisis of 1973 and the greatly underrated financial upheaval that followed, health care institutions and personnel operated under altered conditions. In this chapter, the focus of the story of the 1970s and 1980s shifts from the ideas and practices in the health care system to the interactions of medicine and health with the wider American society. Some of the interactions were stimulated by changes in the culture, many were stimulated by shifts in economics and politics, while others were kindled by elements specifically targeting physicians and their institutions.

Two groups in particular led movements and programs that addressed health care directly: a host of activists and reformers introduced briefly in chapter 11 and leaders from business and government or nonprofit organizations. In 1983 one observer reported: "Outside the profession, there are various efforts to cut medicine down to size: widespread malpractice litigation, massive government regulation, and various attempts by consumer groups and others, to redefine medicine as a trade rather than a profession and the physician as a morally neutered technician for hire under contract."[1] What that observer did not say was that critics of health care often were just making physicians and health care institutions scapegoats for the failure of the whole culture to provide ways to deal adequately with human suffering, pain, and death.

Medicalization and Demedicalization

Another background factor was also operating. In the 1970s sociologists and then popular writers discovered *medicalization*. In various cultures, and particularly that of the United States, opinion leaders had reconceptualized or construed ordinary problems or stages in life as medical problems that physicians could deal with, or were expected to deal with. Some thinkers perceived the expanding authority of medicine as a conspiracy of physicians to gain power and business. Others viewed the change as secularization of life or technological utopianism. But there was no question that many areas of life moved at least partially into the medical realm, and a large number of Americans expected doctors to deal with these problems (a job many practicing physicians did not welcome). As one physician complained in 1988, "Uncle Sam has become Uncle Sam, M.D. Every infirmity and every ache seem to merit treatment. We have medicalized a whole range of human miseries and misfortunes that in the past were outside the doctor's jurisdiction." A sociologist's list in 1992 included social and individual problems: besides criminal behavior and children's misbehavior, "madness, alcoholism, homosexuality, opiate addiction, hyperactivity and learning disabilities, . . . eating problems, . . . child abuse, compulsive gambling, infertility, and transexualism." Also medicalized were natural life processes: "sexuality, childbirth, child development, menstrual discomfort (PMS), menopause, aging, and death."[2]

It is true that some physicians, particularly a number of pediatricians and psychiatrists, wanted to use their medical expertise to improve the world generally, and the interests of many of them, along with some general practitioners, extended beyond the biological aspects of patients' lives to include the psychological and social aspects as well.[3] One of the attractions of specialty practice, however, was that the narrow specialist usually did not have to deal every day with borderline areas such as delinquency, dyslexia, obesity, and sexuality. Consciousness of medicalization facilitated the growth of least two medical/nonmedical areas that became very important in health care in the 1970s and 1980s: pain management, for both acute and chronic patients, and the hospice movement, for end-of-life care.[4]

Yet after the 1960s, observers could also point to less medicalization in society and even to specific areas that moved out of the medical arena and so were "demedicalized." There were technical items, such as take-home pregnancy tests. There was, less exclusively, medical participation in births. "Natural childbirth," typically Lamaze, often met with vigorous physician opposition, in this instance a sign of active demedicalization.[5] Women's groups showed that artificial insemination for couples with fertility problems, which for years had been carried out by physicians as a laboratory procedure, could be effected easily by anyone, without medical assistance, by using a common kitchen utensil, the turkey baster. Ironically, in the 1980s and after, artificial—laboratory—conception returned some reproduction to medicine.[6] Or there were reclassifications of disease. The most notorious example was the way in which the American Psychiatric Association, by a narrow vote, in 1974 declared that homosexuality was not a disease, a decision that damaged some public trust in physicians because the blatant intervention of politics obscured the technical considerations involved in the decision. Physicians had always discussed politicosocial agendas, but in this case, as in the abortion debates, the discussions appeared prominently in the public media.[7]

Finally, in just three decades, the whole society demedicalized much behavior by mentally ill people, who were pushed out of hospitals and into the community to be dealt with as welfare cases or, in the majority of instances, by the criminal justice system. This "deinstitutionalization" was demedicalization on a staggering scale. The medical specialty psychiatry, members of which had led aspects of medicalization, was to some extent superseded, both functionally and economically, by nonmedical psychologists, social workers, clergy, and, most often, police.

Lawyers and courts in particular preempted physicians' decision making. In 1974, for example, a judge ordered that a severely deformed newborn infant undergo surgery, regardless of the parents' wishes or the physicians' advice. After the surgery, the baby, as predicted, died anyway. The judge, as an outraged medical editor pointed out, had without qualifications "engaged in the practice of medicine," with results for all to see.[8]

The assault on medicine, which had intensified by the end of the 1980s, started out with criticism designed to improve health care. It is ironic that ultimately the attacks demedicalized important elements in the American health system. That is, nonmedical figures and groups challenged the authority of physicians and their associates in matters of illness and care. In addition, some Americans, such as disabled people in the independent living movement, wanted to operate entirely outside the "medical presence" and outside a medicalized identity and medicalized institutions, as did many of those in the self-care movements (discussed below). Such challenges to health care workers and institutions came unexpectedly, and from many areas of society.

The Health Care Crisis

The story begins in 1968–70, when many opinion leaders began to talk about what writers in a newsmagazine in 1969 called a "growing crisis in health care."[9] This crisis discourse continued well into the 1970s.

The new analysts built on reformers' critiques from earlier years, but a general consensus emerged from a number of publications and official reports both inside and outside medicine. Ironically, it was the defenders of medicine who listed clearly the flaws in medical services cited in crisis discussions in order to refute each point and show "what's right with American medicine" rather than what was wrong and precipitating the crisis.[10]

The basic elements making for a crisis were, first and foremost, the rising cost of medical care; second, a shortage of doctors; and third, maldistribution of, or lack of access to, medical care. One writer, taking another approach, listed disappointed expectations from American health care: "premature death and disability. . . . unnecessary sufferings and poverty. . . . and above all. . . . the disgraceful discrepancy between what modern scientific health care can be and what it is."[11] Few technical elements raised concern, except for the increasing number of antibiotic-resistant strains of infectious diseases and the overuse of technology and surgery, particularly as so many physicians learned to practice *defensive medicine*, a term still well understood generations later to describe unnecessary procedures ordered to protect physicians and institutions in case some patients contemplated suing them.

One widely repeated factoid created major consternation and sharp doubt about the actual care delivered by American medicine: infant mortality statistics showed that the United States ranked, not first, but nineteenth (or sometimes fifteenth or twenty-fourth) internationally. This single, shocking statistic discredited claims of excellence by leaders in the U.S. medical establishment and made people ask where the extravagant expenditures on health were going. "Sixties" radicals attacked American medicine from one direction, but many other leaders, including President Richard Nixon, also spoke of the crisis in American health care. In 1970 an assistant secretary in the Department of Health, Education, and Welfare told a Harvard Club meeting, "The deepening health care crisis . . . is a matter of grave concern." Later that year, the businessman Thomas J. Watson Jr. referred to the United States as "the home of the free, the home of the brave, and the home of a decrepit, inefficient, high priced system of medical care."[12]

Eventually, well-informed commentators began to discern two aspects to the crisis. First, it was not primarily about scientific medicine, but about personal experience with the health care system. As one early official commission summarized the problem: "Long delays to see a physician for routine care; lengthy periods spent in the well-named 'waiting room' and the hurried and sometimes impersonal attention in a limited appointment time; difficulty in obtaining care on nights and week-ends except through hospital emergency rooms; unavailability of beds in one hospital while some beds are empty in another; reduction of hospital services because of a lack of nurses."[13]

The second aspect of the health care crisis was that the expectations of both users and leaders in health care continued to rise. Most people, including physicians, tended to recognize and appreciate impressive instrumentation. And there was, not least, the "Marcus Welby syndrome," named for the title character in the popular television program *Marcus Welby, M.D.* (number one in popularity in 1970), which emphasized the doctor's personal talents, abilities, and power. Marcus Welby, like his popular culture predecessor, Dr. Kildare, had few medical failures. Indeed, most of the 30 [!] "medical entertainment" shows on television from 1954 to 1977 featured "a generally young, handsome, idealistic and charismatic doctor who is counterbalanced by an older, more experienced doctor who acts as his mentor."[14]

As the president of the Association of American Medical Colleges put it as early as 1969,

Marcus Welby, M.D. was an extremely successful television drama that ran for several years and helped form one image of contemporary physicians, an image ably portrayed by the popular movie star Robert Young. Marcus Welby became a symbol of a type of personal doctoring that was rapidly fading out even as the television program reached the peak of its popularity in 1970. *Look*, 23 Mar. 1971, cover.

People expect more from the medical profession because they know it can do more for them. People have more money to spend on health. We have an elaborate system of private and public health insurance and Government benefits that makes it possible for more people to pay for medical care. That increases the demand for services. The problem is that we have no efficient and effective mechanism for delivering the health services that people need and can pay for.[15]

Talk about a doctor shortage ended in 1972, replaced by fears of a possible surplus. In the view of one analyst in 1976, "Regardless of whether we have too many doctors or too few, the real problem is that we do not have enough doctors where we need them and those we do have are too often the wrong kind." Among policymakers, she continued, "It dawned on everyone that the true but often unstated purpose behind a decade's health manpower legislation was not just to produce more doctors but to produce doctors who would help redress social injustices that left poor people and country folk without access to medical care."[16] Of course, despite the programs, most physicians had still ended up in the suburbs, not in inner cities or isolated rural communities.

A number of opinion leaders, trying to make economic sense of the crisis, held that the coming of Medicare and Medicaid had increased the demand for medical services and driven the prices up. And since the paying hospital patients were now mostly covered by insurance or Medicare, any attempt to save money by cutting Medicaid immediately increased hospital costs by bringing into the hospitals more nonpaying patients, which inflated the cost of health insurance—another version of the tax on the sick to provide health care for the poor.[17]

Mistaken Assumptions about Steady Reform

Underlying all the official and unofficial statements and analyses was the widely shared assumption that Congress would soon enact some form of universal health coverage to even out services and make them more efficient so that health care would reach more people and cost less. This assumption continued for some years, even though the last serious attempt occurred in 1974, when the effort simply fell apart among the political factions in Congress. Despite the persistent efforts of rationalists and some dedicated reformers, compulsory health insurance or any other universal health coverage never again made it onto the political agenda in a serious way.[18] Instead, the various health lobbies continued to advocate specific, rather than general, federal health programs.[19] Policy analysts assumed that further incremental measures would extend health insurance of some kind to first one population group and then another, as had been occurring since the 1930s.[20]

During the Reagan administration, in the 1980s, however, many programs suffered severe cutbacks, with permanent, not temporary, cessation of services. Not least was the notorious administration hostility to environmental protections. For specific health programs, federal defunding made its way to the state and local levels, particularly in public health. Nutrition programs were devastated. Maternal and child health programs were scaled back in a striking parallel to events in the late 1920s (described in chapter 7), when anti-Progressives began to strangle the Sheppard-Towner Act. In 1979 there were about 219 public health workers per 100,000 population. Two decades later the ratio had fallen to 160 per 100,000. Perhaps the most consequential cut was aborting the neighborhood health center movement. Beginning in the mid-1960s the centers had had both success and promise in upgrading care for the poor and the uninsured, even after the centers were ghettoized when Congress prevented them from taking any but welfare cases after 1969.[21] Following the Reagan cutbacks, there was an increase in the percentage of low-birth-weight babies, a sign of general health problems.[22] The obvious inhumanities of the Reagan period led Congress in the late 1980s to extend a number of coverages, adding up to more incremental increases in government insurance and a confusion of trends in policies affecting vulnerable populations.[23]

The Health Care "System"

Meanwhile, early in the 1970s, with almost no one's noticing, analysts had begun writing less about the health care crisis and more about "the American health care system." Courses and textbooks on the subject started appearing. The health care system, however flawed, was extremely complex. It was all one could do to understand how the different parts worked, much less how they worked together: in 1972, one could count "125 identified health occupations . . . , with some 250–300 secondary or specialty designations." As the authors of one book commented, they had written an anatomy of American health services: "It stops short of being physiology, that is, how the system works. It stops short of being pathology, that is, how and where the services fail to work. . . . Social scientists refer to it as a 'system' . . . . Reformers refer to it as a 'non-system' implying that it is chaotic and in need of reorganization along more rational lines." But as another expert argued, each policy action determined the next action, and so they were in fact all linked by historical sequence into a system. For example, in the midst of the Reagan cuts, Congress enacted a successful Orphan Drug Act to fill the gaps where the pharmaceutical industry was finding it unprofitable to devise drugs for relatively rare diseases. Altogether, one could identify the elements in the system and also the steps toward general coverage: federal governmental direct medical care for veterans, service personnel and their families, the aged, and the poor, with huge tax subsidies for insurance offered through employers. This arrangement left "only" between 30 million and 37 million people with no coverage.[24]

In the end, it was free-market economists who argued most successfully that there was rationality in the system and that the parts implicitly could work together. It took some time, but they began to apply a "naturally" adjusting market model to what they called the health care "industry" and health care delivery.

Meanwhile, successive editions of the health care systems textbooks suggested how new elements entered the system from the early 1970s to the late 1980s. The sudden appearance of ethics and legal or malpractice concerns was a new topic. Then authors had to add coverage of attempts to control costs with health maintenance organizations, preferred provider organizations, independent practice associations, and, in the 1980s, diagnosis related groups (DRGs, discussed below). Meanwhile, concern continued to grow about cost containment and competition among health care "providers."[25]

Redefining Elements in Health Care

Although the major health care elements seemed to stay in place, with attempts at economic competition beginning in the 1970s, the functions, if not the identities,

The Health Care Network

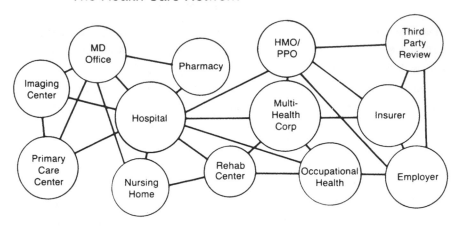

By the 1980s the hierarchical regionalism model no longer adequately described the working of the American health care system. Now a network model, including financial entities as well as the hospital, better represented the health care system. Russell C. Coile Jr., *The New Hospital: Future Strategies for a Changing Industry* (Rockville, MD: Aspen, 1986), 11. Reprinted with the generous permission of Wolters Kluwer Law & Business.

of each element shifted. Physician, patient, health insurer, care, and perhaps even disease were relabeled or redefined from some point of view.

Various circumstances worked to bring the "market" model into health policy. Most obvious was the alarmingly escalating health care costs. By the end of the 1980s, experts tended to blame technological innovation—the price of "progress"—and the growing numbers of old people. Often overlooked was the general price inflation in the United States, which could affect providers of care acutely, as when nursing homes had to purchase food and other everyday items. A second circumstance was the cultural change that made reducing government control over markets popular and acceptable, symbolized by the election of Ronald Reagan in 1980.[26]

In addition, changes within economics moved many theorists to teach that health care was a commodity and that purchasing of health care services and also health care insurance could be controlled by market forces. In 1982 a hostile observer of the coming of procompetition forces in Washington reported that "some political authorities derided the notion that there is anything 'spiritual' about the doctor/patient relationship. Their implication was that professional medical service, like any commodity, can be weighed, prepackaged, precisely 'costed'—and delivered on demand to any destination with predictable effect."[27]

The change to a market interpretation of medicine and health represented a major shift away from those who believed that health care involved a primary responsibility of the nation and the national, state, and local governments or from those who believed that equal access to equal health care was a human right. Moreover, because there was in fact no consensual basis for determining a general level of "need" for health care, it had become morally acceptable to dominant groups to treat demand for medical services and drugs as something that was a personal whim, a whim that could be denied or, if one had enough money, indulged.[28] But as free market advocates pointed out, if the trends in the growth of health care were projected into the future, within a few decades the entire national income would be devoted to health care and the entire population would work providing health care.

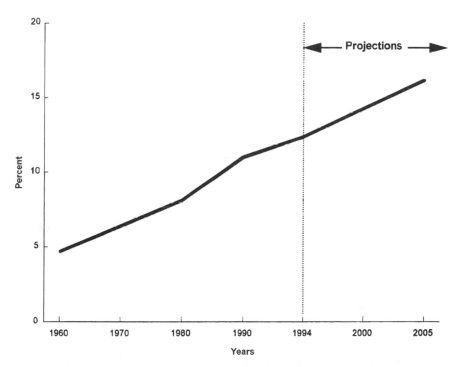

Personal health care expenditures as a percentage of gross domestic product, 1960–2005. Because it was possible to graph increasing health care expenditures as almost a straight line moving steeply upward from 1994 into the future, critics of constantly rising health costs could figure out, simply by extending the projection already made in this graph in 1996, that in a given number of years 100 percent of U.S. expenditures (and 100 percent of U.S. workers) would be devoted to the "health care industry." "Personal Health Care Expenditures: CYs 1960–2005," *Health Care Financing Review*, 1996 suppl., 10.

Reconfiguring Health Insurance

The shift to a market outlook transformed health insurance. As I have suggested, Americans were coming to accept health insurance as a normal part of at least middle-class life. In 1989, out of 248 million people, 217 million had significant public or private health insurance and 31 million had no coverage.[29] Those covered by Medicaid were often inadequately served, however, and the record of the failure of anyone to provide for the uninsured and the poor in general made the lagging international rankings of health and health care in the United States more and more embarrassing.[30]

The story of Blue Cross and Blue Shield shows how the shift to market thinking changed basic institutional arrangements. Originally the then nonprofit Blues had obtained special legislation that permitted them to have exclusive territories within which they would cover significant parts of the population. Most of their coverage came from employers, who could insure all their workers with one simple payment and have the advantage of a workforce that was getting medical care. Because of the large population covered, those who were more prone to sickness were subsidized by the more healthy, which is how any insurance works. The point is that one is never sure into which group one will fall at any time.

Meanwhile, private insurance companies found health insurance lucrative, but they did not work to cover large populations. Instead, they offered group insurance to certain employers, knowing that the active workers covered by the policy would be healthier on average than a more general population that included some already chronically ill or disabled people who were not working. In these and other standard insurance ways, private insurers could "cherry-pick" their subscribers and offer lower rates to selected groups. By restricting the population they covered, private companies undercut the Blues. Eventually, the Blues had to stop offering wide coverage. And then both private insurance companies and the Blues started bargaining with hospitals and group medical providers to get lower costs for covering their policyholders—the same situation that at the turn of the century had led contracting physicians to offer ever lower and more impossible rates to get a group to hire them, but now the bidder was not a solo practitioner, but a health care entity such as a hospital or physician group. In 1986 the Blues lost their tax exemption and became just additional players in the private insurance marketplace. Moreover, as some very large employers began to provide their own health insurance for their employees, there was even more direct competitive pressure.[31]

In actual practice in the 1970s, private insurers, the Blues, and Medicare and Medicaid all incurred much unfavorable publicity by sometimes denying individuals' claims for payment. That left the individual, acting as a customer or employee or citizen (none of which roles provided leverage), with a possibly large bill for services a doctor had ordered. The usual reasons given for denying payment were: "a diagnostic," not treatment, procedure; "a pre-existing condition" not covered by the insurance policy; "merely custodial," that is, without any therapeutic goal; and "not medically necessary."[32] Those without extra financial resources were hit the hardest and usually had the least ability to appeal. The problem was, again, with the health care system, not medicine. By the end of the 1980s it was clear that "the market" applied to large organizations and businesses, not individual patients. Patients either were locked into an employer's insurance or could choose only between insurers, not providers. That arrangement simply did not work well. In the words of two free-market experts, "The market for health insurance does not naturally produce results that are fair or efficient."[33]

Trying to Control Costs and the Advent of DRGs

From the early 1970s to the late 1980s Congress passed a number of adjustments designed to control health care costs. These ultimately added up to a major shift in national policy, indeed helping to bring market forces into health care but often regulating it too. In 1973, responding to the obvious success of the Kaiser Permanente health maintenance organization, Congress initiated a series of measures to encourage the formation of HMOs and then other groups of health care providers, such as physician groups (PPOs), that could compete with one another. That is, doctors had to offer insurance companies lower rates for health services. Then in the 1980s, for-profit contractual organizations, using the HMO structure and laws, expanded greatly, not least because they could and did restrict care given to patients. The Reagan administration aggressively encouraged privately owned HMOs in the hope that speculators would invest and so reduce the need for public funds for health care. The for-profit health "providers" of course restricted whom they would treat (cherry picking). Congress also shored up the implicit custom of providing health insurance through employment by providing COBRA insurance, which allowed an employee leaving an employer to continue his or her insurance for a period of time.[34]

In 1983, in the midst of all these adjustments, a fundamental shift was effected through a technical change in the way Medicare paid for services: hospitals, as a first step, were to be compensated according to how each patient fitted into a schedule of charges, categorized as the diagnosis related groups (DRGs).

Previously, Medicare had compensated hospitals and physicians on a cost-plus basis, whatever costs an accountant could list, plus 2 percent. "Providers," that is, physicians and hospitals together, thus ran up the charges to the point that Medicare funding was threatened. But quietly, through an amendment to a bill in 1983,

Congress, using DRGs, moved the federal government to reward lower rather than higher charges. If hospital administrators could cut costs, the hospital would make more money, because it would still receive a set amount for a patient with a particular diagnosis, regardless. A hip replacement, for example, brought in a set amount, whether the patient was in the high-cost recovery room for twelve hours or seventy-two.[35]

As there had been occasionally in the colonial period and the nineteenth century, as of 1983 there was a fee bill with a set charge for a particular medical condition. That might seem reasonable, but hospitals still had to deal with "outlier" patients, difficult cases for which the costs were much higher than the set fee. Hospitals that served many charity cases were implicitly told to change their missions to balance their books.[36]

From the point of view of health care "providers" such as physicians, the government was now controlling rates, flagrantly breaching the fee-for-service system, in which "providers" could charge almost anything they wanted. Most importantly, DRGs constituted a direct assault on physicians' autonomy, restricting a doctor's freedom to prescribe treatment she or he believed medically indicated. The government would no longer pay for it. Many people at the time viewed the change as revolutionary.[37] The same government that was bringing "the market" into health care was now also regulating fees and, implicitly, rationing care (through the agency of a third-party administrator who could deny or cut services to any patient to save money), while appearing not to take any responsibility for that action. People in health care soon adjusted to the new system, however, and private insurers, with their own schedules of charges, generally followed the government's lead in the method of determining payment.[38]

The pressure to cut costs put the ordinary physician in a difficult position. As a 1986 analyst wrote, "Costs will best be contained by direct disincentives to overutilization of services by physicians. . . . Physicians are comfortable either doing everything that can possibly be done for patients or sending the patients to another source of care; conscientious physicians may find it impossible to function under a system in which they are expected to serve truly as gatekeepers for the rationing of medical care." She added, perhaps unnecessarily, that with DRGs "the moral imperatives of the physician's oath are violated by direct disincentives," putting costs before patient welfare and using economic motivations to influence professional judgments.[39]

Economic Identities for Doctor and Patient

Perhaps the most startling aspect of the marketization and demedicalization of health care decisions was the changes in the identities of doctor and patient. The doctor became a health care "provider," and the patient, a health care "consumer." These were not insignificant changes in language but real shifts in social roles. In the words of one doctor, "Every time a patient is referred to as a health-care consumer, another angel dies."[40]

In the first half of the twentieth century, occasionally an economist would refer technically to "consumers" of medical services. The lawyers who drew up Medicare in 1965 referred to doctors as "providers," obviously to avoid defining exactly who could be paid for services. The consumer protection advocates of the interwar years dealt mostly with products, such as patent medicines, not services. In the 1960s and 1970s new and radical consumer advocates wanted to destroy the authority of the medical profession and bring "power to the patient."[41]

The issue was the "medical monopoly" of licensed physicians. Over the years, growing legions of conscientious practitioners attempted to explain to patients what their choices and risks were. Nevertheless, in a world of rapid technological change and highly trained specialists, the idea of any widespread, fully informed, *independent* "consumer choice" was silly. Beginning in the 1970s, however, economists associated with antiregulatory ideologues appropriated the language of consumer protection and took the antimonopoly, free-market line against the "medical monopoly." Beginning in 1975 even the courts joined in, and especially after 1982 they tried to prevent any attempt of physician organizations to restrict physician advertising or any other marketing.[42] Physicians who believed that professionalism included at least some idealism were particularly upset. An officer in an Alabama medical organization interpreted the first of the court decisions as follows: "Anything goes and professional organizations are enjoined against doing anything about it. Ethics, discipline and order shall henceforth be considered in restraint of trade." Who would benefit, he asked, except "hucksters and charlatans"?[43]

Like 1960s radicals, the free-market economists attacked the embodiment of the medical monopoly, the professional organizations, especially the AMA. Only by destroying the monopoly could costs be lowered, according to free-market enthusiasts. Ideologue lawyers in the Reagan administration and after went to extremes, encouraged by the courts. There was even one private suit in which a supposedly competitive insurance entity tried to destroy the nonprofit Marshfield Clinic, which had successfully brought big-city specialist medicine to a large area of rural Wisconsin. Working particularly through the Federal Trade Commission, ideologues used the rhetoric of consumer rights and consumer choices to undermine legal privileges of the medical profession.[44] Given the tactics medical organizations had earlier used against group practice and prepay plans, their genuinely professional, ethical stance did not win credibility.

But at this point a twist reflected the new realities of the 1970s and 1980s. As I pointed out earlier, the "consumers" who were empowered were not the individual patients, who were mostly limited to choosing health plans, if they were allowed to do that. Instead, the "consumers" in free-market health care were the major purchasers of health care services, namely, the increasingly for-profit insurance companies and "provider" groups and, especially with Medicare and Medicaid, the federal and state governments. The expectation was that freeing the market from restrictions would enable these large purchasers to hold down costs, thus solving one of the major problems in the health care "crisis."[45]

Because by the 1980s most of the population, 85 percent or more, had substantial coverage under either government or private health insurance, the majority of patients were demoted further, from consumers or purchasers of health care to mere claimants in insurance arrangements.[46] The new free-market competition, aimed at large providers, did not just ignore the individual citizen. By insisting that health care was an ordinary commodity, large providers subjected actual users or receivers of health care to smaller packages and exclusion—the common way of reducing the price of any commodity. In the end, then, introducing "competition" and excluding those without insurance served to reinforce the traditional two tiers of the American health care system, one for the well off and one for the less affluent.[47]

Privatizing Health Care, at Least Partially

Particularly in the 1980s, many commentators were alarmed by the number of private business organizations moving in and contracting to perform health care functions. Beyond the groups of physicians who incorporated themselves for tax and business purposes, major chains of for-profit hospitals and corporations hired physicians and other health care personnel to serve the public. For a time at the high point of profit-making health care in the 1980s, turning all of health care into entrepreneurial, profit-making activity seemed to be the way events were trending.[48]

Insurers had the motive, and with computers the tools, to bring conventional business management pressures on doctors and hospitals that were furnishing individual services. Furthermore, whether a health care corporate entity using business models was investor owned or a nonprofit, the result was significant erosion of health care professionals' autonomy and a shrinking demand for hospital beds. By 1986 the average hospital stay in the United States for most conditions was shorter than in any other country in the world.[49]

Before the 1990s, however, the private sector was still just a part, albeit an important one, of the total health care system, sharing with the still dominant non-profits the problems of capital, management, technology, and personnel. And it was turning out that size did not make for more efficiency any more than did business ownership. Indeed, for hospitals, large corporate organization did not bring lower treatment costs, but it did bring higher management costs.[50]

The final irony of the 1980s was that there was one kind of health insurance that in practical terms was universal: catastrophic health coverage. A person who had no money or insurance, or whose insurance had run out, would present himself or herself at the emergency room and receive care. It might not be the same care that an insured person would receive, or it might be even better. The ER professionals just did their jobs. Rationing was still being sorted out by the social entities struggling over who would pay for what, directly or indirectly, with the hospital usually in the middle. Analysts figured that regardless of the means, the federal government was paying about half of all health care costs directly and a good deal more in hidden tax subsidies and other kinds of subsidies.[51]

Subverting and Appropriating Physician Authority with "Ethics"

In these same years, additional elements in the culture worked to appropriate physicians' authority. The figure of the physician in this complex and contentious health care "system" became considerably more conspicuous an issue in the United States. The subverters of physician authority had two approaches. One approach was to attack the role and authority of physicians directly. Even among intellectuals, postmodernists were "deconstructing" any expertise. In a second approach, many laypeople attempted to make decisions—and not just economic decisions—that had customarily been relegated to physicians. A host of amateurs, beyond lawyers and judges, therefore tried to play doctor, in a process of further demedicalization. And insofar as they could not do without physicians, they tried to impose controls on physicians who had been using recognition of professional status to enjoy autonomy—specifically not to have such people question their judgment.

Another facet of the critique of medicine was on one level a replay of the old conflict in American society between science and religion. This time, biomedical scientists and clinicians represented science, and the religionists were "humanists," particularly Roman Catholic humanists, who quickly attracted others from religion and humanism who had been sidelined by the growing prestige of science and the secularization of American culture. All these humanists were put on alert by the medical technology that increasingly determined not only the end of life but also the beginning, starting not just with conception but with tinkering with the newly discovered (in 1953) DNA.[52] Those sounding alarms were not traditional experts. Indeed, over the years professional philosophers had largely abandoned the field of ethics, and the handful who were technically competent in the subject were quickly swamped by self-appointed *ethicists* with their own agendas and a new word coined in 1970, *bioethics*.[53]

In 1966 Henry Beecher, a medical leader from Boston, in a landmark paper had pointed out that because of the hugely expanded scale of research, ethical standards were endangered: "There is reason to fear that these requirements and these resources may be greater than the supply of responsible investigators." In other words, in the huge biomedical research effort, there was unethical experimentation and even fraud in small amounts. Beecher himself had no problem discerning what was ethical, but he recognized that occasionally one of his now very numerous colleagues might not have that ability.[54]

The actual organization of a movement was initiated by a group of religionists who wanted to change some of the operating values of physicians. In the late 1960s this small, shrewd group made two major decisions. First, they would not work for social justice or anything that might smell of social reform. Indeed, they ultimately were able to coopt large numbers of well-meaning doctors

The Hastings Center, in Garrison, New York, a remarkably well-funded think tank for bioethics, symbolized the success of the bioethics movement in critical areas of American society. Scholars at the Hastings Center were influential in shaping American public opinion. Photo reproduced through the generous courtesy of the Hastings Center.

and academics by labeling their approach "medical humanities." Second, they would infiltrate the medical establishment, especially medical education. In this strategy they were successful, gaining remarkable secular and even medical funding.[55]

For generations, physicians had used ethical statements and guidelines to bolster their standing and carve out a protected area for medical professionals in society. That is, they would police the behavior of their colleagues so that other elements of society need not intrude where they were unqualified and did not know either the medicine or the ethical customs. Particularly as the 1960s general challenge to authority got under way, a series of incidents, played up by journalists, led to the suggestion that perhaps outsiders ought to intervene in what doctors and scientists did. The Pill (1960), available only by a physician's prescription, instigated questioning. Then there were transplants of a part of one human to another, and dialysis, which initially could not be available to everyone and had to be rationed openly. In 1968, life support, as in the Quinlan case, was so out of control that physicians had to formulate an additional definition of death, "brain dead," a formulation that was sometimes rejected by competitors from that other profession, religion.

In the 1970s and 1980s private financial support and journalistic interest gave the bioethicists a growing importance in public discussions that they did not have in other cultures. Questions of technical differences between life and death heated up even further with the abortion debate (*Roe v. Wade* was in 1973) and also with causes célèbres of newborns for whom continued life support was debatable on many moral grounds. In 1972 the educated public learned of the Tuskegee syphilis study, carried out for many years, beginning in 1932, by officials of the U.S. Public Health Service on an unwitting population to discover what would happen if syphilis went untreated. It was a study in which racist thinking appeared originally to have been a factor even beyond questionable standards for human experimentation.[56] And over the years there were many other cases in which what had happened in the past was condemned by later generations, who invented new standards, or in which people suddenly decided to start applying dogmatic beliefs to medical practice and investigation.[57]

The implicit, if not explicit, strategy of the bioethicists was to undermine trust in medical professionals and to shift that trust to philosophers, religionists, lawyers, and political officeholders. All these nonmedical personnel had recourse to general principles and rules and procedural prescriptions that they advocated, telling health care personnel and anyone else how to apply those prescribed rules to individual cases.[58] Occasionally ethicists as well as economists referred to "patient rights" to further their agendas. Patient rights, however, had a number of other sources, including genuine personal grievances, institutional public relations, and, once more, rebellion against "authority."[59]

The Malpractice Crises

Part of the original "crisis" in health care was the "malpractice crisis" of the 1970s, which was followed by a second malpractice crisis in the mid-1980s. In each malpractice peak, patients who believed that they had suffered from incompetence or unacceptable actions on the part of physicians and/or medical institutions used the power of the courts, as had been traditional, to do some of the policing of the medical profession that physician organizations had been unwilling or unable to do. The result in the 1970s and 1980s, however, was that lawyers, working on a contingency basis (typically they received a third of any award), were greatly encouraged to initiate suits.

Technology especially fueled lawsuits in the 1970s. And as lawsuits increased by as much as 20 percent per year, firms that insured doctors and hospitals against malpractice claims had to raise their rates. In Washington State a low-risk physician who had been paying $527 (perhaps 2 percent of his or her earnings) a year suddenly was being charged $1,287 (more than 5 percent of net earnings), and that increase was mild compared with those in many areas and most specialties. The doctors whom insurers hit with major increases became frantic. State legislatures hastened to pass numerous measures to control claims and make sure that physicians were insured. In a few areas, doctors went "on strike," refusing to perform nonemergency surgery until the legislators acted.[60]

The immediate malpractice crisis receded, but a decade later the second malpractice crisis came, caused by the extraordinary rates that were necessary to cover claims. Early in the 1980s a jury awarded a Florida man with brain damage more than $12 million, an enormous sum at the time. Especially for surgical specialties, malpractice premiums reached very high levels, some increasing by more than five times in a very few years. Physicians represented the situation as intolerable and blamed the insurance premiums for doctors' and hospitals' higher charges. In 1985 one in ten physicians was the target of a malpractice action. The number of claims rose from 3.3 per hundred physicians before 1978 to 8 per hundred from 1978 to 1983. And for physicians, being sued was not just ordinary business; it was personally devastating and frightening. They blamed the legal system, in which hungry lawyers would take cases and initiate extravagant claims. Many people at the time believed that the coming of "no-fault" auto insurance had turned loose legions of predatory attorneys, who rushed into the area of medical malpractice.[61] Lawyers pointed out that physicians in fact did make serious and unnecessary errors. Defenders of lawyers challenged physicians to clean up their act so the courts did not have to. Physicians pointed out that MDs won 80-95 percent of the actions against them, so cases tended to be attempts at extortion, and that grievances could be settled otherwise.[62] Meanwhile, notorious cases in the news damaged the physician image and intensified the greed among those hoping to profit from malpractice lawsuits.

Effects of the Malpractice Crises

These bitter arguments continued without much change for decades, into the twenty-first century. Yet the malpractice crises of the 1970s and 1980s did have significant effects. First, physicians improved the policing of their own ranks and obtained state legislation enabling them to do so more effectively, for example, protecting a physician from lawsuits when he or she voted to sanction a colleague. Second, close observers figured out that most suits grew out of defective communication between doctors and their patients. Practitioners therefore received a lot of advice about informed consent and, once more, as in the 1950s, about how not to antagonize patients. For many physicians, however, the doctor-patient relationship would never again be characterized by the same level of trust. "I look at every patient as a potential malpractice suit when she first enters my office," wrote an obstetrician in 1975.[63]

The third effect was greatly to increase defensive medicine with its unnecessary testing and paperwork. In poll after poll physicians admitted that to protect themselves they ordered extra tests and procedures (76 percent did in one 1977 poll) or that they unnecessarily hospitalized patients because it would offer a better defense in a malpractice suit. In 1977 Leon Eisenberg wrote, "Today, most doctors order skull x-rays 'to be on the safe side' lest they be sued if an insignificant linear fracture should subsequently be detected. It costs more to order films and it exposes the patient to radiation. But it takes a courageous physician *not* to do it."[64]

A fourth effect of the malpractice publicity was the formation of new consumer groups, which functioned to inform "the public" how to distinguish between health care "providers" (physicians and hospitals) that were technically questionable and those that were more competent. The individual (presumably not locked in by insurance) could then choose a doctor as one would "a technician (or a refrigerator)," the participant-observer Louise Lander wrote in 1978. The system would encourage physicians to advertise their qualifications. The end result was that everyone could blame the patient-consumer victim for receiving flawed medical care, because the consumerist approach, endorsing a medical marketplace "frees the practitioner and the medical institution from an obligation to their patients any greater than, or different from, the obligation of a retail store to its customers." Consumer protection thus again played into the hands of free-market extremists.[65]

Critics of Doctors from Inside Medicine and Outside

With both conservatives and radicals attacking physicians as the convenient symbols of health care, the doctors felt very much besieged. One observer as late as 1984 suggested that hostile forces, including academia, law, and capital, were "encircling" medicine.[66] Many practitioners were discouraged by the loss of respect. Moreover, from within medicine a significant segment of physicians joined in the 1960s assault on authority, often criticizing and attacking both bureaucrats and established colleagues who did not immediately correct faults and errors in the system. A number of physicians attended the legendary Woodstock festival to make a cultural statement (ironically, they ended up spontaneously performing professional duties). Countless others cared for drug addicts and other impoverished fringe and radical people shunned by most doctors, as at David E. Smith's Haight-Ashbury Clinic in San Francisco. A stream of medical student activists performed community service but also pressured medical schools to recruit minorities and to accommodate the poor as well as the elite, all the while raising consciousness and scandalizing established professionals with their rebelliousness and egalitarian demands in the face of what one group called the "crises in community health and social justice." The reformers did make a difference. As one later observed, the ideas of the 1960s survived better than the activist organizations set up at the time.[67]

The growing number of emergency room specialists, for example, felt themselves to be marginalized by colleagues and also to be victims, in this case victims of hospitals and insurance companies, who tried to pay them less than other doctors, and colleagues who denied that ER physicians had special competence. In 1968 some younger emergency room physicians in Michigan gathered together and started organizing. They grandiosely founded the American College of Emergency Physicians, operating out of a basement storage room in a medical society building. Because of the turbulent times, they were able to get attention and attract others, in defiance of both organized surgery and general practice. Later that year they organized a national meeting, and in 1969, when to their own amazement and everyone else's they were attracting many members, they held their first scientific meeting to affirm that emergency room medicine was a regular medical specialty. Their own board was approved less than ten years later. The parallel between these specialist rebels and other groups in The Sixties would be hard to miss.[68]

For the ordinary physician, whether in private or group practice or holding a salaried position with an organization, outsiders' attacks on doctors and the medical profession in general were dismaying, especially because it was at just this time that moderates with some sense were finally, albeit very slowly, taking control of the American Medical Association. Pickets at the AMA meetings in 1969

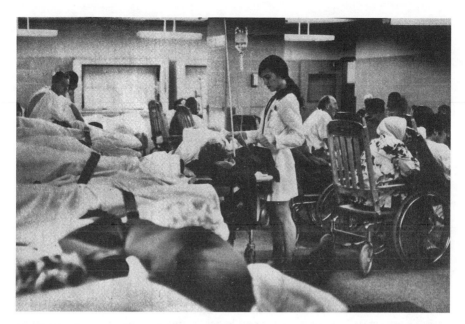

Dr. Shirley Roy, a resident physician at Cook County Hospital in Chicago, stands among patients, mostly on gurneys, who were crowded into the hospital and waited for long periods to be x-rayed. She was one of those who became an "activist," as she joined other residents and interns in 1970–71 in publicly demanding better and safer conditions for patients. Jack Star, "Cook County Hospital: The Terrible Place," *Look*, 18 May 1971, 26.

denounced the organization as the "American Murder Association" because so many Americans did not receive proper health care. The next year, protesters at the meetings denounced racism and sexism in medicine as well as the AMA's ties to the military-industrial complex, indicating vividly how critiques of physicians and the health care system were part of the wider Vietnam War–era protest and social critique.[69] One reformer wrote, "The obsolescence of the American physician today is manifold, a product of his archaic education, his inappropriate orientation to disease and to people, the economic (fee-for-service) and societal (one-to-one) framework of the 'physician-patient relationship,' the traditional notion of a patient-centered rather than a community-centered responsibility."[70]

Doctors, nurses, and other figures in the health care system were vulnerable to attack because they embodied the authority, conservatism, and insensitivity that many younger physicians and patients of all ages recognized in the health care personnel they encountered. Senior physicians, especially in the upper reaches of the regional hierarchies and specialties, were in fact typically wealthy and enjoyed unusual social privileges and recognition. Many of them responded by acting, perhaps unwittingly, in arrogant and authoritarian ways, a style imitated by still others. Wrote one journalist with colorful exaggeration, "The image of the selfless doctor has been gradually changing in the American mind, to that of a callous, overpaid, high-living Cadillac-riding businessman. . . . Dr. Kildare has dissolved into Dr. Scrooge."[71] Already in 1968, a columnist reported that "most doctor jokes are about how much money they make," and the writer described a current cartoon in which a physician is speaking to his chauffeur, who has drawn a Rolls-Royce up to the door of the house: "No, no, Casey, bring the Ford. You know I'm visiting patients today."[72]

An awful additional truth was that, according to the official AMA code, physicians had no ethical duty to treat anyone they did not choose to treat. As a woman living in rural Mississippi noted after a local community health center opened in time to relieve her cataracts and save her from blindness, "Town doctors here in Gunnison, well they don't take up with us poor folks." Many physicians simply refused to take Medicaid patients when the compensation levels fell too low to suit them or to sustain a practice, for example, when they did not even pay overhead office costs.[73]

The flaws people found in doctors and the medical system, as portrayed in a widely known novel of 1970, *The House of God*, of course reflected the flaws critics found in society in general.[74] The professions, and especially the model profession, medicine, had gained social authority in part because professionals claimed to be persons of good character who could be trusted. Showing professionals, including biomedical clinicians and scientists, to be flawed human beings damaged their image and their authority.[75]

In actual practice, physicians, secure in their insulated medical communities, could appear insensitive to the social responsibility that professionals were supposed to exhibit. Beyond the flagrantly selfish political maneuvering of medical organizations, it was in fact difficult to find overworked urban specialists who also took time to provide a significant amount of uncompensated medical care. They were simply too busy with demanding pay patients to put in weekly hours in a free clinic. With many conspicuous exceptions, most physicians acted as if they were unaware of the concept of broad social accountability, unless, of course, they were social Darwinists. In 1986 a statesman of medicine tied developments in health care to those in the general society: "In this depersonalized society, the well-being of our citizens is being equated with the goods and chattels of the marketplace." Only pediatricians as a professional group consistently engaged in acts of public responsibility.[76]

Finally, some talented social commentators made a case that all the work that physicians did was futile or even counterproductive. In the late 1970s the lay critic Ivan Illich contended that "the medical establishment has become a major threat to health." People were living longer, and most deaths were associated with old age or accidents, not contagious diseases, he argued. "There is no evidence of any direct relation between this mutation of sickness and the so-called progress of medicine." Instead, he asserted, medicine caused illness.[77] Like Illich, a number of intellectuals portrayed "diseases" as entities invented by physicians and foisted on the public, and they could cite studies showing that health in the Western world had improved in spite of medicine, not because of it. Some critics could admit that perhaps public health was effective, but they could not say the same for medical practice. Indeed, it was with the help of Illich and others that the term *iatrogenic disease*, referring to an illness caused by physician action, spread beyond medicine. And, appropriately for the time period, statistics revealing a high number of "adverse drug reactions" made medicine itself part of the newly discovered toxic environment.[78]

The theme of rebellion against authority reverberated in many representations of medicine. The authority could be an individual assertive physician or representatives of bureaucracies and social institutions. A generation of medical students or those who would be attracted to medicine grew up watching *M*A*S*H*, a television show that ran from 1972 to 1983, about doctors (mostly remarkably gifted surgeons) in the Korean War of the 1950s—although most people understood that the show was about the world of the Vietnam War and the lingering rebelliousness and other issues of the 1970s. *Marcus Welby, M.D.*, which ran in the same period, like the 1960s program *Dr. Kildare* recalled a past era when a romanticized solo practitioner was heroic and successful. By the 1970s, however, Dr. Welby was no longer the conscious role model for many younger people going into medicine.[79]

Self-Help and Self-Care

Public rejection of physicians was facilitated by two other movements in the environmental era, self-help and self-care, which extended parts of medical consumerism. Self-help made the individual responsible for being informed about how to stay healthy. "The new patient" often came into the doctor's office with up-to-date medical information that the unfortunate practitioner might not yet be aware of. As the Washington, DC, internist Michael Halberstam commented in the 1970s, "I once sat down to watch four weeks' worth of medical shows on TV and came away astounded at the sophistication required of their audiences. One might guess that the American people know more about the workings of the medical profession than they do about any single occupational group, with the possible exception of the Mafia." He went on to say that this was "not a false sophistication; the questions that lower middle-class patients ask reflect a real awareness of disease and the techniques available to treatment."[80] The well-informed, health-conscious patient who could and did challenge the doctor created profoundly new circumstances for the individual practitioners of medicine—or perhaps a reversion to the time when any literate person was entitled to act as a medical authority.[81]

Self-help advocates sought to produce more than just an informed, challenging patient. The individual, in the words of one advocate, would be converted into "the first level of a total health care team where there would be appropriate skills assigned to the health consumer. He would do the work somewhere between that of a physician's assistant or a nurse and someone who knows nothing." Self-help thus rolled easily into the then trendy holistic health movement, emphasizing diet and other personal programs, with medical intervention in a subordinate role. Advocates put patient/consumer autonomy ahead of dependence on medical expertise by substituting a person's self-knowledge for scientific medical knowledge. In the form of patient groups, which offered information and practical mutual advice as well as comfort, self-help groups performed valuable social functions, particularly in a population increasingly burdened by sufferers from chronic disease, whose attempts to deal with their personal situation taxed even the most concerned specialists and caregivers.[82]

Self-care, in contrast to self-help, was the entire rejection of modern medical care. This could take the form of practicing folk medicine or any other type of healing on one's self and one's family. Many rebellious people who practiced traditional crafts and sang folk music glorified folk health wisdom and folk remedies as "natural."

Women's Movements

Both self-help and regular medicine were powerfully affected and driven by the sudden appearance of an assertive women's health movement, which included both organization and doctrine. Historians have long shown that by the late 1960s a number of women active in the antiwar, civil rights, and other movements began to question why they, as women, should not have the same consideration and rights as men. They also advocated special recognition for the ways in which women experienced their bodies, a curious reversion to nineteenth-century emphases on the uniqueness of women's bodies, but now for purposes of liberation, not oppression. By the mid-1970s, among both activists and those whose consciousness was being raised, a number of groups of women in different localities and with different agendas had spontaneously come together to form a powerful women's health movement. Initially, many women were stimulated by facts concerning abortion and birth control, such as the fact that in the ten years after *Roe v. Wade*, abortion-related deaths of women declined by 73 percent.[83] More generally, however, many women concluded that the *natural* functions of women's bodies were overmedicated and that too many women had had unpleasant experiences as patients, particularly as gynecological patients.

One group, in Boston, started comparing experiences and in 1969–71 produced the first editions of the legendary book *Our Bodies, Ourselves*, which eventually sold millions of copies and made untold numbers of women feel empowered by the experiences the Boston collective shared. Another group, in Los Angeles, learned from one of their members how to demystify what was going on in their bodies. Using a speculum, a flashlight, and a mirror, she examined her own cervix. By breaking taboos—touching their bodies and using medical instruments themselves—these women felt extraordinarily liberated. In Chicago, for a short time before *Roe v. Wade*, a group was quietly offering abortions, many performed carefully by the women. And everywhere, women's clinics sprang up, sometimes as part of free clinics. Many of those that initially focused on abortion soon

The members of the women's collective that wrote and produced *Our Bodies, Our Selves* in 1969–71 are shown here in a formal portrait, a recognition that they had achieved something significant. These women spoke for and about women's health experiences at a critical moment. Reprinted from Wendy Kline, *Bodies of Knowledge: Sexuality, Reproduction, and Women's Health in the Second Wave* (Chicago: Univ. of Chicago Press, 2010), 19, with the permission of Phyllis Ewen. Photo copyright by Phyllis Ewen.

expanded to cover women's health more generally. The groups were often very political, and a coalition came together in 1975 formally embodying the political arm of the women's health movement.[84]

In conjunction with other radical thinking at the time and the dawning "age of narcissism" or the "me decade" in 1970s culture in general, most leaders in the women's health movement wanted to use the authority of their collective subjective experiences against the dehumanizing, conformist science they were encountering in the doctor's office and the hospital. They also advocated consideration and respect for women, whether as patients undergoing pelvic examinations or as physicians and paramedicals. Women's clinics and advocacy were especially hard hit by cuts at both the state and federal levels during the 1980s.

But the impact of the movement persisted, particularly, as one historian put it, the impact of thinking based on "the plurality of individual experiences (every woman's experience is different and just as valid as every other woman's, because each individual is the authority over her own body)." The general practice of medicine was never the same after years of consciousness raising. Especially affected were medical students and younger physicians, and in response to pressure, academic institutions and hospitals often altered official policy in ways that over time

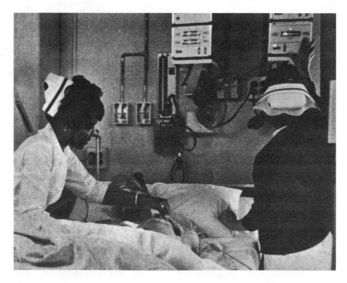

Nurses in the late twentieth century worked with a variety of complex technologies, in addition to managing patients and more traditional medical devices. Technology was altering the roles of both nurse and patient. George I. Lythcott, "The Health Services Administration: Improving the Access to Health Care of the Nation's Underserved," *Public Health Reports*, 93 (1978), 640.

changed how not only women but all patients were treated—a profound shift, particularly for the technological cowboys among hospital residents.[85]

The new feminism introduced additional change to the rapidly growing area of nursing. As noted in earlier chapters, nursing started out as a women's occupation, based on the idea that women naturally would take orders from men—doctors—and perform domestic chores and serve as moral leaders. Feminism in the 1970s galvanized the long, tedious campaign to make nurses educated professionals, experts with autonomy and decision-making functions in technical medical care, rather than fulfilling some former ideal of womanhood.[86]

Attacks on the Hospitals

Throughout the technological and environmental medicine eras of the post–World War II decades, hospitals and medical schools, even when under attack, remained constant features of the social landscape. Even so, external forces from legislation, finances, and society in general, including, naturally, movements from The Sixties, changed how hospitals delivered health care and simultaneously served as learning sites for medical students and residents. Traditionalists did not like most of the changes and often spoke out. One medical school faculty member complained of many medical students he encountered in 1973: "Their slovenly appearance with dirty, long hair touching wounds, with offensive body odors, dirty hands and nails, and nondescript and inappropriate clothing belies their professional medical goals."[87]

Hospitals proved to have enormous staying power as major social institutions, even as Hill-Burton funding was phased out in the mid-1970s. As late as 1980, hospitals took 40 percent of the medical care dollar.[88] The bureaucratic hospital was an easy target for critics. As one observer put it in 1979, "The hospital appears to be a source of uncontrolled inflationary pressure, an instrument of bias and sexual oppression, or an impersonal monolith—managing in its several ways to dehumanize rich and poor at once, if not alike." The place of hospitals in the health care system was beginning to change, however, under pressure not just from critics but from market and structural forces as well.[89] As two analysts wrote in 1989, "The market for the mainstay of the hospital product line—inpatient hospital services—is in the decline stage of its life cycle."[90] Meanwhile, administrators tended to follow the advice of management consultants: cut personnel, demand more output from employees, "look at health care as a commodity," apply management engineering (efficiency analysis), take advantage of scale, and above all keep costs lower than the competition's, regardless of sales.[91]

On another level, where once the basic structure of hierarchical regionalism had been organized around hospitals, now obviously the hierarchy was dissolving. The teaching hospitals, to which community hospitals referred patients needing very advanced specialized care, lost their uniqueness. Highly advanced specialists, trained at the teaching hospitals, along with their requisite technology, began to become available in community hospitals and in independent practice arrangements. A California MD administrator reported in 1984, "As trainees have left training programs, they have migrated to community hospitals and no longer refer their patients to their old chiefs. In fact, they try to do their old chiefs in." Highly remunerative procedures such as heart catheterization effectively moved to the suburbs, leaving many great teaching hospitals with a staggering percentage of unprofitable patients with unprofitable afflictions.[92] The hierarchy thus was flattened and no longer represented how the health care system worked.

Particularly in the 1970s, profit had become a major factor in hospital survival, and from 1980 to 1983, 949 hospitals actually closed.[93] Others were purchased by ambitious, expanding hospital systems, which, as noted above, could be either nonprofit or now, to the alarm of many, for-profit corporations that touted managerial ability as the factor that could reorganize and streamline medical care. Administrators of traditionally charitable, service-oriented institutions found that to do good in the world, they would have to do well in competition with other "providers" of health care. As business models became dominant, physicians involved in hospitals found their professionalism challenged. What if for-profit health care did represent the future? One hospital chain corporation in 1976 reported that "Your Company is a major participant in one of the fastest growing industries in the country"—at that time a better investment than computers and electronics.[94]

The for-profits did persist, but often, it turned out, in alliance with the nonprofits, and in the end the for-profits did not become dominant. Regardless of free-market theory, they were no more efficient than nonprofits in delivering health care. As was pointed out at the time, health care was not an appropriate arena for free-market competition, a fact the proponents of "competition" simply ignored.[95]

At the same time, multiple-hospital systems proliferated. At the beginning of the 1980s already more than one-fourth of the hospitals were part of a system, not just the Veterans Administration, Catholic charities, or Kaiser Permanente but often a regional system. Some units remained charitable, while other units in the same system incorporated, whether or not for profit. Hence, as one scholar pointed out at the time, one could run across Sisters of Sorrowful Mother Management or Alexian Brothers Health Management Inc. (She apparently missed Little Sisters of the Poor Health Care, the ambiguity of whose title must have escaped the good people involved.) Altogether, the health care system at all levels included many conflicting sectors and became far more complex and unpredictable than anyone had anticipated. The disruption from both economics and The Sixties phenomena was particular upsetting to the Catholic hospitals, which were becoming secularized and challenged, so that they had to rethink their mission.[96]

Underlying the hospitals' problems was one overriding fact: the traditional functions of the hospital were dissipating. Where once long bed rest and observation had served both diagnosis and recuperation, now there was early ambulation. Moreover, the technology for both diagnosis and treatment, which had been concentrated in the hospital, was now dispersed, as noted earlier, even into malls, where less invasive surgery, for example, could flourish. Profit-seeking was not good for hospitals and probably inappropriate.[97]

New Symbols for the Hospitals: The ER and the ICU

The public face of the hospital for many people was the emergency room, the use of which, as noted above, grew at an astonishing rate after World War II. At first it seemed contradictory that the emergency room should have to deal with patients whose complaints were not emergencies by any definition. Soon it became a cliché that for many poor or marginal or just isolated or socially inept people the ER served as the primary care provider in lieu of a family doctor or clinic. And the ER had the additional advantage of being available 24/7. Beginning in 1986, when it became illegal to turn patients away from emergency rooms, hospital consolidations led to more overcrowding in ERs. Yet for many Americans, even many with insurance, the ER was the entry point into mainstream medical care.[98]

Until they got used to it (which they had by at least the 1970s), ER staff members, who were prepared to treat trauma, were often bemused by people's coming to the ER because they were constipated or had a headache or sore throat. For hospitals, the ER lost a lot of money, except when it recruited long-stay patients who were severely injured or very sick. Nevertheless, the ER served a very important symbolic purpose, as well as functioning as the last safety net in the health care system—even when it was misused. The number of ER patient visits increased from 28.7 million in 1965 to 92.1 million in 1990.[99]

The other new public face of general hospitals was the intensive care unit, or ICU (often confused with or combined with coronary care or other specialty units providing intensive care). As noted in chapter 10, by the late 1960s ICUs were becoming universal, even in fairly small hospitals. By the 1970s it was clear that the over-the-top technologies and the enormous number of personnel devoted to ICUs were extremely expensive and might account for a substantial part of the rise in hospital costs. But as one investigator commented, "So what if it [the ICU] is expensive? If it does a lot of good, it is probably well worth the money." That same investigator, disconcertingly, ended up largely skeptical that the outcomes justified the costs. Hers was a lonesome voice, however, against the faith and hope of both patients and staff, faith and hope that justified the rapid growth of those new icons of technological medicine and care.[100]

Changes were also taking place in hospital operations. Unions continued to be a challenge in some localities, especially when, occasionally, even the medical staff organized. And when groups of physicians contracted for services with the hospital, relationships became very complex. Peer review and lawsuits continued to convulse hospital policies. Federal regulation of physician practice and hospital operation also complicated internal operations. Who, for example, was supposed to enforce ethical injunctions? Then in the mid-1980s, DRGs, which, as noted above, gave rise to alarmist language in many medical journals, changed the relationship between staff physicians and the hospital. In the end, physicians tended to make their peace with the new system. In their new role physicians were gatekeepers, always under pressure from administrators to save money and discharge patients "quicker and sicker."[101]

Medical Education and Pressures to Change

All the while, the teaching hospitals became ever more central as a site for medical education. It was in the hospitals, as a British observer pointed out in 1984, that "American medical students appear to be fed a rich diet of high technology, to which most of them become addicted for the rest of their professional lives."[102] It was in the hospitals that the patient mix determined what clinical exposure students and residents would get, and it was also in the hospitals that they would learn to interact with paramedical personnel. Thus, to a remarkable extent, the history of medical education in this era is the history of teaching hospitals.[103]

The formal pattern of medical education still had not changed from the Johns Hopkins turn-of-the-twentieth-century model, referred to by one medical critic as "the Flexnerian dinosaur type," in which the last two of four years were spent in clinical training, chiefly in the hospital.[104] By 1969, however, medical school was followed by a hospital residency of not one but three or four years. In this new pattern, the curriculum in the two preclinical years of medical school was increasingly dominated by pathophysiology under a variety of names. And then for the next six years there was clinical and specialty training.

This familiar pattern continued the intense socialization of medical students into a community of all-medical personnel (see chapter 9). In medical school, young people shared with other students portentous encounters with suffering and death, intensifying social bonds formed in class, work, and recreation. Around 1970, however, it was possible to detect signs of change in medical education. One medical educator eyewitness in 1974 listed "advanced placement at all levels; early exposure to clinical medicine; more opportunities for elective study; self-instruction with the aid of audiovisual technology and computers; and abbreviation of the total period of training."[105]

The younger people coming into medicine in the 1970s and 1980s, even when not radical, did gradually reshape the medical schools and eventually the profession. They were very technologically oriented and yet, on the whole, more sensitive than their predecessors to the doctor-patient relationship. A number of the new students were aware of social inequities and tended to take a more egalitarian approach to patients than their predecessors. Some even wanted to work in community medicine. Medical graduates now very often accepted as normal the idea of a professional who could work in a bureaucracy or at least in a group, even when they were caught up in administrators' demands to process more patients. For example, the physician leader of a hospital unit could no longer just terminate an unsatisfactory employee. Instead, the leader had to fill out forms documenting the objectionable performance. For many MDs, the adjustment to bureaucratic procedure was difficult or unacceptable, even if it did not affect how they cared for any particular patient.[106]

Meanwhile, state legislatures, national voluntary bodies, and the federal government, using the power of funding, tried to influence the curriculum of medical schools to solve various problems in the health care system. Medical ethics became a required element, for example, and primary care training was privileged in a number of ways—as if any of this would affect the hundreds of thousands of physicians already trained and working. Direct federal aid to medical students (and those in allied fields such as dentistry, podiatry, and public health) faded out in the 1970s.[107]

When federal funding for research leveled off or even decreased, and patients referred for advanced care started going elsewhere in the community, the teaching hospitals became decidedly more dependent on patient load for income. Not only did the hospitals need more medical residents to care for the load but the students and residents saw a different mix of patients. In 1965, 16 percent of the discharges were aged sixty-five or older; by 1985, 30 percent. And by the 1980s the teaching hospitals, with 5.67 percent of the acute beds in the country, were furnishing 47.2 percent of the free care.[108]

Confusing Health Care with Welfare

It was at that point that one of the major confusions in the health care system, alluded to above, became clear. Up until the mid-twentieth century there were two, largely separate pathways for unfortunate Americans: the welfare route and the medical route. Disabled people generally fell into the welfare category, especially under provisions of Social Security. When chronic diseases became dominant in the United States and, often in a connected way, aged populations expanded rapidly, the familiar categories broke down. In 1972 Congress extended Medicare to disabled people supported under Social Security. At the same time, disabled people became eligible for supplementary income. Where once a person disabled by a serious chronic illness, for example, multiple sclerosis, would have become dependent and been financially ruined by medical bills, now he or she still had income and the social status that came with it.[109] The social role of the chronically ill as welfare cases was significantly changed. In 1988 an observer described what was happening: "We have . . . made the sick person more visible and more acceptable by substituting illness criteria for means criteria in distributing the benefits of social programs. There has been a trend toward reducing governmental welfare and poverty programs and increasing the assistance to the sick and disabled; being sick or impaired will bring you public assistance, but being poor won't." A doctor's order, for example, could get a person into public housing when even desperate need could not.[110]

Medicare should have covered the medical side of the aged, and Medicaid the medical care of welfare cases—poor children and adults who could not work and who had welfare or Social Security support. Instead, nursing homes took over the care of extreme cases of both kinds, welfare and medical. From 1980 to 1996 expenditures for nursing home care increased by a factor of five. In 1995 there were 16,389 nursing homes with 1.75 million beds.[111]

Nursing homes primarily for the aged were a continuing problem. No one knew whether they were medical or welfare institutions. The homes were problematic in addition because the administrators hired from the very bottom of the labor pool. But nursing homes could be very profitable. At least three-quarters were owned by entrepreneurs, and people continued to speak of "grey gold." Greedy for-profit operators produced many scandals and received consistently low scores on quality of care. A major finding was that in many nursing homes no actual nurses were on duty. Competing nonprofits also required close supervision by state governments to prevent abuses—another outstanding example of the need for government regulation of health care. At the same time, by the 1970s and 1980s, nursing home operators had developed powerful lobbies. One thing was certain: a significant part of the population—a million and a half residents plus their families—depended on nursing homes. At the end of the 1980s, 20–25 percent of all elderly people could expect to spend time in a nursing home, typically in their last days.[112]

Disability, another boundary problem for medicine, became acute in the years 1968–88. Some experts measured illness and health in terms of how many people were disabled from work for how many days. Disability was also the main factor in welfare eligibility. The now dominant chronic diseases, including impairment of sensory and physical abilities, disabled people, particularly old people, for long periods or permanently. In 1972 the federal government moved disability into Social Security, and thus it was that over a period of time decisions about who was disabled came to be made by physicians rather than by welfare workers.[113] Simultaneously, it was groups of disabled people who, after the 1960s, organized and demanded social recognition. They wanted to function as citizens, not as either welfare or medical cases, which caused problems and confusion about definitions and status into the twenty-first century.[114]

Public Health

By the 1970s and 1980s, observers were commenting more frequently on "an increasingly fragmented public health infrastructure," an intense form of the fragmentation of the whole field of medicine, which, as noted above, had become noticeable by the 1960s. Public health now included trying to change personal health. And above all, environmental factors gained an enormous amount of attention.[115] The various kinds of public health activities were united only in a general commitment to protect the public's health—and also enhance levels of health. Professionals complained that no one could master all aspects of public health.

Three trends marked public health. First, public health leaders' concerns expanded to include many new areas, in particular protecting the population from chronic as well as acute infectious diseases. Second, they focused much more intensely on special groups at risk, beyond the traditional women and children and occupational groups: now it was urban populations, various ethnic populations, and particular age groups, especially the elderly. It was the introduction of these risk factor groups that led, in the 1970s, to the fading out of general, that is, undifferentiated, population screening programs. Physicians could screen an individual patient in an annual examination.[116] Using risk factors, clinicians could now think in terms of predicting health and diagnosing illnesses in individual patients even before there were symptoms. Epidemiologists were calculating how risk factors could suggest causes of disease, as in lung cancer and smoking.[117]

Third, the influx of environmentalist thinking provided an additional stimulus to broaden the work of public health workers even further. As the authors of a 1973 public health textbook put it,

> Public health today also includes *all* the hazards to health residing in the environment—the microbiologic, physical, and chemical dangers in the atmosphere, in water supplies, and in the soil. It includes problems of *population growth*, the rate of which is a matter of international crisis. Public health, finally, includes the *social and behavioral aspects of life*—endangered by contemporary stresses, addictive diseases, and emotional instability.[118]

Within a short time, the social factors that these authors listed separately from the physical environment became generally understood, in public health and more generally, as part of each American's environment.[119] The authors of the well-known Alameda County (California) study in 1983, for example, added to unhealthful habits another dimension already under investigation: social networks as possible determinants of health, especially in older people. "Evidence is mounting that many of the physiological changes we commonly think of as part of the 'natural aging process' are to a large degree environmentally determined," and hence the elderly constituted a "risk group."[120]

In 1979–80, an important government report, *Healthy People*, was issued. The report contained much that was traditional public health, such as disease prevention and health education, but also provides evidence of a major shift toward blaming people for their own bad health. The authors coupled "promoting health" with "preventing disease." The first of the specific targets of action was high blood pressure, identified as "perhaps the most potent of the risk factors for coronary heart disease and stroke." Other targets were more general: services for family planning, pregnancy and infant health, immunization, and STDs. Then there were protective measures, all standard in the age of environmentalism, aimed at toxins, occupational diseases, injury control, water fluoridation, and infectious disease epidemiology. Finally, there was active "health promotion," in which public health educators targeted smoking and alcohol and drug use; nutrition; physical fitness and exercise; and "control of stress and violent behavior."[121]

This program promoting health included the usual middle-class recommendations, but with emphases that would become even more distinctive over time, such as fitness, losing weight, cutting back on salt, and avoiding tobacco. The word *wellness*, which was not in most dictionaries, became standard only in the 1980s, along with the "health belief" education model.[122] The balance was definitely tilted toward maintaining health, not, as it might have been earlier, avoiding disease. One commentator named the 1980s the "wellness decade." Public health authorities for many years after used the *Healthy People* report to define the goals of public health, in effect finding unity in the report rather than in the many different activities that now constituted public health.[123]

There was one major shift: public health educators less often tried to reach the whole general public. Instead, they tried to reach defined audiences for health promotion and disease prevention, such as those in certain business firms, in educational institutions, or in service categories, such as cafeterias.[124] The educators were, of course, just following the business finding that niche merchandizing could produce more effect for a given amount of money.

Public health continued to be filtered largely through political institutions, but the sums involved on all levels of government were enormous. The support went to many different kinds of experts, from geneticists and statistical epidemiologists to inspectors in the field, laboratory technicians, and educators. Major infrastructures, such as water and sewage systems, were typically in the public utilities budgets rather than the public health budgets, although much public health expertise underlay those expenditures.

A New Meaning for Public Health Education

Attempts at controlling behavior were contentious, as in nutrition education. Occasionally a measure would have an effect. The state and then federal mandate to raise the drinking age to twenty-one was accompanied by a startling 34 percent decline in teenage automobile fatalities between 1982 and 1987. Laws requiring motorcyclists to wear helmets had demonstrable positive effects, but libertarian motorcyclist lobbying efforts succeeded in getting many of the laws repealed beginning in 1976, and the death toll climbed. Despite the swine flu inoculation failure, vaccines for hepatitis B and a type of meningococcal infection were developed, with significant protection for children. Protection of adults did not receive comparable attention, except for programs targeting the ever-increasing population of the elderly.[125]

The public health education efforts in particular fell in with the campaign to blame individuals for any bad health that they suffered. The most formal public statements included "lifestyle" as the major cause of preventable ill health. The lists of hazards tended to become standard, such as an authoritative one in 1987: "tobacco, alcohol, injury risks, high blood pressure, overnutrition, and gaps in primary prevention." The moral tone was reminiscent of that of the reformers of the first half of the nineteenth century and the uplifters of the Progressive era.[126]

But now there was statistical justification for saying that people with cancer and heart disease had chosen to smoke. In 1977, in an often cited declaration, the leading clinician-administrator John H. Knowles called for a major program to get people to take responsibility for their own health by watching their diet, exercising, and following other courses of self-denial. He decried the implicit social Darwinism that suggested that people who had "bad habits" got what they deserved, namely, bad health and death. Yet he also called for social responsibility on the part of Americans, in place of their imagined "individual rights" to "sloth, gluttony, alcoholic intemperance, reckless driving, sexual frenzy, and smoking"—all of which caused health problems and alarming statistics. At almost the same time, it was possible to deny that there was any "right to health care." One journalist spelled out the free-market assumption that everyone had free will to make choices and warned that "today, insurance plans spread the burden of paying for illness: the prudent and dutiful are paying heavily for the irresponsible."[127]

Behavior was becoming a medical problem in one more way. In 1977 the top medical official of the federal government declared that "we have entered a new era in nutrition, when the lack of essential nutrients no longer is the major nutritional problem facing most American people. Evidence suggests that the major problems of heart disease, hypertension, cancer, diabetes, and other chronic disease are significantly related to diet."[128]

A man presumably buying food from a food truck parked in front of an imposing building that symbolized the dignity and power of the U.S. government, 2000. The food truck temptingly offered for sale almost everything government health experts at the time advised against consuming. Marion Nestle and Michael F. Jacobson, "Halting the Obesity Epidemic: A Public Health Policy Approach," *Public Health Reports*, 115 (Jan./Feb. 2000), 15. Reproduced with the generous permission of the Association of Schools of Public Health.

The campaign to make "healthy choices" and choose a healthful style of life had gained great momentum by the 1980s. Many Americans engaged in some sort of body worship appropriate for the age of narcissism. A number joined exercise or dietary congregations like Weight Watchers to reinforce their healthful ways. Health promotion experts could claim convincingly that enough Americans had actually changed their behavior sufficiently to reduce the rate of deaths from heart disease and cancer. One witness reported this impression in 1988:

> The streets are alive with the sound of pounding sneakers, with the grunts of weight training, the hard-rock music of aerobics classes. People are swimming down pool lanes, cycling along park paths and country roads, running on beaches and back streets. "Getting in shape" and "working out" are sacred themes of contemporary life. Strength, endurance, and fitness have become even closer to godliness than cleanliness. Health clubs are cathedrals built to sanctify the body; weight rooms and tracks have become our temples; athletic paraphernalia are our liturgical articles.[129]

All this health enthusiasm fitted in easily with various kinds of environmental thinking and advocacy, such as reducing environmental exposure to cigarette smoke and television. It was easy to use environmental factors and health campaigns as forms of demedicalization, as one 1976 writer hinted: "Environmental health, mental health, school health, and community health are all examples of the explosion of the health concept into a multiprofessional construct.... Most health and sick behaviors are learned [and] consequently can be influenced by environmental events." In a time of chronic diseases with multiple causations, a "second revolution in health" was not beyond credibility. Yet some commentators at the time questioned making individuals, rather than a society that nourished powerful antihealth elements, such as advertising and poverty, responsible—or, again, "blaming the victims."[130]

Perspectives on Health Care at the End of an Era

In 1989 two sociologists summarized one version of how completely health care embodied the social organization that was, along with technology, the other main stream in modernization. Instead of the doctor-patient relationship idealized in health care of the past, "the physician-patient dyad has evolved into a complex network of patients, providers, third-party insurers, government agencies, private entrepreneurs, and national corporations. Patients' decisions about physicians or other providers may be limited by their employer's health plan. Physicians' allocation of medical resources may be constrained by features of a third-party insurer or the protocol of an employer."[131] Yet in the years around the turn of the twenty-first century it was indeed the doctor-patient interaction that highlighted the opening of a new era, beyond modernization (see chapter 14).

Meanwhile, in 1990 Eli Ginzberg, a veteran analyst of the health system, took a step back from immediate problems and offered a commonsense comment: "The U.S. health-care system must be meeting societal goals; otherwise we wouldn't be pumping so much money into it." The public in general (not including lower-income groups), he argued, had a lot of disposable income, and they chose to spend it on health care. After all, in the 1980s, 75–85 percent of Americans polled stated that they were "completely satisfied" with their health care. All levels of government devoted much funding to the system. Industrial disputes at that time were typically centered around health insurance. "Reduced to basics, [the goals of Americans] amount to nothing more or less than a healthier, longer life for the individual with less disability, pain and incapacity."[132]

What Ginzberg did was to point out the logical end point of trying to apply the free market to health care. The free market was not designed to control prices or reduce national spending on health care. Pretending that it would do either was either naïve or duplicitous. The free market was to give those with money the choice to spend as much or as little as they wished for whatever personal health care and medical technology they might think they wanted. Those wants had driven total national expenditures very high.

Because of the high priority of health care after the mid-1970s, some of the best brains in the country tried for two decades and more to figure out what policies would work best to control prices and yet deliver good levels of health care to the maximum number of people. They worried about various kinds of organization and cooperation, about home health care for the elderly, and about many other policy questions. They worried about different kinds of financing and particularly about how much central control was necessary, as opposed to state-level control and spontaneous competition, including competition among health units, such as nonprofit hospitals, that previously had not embraced frankly entrepreneurial competition. Almost every local scheme had some successes and some failures as each type of unit tried to pass unprofitable health care arrangements onto other units. So it was that most observers in the 1980s believed that free-market competition was beginning to play a large role in health care, which for a long time had been dominated by professional restraint and trust and a great deal of formal and informal charity. In 1985 the sociologist David Mechanic concluded, "It is evident that we are living in an era of simultaneous rationing and commercialization."[133] By the 1990s at least some of the effects of a drift toward market control became more obvious. They were summed up in the phrase that physician groups, especially, successfully demonized: *managed care.*

As we shall see in chapter 13, managed care would bring to the surface two major problems. The first was the continuing, wildly disproportionate distribution and possibly conscious rationing of the still limited resources of health care. "We . . . move 90-year-olds to intensive care units and restart their hearts 20 or 30 times before they die, but don't give health insurance to 31 million people," wrote a critic in 1994.[134] The second was the way the system affected direct caregivers, especially those in the home. Indeed, it would soon be possible to talk of the "informalization" of caregiving, which mostly fell to women, including women who also worked outside the home. By the end of the 1980s well-informed observers were already talking about "a major transformation of the health care system" or even "the health care revolution."[135]

PART IV

The End of One Epoch and the Beginning of Another

By the end of the 1980s, it was apparent that another era was beginning. It was not difficult to see that this was the era of genetic medicine, and with it personalized medicine. And to a considerable extent, one could project into this new era the persistence of most of the institutions and legacies of the past, from the colonial beginnings and traditions to the environmental era. As one expert observed at the beginning of the new era, in the time of environmental concern chemicals had tended to replace microbes as the most feared sources of disease, but taking center stage now were degenerations at various levels: of organs, of joints, of nerves, and particularly of DNA as found in cancers.[1]

Historians writing in the era of genetic medicine noticed especially the borderlands of health care. The promise of genetic engineering to remake a person built upon the long struggle to use medical techniques and institutions not just to cure diseases but to move humans toward a state of perfection. Earlier, eugenicists had tried by selective breeding to improve lines of physical heredity. Surgeons had long attempted to restore functions to imperfections of the body, not least, for example, clubfoot. But where was one to stop? Surgeons over several generations had moved from physical functioning to social functioning and developed the specialty of plastic surgery, which often shaded into the cosmetic. With physiological medicines, an enormous number of functions could be enhanced or repressed. Reaction and movement could be improved, especially in sports. Substances could add height and muscle to a person (short stature became labeled as a disease). Even (male) sexual performance could be increased, and not just with a placebo effect from fraudulent patent medicines and devices. Individuals pursued mental acuity and subjective comfort using various substances labeled medical or nonmedical, such as alcohol and drugs that were once part of medicine. In a person-centered consumer society, perfection could take many forms. So genetic medicine followed at the margins of health care in assisting humans to achieve their changing personal goals, starting with the prolongation of life.[2]

This new phase, genetic medicine, was not just another era in which familiar health care components persisted, evolved, and shifted. Although for a time genetic medicine was the conspicuous identifier in biomedicine and marked a new era, the nature of health care was changing, so that it was not just a new era but a new epoch that overtook events. With even a preliminary survey of the signs of that dawning epoch, it is possible to recognize how our broad history of health and ill health and medical care in the United States reaches a natural conclusion with the dawn of the new epoch, even if the exact nature of this larger epoch remains unclear.

Sizable though the health care apparatus was at the turn of the twenty-first century, it was still functioning in a postindustrial, consumer, service-industry society. Now, however, the gap between the rich and the poor was growing dramatically, and the working middle class, with income to spend on health care, was shrinking significantly, renewing attention to the low standing of American health in comparison with the level of health in other developed countries and the extent to which low income per se was even more obviously a major factor in the epidemiology of health and disease.[3] Or one could point out that the more the economic resources that went into health care for individuals, the fewer the resources available for public health, welfare, and other activities that would do a great deal more to sustain and raise general levels of healthiness than would private health care for the affluent or lucky.[4]

Moreover, fragmentation in the society was modifying or disrupting bureaucratization, most obviously when production units such as the large, integrated businesses and the great factory production lines no longer dominated the social landscape. This reading of the record therefore suggests that with the new epoch, modernization as such ceased to explain usefully what was happening in society and in the history of medicine and health care. Scholars increasingly often have had recourse to concepts like networks, systems, and globalization.[5] The recent history of health care therefore also requires a different framing. By the twenty-first century, experts could point out that health care institutions, even in a network model, were so disconnected from one another that the system not only was extremely and unnecessarily costly but overall was not delivering the highest quality of care. The fragmenting of the medical profession into specialties was only one aspect of fragmentation in the health care system.[6]

Adding to the confusion, beginning especially in the 1980s, advocates of providing health care on a free-market model also tended to equate fewer federal government mandates with a devolution of regulation to the local and especially the state level. This meant less uniformity across the country but more federal-state cooperation, and cooperation with private for-profit and nonprofit organizations. In practice health care did not fit standard economic and political models. It was just aggravatingly complex.[7]

Shortly after the turn of the twenty-first century, a group of sociologists trying to understand what had happened, and was happening, cited the fading away of the old categories "medical" and "nonmedical" as marking 1985–90 as a time of transition. They confirmed what various social analysts from other viewpoints were also detecting. Somehow, at least for those in the upper levels of the American social hierarchy, biomedical research and health care had changed. Biomedicine had become a defining element in the broader culture. The sociologists who fixed on a transition in 1985–90 could argue plausibly that large-scale biomedical activities and viewpoints altered the activities and viewpoints of opinion leaders—ideas of the self, the body, and also economics and politics. Ivan Illich and other social critics had not foreseen how far medicalization could go under the influence of the "molecularization" of medicine and the shift from sickness and defined diseases to wellness, prevention, and risk groups. Whereas previous critics of medicalization had written in terms of social control, referring to the organizing element in modernization, the new authors were describing social transformation. Indeed, in the eyes of these scholars even economic and political activities were becoming not just medicalized but, consonant with decades of biomedical research, "biomedicalized."[8]

What such observers were trying to express was a sense of qualitative mutation in the health care apparatus. The major theme in surveys of biomedical research was how each quest for knowledge ended up finding new complexity—interesting explanations, but with innumerable cellular, subcellular, chemical, and network factors operating in any bodily function or malfunction. On the individual level, as the historian Charles Rosenberg observed, "We have moved from the kind of intense and visceral fear associated with acute infectious disease . . . to a species of widely-disseminated anxiety spread over time."[9]

Regardless, after the late 1980s, health care grew ever more important economically. There were still recognizable elements functioning as part of health care, but they were functioning somewhat differently than they had when patterns of modernization could help provide an understanding of what was going on.

One new dimension in particular entered the story. How health care was delivered became an important determinant of how both professionals and the general public perceived medicine and illness. As the 1980s gave way to the 1990s, people detected not only the coming of genetic medicine but, almost simultaneously, reaction to the management style called "managed care." But beyond that, there was a quiet transformation under way in the standard medical encounter, the doctor-patient relationship.

It is on the basis of change in this common social encounter, that between doctor and patient, that it is possible to say that beyond genetic medicine and more blatant managerial intrusion into health care, a whole new epoch began in the history of American medicine and health care.[10] Moreover, consumption, lifestyle, and ideas of risk combined with health promotion to shift concern with the body into new areas of conceptualization and of social functioning, a shift that had roots in The Sixties but flowered only beginning in the late 1980s.[11]

Part IV therefore comprises but one full chapter on genetic medicine, followed by a short general concluding chapter. In taking up the era of genetic medicine, it is necessary to suggest the magnitude of the other changes that became socially important from the late 1980s into the twenty-first century. Then I close by rehearsing evidence of the underlying epochal change that was taking place. Appropriately, most of the detail of what happened in the immediate past and into "present times" is left for future historians.

Landmark Dates

1990	Formal beginning of mapping the human genome
1991	American Board of Medical Genetics established
1992	Evidence-based medicine takes center stage
1993	Discovery that alternative medical resources used by one-third of the population
1994	Peptic ulcer formally confirmed to be a bacterial disease
	Defeat of Clinton health reform
1996–97	Mad cow disease recognized as a prion disease in humans
Late 1990s	Managed care backlash
1999	West Nile virus detected in the United States
2000	Realization that human genome sequencing was succeeding
	Institute of Medicine report on error in medicine published
2002	Economic recession leads to rethinking the economics of health care
2003	SARS frightens but bypasses the United States
2004	Obesity overtakes tobacco in lifestyle hazard ratings
2007	Recognition of increase in incidence of diabetes
2010	Affordable Care Act passed

CHAPTER 13

The Era of Genetic Medicine, Late 1980s and After

As early as 2001 the editor of the *Journal of Health Politics, Policy and Law* sensed that a dramatic change was afoot:

> For the past several years we have been experiencing a revolution in medicine, manifest in many forms and along myriad dimensions. It is most readily recognized in the transformation toward a system of competing managed care plans, but that itself rests on the emergence and increased sophistication of evidence-based medicine, at least implicit cost-effectiveness analysis, and clinical practice guidelines that are supposed to summarize the current state of knowledge and evidence. Kindly Marcus Welby, M.D., for all his anecdotal experience, patient knowledge, and accumulated wisdom, gets supplanted by medical data gathering and algorithms not so prone to idiosyncratic bias and untested norms.[1]

For this observer, the most notable recent revolutionary changes involved evidence-based medicine—an attempt to introduce new standards of practice—with implications for both the legal system and the clinic. Closely aligned were changes related to health insurance arrangements, mostly under the heading "managed care," an approach that had dominated an era beginning in the 1990s. Yet within a few more years, evidence, quality, cost, and policy all converged because they were so intertwined.[2]

At almost the same time, however, there was a fundamental disjuncture in medicine, symbolized by the human genome project. In May 2000 Francis Collins, director of the National Human Genome Research Institute, told a private audience gathered to hear that the project was successful, "We have been engaged in a historic adventure. . . . The enterprise we have gathered here to discuss will change our concepts of human biology, our approach to health and disease, and our view of ourselves. This is the moment, the time when the majority of the human genome sequence, some 85 percent of it, looms into view. You will remember this." And Collins was right. It was indeed, as he said, "an astounding time in our history."[3]

This chapter suggests how people experienced what happened from the late 1980s to about 2010. There were expectable trends, including new diseases. But the whole story of change shifted with the coming of genetic medicine and attempts to apply genetic medicine to individual patients. Meanwhile, germ theory took on new guises and new importance even as alternative and commercial medicine persisted. And meanwhile there was the managed care crisis. The actual work of health care continued in the midst of all these legacies of modernization. Earnest workers, along with policymakers at various levels, were still trying in the first decade of the twenty-first century to accommodate insights from genetic medicine, improve health care, and figure out how to pay for it.

Continuing Trends

From 1988 to 2008 the U.S. population grew by more than 23 percent, from 246 million to 304 million. The number of active physicians increased by about 39 percent, from 562,839 in 1989 to 784,199 in 2008, so that the number of physicians per 100,000 population continued to increase. Already more than 30 percent of physicians were female.[4]

From 1990 to 2007 the percentage of medical students who were women increased from 37 percent to 49 percent, and no one was surprised when, in the years that followed, a majority of medical students were female. After a low point in 1991, medical students' interest in primary care increased to a high point in 1999 and then declined and plateaued. The change in the total number of medical students—c. 17,000 to c. 18,000—was not significant, but the percentage of non-

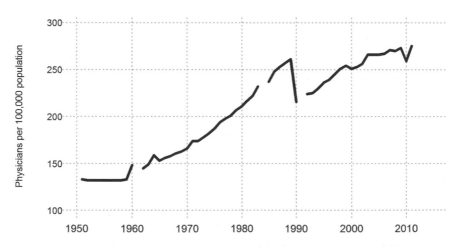

For sixty years, from 1950 to 2010, the ratio of physicians per 100,000 population increased at a generally steady rate. There was more medical care available for each person, on average. The increase in supply did not decrease the price, however. Human as well as economic factors confounded policymakers.

Asian "minority" students finally increased slightly. In 1989 there were 1,666,200 registered nurses; in 2009, 2,538,770. In 2000, 12,211,000 people, or 8.9 percent of the civilian workforce, were working in health services; in 2008 the figure was 15,108,000, or 10.4 percent. Five or six million of the health care workers served in hospitals, and a million and a half or more in nursing homes. The number of beds in hospitals declined from 1.2 million in 1990 to 951,000 in 2008, continuing a long-term trend, as medical procedures were still moving to outpatient sites. Indeed, another 486 hospitals closed from 1987 to 2002, although soon some hospitals were once again reporting that capacities could not accommodate admissions. And just as governmental regulations finally improved the quality of care in nursing homes, health policy leaders launched a campaign to move elderly patients away from institutional care and back to care in the home, where they could be maintained more cheaply with "community" resources, which, translated, meant both low-paid and unpaid labor. Meanwhile, many nursing homes changed to accommodate patients discharged from hospitals but still needing care.[5]

The health record of the country was summarized in an official report in 2009, homing in on data collected in 2007, exactly four centuries after the Jamestown settlement:

In 2007, American men could expect to live 3.5 years longer—and women 1.6 years longer—than they did in 1990. The gap in life expectancy between the black and white populations has narrowed, but it persists. Mortality from heart disease, stroke, and cancer has continued to decline in recent years, although mortality from chronic lower respiratory diseases and unintentional injuries has not. Infant mortality—a major component of overall life expectancy—declined through 2001 and has changed little since then. However, both life expectancy and infant mortality continue to lag behind levels in many other developed countries. . . . Of concern for all Americans is the high prevalence of people with risk factors such as tobacco use, high cholesterol, obesity, and insufficient exercise, which are associated with chronic diseases and conditions such as heart disease, cancer, diabetes, and hypertension.[6]

Unsurprisingly, then, earlier trends continued. This official statement nevertheless contained two notable new elements. First was the special recognition of diabetes as a major fatal disease, along with the traditional big three—heart disease, cancer, and stroke. The proportion of deaths attributed to diabetes increased from 8 percent in 1988–94 to 11 percent in 2005–8. Moreover, diabetes opened the way for biomedical experts to look generally at autoimmune diseases. In this era of genetic medicine, investigators paid particular attention to the probably hereditary side of the internal environment of immune system functioning. A person's

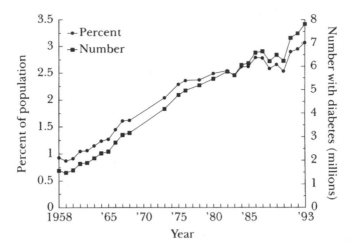

The number of diagnosed diabetes cases in the population and the percentage of the
population suffering from diabetes increased at a surprisingly steady rate from 1958
to 1993, and diabetes had become an alarming national health problem by the turn
of the twenty-first century. *Diabetes in America*, 2nd ed. (Washington, DC: National
Institute of Diabetes and Digestive and Kidney Diseases, 1995), 50.

autoimmune profile helped physicians make diagnoses and especially, and significantly, determined the risk groups into which the person's body fell.[7]

The second new element in the report was noticeably increased concern about
the lack of progress against preventable deaths in the United States. Moreover, it
was obvious that, as in the previous era, social factors were deeply implicated in
many of the unsatisfactory statistics.[8] Most experts continued to blame individual
behavior choices, not television, advertising, or other aspects of the individual's
social and cultural environment.

Statistics showed additional changes in the two decades on either side of the year
2000. There was a striking increase in the percentage of adults taking antidepressant drugs, from 5 percent in 1988–94 to 11 percent in 2005–8. In addition, the percentage of adults on antianxiety medications increased from 4 percent to 6 percent.[9]
Another sign of a marked social trend was that 20 percent fewer people died in
hospitals, while the number of people who died at home increased by 50 percent,
to about a quarter of the total, suggesting (along with a continuing increase in the
number of deaths in nursing homes and hospices) a significant shift in end-of-life
options, or perhaps some demedicalizing of death.[10] Not only were the chronic diseases continuing to increase in importance, but in many cases a person with one
chronic disease was also suffering from one or more additional chronic diseases.

In 2000, by one estimate, chronic conditions accounted for 78 percent of all health expenditures.[11]

Many trends did not diverge greatly from those already in sight before 1990. In 2004 the CEO of a major health care organization stated her perceptions of ongoing problems in the system (by that time even sophisticated commentators used the term *system*):

> The destiny of our healthcare system is controlled to a great degree by legislators. I am concerned that our politicians are not up to the task. . . . The incentives in our system are misaligned. For instance, physicians primarily get paid for keeping the patient in the hospital, but hospitals get paid to get the patients out of the hospital as quickly as possible. Certain diagnostic groups or illnesses are reimbursed more, so there is an incentive to create more of those services whether they are needed or not. Another perverse incentive in the system is that insurance is too high for many to purchase, but at the same time hospitals are required by law to provide care to all who come to their emergency departments. As a result, there is no incentive for many to be insured because they know they will receive care anyway. These types of convoluted incentives prevent us from being more cost effective and progressive. In addition, the industry lags behind in implementing available, new medical and information system technologies that can make healthcare delivery more efficient and can improve care quality.[12]

In these observations it is easy to see a continuation of earlier problems. This CEO still thought in business terms; in particular, she depended on monetary incentives to manage health care and hoped that technology and the profit motive would solve most of the problems.

Emerging Diseases

The relentless march of so-called new or emerging diseases and illnesses continued.[13] One "new" disease involved reconceptualizing the cause behind what a newsmagazine in 1979 labeled an "epidemic of senility." Much earlier in the twentieth century the population of senile elderly people had been substantially shifted from their homes to mental hospitals (often with a diagnosis of "chronic brain syndrome"), but with the deinstitutionalization movement of the 1960s and 1970s they moved to nursing homes or once more were kept at home. Regardless of the site, they increasingly became a problem to both family and society as they deteriorated relentlessly but usually slowly. Early in the 1980s experts continued to refine the definition and conceptualization of "senile dementia," which in popular discourse came to be known as Alzheimer's disease.[14]

With Alzheimer's disease, old age and senility definitively moved into the medical realm. Senile dementia became not an inevitable phase of old age but a disease that might be treated or, because there was probably a genetic factor, prevented. Medicalization also suited the political agendas of those concerned about an aging population. Experts and policymakers could speak in alarming terms about "the slow death of the mind" or "a never-ending funeral." Some believed that, given the shifting demographics, Alzheimer's was not a disease of the month but a disease of the century, one that would increase in incidence dramatically for decades.[15] By the 1990s, biomedical scientists were focusing more insistently on genetic factors in Alzheimer's and the other dementing processes, while the media extended the discussion beyond dementia itself to the damage done to hapless informal "caregivers" and desperate families who were not included in any medical statistics.[16]

Another diagnosis that increased dramatically from the 1980s to the second decade of the twenty-first century was childhood autism. Diagnoses of "autism" replaced those of "mental retardation" and "developmental disability" in a number of cases, but the steeply rising curve of incidence went far beyond developmental disabilities. Many clinicians believed that the increase represented a real increase in incidence and might be parallel to the increase in adult depression. The diagnostic question, however, was irrelevant to the large number of parents of autistic children who, by the twenty-first century, faced burdens of care far beyond what they might reasonably have expected.[17]

Physicians and members of the public alike also continued to be concerned about newly noticed acute and infective diseases. In the 1990s, MRSA, the antibiotic-resistant staph (methicillin-resistant *Staphylococcus aureus*), known for more than a decade as a hospital disease, emerged, offering another model of a way to understand new and dangerous diseases. First, investigators found that extremely dangerous, fast-moving versions had mutated into existence "in the community" and spread not just in hospitals but to a variety of places in the human environment, including interchanges with domestic animals and household pets. Second, using new technologies to characterize genetically different strains and determine what changes in the DNA accounted for resistance to antibiotics, biomedical experts showed the extraordinary adaptability of these staph organisms as they overcame one medication after another. Third, the massive, uncontrolled infections led to substantial alarm, on all levels of knowledge and expertise, about an increase in antibiotic-resistant infections, now including necrotizing fasciitis, or "flesh-eating" disease. The alarm was not just a public health concern. As one clinician complained in 2007, "When we are on rounds, we spend the vast majority of our time going from one patient to the next who are infected with organisms for which we have almost nothing that works, and trying to figure out how to treat them."[18]

Several years into the twenty-first century, a San Francisco infectious disease expert who had been through the AIDS crisis commented concerning MRSA:

> Other than AIDS, this is the biggest thing in the past thirty years. In incidence, in cost, in impact on health care, in how you approach patients. There are so many things where you must assume MRSA could be causing them: Earaches in kids. Sinus infections. Pneumonia. The big three sites for infection are respiratory tract, urinary tract, and skin and soft-tissue, and suddenly half of those skin and soft-tissue infections are resistant. Your workhorse antibiotics, gone. And if you put someone on an ineffective antibiotic, you may make things worse, because it furthers resistance and it prolongs the period when they are infectious.[19]

The Globalization of Health Threats

In the media, most of the infectious disease alarms now signaled another kind of change: the globalization of disease, health, and medicine / public health. The media as well as scientists tracked influenzas as they spread around the world in the 1990s and after. Influenzas could start in local circumstances, as when viruses mutated and passed from birds or pigs or other animals to humans. In 1999 another type of disease, the West Nile virus, originating in Uganda and Egypt and carried by mosquitoes, was detected in New York. The virus spread rapidly to people living in many states. The disease was usually asymptomatic, but in about 1 percent of the cases the virus became neuroinvasive and caused a significant number of deaths, making the outdoor environment in many areas seem potentially deadly.[20]

In 2003 SARS (severe acute respiratory syndrome) spread at first through the local health care system in China (about one-quarter of the known infected people were health care workers). Then experts detected it in nearby travel destinations and soon even in Toronto, Canada, borne by jet airliner passengers who in many cases had merely breathed the same air as a passenger with SARS. While SARS did not reach the United States, in that same year pets from Africa infected pet prairie dogs and brought monkeypox to unsuspecting residents of Middle America. The media portrayed SARS and other diseases as a threat to everyone, in contrast to AIDS, heart disease, and cancer, which reporters suggested were threats mainly to people in risk groups.[21] Repeatedly, there was news of some disease like Ebola fever emerging from a local environment somewhere in the world and spreading in the now globalized society, threatening populations in, not least, the United States. Among health authorities, it was commented bitterly that "the most important disease vector today is the Boeing 747."[22]

In addition to scares of bird flu or swine flu (which resurrected the historical memory of the 1918 flu epidemic), at the beginning of the twenty-first century terrorists and a sharp increase in terrorism made people aware of "bioterrorism," and fear spread that someone would deliberately introduce an infectious disease into the United States. As one group of public health experts recalled, "For the first time in many years, public health infrastructure was viewed as critical by more than those of us in public health."[23] Anthrax was the identified threat at one point, but public health personnel also expected to deal with other diseases and disasters.

Marketers for the pharmaceutical industry continued to construct new diseases or broaden the application of old diagnoses (adults as well as children now suffered from ADHD, attention deficit hyperactivity disorder, and, as noted earlier, normal men suffered from "erectile dysfunction"). Ideas of the ideal body and ideas that discomfort was unnatural and intolerable drove consumers to demand from physicians drugs, surgery, and other physical and functional enhancements, some of which insurance entities would cover and thereby legitimate as "medical." In a similar way, pain, especially back pain, became a major issue in medicine as well as in advertising. Meanwhile, the political and propaganda progress of the oxymoronic idea of "medical marijuana" dramatically and confusingly demedicalized parts of the subject of pain.

Genetic Medicine

The most distinctive mark of the shift into the genetic era was a new way of understanding what allowed disease processes to appear in any particular human organism. With increasing momentum in the 1980s, medical experts attributed more and more diseases to genetic determinants, diseases set off by a variable combination of environmental impingements interacting with each person's preexisting, usually hereditary subcellular constitution—a pattern remarkably reminiscent of nineteenth-century beliefs about pathology, when one's constitution had to stand against miasmas and contagion. In the complexity of modern biomedicine, scientists were attracted by the fact that a person's genetic variants, now often detectable to some degree by instruments, were relatively stable throughout the life of the human organism, so that it was possible to study them. Moreover, it became evident that not just defective genes but variations in the process of DNA replication, transcription, and expression were part of an individual's genetic identity.

One symbolic date for the genetic era was 1991. In that year the American Board of Medical Specialties approved a new board, the American Board of Medical Genetics. Medical geneticists had been organized for years before this recognition came, it is true. But for twenty years previously only one other new board had been approved, the American Board of Emergency Medicine (and that was in 1979).

As noted above, when the study of viruses, subcellular physiology, biochemistry, and genetics came together, visionaries foresaw that it might be possible at some probably distant future time to engineer the basic elements of heredity and life in such a way as to produce humans who would be free of some diseases, especially those diseases presumably caused by "defective," damaged, or mutated genetic material. The visionaries had the same goal as earlier eugenicists: to improve the inherited bases for people's health and well-being. Now, however, the utopians had evidence to suggest that the a person's genetic makeup might be improved, not through long processes of selective breeding, but simply by mechanical tinkering with chemical structures in his or her DNA.[24]

Bringing genetics into medical practice took a very long time. Scientists and the lay public shared a folk belief that heredity determined a person's physical and functional attributes (including anatomy, physiology, and behavior). It was easy to use chromosomes as a more precise way of representing inherited traits. Especially in the 1960s, when scientists were actually visualizing chromosomes with instruments, it became possible to think in terms of equating diseases with the structure and consequent reactions of genes. By 1993 one expert was writing, "By almost universal consensus, cancer is now viewed as resulting from changes in a few key regulatory genes. The disciplines of cancer research have also converged to focus on genes that regulate cell growth and differentiation."[25]

Meanwhile, the combination of genetics and medicine in an early, practical form centered on diagnosing and treating known clinical syndromes. One early achievement was finding a test for infants born with PKU (phenylketonuria), a disorder that without treatment would result in developmental problems and often death. The diagnostic test for PKU in newborns and a dietary regimen that would permit them to attain an often normal adulthood came at the same time, and it appeared that a genetic disease was being controlled by environmental means. The story had a frustrating ending, however, for it turned out that babies who were saved and grew into adults could still carry a genetic makeup that blighted their children. Screening newborns for PKU and other genetic disabilities was required in most states for decades.[26]

Genetic medicine in the 1970s and 1980s had a substantial beginning when investigators developed standard ways of identifying possibly defective or aberrant chromosomes. With amniocentesis and sonography, a pregnant woman could learn about some of the genetic hazards that might show up if her fetus were allowed to proceed to full term. Amniocentesis became practical and widespread in the 1970s as a way of testing fetuses for at least gender and Down's syndrome, with promises for further genetic screening in the future.[27]

Down's syndrome was the first and primary determinant of a woman's choice to continue to term, but PKU and other serious hereditary malformations, such

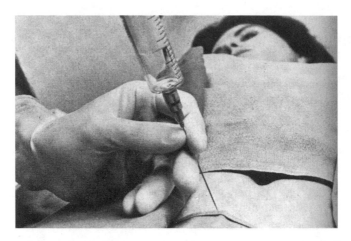

By the 1970s pregnant women commonly underwent amniocenteses, as shown in this photo. This standard procedure safely and with surprising reliability detected genetic abnormalities in fetuses and permitted parents to make informed personal decisions. Theodore Cooper, "Implications of Findings from the Amniocentesis Registry for Public Policy," *Public Health Reports*, 91 (1976), 116.

as congenital hypothyroidism, soon became part of a medicalized pregnancy. By the 1990s, aided by technological developments, a new paramedical occupation, genetic counselor, had joined the medical team. A genetic counselor was someone who could offer a prospective mother examinations and choices based on advanced science and technology. Amidst a storm of practical, moral, and social controversies, increasing numbers of women who could afford to made highly personal decisions concerning whether to have selective abortions.[28]

In all these circumstances medical experts increasingly in the 1980s invoked some hereditary or genetic factor to explain the illnesses they were observing and studying. In 1988, for example, in just one issue (1 September) of the *New England Journal of Medicine* four of the five major articles had to do with probable genetic factors in different diseases.

Reconceptualizing Diseases

Around 1990, however, investigators finally made a great reversal that made use of genetic medicine to reconceptualize diseases in a major way, underlining that they were indeed initiating a new era. Before, the connection between disease and medicine had been made by identifying a syndrome by a pattern of epidemiology or ordinary pathological evidence and then tracing it to the level of a chromosome. Now, however, the process became not to reduce a known disease to genetics but to start with the gene to understand the disease process. For complex, chronic diseases especially, this was a great advantage. Even in clinical practice, some problems, such as dystonias (painful muscle contractions), were being categorized by genetic specifications rather than by signs and symptoms.

Biomedical scientists at the time understood exactly what had happened. As Barton Childs, of Johns Hopkins, wrote, looking back,

> A final and irreversible wedding of medicine and genetics was attained in the suggestion of the gene and its products as the ultimate clues to pathogenesis and perhaps to diagnosis. In the '90s, the diagnostic process that proceeded traditionally from phenotype to protein by way of history, physical signs, and physiological and biochemical expressions, began to be reversed so that the protein mediator of pathogenesis was discovered by way of the gene instead.

Childs, writing in 2002, looked to a time when "the genes and their products constitute the node whence principles of disease originate." It was a way of reducing medicine to biology, but now it was an experimental, laboratory biology.[29]

Moreover, it was subcellular biology. Physiology and virology also were operating on this same biochemical level. Oncogenes, cancer-causing genes, gave way to extreme microlevel mutations and other contingent changes. But it was forty years after the discovery of DNA structure in 1953 before the shift to this start-with-genetics perspective gained traction.

One disease that was paradigmatic of this reversal into "genomic medicine" was cystic fibrosis, which investigators from both sides of the Atlantic found to be coded by a mutated gene. This disease, which was reported for centuries and had been given a clinical description by an American physician, Dorothy Anderson, in 1938, compromises respiratory, gastrointestinal, and reproductive systems and disturbs sweat gland functioning. A common sign is an often fatal difficulty breathing. At first most afflicted children died by the age of two, although eventually careful supportive therapy permitted a number to survive into adulthood. In 1978, investigators identified a specific mutation in a specific gene as the cause of many cases of cystic fibrosis. This was an important discovery for both scientific and practical reasons. It permitted asymptomatic carriers of the gene to be identified, and soon screening for the gene was possible and became widespread in the United States. That in turn permitted couples to make reproductive decisions. Over the years, however, so many (more than a thousand) specific mutations appeared to be involved in cystic fibrosis that the problem took on enormous complexity.[30]

Getting into Genetic Medicine

By the first decade of the twenty-first century investigators were saying that genetic medicine was not just about genes but about proteins that genes produced.

A particular protein, for example, might occur in two very different kinds of cancer. Working on the "nano" level, a scientist might use a nanomarker to test for such proteins and to suggest that in two different individuals with two different diagnoses the same pharmaceutical intervention might be indicated.[31] This recourse to multiple, subtle causes of a set of symptoms bore an eerie resemblance to ideas in the early nineteenth century that many causes operated together to bring on a person's particular illness. Twenty-first-century research was uncovering ever-expanding numbers of factors involved in many diseases, including the hypertension, Alzheimer's, and autism featured prominently in the media.[32]

Investigators working in genetic medicine had three different goals. Two were obvious: diagnosis and potential cure. But the third was to discover what an individual's reaction to a particular drug might be so that it might be used or avoided in that particular case. A whole industry sprang up to find and market "biomarkers" even before their usefulness was completely clear.[33]

Generating much interest in biomarkers in personalized medicine was a movement to take drugs already available or discarded and use them for small groups of patients whose bodies and diseases might respond to any particular chemical in special, genetically determined ways. The biogeneticists hoped, for example, to reduce the number of incidents of adverse drug reactions, still a leading cause of deaths. They also hoped to broaden the possible therapeutic uses of many substances that in exceptional patients could target pathological processes, as in some heart disease. Or to judge the kind of dosage that would be effective. Indeed, informal networks appeared, for example, of family members who shared a particular disease-prone genetic factor and who might learn about some combination of drugs that targeted that factor.[34]

Meanwhile, leaders in genetic medicine had shared with the public their hopes that reengineering chemical structures in the gene might effect a cure in patients whose disease was caused by an identified genetic pathology. Already in 1990, introducing some copies of a normal gene into a four-year-old girl with a genetically immobilized immune system had proved successful.[35] Any progress with this kind of technology, however, was extremely slow. In 1999, when such a tactic, "gene therapy," was tried on a patient who had cystic fibrosis along with other ailments, he died, and the search for a genetic engineering cure was postponed indefinitely. It was soon clear that environmental factors as well as still other genes might be involved in any disease. But the hype and the hope were hard to keep down. The "promise of genetic medicine," which merged research planning into science fiction, was still alive for many diseases, including cystic fibrosis, a decade and more into the twenty-first century. Investigators were identifying contingent genetic factors involved in diseases, but making gene therapy effective and wide spread was still a long way off.[36]

The problem was that for most diseases the genetic factors or the combined genetic and environmental factors were extraordinarily complex, in contrast to the earlier model of one microbe, one medicine. As a leading scientist wrote in 2008, "While genomics has yielded numerous potential new drug targets, it also reveals the complexity of a single cell, not to mention an entire organism. Curing complex diseases with single chemical entities may have been an unrealistic expectation. As a result, current therapies move towards drug combinations to hit diverse biological targets, as seen in the treatment of HIV/AIDS and cancer." He went on to note that not only is each human body complex but so is the immediately interacting biology of that body: "Metagenomics of the bacterial flora in the gut is revealing millions of genes present in the ~10 trillion microbial cells coexisting in the human gut, with yet largely uncharted effects on human diseases, such as obesity and inflammation."[37]

Personalized Medicine

That complexity operating in each human body led to another aspect of the genetic medicine for which the new era was named, the shift to personalizing treatment regimens of patients, the "personalized medicine" to which I have already alluded. That is, each person's genetic profile is unique, and so physicians began to try to adjust therapies to each individual rather than prescribing a standard medication for a standard disease. One author in 1998 described the shift in an article titled "From 'Magic Bullet' to 'Specially Engineered Shotgun Loads': The New Genetics and the Need for Individualized Pharmacotherapy."[38] With the formal launching in 1990 of the quest to map the human genome and its rapid success only a few years later, personalized medicine became a real possibility. By the middle of the first decade of the twentieth century, the costs of genetic mapping were dropping rapidly, and a flood of research seemed to be changing the landscape again. As a popularizer wrote in 2009, "Now, for about the price of a smart phone, you can have your own DNA scanned to learn what bugs might be lurking in your operating code." Well, perhaps "one day" that will happen. But soon enough to alarm all kinds of people who thought about racism, privacy and control, eugenics, and many other concerns.[39]

One model disease for this personalized or boutique therapy reconceptualization was beryllium poisoning. Beryllium and beryllium compounds are not generally found concentrated in the natural environment, but in the mid-twentieth century they became increasingly common industrial substances. Workers suffered a variety of symptoms, both acute and chronic, when they breathed or touched beryllium products. The most notorious exposures came from the first generation of fluorescent light tubes, just after World War II. Workers in the factories were poisoned, but so also, unexpectedly, were children who used discarded lights as toys, typically swords, which broke when they were used for dueling. The variability of symptoms, however, baffled investigators. Some people got vicious irritations of the skin and eyes. Rickets could be produced, and damaged lungs were particularly common. Sometimes the effects, if there were any, were acute, sometimes chronic, sometimes both. Altogether, it was very difficult to sort out the physiological processes that might be involved in this obvious poison. By 1980 one scientist who was concerned about the environmental dangers of beryllium could report that in poisoning, "the immune machinery of the body seems to be in a key role. . . ." "Indeed," he added, "it is widely thought that a beryllium antigen may exist."[40]

In the era of genetic medicine investigators finally figured out that people's reactions to beryllium varied, and many had no reaction at all. It was no wonder that no general measure to counteract the individual effects of the poison emerged, and controlled experiments produced confusion. Soon experts suggested that individuals' immune systems differed and that therefore a person's genetic makeup, including the immune system structure, determined whether he or she would react

Two General Electric engineers examine an early fluorescent light bulb. At first the manufacturing process caused beryllium poisoning among workers. Only in the age of genetic medicine did investigators find a model to explain the clinical patterns they saw in beryllium poisoning. Paul W. Keating, *Lamps for a Brighter America: A History of the General Electric Lamp Business* (New York: McGraw-Hill, 1954), between 184 and 185.

to beryllium, and if so, how. As the authors of a review article in 2007 concluded, "Beryllium studies are paradigmatic for studying the interaction of genetic susceptibility with environmental exposure." Thus what had been assumed to be a general poison came to depend on genetic individuality, a transformation that called for preventive measures, or, when they failed, for personalized medical treatment devised for each victim.[41]

Leaders in both biomedicine and the pharmaceutical industry were quick to see the implications of personalized medicine. The sociologist Andrew Lakoff quotes a 1999 report:

> "Pharmacogenomics"—a science that combines the knowledge and study of genetics with the process of developing new drugs—will enable pharmaceutical companies to create treatments geared to distinct genetic variations of any particular disease. Companies will, in essence, be able to predict which patients are likely to respond to which "suites" of medications. With this capability, a pharmaceutical company will have the opportunity to market to specific patient subgroups.[42]

With the human genome project, it became clear that biomedical scientists wanted to duplicate the germ theory revolution, in which a great many diseases appeared to have demonstrable, mechanical causes. Now the definite, mechanical cause could be traced to a gene or genes interacting in specific ways with environmental impingements. One early proponent spelled out the goal:

> to markedly increase the number of human diseases that we recognize to have major genetic components. We already understand that genetic diseases are not rare medical curiosities with negligible societal impact, but rather constitute a wide spectrum of both rare and extremely common diseases responsible for an immense amount of suffering in all human societies. The characterization of the human genome will lead to the identification of genetic factors in many more human diseases, even those that now seem to be [too] multifactoral or polygenic for ready understanding.[43]

Yet as one NIH expert noted ruefully in 2004, after years of experience, "The path from the identification of a gene to a new drug is well known to be long, expensive, and failure-prone."[44]

Investigators were particularly busy applying genetic studies to complex medical conditions, notably cancers, searching for the "oncogenes" that could trigger or encourage abnormal growth. Or perhaps each tumor would have an individual genetic characteristic susceptible to personalized treatment—as actually happened in 2007 for a twenty-nine-year-old Massachusetts male nonsmoker with a variant lung cancer. Hope continued, supported by a lot of funding, that it might be possible to use the human genome as a key to many baffling and deadly conditions, especially if environmental factors could be tied to gene functioning.[45]

Lifestyle and Environmental Factors in Health

The pharmacogenomist Wolfgang Sadee observed in 2008 that "greater insight into the biology of the human body, and the etiology of disease, enables increasingly accurate prediction of risk and treatment outcome" and that therefore a "focus on early therapy and disease prevention has the potential to transform our health care system." Even genetic medicine thus supported the insight that preventive programs and lifestyle change would best preserve health.[46] And this approach reinforced both longstanding health education efforts and the tendency to blame individuals for their own illnesses. At the beginning of the genetic era, in 1990, two government public health physicians, in a widely cited article, had added up the major environmental causes of mortality and listed them in order of the number of victims: tobacco use, "diet and activity patterns," alcohol, microbial agents, toxic agents (chiefly pollution), firearms, sexual behavior, motor vehicles, and use of illicit drugs. By 2004 it appeared that obesity, now labeled an "epidemic," would oust tobacco as the number one killer in the lifestyle category, and Medicare classified obesity as a disease so that treatment would be compensated. Regardless, out of the whole list, only the microbial agents could be considered "natural" and to some extent beyond human control.[47]

Once again, in a new era, the continuing effects of earlier eras were discernible. Clearly, environmental medicine, with the idea of an immune system interacting with external and internal impingements, was inseparably involved in the maturing of the era of genetic medicine as the complexity of genetic actions became clear. As one expert wrote in 2000, "It is the interaction between individual genetic composition and the environment that determines the most prevalent and important diseases affecting the broader range of humanity."[48]

Of course people's ideas about the environment also expanded after the late 1980s. On the global side was climate change. At the other extreme was the mostly unexplored toxic potential of nanolevel industrial products and processes. And integrating such ever-multiplying, multifarious environmental factors with genomic factors added still one more dimension to the disease processes with which health care scientists worked.[49]

It was equally obvious that in a later version of the era preceding the environmental era, the era of technological medicine, technologies of many kinds were enabling scientists to probe into genetics, map the genome, and suggest the variety in each person's genetic profile, as well as to continue to incorporate more technology into surgery, pharmaceuticals, and diagnostic systems, though not always with happy results.[50] Similarly for the age of antibiotics, the enthusiasm and success of antibiotics against infectious diseases persisted and inspired advocates of genetic medicine. Indeed, the continuing success of antibiotics in muting the effects of infectious diseases helped clear the way for attention to inborn problems with structure and metabolism. And the continuation of both the style and the content of the era of physiological medicine of course marked the era of genetic medicine as well. Investigators were still identifying, and trying to control, the unobvious chemical events and interactions that caused various kinds of illnesses. This model worked as well with genes as it had earlier with organs and chemicals.[51]

Continuing Patterns from Other Earlier Eras

The remnants of the age of surgery and germ theory produced some effects that may not have been obvious, including some very late victories for the germ theory model. Surgeons at the turn of the twenty-first century, for example, were often called on to take preemptive action for asymptomatic women whose genetic testing showed them to be at high risk of developing breast cancer: to remove the breast before a neoplasm developed rather than after. This preventive strategy was particularly plausible, it should still be noted, after two major genetic markers for breast cancer were established in the early 1990s, although the factor of chance, or risk, now was framed for the patient in an extremely complicated scientific and personal context.[52]

Germ theory also continued to prosper as enormous strides were made in virology.[53] Again, the emphasis was on the viruses in the environment that could affect genetic material in the body. As two experts wrote in 2011, "Discoveries made by environmental virologists during the past decade or so have revolutionized our perceptions of the living world. It has become apparent that viruses are the most abundant living entities on Earth, outnumbering their hosts by an order of magnitude. The vast majority of viruses infect microbes, but many infect humans, making each one of us a platform for a complex microbial community."[54]

In 2006 the Food and Drug Administration licensed a vaccine for cervical cancer. This one was different, although still on the model of the many based on germ theory. First, it was based on the connection between a virus, the human papillomavirus, and a type of neoplasm, not an acute disease. Second, the connection was based on reducing a statistical risk factor rather than on demonstrating mechanical effects in a laboratory. This program to immunize girls eleven to thirteen years of age led to much controversy for political and social reasons—would this protection encourage early and/or casual sexual activity?—and questions about unproven long-term effectiveness and commercialization.[55]

One late major victory for classic germ theory, however, startled medical practitioners. As already noted, they had struggled for generations to understand and control ulcers of the stomach and the duodenum. During the first half of the twentieth century, theories of localization had led to competing attempts to cure by medication, especially antacids, and by surgery. None was particularly successful. Nor were strategies based on ideas of psychosomatic cause (stress and the like). Then in 1983, clinicians in Australia announced that a newly identified bacillus caused the symptoms and that antibiotics would successfully heal this baffling syndrome, which affected as much as 12 percent of the U.S. population.[56]

This news that gastric ulcer was a classic infectious disease following the germ theory model was so contrary to general medical opinion that it took years for the idea to become fully accepted—a process reminiscent of the shift in belief to bacterial causes of more transparent diseases exactly a century earlier. At first some American practitioners reported their own amazing therapeutic successes in curing peptic ulcer with antibiotics, and finally, in 1992, a randomized, controlled study out of a Veterans Affairs hospital in Texas confirmed that "most peptic ulcers associated with *H. pylori* infection are curable." Altogether, reported the investigators, studies published by that time provided "compelling evidence for the hypothesis that peptic ulcer, either duodenal or gastric, is the end result of a bacterial infection." Finally, in 1994 a National Institutes of Health Consensus Development Panel established the bacterial viewpoint formally and officially in the United States.[57]

A further reincarnation of germ theory came with the general recognition that another kind of microscopic agent might be causing serious human diseases in a way parallel to that of viruses and bacteria. This was the "slow virus," renamed the *prion* in 1982. From the 1950s through the 1970s, investigators noticed that some fatal neurodegenerative diseases, such as Creutzfeldt-Jakob disease and kuru in humans and scrapie in sheep, seemed to have similar pathologies and suggested that they might therefore have similar causes. The ultimate idea of the agent, the prion, which became generally accepted only in the 1990s, was emblematic of the genetic era, because it resulted from narrowly biochemical research on the probable agent, not from studying symptoms of the varieties of the diseases.[58]

The prion jumped into the news in the late 1980s, when the so-called mad cow disease, or bovine spongiform encephalopathy, hit British agriculture. Then, beginning in 1996–97 it became clear that some people who had consumed the infected beef had become fatally ill with a variant of Creutzfeldt-Jakob disease. Scores more victims were identified, and still more were expected, including those who received blood from infected donors. Americans, it turned out, were not affected, but the mysterious agent got a lot of publicity. Some was very colorful, beyond the hundreds of thousands of British cattle that were slaughtered. Kuru, found in Papua New Guinea, was traced by an American investigator to cannibalism when working with others he studied residents of New Guinea who had consumed the brains of infected people.[59]

Research on prions over a generation had been heavily transatlantic, but finally, in the 1990s, the Nobel Prize winner Stanley Prusiner, of the University of California, San Francisco, persuaded most people that at least one class of diseases was infectious, passed on by a protein that did not contain DNA but in some way was self-replicating. Investigators immediately began to speculate about other neurological diseases that might have had genetic elements but certainly had mysterious onsets and pathophysiological processes, such as Alzheimer's or Parkinson's. Was it possible that this new version of germ theory, properly integrated into genetic constitutions, would add another dimension to understanding diseases? Moreover, the ease with which prions traveled from one species to another was ominous.[60] The influence of not just germ theory but all the earlier eras was alive and well in the era of genetic medicine.

Alternative and Commercial Health Care

Not least of aspects of the past that persisted were alternative therapies, self-treatment, and commercial exploitation.[61] One of the shocks of the 1990s was the publication of a survey showing that residents of the United States continued to personalize their own health care to a remarkable extent by relying on not just regular medical doctors (MDs or DOs) but also alternative therapies—relaxation techniques, chiropractic, massage, imagery, spiritual healing, commercial weight-loss programs, lifestyle diets, herbal medications, megavitamin therapy, self-help groups, biofeedback, hypnosis, homeopathy, acupuncture, and prayer. The use of alternative therapies crossed income and other demographic lines and also drew on continuing environmental-era fears about toxins. About one-third of the population saw neither a medical nor an alternative provider in 1990, but one-third of the survey respondents resorted to irregulars, whether or not they saw a regular provider. Fifty-eight percent saw a medical doctor, 7 percent saw both a medical and an alternative provider (often without telling their medical doctor), and 3 percent saw only an alternative provider. Alternative providers most often treated back problems, psychological problems, allergies, arthritis, insomnia, strains and sprains, headache, high blood pressure, and digestive problems. A follow-up survey uncovered substantially higher percentages of people who utilized alternative therapies, 42 percent in 1997 versus 33 percent in 1990. By that time, Americans' spending on alternative therapies exceeded $20 billion annually, which was more than they spent on hospitals. In 2007 the amount was about $34 billion.[62]

Some people followed fads and made use of unapproved technologies, now called complementary and alternative medicine. And alternative and folk treatments had always existed alongside regular medicine and were deeply embedded in the cultures of many ethnic and other subgroups. Such practices were often tolerated or patronized because of political or cultural forces or, more recently, the high cost of health care.[63]

Using nonstandard means was not clearly demedicalization of the treatment of illness and discomfort or medicalization of expectable life distresses. As David Hess wrote, "The diseases of modernity are largely . . . the result of genetic predisposition interacting with noninfectious risk factors such as environmental toxins, stress, poor diet, and lifestyle. Biomedicine has been much less successful at treating the 'new' [chronic] diseases than the older, infectious ones."[64] Particularly with the persistence of arthritis and back pain, there was an opening for commercial, ideological, and political exploitation in the twenty-first century, just as cancer for generations had opened the way, and continued to do so, for crank cures and for what appeared to be technological treatments. Most notorious in earlier decades had been Laetrile and Krebiozen, both supposedly able to cure cancer and symbolic of commercial cultism.[65] The Center for Complementary and Alternative Medicine, which the federal government established in 1991, had trouble maintaining credibility in the twenty-first century, when leaders in regular medicine continued to push for randomized, controlled clinical trials and the new standard, evidence-based medicine.

Meanwhile, public health authorities who pushed aggressively for vaccines for prevention of infectious disease or for other measures, such as water fluoridation, encountered pockets of opposition. Public health managers continually worried about "popular resistance," which often had political bases that emphasized supposed individual, rather than public, benefits. By the late 1980s, and continuing for decades after, there were significant political attacks on public health and public health funding.[66] Into the twenty-first century, pockets of resistance to vaccination led to local outbreaks of dangerous diseases, such as rubella.[67]

Health Insurance in Trouble

It is ironic that in the midst of all of these scientific and cultural determinants shaping health care, in the United States the era was defined as much by health insurance as by biomedical science and technology. As early as 1996 one group of experts had concluded that where once technology, demographics, and doctor and hospital supply and decision making had driven change, now flawed and "socially amoral economic forces" were shaping health care.[68]

One reason for the problem was that the bulk of the population held health insurance through employers. That might have been a workable model in an industrial economy dominated by large organizations, typically factories, in which Medicaid and Medicare could provide for most of those not employed. By the 1980s, however, as noted above, the United States was a postindustrial society operating in a global economy. Employees of the growing number of services and small businesses supporting a consumer society, often with low wages, did not have access, at least not reasonable access, to health insurance. Between 1980 and 1990 the number of Americans without health insurance, instead of shrinking, grew by 1 million, from 12.3 percent of the population to almost 15 percent. By 2009 about 50 million Americans were not covered. At some time during any year, almost half of younger and middle-aged adults had periods of no coverage.[69] Moreover, large employers who did offer health insurance—a significant number did not— were already paying an enormous sum, about thirty-six hundred dollars a year (almost 11 percent of the U.S. median household income), for each family covered in 1995.[70]

Policymakers in the American health care system therefore faced two problems in the 1990s: how to control costs of care for middle-income sectors, "the worried middle class"; and how to provide for the 30–50 million people who at any time had no health insurance. Indeed, in the eleven years after 1998 the number of uninsured patients treated in community and other health centers increased by 100 percent. There were pockets in which inadequate insurance clearly produced bad effects, such as the Amerindian population, among whom relative death rates actually started to increase significantly in the 1980s. And lack of coverage had an additional cost: patients who were actively and continuously engaged in their own treatment got much better results from the health care system; those who were not did not.[71]

In 1993–94 the Clinton administration attempted to meet the goals of coverage and cost with a reform act. Economic and political enemies of the plan succeeded in killing the initiative by convincing the frightened middle classes that they would have less care for more cost. The opponents were abetted by those who feared that privileged sectors, rather than just the unprivileged, would have their medical care "rationed."[72] A decade and a half later the Obama administration got the Affordable Care Act, very conservative reform, through Congress, primarily using economic incentives to get the private insurance companies to do less cherry picking of healthy, desirable populations and greatly enlarge their risk pools. Rightly or wrongly, a number of commentators counted the reform as a major change, beyond one of the usual incremental steps toward universal coverage.[73]

Managed Care

The reality policymakers had to deal with in the 1990s and into the twenty-first century was not congressional attempts to mend the system but managed care, which became conspicuous in the mid-1980s and in the 1990s generated a powerful reaction. Managed care was an attempt by insurers to control costs, restricting the procedures and care their contractor organizations would deliver to patients—and of course the insurers bargained with "providers" for reduced fees for treatment the insurers approved. Often there had been little price competition in health care (even, surprisingly, among pharmaceutical suppliers), and business managers introduced some pressure against price increases among both nonprofit and for-profit entities. As late as 1994 an industry publication writer noted that "managed care has been the choice of not only employers and the public, it has also emerged as the choice of politicians pursuing health care reform." The managed care approach became commonplace, but it was 1990 before the rapid increase in the use of the tactic, and subsequent complaints, reached such a level that journalists noticed. In 1991 *Business Week* finally asked, "Insurers vs. Doctors: Who Knows Best?" and could report that fourteen states had already passed laws to regulate denial of patient claims. In 2002 a sociologist reported that "managed care has become the defining term of the current health care system."[74]

By 1998 observers were writing about "the managed care backlash." The evidence was, as the sociologist David Mechanic explained, "the chorus of opposition from physicians and other professionals, negative media coverage, repeated atrocity-type anecdotes, and bashing by politicians." There was, for example, Steven Olsen, the two-year-old who was denied a CT scan (cost: $800) by hospital personnel: he turned out to have a brain abscess. Such were the tragic stories recounted in muckraking reports and court documents. The reality was somewhat different. One physician who administered managed care, acting, according to the model of a family practitioner, as gatekeeper and coordinator of a patient's care, in 1999 got fed up with "all the negativity all the time. . . . Even when I was doing good things in the organization, the negative pressure from the media, from lawyers, from the public was so strong that it was hard to take pride or pleasure from my work."[75]

The tangible form of the backlash was the introduction in state legislatures and Congress of more than a thousand bills for "consumer protection" of those covered by managed care; however, in this case the consumers were not health care consumers as such but health insurance beneficiaries. One survey made it clear that Americans who were well off did not want any changes and opposed regulation or anything that might cost them more. Respondents who wanted protection were those who were fearful of what further insurer restrictions might do to their

"Very scary, Jennifer—does anyone else have an H.M.O. horror story?"

By the late 1990s most people knew of patients who had been denied treatment by a health insurance company or HMO organization, that is, "managed care." The media played up particular cases of what might appear to be outrageous denial as human-interest stories. Reprinted from the *New Yorker*, 19 July 1999, 38, by permission of Condé Nast Licensing.

coverage. People understood that managed care was a form of rationed health care, and most did not like it—even in the face of evidence that Americans were, on the whole, "overtreated." Managed care did in fact produce the first widespread public confrontation with overt rationing. A large share of health plan members now had to phone the plan and get permission for a referral or a procedure, or their doctors had to delay a procedure until it could be approved. Meanwhile, the horror stories in the media instigated fear of what else might happen as the insurance companies increased their profits.[76]

Later Stages of Managed Care

A series of developments in the 1990s brought change to managed care, change beyond the state regulatory statutes. Especially after a Supreme Court decision in 1992, insurers were whipsawed by legal problems. There were lawyers trying to reverse care decisions in individual cases, and there were lawyers seeking damages because patients had been denied one procedure or another. In a 1998 syndicated cartoon, a patient asks a bemused doctor, "What is the health care system coming to when my HMO won't allow you to prescribe unnecessary tests that protect you from frivolous malpractice lawsuits?" Year after year, the courts and the legislatures were changing the rules, trying to find some just way to control events. Within the system, doctors trying to hold on to their jobs made heavy-handed decisions, decisions that would hold prices down and make the physician look good in his or her utilization review but could raise difficulties for the patient. There were executives who put caps on what their companies would pay for a given patient or, as in DRGs, a given illness or procedure, which put pressure on health care "providers" to do less for the patient, and to do it more cheaply. In fact, managed care required physicians to reverse the traditional disconnect of a professional between fees and service. Ideally, a professional offering medical care would treat patients without regard for billing and payment. Under managed care, however, the physician had to consider the cost of any course of treatment first and then do a cost-benefit analysis, keeping in mind the public or business goal of lowering the costs of health care. A medical decision was definitely no longer purely clinical.[77]

For a short time, managed care did hold down price increases, mostly because the length of hospital stays dropped substantially. From 1993 to 1998 the remarkable inflation in health insurance prices did pause temporarily. In the later 1990s, however, managers' price controls began to fail, arguably because of the growing bureaucracy necessary to deny coverage to patients. At the turn of the twenty-first century, 31 percent of the money spent on health care in the United States went to administrative costs (compared with 17 percent in Canada). Private firms had much higher administrative costs than did government programs. Nevertheless, with rising costs and prices, it became clear that competition between insurance providers might put a temporary brake on price increases, but there was no consistent evidence that competition improved or even sustained the quality of care. Indeed, quite the opposite. It turned out that business managers could show savings in the near term but were not making adequate provision for long-term expenses and other future problems. The greatest problem was the rising cost of medical devices and pharmaceuticals (together up more than 300 percent in just the decade 1991–2001), along with the fact that malpractice liability was now sometimes extended to insurers. One response was to try to shift economic risk to physicians by capitation, that is, paying a flat sum to a health care "provider" for all of a patient's care, whatever that care might be, for a period of time—an even more direct re-creation of the hated contract system of a century earlier.[78]

Around the turn of the twenty-first century, however, there was a general shift. First, the number of managed care companies declined as a result of sales or consolidations. Second, insurers started offering a variety of plans from which purchasers could choose, not just the earlier inflexible single plan from a particular insurance provider. This was, as one group of authors suggested, "managed care lite." There would be only one managing entity, but an insured person had choices, even alternative therapies such as acupuncture, chiropractic, and massage. Most distinctively, a person could choose to pay less each month but have fewer coverages of various kinds, such as caps to totals paid for one illness incident or much higher co-pays for seeing a physician. That is, a person formally "insured" in fact had only incomplete, partial insurance, sometimes derisively referred to as "unsurance."[79] The rationing was thus disguised as the consumer's choice. The implicitly or explicitly for-profit system was safe, because if something went wrong, the policyholder could be blamed for choosing a cheap alternative. One did one's own rationing, and quality control receded as an issue, at least for managers of health care. In practice, as with much else in health care, "consumer" (policyholder) choices of insurance plans generally turned out to be irrational in more than one way.[80] Once again, free-market policymakers and for-profit insurers could blame the victim for making bad choices.

The Doctor's Work in the Age of Genetics and Managed Care

The doctor's work continued, of course, but with some modifications. First of all, at the turn of the twenty-first century a new physician figure appeared in great numbers, fulfilling the drive to bureaucratization. This was the *hospitalist*, a physician who worked full time in the hospital as a hospital employee. She or he was fully in charge of patients sent by office-based physicians. The hospitalist had the advantage of knowing how to maneuver the patient through the complexities of the hospital system. Responding to the requirements of managed care, the knowledgeable hospitalist, by providing coordination and continuity of care, could reduce the length and inefficiency of treatment and at the same time maintain quality of care—perfectly responding to the requirements of managed care administrators.[81]

Meanwhile, under the influence of relentless pressure from insurance companies, a new work pattern emerged for the typical primary care physician and many specialists in outpatient offices: the group practice. Most physicians found that a group practice could hold off the insurance company better than an independent practice could. Therefore, ordinary practice consisted in working in the office and seeing as many patients as possible, typically spending not more than fifteen minutes with any one patient. The physician's day lasted basically from eight or nine to five, matching the hours of the office staff. One doctor reported, "I usually try and grab something to eat around midday, and we'll not schedule patients for an hour or so to catch up and give the staff time for lunch. But, I often run late and will finish my morning patients during that time off, which leaves me only half an hour or so to get something to eat, return any urgent phone calls from patients, check labs and other stuff that might've come in." Such practitioners did get to see their families nights and weekends, but they often took work home via computer and continued to average well over fifty hours of work a week. Their pay, by the corporation or group, was based on how many patients they got in and out of the office each day. Beyond that pressure, and pressure from demanding patients who searched out health "information" on the Internet, harassed primary care physicians had a comfortable, often rewarding job.[82] The younger and beginning physicians tended to adapt to business requirements more easily than established colleagues. As one observer put it, they "have introduced information and accounting systems to their practices, developed treatment guidelines and disease management programs, joined group practices and newer organizational arrangements." Moreover, even with many daunting, chronic diseases such as Parkinson's or Alzheimer's, or lupus, epilepsy, arthritis, and Crohn's disease, genetic-era physicians often offered some treatment that could alleviate some of the symptoms.[83]

Health Costs and Payoff

By the first decade of the twenty-first century, a profile for the whole medical system had emerged. Health care expenditures continued to rise. Per capita dollar expenditures were $3,469 in 1993, $4,790 in 2000, and $6,697 in 2005. Yet the rate at which total expenditures rose had moderated substantially, probably in part because so many fewer people were insured. Much of the cost increase, it should be emphasized, came from technologies, including pharmaceuticals. Hospitals offered special services and were less dependent on price competition to gain insurer contracts. In the end, one 2002 analyst concluded,

> In the battle between tight-fisted (and frequently scandalously profiteering) private wholesalers of physician labor, physicians won. This is not to say that physicians haven't been bloodied in the process, with wounded incomes, sunken morale, and perhaps even some market fatalities in terms of physicians opting for early retirement or alternative careers. But as the dust from this fracas begins to settle, it appears that the U.S. health system is heading back on a course of increasing specialization, rising physician incomes, and pressures to increase overall physician supply.[84]

The constant increase in money allocated to health care in the United States was in the end relentless. For example, the national health expenditures in constant

(2000) dollars (billions) for selected years from 1970 to 2006, with a total increase for the period of 564 percent, was as follows:

1970	1980	1990	2000	2006
$272	$469	$875	$1,354	$1,806[85]

As the ethicist Daniel Callahan observed in 2009, technology, including pharmaceuticals, accounted for 50 percent of the costs. In addition, medical intervention for newly defined illnesses increased. One could cut back the costs of technology by decreasing testing and treatment in one of two ways: either reduce demand on the part of both doctors and patients—demand fueled by industry marketing—or frankly ration medical interventions. If one did not undertake such a program, Callahan concluded, one had no right to complain about health care costs.[86]

It was true that much of the funding went into diagnosing and treating old people. Yet each year it became more obvious that individuals who were living longer benefited substantially from medicine. One need only mention how artificial lenses saved an enormous number from blindness, and how joint and other replacements kept aging citizens mobile. A long list of additional measures kept people of all ages functional to a degree that had not been possible only a generation or two earlier. For large parts of the population, the medical system was having remarkable effects, even beyond preventive measures such as inoculations. Beginning in the 1980s and accelerating after the turn of the twenty-first century, among older Americans the incidence of disability continued to decline, even though there were more older people. Altogether, an amazing and increasing percentage of Americans over sixty-five, more than 60 percent, were reporting that their health was good or at least not poor.[87]

The Evidence-Based Medicine Movement

Many people in the decades before and after 2000 spoke and wrote about day-to-day changes in health care institutions and practice. What specially caught the attention of observers writing about the science of medicine was evidence-based medicine. This, as the term suggests, was an attempt to get practitioners to think in terms of actual patient outcomes, that is, in terms of scientific data rather than intuition (clinical sense) or abstract, impersonal pathophysiological mechanisms, or to follow a cookbook model of "protocols" for diseases. A whole field, the *outcomes movement*, had started to come together in the 1960s and 1970s and was well established in the 1980s and 1990s, when evidence-based medicine absorbed it. At first some physicians had tried to chart the procedures that had favorable or unfavorable long-term outcomes. The basic question had been, do suggested treatments actually work? Then systematic inquiries and especially computer-assisted analyses became possible, and some leading experts turned a private conversation among doctors into public information that insurance companies, hospitals, and even patients could use. Above all, as one experienced team put it, evidence-based medicine "places the burden of proof on clinicians and researchers to demonstrate that an intervention results in more good than harm before recommending it to the public." "Data-based decision-making," in whatever form, influenced legislators as well as actual managers of health care.[88]

Finally, because of difficulties in, and contradictions between, different randomized, controlled clinical trials, another tool became prominent: meta-analysis. Meta-analysis was simply pooling and then analyzing all the reports on a particular question from randomized, controlled trials. In this way, analysts used statistics to trump varying, conflicting results from different studies of the same phenomenon. As two experts commented in 2006, "Caches of reviews, many regularly updated and available on the Internet . . . have the potential to identify and speed the adoption of new practices. News and radio journalists have learned how to describe a systematic review or meta-analysis as a new study with new findings."[89]

As one historian has reported, beginning especially in 1992, *evidence-based medicine*, variously interpreted, became a slogan that served to defend medical authority against the assaults of insurance companies and alternative medicine, as well as cranky patients.[90] Data-based decision making in medicine took on an additional dimension, however. Beginning especially in the 1990s, two international movements produced a whole additional base of objective measurements of outcomes as economists and especially mathematical statisticians took meta-analysis far beyond a trendy motto. As one pioneer recalled, "We didn't have any heroes or mentors . . . anywhere in the country to encourage another way of thinking." The new investigators offered large aggregations of strictly reviewed, worldwide studies of actual outcomes of medical interventions. The United States was only one site for global data production. The new data on interventions or clinical trials had to be of high quality, not just a clinician's subjective impression. The data also had to be independent of pharmaceutical firms. And then the mathematicians could go on to use advanced analytic techniques to control for variations in types and sources of data, such as locations, population groups, and patient compliance.[91]

Following the economic recession of 2002, when state governments in the United States ran short of funding, many state, not federal, officials began using the new, much more reliable data on outcomes to control costs for Medicaid, and they set a new standard in medicine in general. The main target of the new "clinical epidemiologists" was big Pharma and similar firms, all of which fought vigorously to derail exposés of ineffective, expensive drugs. Intense opposition to clinical epidemiology also came from single-disease interest groups whose publicists were making claims that went beyond the facts.[92]

Commercialization, Translational Medicine, and Errors

Concern about the continually increasing influence of commercialization in bio-medical science and practice was a conspicuous and strident theme in the literature on medicine at the turn of the twenty-first century. As noted earlier, any number of writers produced exposés of the ways in which marketing personnel within pharmaceutical firms created illnesses such as depression or encouraged the use of commercial products by influencing physicians and through direct advertising to patients. A typical big Pharma firm spent more on marketing than on research. There was about one drug rep for every nine or ten physicians. But commercialization went further. On occasion physicians—who could now advertise—might perform legal but questionable procedures. Indeed, across the spectrum of medical practice there were endless pressures on physicians, who faced mostly unregulated conflicts of interest, to take one action or another for direct or indirect financial gain—to prescribe a particular drug or procedure or contrive to increase or decrease some patient's treatment.[93] Would personnel in biomedical laboratories turn from science enthusiasts into investigators with ties to commercial ventures (as had long occurred in pharmacology)?[94] No one knew where free-market forces would take medicine when financial incentives constituted the main instrument for social and policy manipulation and control.

A great deal of attention and money was devoted to *translational medicine,* a term that first appeared in 2003, although the strategy was already basic in the war on cancer. As one prominent neuropharmacologist wrote, translational medicine grew out of a fear that "the profound advances in biomedical research so rapidly accruing today may never be effectively transformed into meaningful advances in health care for society." Translational medicine was simply an effort to bring scientific findings into everyday clinical practice.[95]

One other notable trend in American health care was a new concern about errors in the system that resulted in harm to patients. In 2000 an Institute of Medicine task force issued a report on the need for patient safety. More than almost any other official report, this one created a widespread movement—in this case to increase patient safety. The analysis and recommendations came from the safety literature, but when they were systematically applied to the health system, the results were startling. The authors of the report suggested, for example, that more people were dying because of "medical errors" than from automobile accidents or breast cancer. As one witness wrote in 2007, "For two decades healthcare organizations have faced dramatic changes in technology, demographics, labor markets, compensation and benefit levels, and malpractice claims and awards. Publication of the Institute of Medicine's study of the consequences of medical errors . . . increased the public's awareness, which created enormous pressure for improving patient outcomes." Moreover, although malpractice claims continued to be a conspicuous problem in the years after the report, it also became clear that lawsuits for damages had not had the hoped-for effect. Indeed, noninvasive surgery ironically gave rise to a remarkable number of malpractice suits. And, once again, private ownership of medical units did not produce quality care. For-profit entities did not match nonprofits in patient safety.[96]

Conclusion

With all of these specific developments, it still was not clear what direction health care would take in the rest of the twenty-first century. As one senior analyst cautioned in 2010, "The current system does not *and cannot* do the job. . . . We are in real danger of destroying the impressive achievements that help to define the American medical care system." Both within and outside the biomedical research community it was hard not to notice that the National Institutes of Health budget was "flattened" beginning in 2003 and that, at the same time, political drives resulted in the proliferation of separately funded NIH institutes.[97]

Yet from a distance, it was striking how remnants of traditional practices and beliefs carried across the Atlantic in the seventeenth century persisted in some recognizable form in the new era of genetic medicine. Modernization had, it is true, brought enormous complexity, along with much effective care. But where the new epoch that coincided with the era of genetic medicine would take health care was an intriguing—and unanswerable—question.

CHAPTER 14

The Recent Past as a New Epoch

In concluding, I wish to suggest that around the turn of the twenty-first century all of health care was entering a new epoch, signaling overarching change. Overshadowing genetic medicine, managed care, evidence-based medicine, and other dramatic developments of these years was the reality of *a different kind of health care*. As seen in chapter 13, genetic medicine and other new approaches in medicine and health care developed even while many concepts and institutions from the previous four hundred years persisted to a remarkable extent. The wider setting in which they all operated, however, belonged to the new epoch. The epoch of modernization had played itself out in the late twentieth century. The new epoch coincided with the era of genetic medicine, but while genetic medicine identified a new era, it did not define the new epoch.

Conceptualizing the new epoch superimposed on the era of genetic medicine requires an altered perspective. One name that has been submitted to cover the new epoch, as I shall explain below, is *desktop medicine*, in which risk groups played a central role. Meanwhile, still other trends suggested that a break from the epoch of modernization was under way in the last decades of the twentieth century. Not least was the astonishing rise of retail medical clinics, staffed by nurse practitioners and other non-MDs and located in large retail stores and pharmacies.[1] Trend spotters also noted a possible end to twentieth-century specialization as many specialist physicians with board certification, such as neurologists, were moving back toward general practice, claiming that they were suited to do primary care as well as specialized medicine. Or investigators around the turn of the twenty-first century were adding new layers of complexity to biological functioning, so that they could speak of "our limited understanding of how the vast majority of genes, proteins, and RNAs work, irrespective of whether they are disease-associated or not."[2] None of these new-epoch developments could have been easily predicted before the late 1980s.

Simultaneously, modernity took on new meanings as globalization had to adjust to environmentalism. On the model of countercolonialism, a number of U.S. population elements reacted against modern medicine by favoring complementary and alternative medical practices found in different parts of a fragmented American culture. At the end, the modernization of health care left a powerful legacy, the details of which continued to shape the future of the eternal problems of health and illness—just as in each era of modernization versions from previous eras continued to define and mold how people offering health care or using it understood what was happening in their time.[3]

Thus one can identify in the long story of medicine and health care in America many elements that continued even as one epoch was ending and another beginning. Originally, and for two and a half centuries after 1607, most of the population had been constantly aware of their bodies and struggled to remain functional under difficult conditions. For these earlier Americans, everyday life included pains, itches, "the miseries," body functions and childbearing, collapse, and, constantly, death all about. Every home often housed people who were incapacitated or even dying. By the late twentieth century, people romanticized their bodies. Advertisements in the media suggested that pain and itching were unacceptable and required some purchase. Disability was becoming socially normalized. Death was distant and unusual in the personal experience of much of the population.

As my narrative suggests, what happened to traditional medicine and health care between the 1880s and the 1990s can be attributed generally to the forces in modernization, especially communication, transportation, and the breaking down of cultural isolation as industrialization, urbanization, "technologicalization," and organization transformed American society. Health care into the late nineteenth century tended to remain local and derivative more than did the rest of American social processes. Typically, however, in any community, medicine represented rationalism as the trained and to some extent professionalized doctor practiced alongside the worst hucksters, who promised health for a price. Then "science," with new meanings, arrived, and physicians started to portray themselves as agents of experimental science, adding another dimension to their rationalism. Where they had once functioned as nurse, priest, and counselor to the sick, doctors now had a new role and soon started wearing the white laboratory coat.

It was clear at the beginning of the twenty-first century that physicians remained the central figures in the health care system. It was also clear that the profession of medicine as such was profoundly fragmented. It was made up largely of specialized practitioners, whose identity was not with medicine as such but with the specialty, even if the specialty was primary care or family practice. An average Medicare patient would see "two physicians and five specialists" each year. Someone with a chronic disease would see on average thirteen physicians a year.[4] Most physicians now were either salaried or worked in a group practice, probably part of a "provider" group furnishing services under a contract with an insurance division or corporation.

Nevertheless, it is necessary to recognize that from the beginning the central figure in health care was a person, usually a physician, who with nurses and other paramedicals undertook to heal other people. By the twenty-first century, physicians were still admired, envied, and endowed with tremendous importance in the minds of patients and in portrayals in the media. Especially in the post–World War II decades, however, physicians found themselves subjected to business and bureaucratic restraints even as they remained symbolically central in the huge health care establishment and also the object of major political and economic drives. Experts and amateurs alike studied and discussed every detail of the ways American physicians functioned.[5] Even as the bureaucracies and organizations of modernization drew in more and more physicians, a substantial core of physicians in the 1990s and after viewed themselves, according to an AMA official, as "physicians in individual practices, the small practice of medicine, small businessmen," resistant to regulation and red tape.[6]

The one thing that traditionally minded physicians retained, in part from the long history of the profession, was the claim that the individual doctor-patient relationship was special, even in a heavily bureaucratic setting. Embedded in the doctor-patient cliché was the further contention that a good physician by training and experience had a special "clinical sense" that no formula or machine, not even a computer, could replicate.[7]

The "personalized" medicine of the genetic era drew on this traditional claim. Just as nineteenth-century homeopathic physicians had chosen from their kit of medicines a combination of diluted doses specially suited to each individual patient, so out of biochemistry and experimentation one could suggest that there was a combination of medications with biomarkers that would suit the genetic makeup and peculiar pathology of each modern patient.

Yet the claim of a clinical sense went beyond any mechanical process. It is easy to ascribe the clinical sense to the socially constructed priestly role that came with the designation of healer or physician. Or perhaps it can be reduced to so-called tacit knowledge.[8] It is when one turns to the patient role, however, that it is possible to see more clearly the power of a social role to enter the realm of illness, science or no.

It was, therefore, a changed pattern in the usual doctor-patient medical encounter that definitively signaled the advent of a new epoch in the history of American medicine and health care at the end of the twentieth century. The change was comparable to that initiated with modernization in the late nineteenth century, when physicians turned from the question of how the individual patient's constitution conformed to nature to that of how the patient compared with other human beings, that is, what was normal, a factor usually measured with technology. Moreover, that new determination soon came in terms of distinct diseases based on laboratory as well as personal and observational data.

Around the year 2000 some acute observers hinted that a change was under way. As I have suggested already, these observers might report on the specifics of health care practice or on major trends and patterns. From different viewpoints, they detected fundamental changes in American medicine and health care.[9]

Moreover, there existed a template to show how fundamental the change was.[10] In the mid-twentieth century an American sociologist, Talcott Parsons, had identified the social role of "the sick person." It was a role one could see every day in medicine and in all cultures that were available for study around 1950. Even historians eventually used this surprisingly universal, socially prescribed role to communicate what happened when someone fell ill: he or she acted out the role of being sick.

Parsons divided into four steps the process by which a person assumed the sick role. First, the person who may be ill or injured has to detect from signs or symptoms—pain, fever, bodily malfunction, general malaise, weakness, dizziness— that his or her status is not normal. One's immediate associates may also remark on evidence such as coughing, paleness, bleeding, or limping. Second, one communicates to one's immediate associates, typically family or coworkers, that one is taking on the sick role and receives from those associates recognition of a change in one's role—a change for which the ill person is not responsible. When one assumes the new role—the sick role—one involuntarily abandons the ordinary roles and responsibilities of life, such as going to work or school or performing domestic duties. At that point, two things have to happen: (1) someone in the social environment, typically a boss or a family member, has to ratify that the ill person is not responsible for the illness and can indeed be excused from ordinary responsibilities; and (2) the sick person has to try to get well and resume his or her normal responsibilities. Further, the sick person is obliged to seek socially approved assistance in trying to leave the sick role and once again take up his or her usual role in society. The socially approved assistance typically comes from a witch doctor or a physician, with whom the sick person is obliged to cooperate in the effort to get well. The sick role thus defines what a sick person does and what the physician's role is.

At the end of the twentieth century, with hardly anyone's noticing, the classic sick role often failed to work in the United States. To begin with, while the old signs, symptoms, and subjective feelings of illness still operated, many medical conditions could not be detected by the individual afflicted person. The illness was "sub-clinical," as in early stages of breast or prostate cancer. Asymptomatic conditions were, as noted above, increasingly recognized in environmental medicine. The existence of "conditions" depended largely on laboratory readings or other technological inputs. Moreover, it had become difficult to communicate one's feeling of illness and to obtain appropriate recognition for one's sick role. Many businesses refused to distinguish any longer between involuntary sick leave and voluntary vacation time.[11] It was often notoriously difficult to obtain disability or worker compensation, and health insurance claims were often disputed, especially under managed care. Or people could be discharged from hospitals after only a very short time.

In addition, normal social support for the sick role often disappeared. People whose functioning was limited denied that they were "sick." Instead, the disabled were demanding rights and recognition of their status as persons. Deaf people refused cochlear implants because to accept them would be to admit that deaf people were deficient. It became increasingly difficult for a person to claim that he or she had a deviant or abnormal status at all. To top it all off, as described earlier, many people who felt ill or could not function fully became villains who were responsible for their own bad health because they smoked, drank, ate too much, failed to exercise, and in general brought on themselves "preventable" conditions. It was at this same time that, in concert with free-market thinking, the sick person became "a consumer." Consumers, who could make commercial choices, were not generally excusable or subject to pity or charity, as a sick person should be.

Most significantly, as some astute observers also noticed, the physician's function was changing. It was when both the sick person and the doctor had altered expectations and roles that it became obvious that a different epoch had opened in the history of American medicine and health care.

In part, physicians were responding to the increased prevalence of chronic disease as opposed to acute disease. They had, again with hardly anyone's noticing, begun to work with a patient "at-risk role," as opposed to the "sick role." In the new practice of medicine, a patient went to a clinic not because of symptoms so much as for screening or because that patient knew that he or she was a member of a demographic group at risk for some disease, such as diabetes, depression, breathing problems, or heart failure.

In this new health care, physicians would treat a patient for an illness that she or he might get, such as coronary disease based on too high a level of cholesterol, or exposure to environmental or genetic elements associated with other chronic diseases, such as diabetes. In one sense, a "risk group" offered a means for moving social concerns to an individual level, parallel to applying a free-market, consumer model to health care. That is, it was an individual's risk profile, a concern of a clinician, not the risk existing in the community, which might be a concern of a public health worker. And an individual's risks could also translate to another form of "patient experience."[12]

While this revolution had long been gestating among experts, the first sign that a person without symptoms might fall into a risk category was the popular and medical conclusion in the 1950s that there was a robust statistical correlation between smoking and ultimately developing lung cancer. And in the 1960s the Framingham study explicitly introduced "risk factors" for coronary heart disease.[13] In addition, subsequent advances in technology and laboratory determinations, particularly imaging, enabled physicians and their technician colleagues to do something for a patient who did not yet perceive that he or she was ill.

Altogether, the insurance model of risk groups was combined with what one historian calls "preventive pharmaceuticals." For half a century after the 1950s, physicians often prescribed products of the pharmaceutical industry in order to prevent illnesses, not cure them. With great, uncounted social and medical costs, this historian continues, health had been translated into risk:

> Pharmaceuticals have become increasingly important to how we live our lives and how we understand both chronic disease and healthy living. In addition to changing our conceptions of disease, the widespread practice of risk reduction through long-term pharmaceutical consumption has reshaped the experience of patienthood, the ethical priorities of medical practice, and the political economy of health and medicine.[14]

It was in 2010 that the physician Jason Karlawish named the recent style of practice *desktop medicine*, as opposed to bedside medicine. The doctor, using a computer, took what he or she could learn about an individual patient and compared it with networks of large data sets to see what risk groups the patient would fall into. "The clinician," Karlawish explained, "then uses these risk factors to determine whether the patient is at sufficient risk to recommend treatment." Only then would a patient have the opportunity to begin medical rituals of taking pills and undergoing periodic monitoring.[15] In desktop medicine, the patient did not have to feel ill and did not have to ask to be excused from work, school, or household duties. He or she could, however, take on the role of consumer and decide which treatment to undergo—or decide not to undergo any treatment at all.

The presumption in desktop medicine was that the patient was asymptomatic or that any immediate complaint of the patient was secondary and should be integrated into life-extending desktop medicine regimens. Of course, when it was appropriate, the physician assumed that bedside medicine continued as before: one applied standard treatments to fractures and clearly discernible acute illnesses. The

A doctor practicing desktop medicine with a patient in the early twenty-first century.

pattern was therefore the same as in the eras of modernization of health care: carryover from the old was embodied and carried on in the new. But for the physician, it was easier successfully to treat risk factors than to treat diseases.[16] The new, desktop medicine did change perceptions and behavior, but it also added another layer of complexity to health care efforts.

With desktop medicine, or whatever the new epoch will be named, once again the eras and epochs in the history of American health care were being defined by the doctor-patient relationship, even though technology and millions of workers extended the effects of that relationship. Both doctor and patient found risk groups and desktop medicine unfamiliar and difficult to deal with.[17] Moreover, part of the ordinary, everyday context of the new epoch was the unwelcome idea that a person's body was put at risk by his or her lifestyle—habits of consumption and consumption-related behavior such as smoking, drinking, unhealthful eating, and lack of exercise, reflecting altered thinking about how one's social existence could affect one's bodily existence.[18]

The narrative thus cuts off four hundred years after 1607. The idea of progress, which was fundamental to the modernization of health care, suggests that humans can control events, specifically events in health and sickness. I need not rehearse the critiques of that naïve presumption. Indeed, just as a new epoch began in the 1990s, it was possible that Americans' subjective judgment of their own healthiness was actually turning negative.[19]

Yet the overall direction of events, as recorded in the foregoing chapters, disconcertingly suggests that, beyond the background of decreasing malnutrition, contamination, and overcrowding, Americans' deliberate health efforts succeeded in substantially extending many human lives and expanding human comfort—and not just in the United States. Those efforts of course did not eliminate evil from the world, not even excesses in attempts at therapy and care, much less ignorance, pretension, and knavery. The historical record summarized in this book is riddled with human frailties, failures, and missteps. Physicians and other health care workers unwittingly (and sometimes wittingly) acted as agents for oppressive elements in the society and culture. Very frequently they mistook pride for professionalism. And it is possible to read the narrative in such a way as to raise the question whether the whole, ultimately gigantic effort was not misdirected.

Yet it would be hard to deny the existence there also of a story of human aspiration and ingenious technical constructions. Perhaps without anyone's intending it, the story can be inspiring. And inspiring in various areas: science and learning, caring and compassion. Even suffering and dying. History helps us become more deeply thoughtful and concerned as we deal with the present and the dark complexities of the future.

NOTES

Introduction

1. *Modernization* to refer to a concept came into use only after the process was well established. It is ironic that the term *modern medicine* did not appear in the titles of articles in either medical journals or general magazines in the United States until very late, in the 1890s and the first decade of the twentieth century. The first American book with the phrase *modern medicine* in the title was not published until 1900—Julius L. Salinger and Frederick J. Kakeyer, *Modern Medicine* (Philadelphia: W. B. Saunders, 1900). Before that, Americans in medicine or outside of it thought in terms of "new" or "recent" or the equivalent of up-to-date, as well as in terms of the idea of progress as developed in the seventeenth and eighteenth centuries.

2. See Roger Cooter, "Medicine and Modernity," in *The Oxford Handbook of the History of Medicine*, ed. Mark Jackson (Oxford: Oxford Univ. Press, 2011), 100–116; and, e.g., Paul Weindling, "Medicine and Modernization: The Social History of German Health and Medicine," *History of Science*, 24 (1986), 277. Even though the concept of modernization has changed in the last half century, refined and of course refuted by theorists, for ordinary purposes the basic idea is useful for summarizing historical processes, especially since modernization embodied cultural as well as social and economic change. See, e.g., Jerrold Siegel, *Modernity and Bourgeois Life: Society, Politics, and Culture in England, France, and Germany since 1750* (New York: Cambridge Univ. Press, 2012). In spite of the disdain of some purists, leading thinkers in the twenty-first century are still finding new understandings of the idea of modernity/modernizing that make my framing of this book for a general reader appropriate. See esp. David Hess, "Technology, Medicine, and Modernity: The Problem of Alternatives," in *Modernity and Technology*, ed. Thomas J. Misa, Philip Brey, and Andrew Feenberg (Cambridge, MA: MIT Press, 2003), 279–302, and other essays in that volume.

3. See esp. Richard D. Brown, *Modernization: The Transformation of American Life, 1600–1865* (New York: Hill & Wang, 1976); or Lewis Perry, *Childhood, Marriage, and Reform: Henry Clarke Wright, 1797–1870* (Chicago: Univ. of Chicago Press, 1979).

4. K. Codell Carter, *The Decline of Therapeutic Bloodletting and the Collapse of Traditional Medicine* (New Brunswick, NJ: Transaction, 2012), esp. 58–59.

5. Michael Bliss, *The Making of Modern Medicine: Turning Points in the Tradition of Disease* (Chicago: Univ. of Chicago Press, 2011), 1. For the present discussion, much of the literature on modernization, such as that equating modernization to imperialism or Americanization, is not useful. I am also ignoring the idea that "modern" history began in about

1500; see, e.g., John Lukacs, *The Passing of the Modern Age* (New York: Harper & Row, 1970), chap. 2.

6. See, e.g., David Mechanic, "The Growth of Medical Technology and Bureaucracy: Implications for Medical Care," *Milbank Memorial Fund Quarterly / Health and Society,* 55 (1977), 61–78.

7. Epochs and eras are of course only approximate, with many exceptional developments at any time. Rigidly philosophically minded people may not easily conceptualize these rough historical constructions.

8. Scholars may find the narrative in the first parts of this book somewhat familiar. The general narrative was established many generations ago and amended later by numerous fine historians. This book differs, perhaps, in its emphasis on how traditional medicine was and on how physicians and patients tended to look to the past even in the face of innovation that was becoming obvious and even exciting in the mid-nineteenth century and after.

9. The eras initially tended to be defined by medical science. As Steve Sturdy shows in "Looking for Trouble: Medical Science and Clinical Practice in the Historiography of Modern Medicine," *Social History of Medicine,* 24 (2011), 739–757, science and health care were functionally interdependent, interacting activities. In common parlance and conceptualization, *science* came to mean just "new."

10. See, e.g., Richard Freeman, *The Politics of Health in Europe* (Manchester: Manchester Univ. Press, 2000), esp. 26–31.

Part I. Establishing and Nurturing Traditional Medicine and Ideas about Health

1. See, e.g., John P. Harrison, *The Benefits Accruing to Society from the Medical Profession: An Introductory Lecture* (Cincinnati: R. P. Donogh, 1843), 3–5.

2. See Robley Dunglison, *The Practice of Medicine: A Treatise on Special Pathology and Therapeutics,* 3rd ed., 2 vols. (Philadelphia: Lea & Blanchard, 1848), 1:v, vii; or, for example, Nathaniel Chapman, *A Compendium of Lectures on the Theory and Practice of Medicine,* ed. N. D. Benedict (Philadelphia: Lea & Blanchard, 1846).

3. Eric J. Cassell, "Ideas in Conflict: The Rise and Fall (and Rise and Fall) of New Views of Disease," *Daedalus,* Spring 1986, 22–23; K. Codell Carter, *The Decline of Therapeutic Bloodletting and the Collapse of Traditional Medicine* (New Brunswick, NJ: Transaction, 2012); John Harley Warner, *The Therapeutic Perspective: Medical Practice, Knowledge, and Identity in America, 1820–1885* (1986; reprint, Princeton, NJ: Princeton Univ. Press, 1997); Elaine G. Breslaw, *Lotions, Potions, Pills, and Magic: Health Care in Early America* (New York: New York Univ. Press, 2012).

Chapter 1. Health and Disease in a Land New to Europeans

1. Herbert S. Klein, *A Population History of the United States* (New York: Cambridge Univ. Press, 2004), chap. 1.

2. David S. Jones, *Rationalizing Epidemics: The Meanings and Uses of American Indian Mortality since 1600* (Cambridge, MA: Harvard Univ. Press, 2004), 32.

3. Conevery Bolton Valenčius, *The Health of the Country: How American Settlers Understood Themselves and Their Land* (New York: Basic Books, 2002).

4. See, e.g., David Dary, *Frontier Medicine: From the Atlantic to the Pacific, 1492–1941* (New York: Alfred A. Knopf, 2008).

5. Roy Porter and Dorothy Porter, *In Sickness and in Health: The British Experience, 1650–1850* (London: Fourth Estate, 1988), 149.

6. Ibid., 22–23.

7. Thomas P. Hughes, *Medicine in Virginia, 1607–1609* (Williamsburg: Virginia 350th Anniversary Celebration Corporation, 1957), 26.

8. Quoted in Wyndham B. Blanton, *Medicine in Virginia in the Seventeenth Century* (Richmond: William Byrd, 1930), 42.

9. Quoted in Valenčius, *Health of the Country*, 24.

10. Gerald N. Grob, *The Deadly Truth: A History of Disease in America* (Cambridge, MA: Harvard Univ. Press, 2002), 49–53.

11. Peter H. Wood, "The Impact of Smallpox on the Native Population of the 18th Century South," *New York State Journal of Medicine*, 87 (1987), 36.

12. Grob, *Deadly Truth*, 38–47, quotation from 45.

13. David S. Jones, "Virgin Soils Revisited," *William and Mary Quarterly*, 60 (2003), 723, 733; Linda Nash, *Inescapable Ecologies: A History of Environment, Disease, and Knowledge* (Berkeley: Univ. of California Press, 2006), 20–24.

14. Suzanne Austin Alchon, *A Pest in the Land: New World Epidemics in a Global Perspective* (Albuquerque: Univ. of New Mexico Press, 2003), esp. 107.

15. Grob, *Deadly Truth*, 37–47; Jones, "Virgin Soils Revisited," 703–742; James C. Riley, "Smallpox and American Indians Revisited," *Journal of the History of Medicine and Allied Sciences*, 65 (2010), 445–477; Affra Coming, quoted in Peter McCandless, *Slavery, Disease, and Suffering in the Southern Lowcountry* (New York: Cambridge Univ. Press, 2011), 25.

16. Jones, *Rationalizing Epidemics*, chap. 1.

17. Elizabeth A. Fenn, *Pox Americana: The Great Smallpox Epidemic of 1775–82* (New York: Hill & Wang, 2001).

18. J. Worth Estes, "The Practice of Medicine in 18th-Century Massachusetts: A Bicentennial Perspective," *New England Journal of Medicine*, 305 (1981), 1040; Mary J. Dobson, "Mortality Gradients and Disease Exchanges: Comparisons from Old England and Colonial America," *Social History of Medicine*, 2 (1989), 259–297.

19. See Flurin Condrau and Michael Worboys, "Second Opinions: Epidemics and Infections in Nineteenth-Century Britain," *Social History of Medicine*, 20 (2007), 147–158.

20. J. Worth Estes, "Therapeutic Practice in Colonial New England," in *Medicine in Colonial Massachusetts, 1620–1820*, ed. J. Worth Estes, Philip Cash, and Eric H. Christianson (Boston: Colonial Society of Massachusetts, 1980), 307–311; Kenneth F. Kiple, ed., *The Cambridge World History of Human Disease* (Cambridge: Cambridge Univ. Press, 1993), contains many essays relevant to colonial America.

21. John Duffy, *Epidemics in Colonial America* (Baton Rouge: Louisiana State Univ. Press, 1953), 164–179, quotation from 164–165; and Grob, *Deadly Truth*, 78–80, quotation from 79–80.

22. See previous note.

23. Duffy, *Epidemics in Colonial America*, 179–183.

24. Ibid., 113–137; Grob, *Deadly Truth*, 81–83.

25. Riley, "Smallpox and American Indians Revisited."

26. Duffy, *Epidemics in Colonial America,* 17.

27. Archibald L. Hoyne, quoted in Berton Roueché, *Eleven Blue Men, and Other Narratives of Medical Detection* (Boston: Little, Brown, 1953), 101.

28. Claude Edwin Heaton, "Medicine in New York during the English Colonial Period," *Bulletin of the History of Medicine,* 17 (1945), 23–25.

29. Elizabeth A. Fenn, "Biological Warfare in Eighteenth-Century North America: Beyond Jeffery Amherst," *Journal of American History,* 86 (2000), 1552–1580.

30. Simon Finger, *The Contagious City: The Politics of Public Health in Early Philadelphia* (Ithaca, NY: Cornell Univ. Press, 2012), 91.

31. Darrett B. Rutman and Anita H. Rutman, "Of Agues and Fevers: Malaria in the Early Chesapeake," *William and Mary Quarterly,* 3rd ser., 33 (1976), 31–60; Margaret Humphreys, *Malaria: Poverty, Race, and Public Health in the United States* (Baltimore: Johns Hopkins Univ. Press, 2001), esp. chap. 1.

32. Humphreys, *Malaria,* 25, quoting "Dying in Paradise: Malaria, Mortality, and the Perceptual Environment in Colonial South Carolina," by H. Roy Merrens and George D. Terry, *Journal of Southern History,* 50 (1984), 449.

33. Erwin H. Ackerknecht, *Malaria in the Upper Mississippi Valley, 1760–1900* (Baltimore: Johns Hopkins Press, 1945); Humphreys, *Malaria.*

34. K. David Patterson, "Yellow Fever Epidemics and Mortality in the United States, 1693–1905," *Social Science and Medicine,* 34 (1992), 855–865.

35. Duffy, *Epidemics in Colonial America,* 138–163; Margaret Humphreys, *Yellow Fever and the South* (New Brunswick, NJ: Rutgers Univ. Press, 1992), quotation from 6.

36. Grob, *Deadly Truth,* 84–86; Wyndham B. Blanton, *Medicine in Virginia in the Eighteenth Century* (Richmond: Garrett & Massie, 1931), 89.

37. Duffy, *Epidemics in Colonial America,* 214–222, quotation from 217; Lord Delaware [De La Warr], quoted in Blanton, *Medicine in Virginia in the Seventeenth Century,* 63.

38. Richard Harrison Shryock, *Medicine and Society in America, 1660–1860* (New York: New York Univ. Press, 1960), 92–93.

39. Grob, *Deadly Truth,* 105; Charles E. Rosenberg, *The Cholera Years: The United States in 1832, 1849, and 1866* (Chicago: Univ. of Chicago Press, 1962).

40. Rosenberg, *Cholera Years,* 3.

41. See, e.g., Dary, *Frontier Medicine,* chap. 6.

42. [John Tennent], *Every Man His Own Doctor: Or, The Poor Planter's Physician; Prescribing Plain and Easy Means for Persons to Cure Themselves of All, Or Most of the Distempers, Incident to This Climate, and With Very Little Charge, the Medicines Being Chiefly of the Growth and Production of This Country,* 3rd ed. (Williamsburg, VA: Wil. Parks, 1736), 4–5.

43. William D. Snively Jr. and Louanna Furbee, "Discoverer of the Cause of Milk Sickness," *JAMA,* 196 (1966), 1055–1060, quotations from 1056 and 1057; William D. Snively, "Mystery of the Milksick," *Minnesota Medicine,* 50 (1967), 469–476.

44. See previous note.

45. Ibid.

46. Shryock, *Medicine and Society in America,* 96.

47. Larry L. Burkhart, *The Good Fight: Medicine in Colonial Pennsylvania* (New York: Garland, 1989), 34–38.

48. Stephanie E. Smallwood, *Saltwater Slavery: A Middle Passage from Africa to American Diaspora* (Cambridge, MA: Harvard Univ. Press, 2007), esp. 135–152; Herbert S. Klein, *The Atlantic Slave Trade* (Cambridge: Cambridge Univ. Press, 1999), chap. 6; James A. Rawley and Stephen D. Behrendt, *The Transatlantic Slave Trade: A History*, 2nd ed. (Lincoln: Univ. of Nebraska Press, 2005), chap. 12; McCandless, *Slavery, Disease, and Suffering*, 47–50.

49. Wyndham B. Blanton, "Epidemics, Real and Imaginary, and Other Factors Influencing Seventeenth Century Virginia Population," *Bulletin of the History of Medicine*, 31 (1957), 462; Richard A. Meckel, "Levels and Trends of Death and Disease in Childhood, 1620 to the Present," in *Children and Youth in Sickness and in Health: A Historical Handbook and Guide*, ed. Janet Golden, Richard A. Meckel, and Heather Munro Prescott (Westport, CT: Greenwood, 2004), 4–8.

50. Thomas Jefferson Wertenbaker, *The First Americans, 1607–1690* (New York: Macmillan, 1927), 185–186.

51. Ibid.

52. Shryock, *Medicine and Society in America*, 96.

53. David Freeman Hawke, *Everyday Life in Early America* (New York: Harper & Row, 1988), 59, 65.

54. Judith Walzer Leavitt, *Brought to Bed: Childbearing in America, 1750–1950* (New York: Oxford Univ. Press, 1986), 14–16.

55. Wertenbaker, *First Americans*, 182–184.

56. Rutman and Rutman, "Of Agues and Fevers," 49.

57. Janet Carlisle Bogdan, "Childbirth in America, 1650–1990," in *Women, Health, and Medicine in America: A Historical Handbook*, ed. Rima D. Apple (New Brunswick, NJ: Rutgers Univ. Press, 1992), 102–107; Richard W. Wertz and Dorothy C. Wertz, *Lying-In: A History of Childbirth in America* (New York: Free Press, 1977), chap. 1.

58. Duffy, *Epidemics in Colonial America*, 241–242; McCandless, *Slavery, Disease, and Suffering*, esp. 50–55.

59. William Fitzhugh, quoted in Darrett B. Rutman and Anita H. Rutman, "Non-Wives and Sons-in-Law: Parental Death in a Seventeenth-Century Virginia County," in *The Chesapeake in the Seventeenth Century: Essays on Anglo-American Society*, ed. Thad W. Tate and David L. Ammerman (New York: W. W. Norton, 1979), 168.

60. James H. Cassedy, *Medicine in America: A Short History* (Baltimore: Johns Hopkins Univ. Press, 1991), 4.

61. Grob, *Deadly Truth*, 87; Billy G. Smith, *The "Lower Sort": Philadelphia's Laboring People, 1750–1800* (Ithaca, NY: Cornell Univ. Press, 1990), 59–61; McCandless, *Slavery, Disease, and Suffering*.

62. Blanton, "Epidemics, Real and Imaginary," 454–455.

63. See, e.g., Kenneth F. Kiple and Virginia H. Kiple, "Deficiency Diseases in the Caribbean," in *Health and Disease in Human History: A "Journal of Interdisciplinary History" Reader*, ed. Robert I. Rotberg (Cambridge, MA: MIT Press, 2000), 238–248.

64. Estes, "Therapeutic Practice in Colonial New England," 306–307, 312.

65. James H. Cassedy, *Demography in Early America: Beginnings of the Statistical Mind, 1600–1800* (Cambridge, MA: Harvard Univ. Press, 1969), chap. 7.

66. Estes, "Therapeutic Practice in Colonial New England," 289–379.

67. Klein, *Population History of the United States*, esp. 52–55; Rutman and Rutman, "Of Agues and Fevers," 49–50; Dobson, "Mortality Gradients."

68. Kathleen M. Brown, *Foul Bodies: Cleanliness in Early America* (New Haven, CT: Yale Univ. Press, 2009); Shryock, *Medicine and Society in America*, 90–91, quoting Cecil K. Drinker, *Not So Long Ago: A Chronicle of Medicine and Doctors in Colonial Philadelphia* (New York: Oxford Univ. Press, 1937), 29.

69. Shryock, *Medicine and Society in America*; Burkhart, *Good Fight*, esp. 52; Hawke, *Everyday Life*, esp. 72–73.

70. Hawke, *Everyday Life*, 74–80; Shryock, *Medicine and Society in America*, 88–90; Elaine G. Breslaw, *Lotions, Potions, Pills, and Magic: Health Care in Early America* (New York: New York Univ. Press, 2012), 62–63.

71. McCandless, *Slavery, Disease, and Suffering*.

72. John B. Blake, *Public Health in the Town of Boston, 1630–1822* (Cambridge, MA: Harvard Univ. Press, 1959), esp. 74; Andrea A. Conti and Gian Franco Gensini, "The Historical Evolution of Some Intrinsic Dimensions of Quarantine," *Medicina nei secoli*, 19 (2007), 173–187; Finger, *Contagious City*, 35–38 and chap. 9.

73. Quoted in Blake, *Public Health in the Town of Boston*, 4.

74. John Duffy, *The Sanitarians: A History of American Public Health* (Urbana: Univ. of Illinois Press, 1990), chaps. 1–3; Finger, *Contagious City*, 72; Nathaniel Potter, "On the Epidemic Distempers of the Year 1802," *Medical Repository*, 6 (1802), 353.

75. Duffy, *Sanitarians*, chaps. 1–3.

Chapter 2. Traditional Treatment and Traditional Healers

1. Francis R. Packard, *History of Medicine in the United States*, 2 vols. (1931; reprint, New York: Hafner, 1963), 1:10–15, quotation from 11.

2. Robert C. Black III, *The Younger John Winthrop* (New York: Columbia Univ. Press, 1966), 169.

3. John B. Blake, "The Compleat Housewife," *Bulletin of the History of Medicine*, 49 (1975), 30.

4. Genevieve Miller, "A Physician in 1776," *JAMA*, 236 (1976), 29.

5. Letter signed by James Craik and Elisha C. Dick, in the *Times* (Alexandria, VA), reprinted in *Medical Repository*, 3 (1800), 311–312.

6. K. Codell Carter, *The Decline of Therapeutic Bloodletting and the Collapse of Traditional Medicine* (New Brunswick, NJ: Transaction, 2012).

7. Wyndham B. Blanton, *Medicine in Virginia in the Eighteenth Century* (Richmond: Garrett & Massie, 1931), 6.

8. Miller, "Physician in 1776," 29.

9. John A. Lanzalotti, MD, personal communication.

10. Carter, *Decline of Therapeutic Bloodletting*, esp. chaps. 1 and 2.

11. See, e.g., Anne Harrington, "The Placebo Effect and Alternative Medicine: Reimagining the Relationship," in *Alternative and Complementary Treatment in Neurological Illness*, ed. Michael I. Weintraub and Marc S. Micozzi (Philadelphia: Churchill Livingstone, 2001), 151–155.

12. A mnemonic encapsulates much of traditional therapy: "Clyster, blister, bleed. Vomit, purge, and sweat. And dose, dose dose!"

13. J. Worth Estes, "Patterns of Drug Usage in Colonial America," *New York State Journal of Medicine*, 87 (1987), 37–45.

14. Roy Porter, "The Patient in England, c. 1660–c. 1800," in *Medicine in Society: Historical Essays*, ed. Andrew Wear (Cambridge: Cambridge Univ. Press, 1992), 95.

15. Estes, "Patterns of Drug Usage in Colonial America."

16. Erwin H. Ackerknecht, *Malaria in the Upper Mississippi Valley, 1760–1900* (Baltimore: Johns Hopkins Press, 1945), 102; Elaine G. Breslaw, *Lotions, Potions, Pills, and Magic: Health Care in Early America* (New York: New York Univ. Press, 2012), 37–38.

17. John B. Blake, "The Inoculation Controversy in Boston: 1721–1722," *New England Quarterly*, 25 (1952), 489–506; James H. Cassedy, *Demography in Early America: Beginnings of the Statistical Mind, 1600–1800* (Cambridge, MA: Harvard Univ. Press, 1969), 131–138.

18. Peter McCandless, *Slavery, Disease, and Suffering in the Southern Lowcountry* (New York: Cambridge Univ. Press, 2011), 204–222; Packard, *History of Medicine in the United States*, 1:82–89; Reginald Fitz, "The Treatment for Inoculated Small-Pox in 1764 and How It Actually Felt," *Annals of Medical History*, 3rd ser., 4 (1942), 110–113; J. Worth Estes, *Hall Jackson and the Purple Foxglove: Medical Practice and Research in Revolutionary America, 1760–1820* (Hanover, NH: Univ. Press of New England, 1979), 15–34; Stephen Wickes, *History of Medicine in New Jersey, and of Its Medical Men, From the Settlement of the Province to A.D. 1800* (Newark, NJ: Martin R. Dennis, 1879), 68–69; Sara Stidstone Gronim, "Introducing Inoculation: Smallpox, the Body, and Social Relations of Healing in the Eighteenth Century," *Bulletin of the History of Medicine*, 80 (2006), 247–268; Morris H. Saffron, *Surgeon to Washington: Dr. John Cochran, 1730–1807* (New York: Columbia Univ. Press, 1977).

19. Reginald Fitz, "Something Curious in the Medical Line," *Bulletin of the History of Medicine*, 11 (1942), 239–264; Morris C. Leikind, "The Introduction of Vaccination into the United States," *Ciba Symposia*, 3 (1942), 1114–1124; Philip Cash, *Dr. Benjamin Waterhouse: A Life in Medicine and Public Service (1754–1846)* (Sagamore Beach, MA: Science History Publications, 2006), chap. 7.

20. John B. Blake, *Benjamin Waterhouse and the Introduction of Vaccination: A Reappraisal* (Philadelphia: Univ. of Pennsylvania Press, 1957); Cash, *Dr. Benjamin Waterhouse*, chaps. 8–14; McCandless, *Slavery, Disease, and Suffering*, 220–225.

21. Estes, *Hall Jackson*, quotations from 217 and 218.

22. James E. McWilliams, *A Revolution in Eating: How the Quest for Food Shaped America* (New York: Columbia Univ. Press, 2005), 91–92; Larry L. Burkhart, *The Good Fight: Medicine in Colonial Pennsylvania* (New York: Garland, 1989), chap. 3; McCandless, *Slavery, Disease, and Suffering*.

23. Herbert C. Covey, *African American Slave Medicine: Herbal and Non-Herbal Treatments* (Lanham, MD: Lexington Books, 2007); Francisco Guerra, *American Medical Bibliography, 1639–1783* (New York: Lathrop C. Harper, 1962), 15, 298–300.

24. George Edmund Gifford Jr., *Medicine and Science in Early America*, ed. Dorothy I. Lansing (Devon, PA: Anro, 1982), [323].

25. John S. Haller Jr., *American Medicine in Transition, 1840–1910* (Urbana: Univ. of Illinois Press, 1981), 67 and notes on 351–352.

26. Burkhart, *Good Fight*, 172–173; J. Worth Estes, "Therapeutic Practice in Colonial New England," in *Medicine in Colonial Massachusetts, 1620–1820*, ed. J. Worth Estes, Philip Cash, and Eric H. Christianson (Boston: Colonial Society of Massachusetts, 1980), 314.

27. Gifford, *Medicine and Science in Early America*, [321].

28. See, e.g., Katherine Bankole, *Slavery and Medicine: Enslavement and Medical Practices in Antebellum Louisiana* (New York: Garland, 1998), chaps. 10–13.

29. Quoted in Thomas A. Horrocks, "Rules, Remedies, and Regimens: Health Advice in Early American Almanacs," in *Right Living: An Anglo-American Tradition of Self-Help Medicine and Hygiene*, ed. Charles E. Rosenberg (Baltimore: Johns Hopkins Univ. Press, 2003), 120–121.

30. Joseph Ioor Waring, *A History of Medicine in South Carolina, 1670–1825* (Charleston: South Carolina Medical Association, 1964), 6.

31. William N. Fenton, "Contacts Between Iroquois Herbalism and Colonial Medicine," *Annual Report of the Board of Regents of the Smithsonian Institution*, 1941, 507.

32. Todd L. Savitt, *Medicine and Slavery: The Diseases and Health Care of Blacks in Antebellum Virginia* (Urbana: Univ. of Illinois Press, 1978), 149, 174; Covey, *African American Slave Medicine*, esp. 15–18.

33. Horrocks, "Rules, Remedies, and Regimens," 112–146; Thomas A. Horrocks, *Popular Print and Popular Medicine: Almanacs and Health Advice in Early America* (Amherst: Univ. of Massachusetts Press, 2008).

34. Blanton, *Medicine in Virginia in the Eighteenth Century*, 217.

35. Guerra, *American Medical Bibliography*, 463; *Boston News-Letter*, 27 Sept.–4 Oct. 1708, 4.

36. Charles E. Rosenberg, "The Book in the Sickroom: A Tradition of Print and Practice," in *"Every Man His Own Doctor": Popular Medicine in Early America*, by Charles E. Rosenberg and William H. Helfand (Philadelphia: Library Company of Philadelphia, 1998), 1–2; [John Tennent], *Every Man His Own Doctor: Or, The Poor Planter's Physician; Prescribing Plain and Easy Means for Persons to Cure Themselves of All, Or Most of the Distempers, Incident to This Climate, and With Very Little Charge, the Medicines Being Chiefly of the Growth and Production of This Country*, 3rd ed. (Williamsburg, VA: Wil. Parks, 1736), 45.

37. Charles E. Rosenberg, "Medical Text and Social Context: Explaining William Buchan's *Domestic Medicine*," *Bulletin of the History of Medicine*, 57 (1983), 22–42.

38. A. Wesley Hill, *John Wesley among the Physicians: A Study of Eighteenth-Century Medicine* (London: Epworth, 1958), 111.

39. Blake, "Compleat Housewife," 37–42.

40. Lamar Riley Murphy, *Enter the Physician: The Transformation of Domestic Medicine, 1760–1860* (Tuscaloosa: Univ. of Alabama Press, 1991), quotation from 9.

41. Lucinda McCray Beier, *Sufferers and Healers: The Experience of Illness in Seventeenth-Century England* (London: Routledge & Kegan Paul, 1987), 246.

42. William Byrd, quoted in Blanton, *Medicine in Virginia in the Eighteenth Century*, 182.

43. Thomas Broman, "The Semblance of Transparency: Expertise as a Social Good and an Ideology in Enlightened Societies," *Osiris*, 2nd ser., 27 (2012), 188–208.

44. William Byrd, quoted in Blanton, *Medicine in Virginia in the Eighteenth Century*, 181.

45. Genevieve Miller, "Medical Education in the American Colonies," *Journal of Medical Education*, Feb. 1956, 83–84.

46. Waring, *History of Medicine in South Carolina*, 30.

47. Whitfield J. Bell Jr., "Medical Practice in Colonial America," *Bulletin of the History of Medicine*, 31 (1957), 442.

48. Jacques M. Quen, "Elisha Perkins, Physician, Nostrum-Vendor, or Charlatan?," ibid., 27 (1963), 159–166.

49. Blanton, *Medicine in Virginia in the Eighteenth Century*, 209.

50. Alexander Hamilton, quoted in Bell, "Medical Practice in Colonial America," 443.

51. Philip Cash, "The Professionalization of Boston Medicine, 1760–1803," in Estes, Cash, and Christianson, *Medicine in Colonial Massachusetts*, 71.

52. Bell, "Medical Practice in Colonial America," 448.

53. William Frederick Norwood, *Medical Education in the United States before the Civil War* (Philadelphia: Univ. of Pennsylvania Press, 1944), 24.

54. Rebecca J. Tannenbaum, *The Healer's Calling: Women and Medicine in Early New England* (Ithaca, NY: Cornell Univ. Press, 2002); Ellen G. Gartrell, "Women Healers and Domestic Remedies in 18th Century America: The Recipe Book of Elizabeth Coates Paschall," *New York State Journal of Medicine*, 87 (1987), 23–29.

55. Laurel Thatcher Ulrich, *A Midwife's Tale: The Life of Martha Ballard, Based on Her Diary, 1785–1812* (New York: Alfred A. Knopf, 1992).

56. Barnes Riznik, *Medicine in New England, 1790–1840* (Sturbridge, MA: Old Sturbridge Village, 1965), 9–10; Bell, "Medical Practice in Colonial America," 445; Tannenbaum, *Healer's Calling*.

57. Jane B. Donegan, *Women & Men Midwives: Medicine, Morality, and Misogyny in Early America* (Westport, CT: Greenwood, 1978), 91; Packard, *History of Medicine in the United States*, 1:50.

58. Joseph F. Kett, *The Formation of the American Medical Profession: The Role of Institutions, 1780–1860* (New Haven, CT: Yale Univ. Press, 1968), 7.

59. John Watts, quoted in Gronim, "Introducing Inoculation," 264.

60. Bell, "Medical Practice in Colonial America," 444.

61. Richard Harrison Shryock, *Medical Licensing in America, 1650–1965* (Baltimore: Johns Hopkins Press, 1967), 3.

62. Genevieve Miller, "Medical Apprenticeship in the American Colonies," *Ciba Symposia*, 8 (1947), 502–510, quotation from 508.

63. Miller, "Medical Education in the American Colonies," 85–90; George F. Sheldon and Mary Jane Kagarise, "John Hunter and the American School of Surgery," *Journal of Trauma, Injury, Infection, and Critical Care*, 44 (1998), 13–40; Simon Finger, *The Contagious City: The Politics of Public Health in Early Philadelphia* (Ithaca, NY: Cornell Univ. Press, 2012), 79.

64. Norwood, *Medical Education in the United States*; Martin Kaufman, *American Medical Education: The Formative Years, 1765–1910* (Westport, CT: Greenwood, 1976), chap. 2.

65. Richard Harrison Shryock, *Medicine and Society in America, 1660–1860* (New York: New York Univ. Press, 1960), 9; Eric H. Christianson, "The Medical Practitioners of Massachusetts, 1630–1800," in Estes, Cash, and Christianson, *Medicine in Colonial Massachusetts*, 54–55.

66. Eric H. Christianson, "The Emergence of Medical Communities in Massachusetts, 1700–1794: The Demographic Factors," *Bulletin of the History of Medicine*, 54 (1980), 66, 69–70.

67. Blanton, *Medicine in Virginia in the Eighteenth Century*, 207–208.

68. Whitfield J. Bell Jr., "Medicine in Boston and Philadelphia: Comparisons and Contrasts, 1750–1820," in Estes, Cash, and Christianson, *Medicine in Colonial Massachusetts*, 168.

69. Bell, "Medical Practice in Colonial America," 449–453; Christianson, "Medical Practitioners of Massachusetts," 51–54.

70. Blanton, *Medicine in Virginia in the Eighteenth Century*, 214–215.

71. Bell, "Medical Practice in Colonial America," 445, 449–453; Estes, "Therapeutic Practice in Colonial New England," 345–363.

72. Janet Carlisle Bogdan, "Childbirth in America, 1650–1990," in *Women, Health, and Medicine in America: A Historical Handbook*, ed. Rima D. Apple (New Brunswick, NJ: Rutgers Univ. Press, 1990), 108–112; Donegan, *Women & Men Midwives*, esp. chap 4.

73. William H. Williams, "The Early Days of Anglo-America's First Hospital: The Pennsylvania Hospital, 1751–1775," *JAMA*, 220 (1972), 115–119; Burkhart, *Good Fight*, 275–288; William H. Williams, "The 'Industrious Poor' and the Founding of the Pennsylvania Hospital," *Pennsylvania Magazine of History and Biography*, 97 (1973), 431–443.

74. John Duffy, ed., *The Rudolph Matas History of Medicine in Louisiana*, 2 vols. (Baton Rouge: Louisiana State Univ. Press, 1958–1962), vol. 1, chap. 4.

75. William G. Rothstein, *American Medical Schools and the Practice of Medicine: A History* (New York: Oxford Univ. Press, 1987), 21–24, 44–48.

76. Heaton, "Medicine in New York," 10–12.

77. Blanton, *Medicine in Virginia in the Eighteenth Century*, 372.

78. Burkhart, *Good Fight*, 274–288.

79. Quoted in Parnel Wickham, "Idiocy in Virginia, 1616–1860," *Bulletin of the History of Medicine*, 80 (2006), 689.

80. See, e.g., Louis C. Duncan, *Medical Men in the American Revolution* (New York: Augustus M. Kelley, 1970); Oscar Reiss, *Medicine and the American Revolution: How Diseases and Their Treatments Affected the Colonial Army* (Jefferson, NC: McFarland, 1998); and Whitfield J. Bell Jr., *John Morgan, Continental Doctor* (Philadelphia: Univ. of Pennsylvania Press, 1965).

Chapter 3. The Beginnings of Change in Traditional Health Care

1. Elaine G. Breslaw, *Lotions, Potions, Pills, and Magic: Health Care in Early America* (New York: New York Univ. Press, 2012), chap. 9.

2. John S. Haller Jr., "The United States Pharmacopoeia: Its Origin and Revision in the 19th Century," *Bulletin of the New York Academy of Medicine*, 58 (1982), 480–492; George B. Wood and Franklin Bache, *The Dispensatory of the United States of America* (Philadelphia: Grigg & Elliot, 1833); John S. Billings, "Literature and Institutions," in *A Century of American Medicine, 1776–1876*, by Edward H. Clarke et al. (Philadelphia: Henry C. Lea, 1876), 303.

3. Richard Malmsheimer, *"Doctors Only": The Evolving Image of the American Physician* (Westport, CT: Greenwood, 1988); Stephanie P. Browner, *Profound Science and Elegant Literature: Imagining Doctors in Nineteenth-Century America* (Philadelphia: Univ. of Pennsylvania Press, 2005), esp. chap. 4; Michael Brown, *Performing Medicine: Medical Culture and Identity in Provincial England, c. 1760–1850* (Manchester: Manchester Univ. Press, 2011).

4. Steven J. Peitzman, "'I Am Their Physician'. Dr. Owen J. Wister of Germantown and His Too Many Patients," *Bulletin of the History of Medicine*, 83 (2009), 267.

5. See, e.g., Malmsheimer, *"Doctors Only,"* 63.

6. Charles E. Rosenberg, prologue to *The Structure of American Medical Practice, 1875–1941*, by George Rosen, ed. Charles E. Rosenberg (Philadelphia: Univ. of Pennsylvania Press, 1983), 5.

7. Elizabeth Barnaby Keeney, "Unless Powerful Sick: Domestic Medicine in the Old South," in *Science and Medicine in the Old South*, ed. Ronald L. Numbers and Todd L. Savitt (Baton Rouge: Louisiana State Univ. Press, 1989), 282.

8. Marie Jenkins Schwartz, *Birthing a Slave: Motherhood and Medicine in the Antebellum South* (Cambridge, MA: Harvard Univ. Press, 2006).

9. Todd L. Savitt, *Race and Medicine in Nineteenth- and Early-Twentieth-Century America* (Kent, OH: Kent State Univ. Press, 2007), esp. chap. 7; Steven Stowe, "Conflict and Self-Sufficiency: Domestic Medicine in the American South," in *Right Living: An Anglo-American Tradition of Self-Help Medicine and Hygiene*, ed. Charles E. Rosenberg (Baltimore: Johns Hopkins Univ. Press, 2003), 147–169; Herbert C. Covey, *African American Slave Medicine: Herbal and Non-Herbal Treatments* (Lanham, MD: Rowman & Littlefield, 2007).

10. Todd L. Savitt, *Medicine and Slavery: The Diseases and Health Care of Blacks in Antebellum Virginia* (Urbana: Univ. of Illinois Press, 1978); Stephen C. Kenny, "'A Dictate of Both Interest and Mercy'? Slave Hospitals in the Antebellum South," *Journal of the History of Medicine and Allied Sciences*, 65 (2010), 1–47.

11. John Harley Warner, "The Idea of Southern Medical Distinctiveness: Medical Knowledge and Practice in the Old South," in Numbers and Savitt, *Science and Medicine in the Old South*, esp. 204–205; Savitt, *Race and Medicine*; Kenny, "Dictate of Both Interest and Mercy."

12. Richard Harrison Shryock, *Medical Licensing in America, 1650–1965* (Baltimore: Johns Hopkins Press, 1967); Shryock, *Medicine in America: Historical Essays* (Baltimore: Johns Hopkins Press, 1966), 149.

13. Guenter B. Risse, "From Horse and Buggy to Automobile and Telephone: Medical Practice in Wisconsin, 1848–1930," in *Wisconsin Medicine: Historical Perspectives*, ed. Ronald L. Numbers and Judith Walzer Leavitt (Madison: Univ. of Wisconsin Press, 1981), 28–29; Eugene Perry Link, *The Social Ideas of American Physicians (1776–1976): Studies in the Humanitarian Tradition in Medicine* (Selinsgrove, PA: Susquehanna Univ. Press, 1992), 17; E. Brooks Holifield, "The Wealth of Nineteenth-Century American Physicians," *Bulletin of the History of Medicine*, 64 (1990), 79–85.

14. William Frederick Norwood, "The Early History of American Medical Societies," *Ciba Symposia*, 9 (1947), 762–772; W. B. McDaniel II, "A Brief Sketch of the Rise of American Medical Societies," in *History of American Medicine: A Symposium*, ed. Felix Marti-Ibañez (New York: MD Publications, 1959), 134–139; Creighton Barker, "The Origin of the Connecticut State Medical Society" and other essays in *The Heritage of Connecticut Medicine*, ed. Herbert Thoms (New Haven: Connecticut State Medical Society, 1942), 1–30; Philip Cash, "The Phoenix and the Eagle: The Founding of the Boston and Massachusetts Medical Societies in 1780 and 1781," *New England Journal of Medicine*, 305 (1981), 1033–1039; Peter Dobkin Hall, *The Organization of American Culture* (New York: New York Univ. Press, 1982), chap. 7; Joseph F. Kett, *The Formation of the American Medical Profession: The Role*

of Institutions, 1780–1860 (New Haven, CT: Yale Univ. Press, 1968); William G. Rothstein, *American Physicians in the Nineteenth Century: From Sects to Science* (Baltimore: Johns Hopkins Univ. Press, 1972).

15. George H. Daniels, *American Science in the Age of Jackson* (New York: Columbia Univ. Press, 1968), 231–232; Myrl Ebert, "The Rise and Development of the American Medical Periodical, 1797–1850," *Bulletin of the Medical Library Association*, 40 (1952), 243–276; Richard J. Kahn and Patricia G. Kahn, "The *Medical Repository*—The First U.S. Medical Journal (1797–1824)," *New England Journal of Medicine*, 337 (1997), 1926–1930.

16. See, e.g., Michael Sappol, "The Odd Case of Charles Knowlton: Anatomical Performance, Medical Narrative, and Identity in Antebellum America," *Bulletin of the History of Medicine*, 83 (2009), 460–498.

17. James H. Cassedy, "The Flourishing and Character of Early American Medical Journalism, 1797–1860," *Journal of the History of Medicine and Allied Sciences*, 38 (1983), 135–150, quotation from 139; Robley Dunglison, *The Practice of Medicine: A Treatise on Special Pathology and Therapeutics*, 3rd ed., 2 vols. (Philadelphia: Lea & Blanchard, 1848), 1:vii.

18. Wyndham B. Blanton, *Medicine in Virginia in the Eighteenth Century* (Richmond: Garrett & Massie, 1931), 399–400.

19. George Rosen, *Fees and Fee Bills: Some Economic Aspects of Medical Practice in Nineteenth Century America* (Baltimore: Johns Hopkins Press, 1946), 2–5; William Moll, "Medical Fee Bills," *Virginia Medical Monthly*, 93 (1966), 657–664.

20. *Boston Medical and Surgical Journal*, 18 (1838), 50–51.

21. Henry Burnell Shafer, *The American Medical Profession, 1783–1850* (New York: Columbia Univ. Press, 1936), 237–240; James G. Burrow, *AMA: Voice of American Medicine* (Baltimore: Johns Hopkins Press, 1963), chap. 1.

22. Shryock, *Medical Licensing in America*, 23–27; Kett, *Formation of the American Medical Profession*, 13–31.

23. John S. Haller Jr., *The People's Doctors: Samuel Thomson and the American Botanical Movement, 1790–1860* (Carbondale: Southern Illinois Univ. Press, 2000), 129; Fanny J. Anderson, "The Doctor and the Newspaper in the Territory of Michigan, 1817–1837," *Journal of the History of Medicine and Allied Sciences*, 2 (1947), 20–21.

24. Martin Kaufman, *American Medical Education: The Formative Years, 1765–1910* (Westport, CT: Greenwood, 1976), chap. 3; William Frederick Norwood, *Medical Education in the United States before the Civil War* (Philadelphia: Univ. of Pennsylvania Press, 1944); Genevieve Miller, "Medical Education in the American Colonies," *Journal of Medical Education*, Feb. 1956, 82–94; George H. Yeager, "Medical Schools of Southern United States, 1779–1830," *Annals of Surgery*, 171 (1970), 623–640.

25. Norwood, *Medical Education in the United States*; Rothstein, *American Physicians in the Nineteenth Century*, 93.

26. Norwood, *Medical Education in the United States*, 276.

27. Rothstein, *American Physicians in the Nineteenth Century*, 98.

28. Kett, *Formation of the American Medical Profession*, chap. 3; Alfred Stillé, quoted in Kenneth Allen De Ville, *Medical Malpractice in Nineteenth-Century America: Origins and Legacy* (New York: New York Univ. Press, 1990), 73–74; Shafer, *American Medical Profession*, 46–37.

29. Daniel Drake, *Practical Essays on Medical Education, and the Medical Profession, in the United States* (Cincinnati: Roff & Young, 1832), esp. 47–49 and 51.

30. Kaufman, *American Medical Education*, chaps. 5–7; Simon Baatz, "The Campaign for Reform: Medical Education and Professional Improvement in Antebellum America," *Transactions and Studies of the College of Physicians of Philadelphia*, 21 (1999), 118–148.

31. Norwood, *Medical Education in the United States*, 399; Michael Sappol, *A Traffic of Dead Bodies: Anatomy and Embodied Social Identity in Nineteenth-Century America* (Princeton, NJ: Princeton Univ. Press, 2002).

32. David C. Humphrey, "Dissection and Discrimination: The Social Origins of Cadavers in America, 1760–1915," *Bulletin of the New York Academy of Medicine*, 49 (1973), 822.

33. Shafer, *American Medical Profession*, 77; Shryock, *Medicine in America*, 154; Pliny Earle, *The Memoirs of Pliny Earle, M.D.*, ed. F. B. Sanborn (Boston: Damrell & Upham, 1898), 56; Steven M. Stowe, introduction to *A Southern Practice: The Diary and Autobiography of Charles A. Hentz, M.D.*, ed. Stowe (Charlottesville: Univ. Press of Virginia, 2000), 21.

34. Benjamin Rush, *Sixteen Introductory Lectures to Courses of Lectures upon the Institutes and Practice of Medicine* (Philadelphia: Bradford & Innskeep, 1811), 69, 147; Carl Binger, *Revolutionary Doctor: Benjamin Rush, 1746–1813* (New York: W. W. Norton, 1966).

35. Benjamin Rush, *Medical Inquiries and Observations*, 3rd ed, 4 vols. (Philadelphia: Johnson & Warner et al., 1809), 3:269, 312; Paul E. Kopperman, "'Venerate the Lancet': Benjamin Rush's Yellow Fever Therapy in Context," *Bulletin of the History of Medicine*, 78 (2004), 539–574.

36. Charles E. Rosenberg, "The Therapeutic Revolution: Medicine, Meaning, and Social Change in Nineteenth-Century America," in *The Therapeutic Revolution: Essays in the Social History of American Medicine*, ed. Morris J. Vogel and Charles E. Rosenberg (Philadelphia: Univ. of Pennsylvania Press, 1979), 5–8.

37. Nathan Smith, quoted in *Improve, Perfect, and Perpetuate: Dr. Nathan Smith and Early American Medical Education*, by Oliver S. Hayward and Constance E. Putnam (Hanover, NH: Univ. Press of New England, 1998), 144.

38. Guenter B. Risse, "Calomel and the American Medical Sects During the Nineteenth Century," *Mayo Clinic Proceedings*, 48 (1973), 57–64, esp. 59.

39. James C. Whorton, *Inner Hygiene: Constipation and the Pursuit of Health in Modern Society* (New York: Oxford Univ. Press, 2000).

40. Ibid., esp. 59 and 60.

41. Alex Berman, "The Heroic Approach in 19th Century Therapeutics," *Bulletin of the American Society of Hospital Pharmacists*, 11 (1954), 320–327.

42. John B. Brown, "Case in which the Oil of Turpentine Was Employed with Success in Taenia," *New England Journal of Medicine and Surgery*, 1 (1812), 269–270; Eric Howard Christianson, "The Search for Diagnostic and Therapeutic Authority in the Early American Healer's Encounter with 'The Animals which Inhabit the Human Stomach and Intestines,'" in *Folklore and Folk Medicines*, ed. John Scarborough (Madison, WI: American Institute of the History of Pharmacy, 1987), 62–83.

43. Shryock, *Medicine in America*, 150–151, quotation from 151; Kett, *Formation of the American Medical Profession*, chap. 2.

44. John Duffy, ed., *The Rudolph Matas History of Medicine in Louisiana*, 2 vols. (Baton Rouge: Louisiana State Univ. Press, 1958–1962), 2:88–90.

45. Shafer, *American Medical Profession*, 213–214; David L. Cowen, *Medicine and Health in New Jersey: A History* (Princeton, NJ: D. Van Nostrand, 1964), 69; Kett, *Formation of the American Medical Profession*, chap. 1 and app. 1; Haller, *People's Doctors*, 131; Toby A. Appel, "The Thomsonian Movement, the Regular Profession, and the State in Antebellum Connecticut," *Journal of the History of Medicine and Allied Sciences*, 65 (2010), 153–186.

46. Rosenberg, "The American Medical Profession: Mid-Nineteenth Century," *Mid America*, 44 (1962), 169.

47. Harris L. Coulter, *Divided Legacy: A History of the Schism in Medical Thought*, 3 vols. (Washington, DC: Wehawken, 1973), 3:144; Worthington Hooker, quoted in ibid.

48. See, e.g., William Barlow and David O. Powell, "To Find a Stand: New England Physicians on the Western and Southern Frontier, 1790–1840," *Bulletin of the History of Medicine*, 54 (1980), 386–401; and John S. Haller Jr., *American Medicine in Transition, 1840–1910* (Urbana: Univ. of Illinois Press, 1981), 213.

49. Elizabeth Blackwell, *Pioneer Work in Opening the Medical Profession to Women: Autobiographical Sketches* (London: Longmans, Green, 1895).

50. Rothstein, *American Physicians in the Nineteenth Century*, pts. 3 and 4; Norman Gevitz, ed., *Other Healers: Unorthodox Medicine in America* (Baltimore: Johns Hopkins Univ. Press, 1988).

51. Haller, *People's Doctors*, esp. chap. 1, quotation from 29; Kett, *Formation of the American Medical Profession*, chap. 4.

52. Haller, *People's Doctors*; James Harvey Young, *The Toadstool Millionaires: A Social History of Patent Medicines in America before Federal Regulation* (Princeton, NJ: Princeton Univ. Press, 1961), 52; Anderson, "Doctor and the Newspaper," 45.

53. Kett, *Formation of the American Medical Profession*, chap. 5; Haller, *People's Doctors*; Rothstein, *American Physicians in the Nineteenth Century*, chap. 7.

54. See previous note.

55. Haller, *People's Doctors*, 102–104; John S. Haller Jr., *Medical Protestants: The Eclectics in American Medicine, 1825–1939* (Carbondale: Southern Illinois Univ. Press, 1994); Haller, *Kindly Medicine: Physio-Medicalism in America, 1836–1911* (Kent, OH: Kent State Univ. Press, 1997).

56. See esp. Kett, *Formation of the American Medical Profession*, chap. 4.

57. Haller, *Medical Protestants*, esp.164–165.

58. John S. Haller Jr., *The History of American Homeopathy: The Academic Years, 1820–1935* (New York: Pharmaceutical Products Press, 2005).

59. Martin Kaufman, *Homeopathy in America: The Rise and Fall of a Medical Heresy* (Baltimore: Johns Hopkins Press, 1971), chaps. 2 and 3, quotation from 30.

60. Coulter, *Divided Legacy*, 3:113; Haller, *History of American Homeopathy*, 113–120.

61. Kaufman, *Homeopathy in America*; Coulter, *Divided Legacy*, 3:125.

62. Naomi Rogers, "American Homeopathy Confronts Scientific Medicine," in *Culture, Knowledge, and Healing: Historical Perspectives of Homeopathic Medicine in Europe and North America*, ed. Robert Jütte, Guenter B. Risse, and John Woodward (Sheffield, UK: European Association for the History of Medicine and Health Publications, 1998), 38; Haller, *History of American Homeopathy*; Paul Starr, *The Social Transformation of American Medicine* (New York: Basic Books, 1982), 99.

63. Kett, *Formation of the American Medical Profession*, 161, 163–164; Coulter, *Divided Legacy*, 3:118.

64. Jütte, Risse, and Woodward, *Culture, Knowledge, and Healing*, esp. Eberhard Wolff, "Sectarian Identity and the Aim of Integration: Attitudes of American Homeopaths Towards Smallpox Vaccination in the Late Nineteenth Century," 217–250; John S. Haller Jr., *History of American Homeopathy: From Rational Medicine to Holistic Health Care* (New Brunswick, NJ: Rutgers Univ. Press, 2009), esp. 94–101.

65. Jane B. Donegan, *"Hydropathic Highway to Health": Women and Water-Cure in Antebellum America* (Westport, CT: Greenwood, 1986), esp. 3–7.

66. Ibid.; Marshall Scott Legan, "Hydropathy in America: A Nineteenth Century Panacea," *Bulletin of the History of Medicine*, 45 (1971), 267–280; Susan E. Cayleff, "Gender, Ideology, and the Water-Cure Movement," in Gevitz, *Other Healers*, 82–98.

67. John D. Davies, *Phrenology: Fad and Science; A 19th-Century American Crusade* (New Haven, CT: Yale Univ. Press, 1955); Madeleine B. Stern, *Heads & Headlines: The Phrenological Fowlers* (Norman: Univ. of Oklahoma Press, 1971).

68. James H. Cassedy, "Why Self-Help? Americans Alone with Their Diseases, 1800–1850," in *Medicine without Doctors: Home Health Care in American History*, ed. Guenter B. Risse, Ronald L. Numbers, and Judith Walzer Leavitt (New York: Science History Publications, 1977), 34.

69. Young, *Toadstool Millionaires*.

70. Ibid., 32.

71. Madge E. Pickard and R. Carlyle Buley, *The Midwest Pioneer: His Ills, Cures, and Doctors* (Crawfordsville, IN: R. E. Banta, 1945), 269.

72. Young, *Toadstool Millionaires*; L. C. Butler, "The Decadence of the American Race, As Exhibited in the Registration Reports of Massachusetts, Vermont, and Rhode Island— The Cause and the Remedy," *Saint Louis Medical Reporter*, 2 (1867), 486–487; Louis D. Vottero, unpublished paper.

73. Adelaide Hechtlinger, *The Great Patent Medicine Era, Or, Without Benefit of Doctor* (New York: Grosset & Dunlap, 1970), 173.

74. George M. Kober, "The Progress and Tendency of Hygiene and Sanitary Science in the Nineteenth Century," *Medical Record*, 59 (1901), 903.

75. See, e.g., Charles E. Rosenberg, "Catechisms of Health: The Body in the Prebellum Classroom," *Bulletin of the History of Medicine*, 69 (1995), 175–197.

76. John C. Burnham, "Change in the Popularization of Health in the United States," ibid., 58 (1984), 185 (quotation); James C. Whorton, *Crusaders for Fitness: The History of American Health Reformers* (Princeton, NJ: Princeton Univ. Press, 1982), chaps. 1–4; Stephen Nissenbaum, *Sex, Diet, and Debility in Jacksonian America: Sylvester Graham and Health Reform* (Westport, CT: Greenwood, 1980); Kathleen M. Brown, *Foul Bodies: Cleanliness in Early America* (New Haven, CT: Yale Univ. Press, 2009).

77. Karen Iacobbo and Michael Iacobbo, *Vegetarian America: A History* (Westport, CT: Praeger, 2004), esp. 71–72.

78. Nissenbaum, *Sex, Diet, and Debility*.

79. Whorton, *Crusaders for Fitness*, 53.

80. Gerald N. Grob, *Mental Institutions in America: Social Policy to 1875* (New York: Free Press, 1973), chap. 3.

81. Rosenberg, "Catechisms of Health," 195.

82. Martha H. Verbrugge, "The Social Meaning of Personal Health: The Ladies' Physiological Institute of Boston and Vicinity in the 1850s," in *Health Care in America: Essays in Social History*, ed. Susan Reverby and David Rosner (Philadelphia: Temple Univ. Press, 1979), 45–66, quotation from 49; Regina Markell Morantz, "Nineteenth Century Health Reform and Women: A Program of Self-Help," in Risse, Numbers, and Leavitt, *Medicine without Doctors*, 75–93.

83. Morantz, "Nineteenth Century Health Reform and Women," 82.

84. Thomas A. Horrocks, *Popular Print and Popular Medicine: Almanacs and Health Advice in Early America* (Amherst: Univ. of Massachusetts Press, 2008).

85. Robert H. Wiebe, *The Search for Order, 1877–1920* (New York: Hill & Wang, 1967); Steven M. Stowe, *Doctoring the South: Southern Physicians and Everyday Medicine in the Mid-Nineteenth Century* (Chapel Hill: Univ. of North Carolina Press, 2004).

86. Stowe, *Doctoring the South*, esp. 4, 104, and chap. 5.

87. John Syng Dorsey, *Elements of Surgery for the Use of Students*, 2 vols. (Philadelphia: E. Parker, 1813), 1:iii; John Eberle, *Treatise on the Practice of Medicine*, 2nd ed., 2 vols. (Philadelphia: John Grigg, 1831), 1:v–vi.

88. Billings, "Literature and Institutions," 294.

89. Russell M. Jones, "American Doctors and the Parisian Medical World, 1830–1840," *Bulletin of the History of Medicine*, 47 (1973), 40–65, esp. 43–51; John Harley Warner, *Against the Spirit of System: The French Impulse in Nineteenth-Century American Medicine* (Princeton, NJ: Princeton Univ. Press, 1998), esp. 38–39, 48, 137–138, 156–157.

90. William Osler, *An Alabama Student and Other Biographical Essays* (1908; reprint, London: Oxford Univ. Press, 1929), quotation from 111.

91. Stanley Joel Reiser, *Technological Medicine: The Changing World of Doctors and Patients* (New York: Cambridge Univ. Press, 2009), 106–111.

92. Ibid., chap. 1.

93. John Harley Warner, *The Therapeutic Perspective: Medical Practice, Knowledge, and Identity in America, 1820–1885* (Cambridge, MA: Harvard Univ. Press, 1986), chap. 1.

94. John Harley Warner, "'The Nature-Trusting Heresy': American Physicians and the Concept of the Healing Power of Nature in the 1850's and 1860's," *Perspectives in American History*, 11 (1977–1978), 291–324; Warner, *Therapeutic Perspective*, chap. 1.

95. Jacob Bigelow, "On Self-Limited Diseases," in *Medical America in the Nineteenth Century*, ed. Gert H. Brieger (Baltimore: Johns Hopkins Press, 1972), 99–100.

96. Ibid., 100, 106; Warner, *Therapeutic Perspective*, esp. 27–36.

97. Oliver Wendell Holmes, *Currents and Counter-Currents in Medical Science, With Other Addresses and Essays* (Boston: Ticknor & Fields, 1861), 38–39.

98. Phyllis Allen, "Etiological Theory in America Prior to the Civil War," *Journal of the History of Medicine and Allied Sciences*, 2 (1947), 489–520, esp. 490–491; Lester S. King, *Transformations in American Medicine from Benjamin Rush to William Osler* (Baltimore: Johns Hopkins Univ. Press, 1991), chap. 4, esp. 84 and 143–144.

99. George B. Wood, *A Treatise on the Practice of Medicine*, 2 vols. (Philadelphia: Grigg, Elliot, 1847); Warner, *Therapeutic Perspective*, esp. chap. 3.

100. K. Codell Carter, *The Decline of Therapeutic Bloodletting and the Collapse of Traditional Medicine* (New Brunswick, NJ: Transaction, 2012).

101. Dale C. Smith, "Gerhard's Distinction Between Typhoid and Typhus and Its Reception in America, 1833–1860," *Bulletin of the History of Medicine*, 54 (1980), 368–385.

102. Thomas E. Keys, *The History of Surgical Anesthesia* (New York: Schuman's, 1945), 21–37; Henry Jacob Bigelow, "Insensibility during Surgical Operations Produced by Inhalation," *Boston Medical and Surgical Journal*, 35 (1846), 317; Martin S. Pernick, *A Calculus of Suffering: Pain, Professionalism, and Anesthesia in Nineteenth-Century America* (New York: Columbia Univ. Press, 1985), esp. 21–22; Reiser, *Technological Medicine*, 172.

103. Seale Harris, *Women's Surgeon: The Life Story of J. Marion Sims* (New York: Macmillan, 1950), quotation from 86; Deborah Kuhn McGregor, *From Midwives to Medicine: The Birth of American Gynecology* (New Brunswick, NJ: Rutgers Univ. Press, 1998); Alexa Green, "Working Ethics: William Beaumont, Alexis St. Martin, and Medical Research in Antebellum America," *Bulletin of the History of Medicine*, 84 (2010), 193–216; Reginald Horsman, *Frontier Doctor: William Beaumont, America's First Great Medical Scientist* (Columbia: Univ. of Missouri Press, 1996).

104. Dale C. Smith, "Quinine and Fever: The Development of the Effective Dosage," *Journal of the History of Medicine and Allied Sciences*, 31 (1976), 343–367.

105. Warner, *Therapeutic Perspective*; Warner, *Against the Spirit of System*.

106. Warner, *Therapeutic Perspective*, esp. 3–7 and chap. 4; Carter, *Decline of Therapeutic Bloodletting*; Breslaw, *Lotions, Potions, Pills, and Magic*.

107. See previous note; and John Harley Warner, "Medical Sectarianism, Therapeutic Conflict, and the Shaping of Orthodox Professional Identity in Antebellum American Medicine," in *Medical Fringe and Medical Orthodoxy, 1750–1850*, ed. W. F. Bynum and Roy Porter (London: Croom Helm, 1987), 234–260, quotation from 247.

108. Richard H. Steckel, "Heights and Health in the United States, 1710–1950," in *Stature, Living Standards, and Economic Development: Essays in Anthropometric History*, ed. John Komlos (Chicago: Univ. of Chicago Press, 1994), 153–170.

109. Herbert S. Klein, *A Population History of the United States* (New York: Cambridge Univ. Press, 2004), 102–103. See also esp. Clayne L. Pope, "Adult Mortality in America before 1900: A View from Family Histories," in *Strategic Factors in Nineteenth Century American Economic History*, ed. Claudia Goldin and Hugh Rockoff (Chicago: Univ. of Chicago Press, 1992), 267–296; Michael R. Haines, "Growing Incomes, Shrinking People—Can Economic Development Be Hazardous to Your Health? Historical Evidence for the United States, England, and the Netherlands in the Nineteenth Century," *Social Science History*, 28 (2004), 253; John Komlos, "On the Significance of Anthropometric History," in Komlos, *Stature, Living Standards, and Economic Development*, esp. 214–215; and Komlos, "A Three-Decade History of the Antebellum Puzzle: Explaining the Shrinking of the U.S. Population at the Onset of Modern Economic Growth," *Journal of the Historical Society*, 12 (2012), 395–445.

110. Allen, "Etiological Theory in America," esp. 498; Robley Dunglison, *A Dictionary of Medical Science*, 2nd ed. (Philadelphia: Blanchard & Lea, 1857), 400.

111. Erwin H. Ackerknecht, "Anticontagionism between 1821 and 1867," *Bulletin of the History of Medicine*, 22 (1948), 572.

112. John Duffy, *The Sanitarians: A History of American Public Health* (Urbana: Univ. of Illinois Press, 1990), 21–22; Allen, "Etiological Theory in America," 492–494; Margaret H.

Warner, "Public Health in the Old South," in Numbers and Savitt, *Science and Medicine in the Old South*, 242; Eberle, *Treatise on the Practice of Medicine*, 1:40.

113. Allen, "Etiological Theory in America," 492–496, quotation from 496.

114. Dunglison, *Dictionary of Medical Science*, 992.

115. Margaret Humphreys, *Yellow Fever and the South* (New Brunswick, NJ: Rutgers Univ. Press, 1992); Duffy, *Sanitarians*, 102–105.

116. Martin V. Melosi, "How Bad Theory Can Lead to Good Technology: Water Supply and Sewerage in the Age of Miasmas," in *Inventing for the Environment*, ed. Arthur Molella and Joyce Bedi (Cambridge, MA: MIT Press, 2003), 231–256; Melosi, *The Sanitary City: Urban Infrastructure in America from Colonial Times to the Present* (Baltimore: Johns Hopkins Univ. Press, 2000), 74; E. H. Barton, *The Cause and Prevention of Yellow Fever, Contained in the Report of the Sanitary Commission of New Orleans* (Philadelphia: Lindsay & Blakiston, 1855), 8; anonymous review of *Report of a General Plan for the Promotion of Public and Personal Health*, in *North American Review*, 73 (1851), 118.

117. Alan I Marcus, *Plague of Strangers: Social Groups and the Origins of City Services in Cincinnati, 1819–1870* (Columbus: Ohio State Univ. Press, 1991), 98–99.

118. Ibid.; Ackerknecht, "Anticontagionism," 562–593.

119. Warner, *Therapeutic Perspective*; Breslaw, *Lotions, Potions, Pills, and Magic*; Carter, *Decline of Therapeutic Bloodletting*.

Chapter 4. Setting the Stage for Modern Medicine and Health, 1850s to 1880s

1. Charles E. Rosenberg, "The Therapeutic Revolution: Medicine, Meaning, and Social Change in Nineteenth-Century America," in *The Therapeutic Revolution: Essays in the Social History of American Medicine*, ed. Morris J. Vogel and Charles E. Rosenberg (Philadelphia: Univ. of Pennsylvania Press, 1979), 3–25; John Harley Warner, *The Therapeutic Perspective: Medical Practice, Knowledge, and Identity in America, 1820–1885* (1986; reprint, Princeton, NJ: Princeton Univ. Press, 1997), esp. chap. 7; K. Codell Carter, *The Decline of Therapeutic Bloodletting and the Collapse of Traditional Medicine* (New Brunswick, NJ: Transaction, 2012).

2. Phyllis Allen Richmond, "American Attitudes Toward the Germ Theory of Disease (1860–1880)," *Journal of the History of Medicine and Allied Sciences*, 9 (1954), 437.

3. James C. White, "Dermatology in America," *Archives of Dermatology*, 4 (1879), 1–28, quotation from 4; A. R. Robinson, "Address," in *Transactions of the International Medical Congress*, ed. John B. Hamilton, 5 vols. (Washington, DC: International Medical Congress, 1887), 4:156.

4. See esp. John Harley Warner, *Against the Spirit of System: The French Impulse in Nineteenth-Century American Medicine* (Princeton, NJ: Princeton Univ. Press, 1998), chap. 9.

5. Wyndham D. Miles, *A History of the National Library of Medicine: The Nation's Treasury of Medical Knowledge*, NIH Publication No. 82-1904 (Bethesda, MD: National Library of Medicine, 1982); John B. Blake, ed., *Centenary of Index Medicus, 1879–1979*, NIH Publication No. 80-2068 (Bethesda, MD: National Library of Medicine, 1980).

6. George B. Wood, *A Treatise on the Practice of Medicine*, 5th ed., 2 vols. (Philadelphia: J. B. Lippincott, 1858), 1:vi.

7. Joseph K. Barnes, comp., *The Medical and Surgical History of the War of the Rebellion (1861–65)*, 15 vols. (1870; reprint, Wilmington, NC: Broadfoot, 1990).

8. Ibid., esp. vol. 9.

9. Alfred Jay Bollet, *Civil War Medicine: Challenges and Triumphs* (Tucson, AZ: Galen, 2002).

10. John S. Haller Jr., *Battlefield Medicine: A History of the Military Ambulance from the Napoleonic Wars through World War I* (Carbondale: Southern Illinois Univ. Press, 2011), esp. 23–39.

11. Bollet, *Civil War Medicine*.

12. Ibid., 454.

13. Frederick Law Olmsted, quoted in Ira M. Rutkow, *Bleeding Blue and Gray: The Untold Story of Civil War Medicine* (New York: Random House, 2005), 76–77.

14. Margaret Humphreys, *Intensely Human: The Health of the Black Soldier in the American Civil War* (Baltimore: Johns Hopkins Univ. Press, 2008), esp. 11; James O. Breeden, *Joseph Jones, M.D.: Scientist of the Old South* (Lexington: Univ. Press of Kentucky, 1975), 155.

15. Frank R. Freemon, *Gangrene and Glory: Medical Care during the American Civil War* (Madison, NJ: Fairleigh Dickinson Univ. Press, 1998), 118–119.

16. Michael A. Flannery, *Civil War Pharmacy: A History of Drugs, Drug Supply and Provision, and Therapeutics for the Union and Confederacy* (New York: Pharmaceutical Products Press, 2004), esp. 20 and chap. 4.

17. Bollet, *Civil War Medicine*, 225–226; Warner, *Therapeutic Perspective*, 98–100.

18. Rutkow, *Bleeding Blue and Gray*, chap. 9.

19. Bollet, *Civil War Medicine*, 153, quotation from 203; Breeden, *Joseph Jones, M.D.*, quotation from 206.

20. Bollet, *Civil War Medicine*.

21. Rutkow, *Bleeding Blue and Gray*, 321–323.

22. Howard D. Kramer, "Effect of the Civil War on the Public Health Movement," *Mississippi Valley Historical Review*, 35 (1948), 456.

23. Ibid., 449–452.

24. Alan I Marcus, *Plague of Strangers: Social Groups and the Origins of City Services in Cincinnati, 1819–1870* (Columbus: Ohio State Univ. Press, 1991), esp. 108.

25. J. S. Billings, quoted in Henry I. Bowditch, *Public Hygiene in America: Being the Centennial Discourse Delivered Before the International Medical Congress, Philadelphia, September 1876* (Boston: Little, Brown, 1877), 147.

26. Charles E. Rosenberg, *The Care of Strangers: The Rise of America's Hospital System* (New York: Basic Books, 1987), esp. 219–220; Mary Denis Maher, *To Bind Up the Wounds: Catholic Sister Nurses in the U.S. Civil War* (Westport, CT: Greenwood, 1989), quotation from 102.

27. Susan M. Reverby, *Ordered to Care: The Dilemma of American Nursing, 1850–1945* (Cambridge: Cambridge Univ. Press, 1987), 44; Freemon, *Gangrene and Glory*, 51–60; Jane E. Schultz, *Women at the Front: Hospital Workers in Civil War America* (Chapel Hill: Univ. of North Carolina Press, 2004); Emily K. Abel, "Family Caregiving in the Nineteenth Century: Emily Hawley Gillespie and Sarah Gillespie, 1858–1888," *Bulletin of the History of Medicine*, 68 (1994), 573–599.

28. Philip A. Kalisch and Beatrice J. Kalisch, *The Advance of American Nursing*, 2nd ed. (Boston: Little, Brown, 1986), 101–102.

29. Reverby, *Ordered to Care*, esp. chaps. 1–3.

30. Arleen Marcia Tuchman, *Science Has No Sex: The Life of Marie Zakrzewska, M.D.* (Chapel Hill: Univ. of North Carolina Press, 2006).

31. Charles Rosenberg, *The Cholera Years: The United States in 1832, 1849, and 1866* (Chicago: Univ. of Chicago Press, 1962), esp. chaps. 10–11.

32. Robley Dunglison, *A Dictionary of Medical Science*, 2nd ed. (Philadelphia: Blanchard & Lea, 1857), 306.

33. John Duffy, *The Sanitarians: A History of American Public Health* (Urbana: Univ. of Illinois Press, 1990), esp. 114–115 and 148–149.

34. John H. Ellis, *Yellow Fever and Public Health in the New South* (Lexington: Univ. Press of Kentucky, 1992), figures from 57.

35. Ibid., 6; Martin V. Melosi, *The Sanitary City: Urban Infrastructure in America from Colonial Times to the Present* (Baltimore: Johns Hopkins Univ. Press, 2000), esp. chaps. 4 and 5.

36. Stuart Galishoff, *Safeguarding the Public Health: Newark, 1895–1918* (Westport, CT: Greenwood, 1975), esp. 8 and 54–56.

37. Rosenberg, *Cholera Years*, 135.

38. Richard J. Hopkins, "Public Health in Atlanta: The Formative Years, 1865–1879," *Georgia Historical Quarterly*, 53 (1969), 300–301.

39. Jim Downs, *Sick from Freedom: African-American Illness and Suffering during the Civil War and Reconstruction* (New York: Oxford Univ. Press, 2012); Lynn Marie Pohl, "African-American Southerners and White Physicians: Medical Care at the Turn of the Twentieth Century," *Bulletin of the History of Medicine*, 86 (2012), 178–205.

40. Richard H. Steckel, "Heights and Health in the United States, 1710–1950," in *Stature, Living Standards, and Economic Development: Essays in Anthropometric History*, ed. John Komlos (Chicago: Univ. of Chicago Press, 1994), 153–170; Dora L. Costa and Richard H. Steckel, "Long-Term Trends in Health, Welfare, and Economic Growth in the United States," in *Health and Welfare during Industrialization*, ed. Steckel and Roderick Floud (Chicago: Univ. of Chicago Press, 1997), 47–89.

41. Costa and Steckel, "Long-Term Trends," 51.

42. Gerald N. Grob, *The Deadly Truth: A History of Disease in America* (Cambridge, MA: Harvard Univ. Press, 2002), esp. chaps. 5 and 6.

43. Samuel Abbott, "Public Hygiene in the United States," in *The United States of America*, ed. Nathaniel Shaler, 3 vols. (New York: D. Appleton, 1894), 2:565.

44. Jacqueline Karnell Corn, "Social Responses to Epidemic Disease in Pittsburgh, 1872–1895," *Western Pennsylvania Historical Magazine*, Jan. 1973, 64.

45. James C. Mohr, *Plague and Fire: Battling Black Death and the 1900 Burning of Honolulu's Chinatown* (New York: Oxford Univ. Press, 2004).

46. John S. Billings, "American Inventions and Discoveries in Medicine, Surgery, and Practical Sanitation," *Annual Report of the Smithsonian Institution*, 1892, 614.

47. Sarah Stage, *Female Complaints: Lydia Pinkham and the Business of Women's Medicine* (New York: W. W. Norton, 1979), esp. 91.

48. Sam B. Warner Jr., "Public Health Reform and the Depression of 1873–1878," *Bulletin of the History of Medicine*, 29 (1955), 503–516; Nelson Manfred Blake, *Water for the Cities: A History of the Urban Water Supply Problem in the United States* (Syracuse, NY: Syracuse Univ. Press, 1956).

49. Quoted in Bowditch, *Public Hygiene in America*, 38.

50. Billings, "American Inventions and Discoveries," 619.

51. Elisha Harris, in *Sanitarian*, 2 (1874), 94.

52. Leo F. Schnore, "Statistical Indicators of Medical Care: An Historical Note," *Journal of Health and Human Behavior*, 3 (1962), 133.

53. Guenter B. Risse, "From Horse and Buggy to Automobile and Telephone: Medical Practice in Wisconsin, 1848–1930," in *Wisconsin Medicine: Historical Perspectives*, ed. Ronald L. Numbers and Judith Walzer Leavitt (Madison: Univ. of Wisconsin Press, 1981), 29.

54. Lester S. King, *American Medicine Comes of Age, 1840–1920: Essays to Commemorate the Founding of "The Journal of the American Medical Association," July 14, 1883* (Chicago: American Medical Association, 1984), chap. 10.

55. Francis Delafield, "Chronic Catarrhal Gastritis, With Opening Remarks by the President," *Transactions of the Association of American Physicians*, 1 (1886), 1.

56. James H. Cassedy, "The Microscope in American Medical Science, 1840–1860," *Isis*, 67 (1976), 76–97; Charles E. Rosenberg, "What It Was Like to be Sick in 1884," *American Heritage*, 35 (1984), 23–31.

57. King, *American Medicine Comes of Age*, 45.

58. L. E. Atkinson, "Forms of Typhoid Fever Simulating Remittent Malarial Fever," *Transactions of the Association of American Physicians*, 2 (1887), 209–220, William H. Draper reported on 209, William Osler reported on 230.

59. Quoted in Robert G. Frank Jr., "American Physiologists in German Laboratories, 1865–1914," in *Physiology in the American Context, 1850–1940*, ed. Gerald L. Geison (Bethesda, MD: American Physiological Society, 1987), 13.

60. Quoted in Joseph C. Aub and Ruth K. Hapgood, *Pioneer in Modern Medicine: David Linn Edsall of Harvard* (Cambridge, MA: Harvard Medical Alumni Association, 1970), 9.

61. W. Bruce Fye, *The Development of American Physiology: Scientific Medicine in the Nineteenth Century* (Baltimore: Johns Hopkins Univ. Press, 1987), esp. chap. 1.

62. Warner, *Therapeutic Perspective*, chap. 8.

63. Michael J. Aminoff, *Brown-Séquard: An Improbable Genius Who Transformed Medicine* (New York: Oxford Univ. Press, 2011).

64. Fye, *Development of American Physiology*, esp. chap. 2 and 94.

65. Ibid., esp. chap. 3, quotations from 103 and 125.

66. Ibid., passim, Cathell quoted on 10.

67. See, e.g., John C. Burnham, *How Superstition Won and Science Lost: Popularizing Science and Health in the United States* (New Brunswick, NJ: Rutgers Univ. Press, 1987), esp. chap. 4.

68. John W. Draper, *Human Physiology, Statical and Dynamical; Or, the Conditions and Course of the Life of Man* (New York: Harper & Bros., 1856), quotation from v.

69. Theodore Hough and William T. Sedgwick, *The Human Mechanism: Its Physiology and Hygiene and the Sanitation of Its Surroundings* (Boston: Ginn, 1906), 4.

70. Charles E. Rosenberg, *No Other Gods: On Science and American Social Thought*, 2nd ed. (Baltimore: Johns Hopkins Univ. Press, 1997), 4–7.

71. See, e.g., J. Dorman Steele, *Fourteen Weeks in Human Physiology* (New York: A. S. Barnes, 1872); and H. Newell Martin, *The Human Body*, 5th ed. (New York: Henry Holt, 1888), 180–200.

72. George F. Barker, "The Correlation of Vital and Physical Forces," in *Half-Hours with Modern Scientists: Lectures and Essays*, by Thomas Henry Huxley et al. (New Haven, CT: Charles C. Chatfield, 1872), 60.

73. James P. Morgan, "The First Reported Case of Electrical Stimulation of the Human Brain," *Journal of the History of Medicine and Allied Sciences*, 37 (1982), 51–64; Gregory L. Holmes, "Roberts Bartholow: In Search of Anatomic Localization," *New York State Journal of Medicine*, 82 (1982), 238–241.

74. Warner, *Therapeutic Perspective*, 85–91.

75. Audrey B. Davis, *Medicine and Its Technology: An Introduction to the History of Medical Instrumentation* (Westport, CT: Greenwood, 1981); Warner, *Therapeutic Perspective*, esp. 85–91, 101–102, 156; Guenter B. Risse and John Harley Warner, "Reconstructing Clinical Activities: Patient Records in Medical History," *Social History of Medicine*, 5 (1992), 191–192.

76. Davis, *Medicine and Its Technology*.

77. Lawrence T. Weaver, "In the Balance: Weighing Babies and the Birth of the Infant Welfare Clinic," *Bulletin of the History of Medicine*, 84 (2010), 30–57.

78. David A. Johnson and Humayun J. Chaudhry, *Medical Licensing and Discipline in America: A History of the Federation of State Medical Boards* (Lanham, MD: Lexington Books, 2013), 23–25.

79. George Rosen, *Fees and Fee Bills: Some Economic Aspects of Medical Practice in Nineteenth Century America* (Baltimore: Johns Hopkins Press, 1946).

80. William Allen Pusey, *A Doctor of the 1870's and 80's* (Springfield, IL: Charles C Thomas, 1932), esp. 71; Steven M. Stowe, *Doctoring the South: Southern Physicians and Everyday Medicine in the Mid-Nineteenth Century* (Chapel Hill: Univ. of North Carolina Press, 2004).

81. See esp. Donald E. Konold, *A History of American Medical Ethics, 1847–1912* (Madison: State Historical Society of Wisconsin, 1962); John S. Haller Jr., *American Medicine in Transition, 1840–1910* (Urbana: Univ. of Illinois Press, 1981), esp. chap. 7; James G. Burrow, *AMA: Voice of American Medicine* (Baltimore: Johns Hopkins Press, 1963); and Eric W. Boyle, *Quack Medicine: A History of Combating Health Fraud in Twentieth-Century America* (Santa Barbara, CA: Praeger, 2013).

82. Warner, *Against the Spirit of System*, 347; Martin Kaufman, *Homeopathy in America: The Rise and Fall of a Medical Heresy* (Baltimore: Johns Hopkins Press, 1971), 90.

83. Burrow, *AMA*, 6; Todd Savitt, *Race and Medicine in Nineteenth- and Early-Twentieth-Century America* (Kent, OH: Kent State Univ. Press, 2007), esp. ix; Charles H. Wright, *The National Medical Association Demands Equal Opportunity: Nothing More, Nothing Less* (Southfield, MI: Charro, 1995).

84. Lee Anderson, " 'Headlights Upon Sanitary Medicine': Public Health and Medical Reform in Late Nineteenth-Century Iowa," *Journal of the History of Medicine and Allied Sciences*, 46 (1991), 178–200.

85. James C. Mohr, *Licensed to Practice: The Supreme Court Defines the American Medical Profession* (Baltimore: Johns Hopkins Univ. Press, 2013); Johnson and Chaudhry, *Medical Licensing and Discipline in America*, chap. 1.

86. Norman Gevitz, *The D.O.'s: Osteopathic Medicine in America* (Baltimore: Johns Hopkins Univ. Press, 1982), chaps. 1 and 2.

87. J. Stuart Moore, *Chiropractic in America: The History of a Medical Alternative* (Baltimore: Johns Hopkins Univ. Press, 1993).

88. Warner, *Therapeutic Perspective*, esp. chaps. 8 and 9.

89. Burnham, *How Superstition Won and Science Lost*, esp. 29–31.

90. S. Weir Mitchell, quoted in Fye, *Development of American Physiology*, 83.

91. Kenneth M. Ludmerer, *Learning to Heal: The Development of American Medical Education* (New York: Basic Books, 1985), 12.

92. Martin Kaufman, *American Medical Education: The Formative Years, 1765–1910* (Westport, CT: Greenwood, 1976), chaps. 7 and 8; Fye, *Development of American Physiology*, esp. chaps. 1 and 2; Charles E. Rosenberg, "Between Two Worlds: American Medicine in 1879," in Blake, *Centenary of Index Medicus, 1879–1979*, 8–9.

93. "Medical Teaching in Philadelphia," *Medical Times*, 1 (1871), 218–219; Fye, *Development of American Physiology*.

94. Rosenberg, "Between Two Worlds," 3–18, esp. 14–15.

95. Ludmerer, *Learning to Heal*, esp. chaps. 1–3 and pp. 4–7, 63–69.

96. Thomas Neville Bonner, *American Doctors and German Universities: A Chapter in International Intellectual Relations, 1870–1814* (Lincoln: Univ. of Nebraska Press, 1963); Ludmerer, *Learning to Heal*, chap. 4.

97. Charles E. Rosenberg, "Social Class and Medical Care in Nineteenth-Century America: The Rise and Fall of the Dispensary," *Journal of the History of Medicine and Allied Sciences*, 29 (1974), 32–54; Milton I. Roemer, *Ambulatory Health Services in America: Past, Present, and Future* (Rockville, MD: Aspen, 1981), esp. 17.

98. Warner, *Against the Spirit of System*, esp. 292–293; George Rosen, *The Specialization of Medicine, with Particular Reference to Ophthalmology* (New York: Froben, 1944); David Mechanic, "The Growth of Medical Technology and Bureaucracy: Implications for Medical Care," *Milbank Memorial Fund Quarterly/Health and Society*, 55 (1977), 69–72.

99. Richard Harrison Shryock, *Medicine in America: Historical Essays* (Baltimore: Johns Hopkins Press, 1966), 168–169; Rosen, *Specialization of Medicine*.

100. S. D. Gross, "Surgery," in *A Century of American Medicine, 1776–1876*, by Edward H. Clarke et al. (1876; reprint, Brinklow, MD: Old Hickory Bookshop, 1962), 117–118.

101. George Weisz, *Divide and Conquer: A Comparative History of Medical Specialization* (Oxford: Oxford Univ. Press, 2006), chap. 4.

102. Bollet, *Civil War Medicine*, 227–229.

103. John Shaw Billings, quoted in Weisz, *Divide and Conquer*, 75.

104. Konold, *History of American Medical Ethics*.

105. Ibid., 35–37.

106. M. H. Henry, *Specialists and Specialties in Medicine, Address Delivered Before the Alumni Association of the Medical Department of the University of Vermont, Burlington, June 27, 1876* (New York: Wm. Wood, 1876), 6–7, reprinted in *The Origins of Specialization in American Medicine: An Anthology of Sources*, ed. Charles E. Rosenberg (New York: Garland, 1989).

107. Warner, *Against the Spirit of System*, 332–341.

108. Austin Flint, *Essays on Conservative Medicine and Kindred Topics* (Philadelphia: Henry C. Lea, 1874), esp. 16.

109. Morrill Wyman, "Medicine and Surgery in 1881," *New England Journal of Medicine*, 305 (1981), 1061.

110. Francis H. Williams, "Report on Progress in Therapeutics," *Boston Medical and Surgical Journal*, 121 (1889), 481–483, 508–511; Warner, *Therapeutic Perspective*, chap. 4; Rosenberg, "Therapeutic Revolution," 8–18.

111. Stowe, *Doctoring the South*, esp. 262–265.

112. Risse, "From Horse and Buggy to Automobile," 30–31.

113. See, e.g., Leon S. Bryan Jr., "Blood-Letting in American Medicine, 1830–1892," *Bulletin of the History of Medicine*, 38 (1964), 516–529.

114. See esp. Carter, *Decline of Therapeutic Bloodletting*; and Warner, *Therapeutic Perspective*.

Part II. Medicine and Health in the Age of Science and Modernization

1. Thomas L. Haskell, ed., *The Authority of Experts: Studies in History and Theory* (Bloomington: Indiana Univ. Press, 1984); Ronald G. Walters, ed., *Scientific Authority and Twentieth-Century America* (Baltimore: Johns Hopkins Univ. Press, 1997).

2. See Roger Cooter, "Medicine and Modernity," in *The Oxford Handbook of the History of Medicine*, ed. Mark Jackson (Oxford: Oxford Univ. Press, 2011), 103; and, e.g., David Knight, *The Making of Modern Science: Science, Technology, Medicine and Modernity: 1789–1914* (Cambridge: Polity, 2009), chap. 12.

3. See, e.g., Stephen J. Kunitz, *The Health of Populations: General Theories and Particular Realities* (New York: Oxford Univ. Press, 2007), chap. 1.

4. See, e.g., Martin Daunton and Bernhard Rieger, eds., *Meanings of Modernity: Britain from the Late-Victorian Era to World War II* (Oxford: Berg, 2001), esp. the introduction, 1–21.

5. Daniel P. Levine, "Creativity and Change: On the Psychodynamics of Modernity," *American Behavioral Scientist*, 43 (1999), 225–244.

6. Charles E. Rosenberg, *Explaining Epidemics and Other Studies in the History of Medicine* (Cambridge: Cambridge Univ. Press, 1992), chap. 1.

7. Roland Marchand, *Advertising the American Dream: Making Way for Modernity, 1920–1940* (Berkeley: Univ. of California Press, 1985).

Chapter 5. The Age of Surgery and Germ Theory, 1880s to 1910s

1. Lester S. King, *American Medicine Comes of Age: Essays to Commemorate the Founding of "The Journal of the American Medical Association," July 14, 1883* (Chicago: American Medical Association, 1984), esp. chap. 10.

2. Edward Meeker, "The Improving Health of the United States, 1850–1915," *Explorations in Economic History*, 9 (1972), 353; Gretchen A. Condran and Rose A. Cheney, "Mortality Trends in Philadelphia: Age- and Cause-Specific Death Rates, 1870–1930," *Demography*, 19 (1982), 97–123; David Cutler and Grant Miller, "The Role of Public Health Improvements in Health Advances: The Twentieth-Century United States," ibid., 42 (2005), 1–22.

3. John Komlos, "Shrinking in a Growing Economy? The Mystery of Physical Stature during the Industrial Era," *Journal of Economic History*, 58 (1998), 779–802; Allen M. Brandt, "'Just Say No': Risk, Behavior, and Disease in Twentieth-Century America," in *Scientific*

Authority and Twentieth-Century America, ed. Ronald G. Walters (Baltimore: Johns Hopkins Univ. Press, 1997), 84–85.

4. Gerald N. Grob, *The Deadly Truth: A History of Disease in America* (Cambridge, MA: Harvard Univ. Press, 2002), chap. 8; Margaret Humphreys, *Yellow Fever and the South* (New Brunswick, NJ: Rutgers Univ. Press, 1992), esp. 4.

5. Mortimer Spiegelman and Carl L. Erhardt, "Mortality in the United States by Cause," in *Mortality and Morbidity in the United States*, ed. Carl L. Erhardt and Joyce E. Berlin (Cambridge, MA: Harvard Univ. Press, 1974), 24–26; Andrew Noymer and Beth Jarosz, "Causes of Death in Nineteenth-Century New England: The Dominance of Infectious Disease," *Social History of Medicine*, 21 (2008), 577.

6. Robert S. Lynd and Helen Merrell Lynd, *Middletown: A Study in American Culture* (New York: Harcourt, Brace, 1929), 447.

7. Grob, *Deadly Truth*, chap. 8; Lloyd G. Stevenson, "Exemplary Disease: The Typhoid Pattern," *Journal of the History of Medicine and Allied Sciences*, 37 (1982), 159–181; John Duffy, *The Sanitarians: A History of American Public Health* (Urbana: Univ. of Illinois Press, 1990), chap. 12; Nelson Manfred Blake, *Water for the Cities: A History of the Urban Water Supply Problem in the United States* (Syracuse, NY: Syracuse Univ. Press, 1956), 262–263; C.-E. A. Winslow, *The Life of Hermann M. Biggs, M.D., D.Sc., LL.D., Physician and Statesman of the Public Health* (Philadelphia: Lea & Febiger, 1929), chap. 7; Martin V. Melosi, *The Sanitary City: Urban Infrastructure in America from Colonial Times to the Present* (Baltimore: Johns Hopkins Univ. Press, 2000), esp. 151.

8. S. Josephine Baker, quoted in Samuel H. Preston and Michael R. Haines, *Fatal Years: Child Mortality in Late Nineteenth-Century America* (Princeton, NJ: Princeton Univ. Press, 1991), 31.

9. Jacob A. Riis, *How the Other Half Lives: Studies among the Tenements of New York* ([c. 1890]; reprint, New York: Hill & Wang, 1957).

10. Preston and Haines, *Fatal Years*, esp. chaps. 1 and 3, figures from 22.

11. Natalia Molina, *Fit to Be Citizens? Public Health and Race in Los Angeles, 1879–1939* (Berkeley: Univ. of California Press, 2006).

12. See, e.g., Sheila M. Rothman, *Living in the Shadow of Death: Tuberculosis and the Social Experience of Illness in American History* (New York: Basic Books, 1994); and Katherine Ott, *Fevered Lives: Tuberculosis in American Culture since 1870* (Cambridge, MA: Harvard Univ. Press, 1996).

13. Gretchen A. Condran, "Changing Patterns of Epidemic Disease in New York City," in *Hives of Sickness: Public Health and Epidemics in New York City*, ed. David Rosner (New Brunswick, NJ: Rutgers Univ. Press, 1995), 36–37.

14. Naomi Rogers, *Dirt and Disease: Polio before FDR* (New Brunswick, NJ: Rutgers Univ. Press, 1992); Guenter B. Risse, "Epidemics and History: Ecological Perspectives and Social Responses," in *AIDS: The Burdens of History*, ed. Elizabeth Fee and Daniel M. Fox (Berkeley: Univ. of California Press, 1988), 48–56.

15. Adam Biggs, personal communication based on unpublished research.

16. See, e.g., Frederick Peterson, "Recent Progress in Surgery and Medicine," *Journal of Social Science*, 31 (1894), l–lix.

17. Judy Barrett Litoff, *American Midwives, 1860 to the Present* (Westport, CT: Greenwood, 1978).

18. Morris J. Fogelman and Elinor Reinmiller, "1880–1890: A Creative Decade in World Surgery," *American Journal of Surgery*, 115 (1968), 812–824.

19. See, e.g., Dale C. Smith, "Appendicitis, Appendectomy, and the Surgeon," *Bulletin of the History of Medicine*, 70 (1996), 414–441.

20. Dale C. Smith, "A Historical Overview of the Recognition of Appendicitis," *New York State Journal of Medicine*, 86 (1986), 571–583, 639–647.

21. Thomas P. Gariepy, "The Introduction and Acceptance of Listerian Antisepsis in the United States," *Journal of the History of Medicine and Allied Sciences*, 49 (1994), 167–206, esp. 175.

22. William Goodell, "Four Cases of Ovarian Tumour and One of Fibro-Cystic Tumour of the Womb, Operated on under the Spray," *American Journal of the Medical Sciences*, 78 (1879), 377–386, quotation from 378.

23. Gert H. Brieger, "American Surgery and the Germ Theory of Disease," *Bulletin of the History of Medicine*, 40 (1966), 135–145; Gariepy, "Introduction and Acceptance of Listerian Antisepsis"; "Antiseptic Surgery in America," *Lancet*, 1 (1882), 232.

24. B. A. Watson, quoted in "Surgery," by Francis D. Moore, in *Advances in American Medicine: Essays at the Bicentennial*, ed. John Z. Bowers and Elizabeth F. Purcell, 2 vols. (New York: Josiah Macy, Jr., Foundation, 1976), 2:629.

25. Stephen Smith, "The Comparative Results of Operations in Bellevue Hospital," *Medical Record*, 28 (1885), 427.

26. A. Scott Earle, in *Surgery in America: From the Colonial Era to the Twentieth Century, Selected Writings*, ed. Earle (Philadelphia: W. B. Saunders, 1965), 207; Gariepy, "Introduction and Acceptance of Listerian Antisepsis," esp. 202–204; Justine Randers-Pherson, *The Surgeon's Glove* (Springfield, IL: Charles C Thomas, 1960), chap. 4; Thomas Schlich, "Negotiating Technologies in Surgery: The Controversy about Surgical Gloves in the 1890s," *Bulletin of the History of Medicine*, 87 (2013), 170–197.

27. J. H. McClelland, "Antisepsis," *Transactions, American Institute of Homeopathy*, 1882, 527–528.

28. Thomas G. Benedek, "A Century of American Rheumatology," in *Grand Rounds: One Hundred Years of Internal Medicine*, ed. Russell C. Maulitz and Diana E. Long (Philadelphia: Univ. of Pennsylvania Press, 1988), 165.

29. E. A. Codman, *The Shoulder: Rupture of the Superspinatus Tendon and Other Lesions in and about the Subacromial Bursa* (Boston: T. Todd, 1934), viii–x.

30. Harvey Cushing, quoted in *Harvey Cushing: A Life in Surgery*, by Michael Bliss (Oxford: Oxford Univ. Press, 2005), 8.

31. Robert A. Aronowitz, *Unnatural History: Breast Cancer and American Society* (Cambridge: Cambridge Univ. Press, 2007), chap. 4.

32. John Hunt Shephard, "Surgical Practice: Looking Forward," *California and Western Medicine*, 46 (1937), 376.

33. Edward Preble, "Marvels of Modern Surgery," *World's Work*, 26 (1913), 588–595, quotation from 595.

34. W. Bruce Fye, "The Origins and Evolution of the Mayo Clinic from 1864 to 1939: A Minnesota Family Practice Becomes an International 'Medical Mecca,'" *Bulletin of the History of Medicine*, 84 (2010), 323–357; Helen Clapesattle, *The Doctors Mayo* (1941; reprint, Garden City, NY: Garden City Publishing, 1943), quotation from 411.

35. Owen Wangensteen, quoted in Allen B. Weisse, *Conversations in Medicine: The Story of Twentieth-Century American Medicine in the Words of Those Who Created It* (New York: New York Univ. Press, 1984), 42; Clapesattle, *Doctors Mayo*, esp. 522–524.

36. Clapesattle, *Doctors Mayo*.

37. Joel D. Howell, *Technology in the Hospital: Transforming Patient Care in the Early Twentieth Century* (Baltimore: Johns Hopkins Univ. Press, 1995), 57–68, quotation from 58.

38. E. H. L. Corwin, *The American Hospital* (New York: Commonwealth Fund, 1946), 8.

39. Clara L. Came, "Hospital or Home?," *Good Housekeeping*, Dec. 1903, 555–559; Margaret Sutton Briscoe, "The Hospitalized Child," ibid., 51 (1910), 542–548.

40. "The Hospital Hotel," *JAMA*, 26 (1896), 94.

41. See esp. Charles E. Rosenberg, *The Care of Strangers: The Rise of America's Hospital System* (New York: Basic Books, 1987), chap. 10.

42. Morris J. Vogel, "The Transformation of the American Hospital, 1850–1920," in *Health Care in America: Essays in Social History*, ed. Susan Reverby and David Rosner (Philadelphia: Temple Univ. Press, 1979), 105–116; Emily K. Abel, *Hearts of Wisdom: American Women Caring for Kin, 1850–1940* (Cambridge, MA: Harvard Univ. Press, 2000).

43. Hanna Bloomfield Rubins, "The Case History in Historical Perspective: Time for an Overhaul?," *Journal of General Internal Medicine*, 9 (1994), 220; Charles E. Rosenberg, "Inward Vision and Outward Glance: The Shaping of the American Hospital, 1880–1914," *Bulletin of the History of Medicine*, 53 (1979), 346–391, quotation from 379.

44. Rosenberg, *Care of Strangers*, esp. chaps. 11 and 13.

45. Morris J. Vogel, ed., *On the Administrative Frontier of Medicine: The First Ten Years of the American Hospital Association, 1899–1908* (New York: Garland, 1989).

46. See, e.g., William G. Rothstein, *American Medical Schools and the Practice of Medicine: A History* (New York: Oxford Univ. Press, 1987), 134–138.

47. Susan M. Reverby, *Ordered to Care: The Dilemma of American Nursing, 1850–1945* (Cambridge: Cambridge Univ. Press, 1987), 62.

48. Philip A. Kalisch and Beatrice J. Kalisch, *The Advance of American Nursing*, 2nd ed. (Boston: Little, Brown, 1986), esp. chaps. 5–8.

49. Rosenberg, *Care of Strangers*, 295.

50. Alan M. Kraut and Deborah A. Kraut, *Covenant of Care: Newark Beth Israel and the Jewish Hospital in America* (New Brunswick, NJ: Rutgers Univ. Press, 2007), esp. 3–5; Rosenberg, *Care of Strangers*; Barbra Mann Wall, *Unlikely Entrepreneurs: Catholic Sisters and the Hospital Marketplace, 1865–1925* (Columbus: Ohio State Univ. Press, 2005).

51. Virginia G. Drachman, *Hospital with a Heart: Women Doctors and the Paradox of Separatism at the New England Hospital, 1862–1969* (Ithaca, NY: Cornell Univ. Press, 1984), chap. 5, esp. 130.

52. Regina Markell Morantz-Sanchez, *Sympathy and Science: Women Physicians in American Medicine* (New York: Oxford Univ. Press, 1985), esp. xi, 232; Anne Taylor Kirschmann, "Adding Women to the Ranks, 1860–1890: A New View with a Homeopathic Lens," *Bulletin of the History of Medicine*, 73 (1999), 429–446.

53. George Crile, *An Autobiography*, ed. Grace Crile, 2 vols. (Philadelphia: J. B. Lippincott, 1947), 1:113.

54. William Allen Pusey, *A Doctor of the 1870's and 80's* (Springfield, IL: Charles C Thomas, 1932), 78–79.

55. Phyllis Allen Richmond, "American Attitudes Toward the Germ Theory of Disease (1860–1880)," *Journal of the History of Medicine and Allied Sciences*, 9 (1954), 447.

56. "The Germ Theory of Disease in Its Relation to Therapeutics," *Boston Medical and Surgical Journal*, 112 (1885), 234.

57. Richmond, "American Attitudes Toward the Germ Theory of Disease," 428–554, esp. 444; Nancy Tomes, "The Private Side of Public Health: Sanitary Science, Domestic Hygiene, and the Germ Theory, 1870–1900," *Bulletin of the History of Medicine*, 64 (1990), 509–539.

58. William Belfield, quoted in *Transformations in American Medicine from Benjamin Rush to William Osler*, by Lester S. King (Baltimore: Johns Hopkins Univ. Press, 1991), 176.

59. See esp. King, *American Medicine Comes of Age*, chap. 8; Nancy Tomes, "American Attitudes toward the Germ Theory of Disease: Phyllis Allen Richmond Revisited," *Journal of the History of Medicine and Allied Sciences*, 52 (1997), 17–50; and Stephen J. Kunitz, *The Health of Populations: General Theories and Particular Realities* (New York: Oxford Univ. Press, 2007).

60. Nancy Tomes, *The Gospel of Germs: Men, Women, and the Microbe in American Life* (Cambridge, MA: Harvard Univ. Press, 1998), chap. 1.

61. U. P. Stair, "Is Consumption an Infectious Disease?," *JAMA*, 1 (1883), 425.

62. Richmond, "American Attitudes Toward the Germ Theory," 451–453; Austin Flint, *A Treatise on the Principles and Practice of Medicine; Designed for the Use of Practitioners and Students of Medicine*, 5th ed. (Philadelphia: Henry C. Lea's Son, 1884).

63. Nancy Tomes, "Moralizing the Microbe: The Germ Theory and the Moral Construction of Behavior in the Late-Nineteenth-Century Antituberculosis Movement," in *Morality and Health*, ed. Allan M. Brandt and Paul Rozin (New York: Routledge, 1997), 276–277.

64. Lewis H. Taylor, "The Epidemic of Typhoid Fever at Plymouth, Pennsylvania," *Medical News*, 46 (1885), 541–543; Taylor, "An Epidemic of Typhoid Fever at Plymouth, Pa., (Second Paper)," ibid., 681–686.

65. Tomes, *Gospel of Germs*, chap. 1.

66. G. S. Franklin, "Sanitary Care of Privies," *Sanitarian*, 14 (1885), 442.

67. Daniel Eli Burnstein, *Next to Godliness: Confronting Dirt and Despair in Progressive Era New York City* (Urbana: Univ. of Illinois Press, 2006).

68. See, e.g., Albert L. Gihon, "Health, the True Nobility," *Public Health Papers and Reports*, 7 (1881), 342–352.

69. David Glassberg, "The Design of Reform: The Public Bath Movement in America," *American Studies*, 20 (1979), 5–21, quotation from 9; Howard D. Kramer, "The Germ Theory and the Early Public Health Program in the United States," *Bulletin of the History of Medicine*, 22 (1948), 233–247, esp. 239–240.

70. *Sanitary Era*, 1 (1886), unpaginated.

71. William Henry Welch, *Papers and Addresses*, 3 vols. (Baltimore: Johns Hopkins Press, 1920), 1:568.

72. Joseph Jones, quoted in Margaret Humphreys, "A Stranger to Our Camps: Typhus in American History," *Bulletin of the History of Medicine*, 80 (2006), 280.

73. John S. Haller Jr., *American Medicine in Transition, 1840–1910* (Urbana: Univ. of Illinois Press, 1981), esp. 154–165.

74. King, *Transformations in American Medicine*, 179.

75. See, e.g., Frank S. Billings, "Fourteen Days with Pasteur," *Medical News*, 48 (1886), 90–96; "The Fear of Hydrophobia," ibid., 467; George M. Sternberg, "Pasteur's Method for the Prevention of Hydrophobia," ibid., 449–453; Bert Hansen, "America's First Medical Breakthrough: How Popular Excitement about a French Rabies Cure in 1885 Raised New Expectations for Medical Progress," *American Historical Review*, 103 (1998), 373–418; and Hansen, "New Images of a New Medicine: Visual Evidence for the Widespread Popularity of Therapeutic Discoveries in America after 1885," *Bulletin of the History of Medicine*, 73 (1999), 629–678.

76. King, *American Medicine Comes of Age*, 35–36; Peter C. English, "Diphtheria and Theories of Infectious Disease: Centennial Appreciation of the Critical Role of Diphtheria in the History of Medicine," *Pediatrics*, 76 (1985), 1–9. See also, e.g., Charles T. McClintock, "Immunity," *Columbus Medical Journal*, 21 (1898), 441–448; and Arthur E. Hertzler, *The Horse and Buggy Doctor* (1938; reprint, Lincoln: Univ. of Nebraska Press, 1970), 1–2.

77. Isaac A. Abt, *Baby Doctor* (New York: Whittlesey House, 1944), 65–66.

78. See, e.g., Joseph C. Aub and Ruth K. Hapgood, *Pioneer in Modern Medicine: David Linn Edsall of Harvard* (Cambridge, MA: Harvard Medical Alumni Association, 1970), 19–20; Evelynn Maxine Hammonds, *Childhood's Deadly Scourge: The Campaign to Control Diphtheria in New York City, 1880–1930* (Baltimore: Johns Hopkins Univ. Press, 1999); and W. P. Northrup, "Antitoxin Treatment of Diphtheria a Pronounced Success," *Forum*, 22 (1896), 53–65, quotation from 61.

79. David Linn Edsall, "The Transformation in Medicine," *Southern Medical Journal*, 24 (1931), 1106; Edward H. Kass, "History of the Subspecialty of Infectious Diseases in the United States," in Maulitz and Long, *Grand Rounds*, 97.

80. Nancy Tomes, "Spreading the Germ Theory: Sanitary Science and Home Economics, 1880–1930," in *Rethinking Home Economics: Women and the History of a Profession*, ed. Sarah Stage and Virginia B. Vincenti (Ithaca, NY: Cornell Univ. Press, 1997), 34–54; Tomes, *Gospel of Germs*; JoAnne Brown, "Crime, Commerce, and Contagionism: The Political Languages of Public Health and the Popularization of Germ Theory in the United States, 1870–1950," in Walters, *Scientific Authority and Twentieth-Century America*, 53–81.

81. Samuel G. Dixon, "The Family Physician," *Pennsylvania Health Bulletin*, 3 (Sept. 1909), 9.

82. Vincent J. Cirillo, *Bullets and Bacilli: The Spanish-American War and Military Medicine* (New Brunswick, NJ: Rutgers Univ. Press, 2004).

83. Judith Walzer Leavitt, *The Healthiest City: Milwaukee and the Politics of Health Reform* (Princeton, NJ: Princeton Univ. Press, 1982), chap. 3, quotation from 104.

84. Suellen Hoy, *Chasing Dirt: The American Pursuit of Cleanliness* (New York: Oxford Univ. Press, 1995); Tomes, *Gospel of Germs*, quotation from 108; Tomes, "Private Side of Public Health," esp. 533–537.

85. Sally Smith Hughes, *The Virus: A History of the Concept* (New York: Science History Publications, 1977), 76.

86. Quoted in Tomes, *Gospel of Germs*, 44.

87. See, e.g., Carolyn Thomas de la Peña, *The Body Electric: How Strange Machines Built the Modern American* (New York: New York Univ. Press, 2003).

88. Alan M. Kraut, *Silent Travelers: Germs, Genes, and the "Immigrant Menace"* (New York: Basic Books, 1994).

89. Barbara Gutmann Rosenkrantz, "Cart before Horse: Theory, Practice and Professional Image in American Public Health, 1870–1920," *Journal of the History of Medicine and Allied Sciences*, 29 (1974), 55–62; Elizabeth Fee and Dorothy Porter, "Public Health, Preventive Medicine and Professionalization: England and America in the Nineteenth Century," in *Medicine in Society: Historical Essays*, ed. Andrew Wear (Cambridge: Cambridge Univ. Press, 1992), 254–260.

90. Richard L. Golden, *A History of William Osler's "The Principles and Practice of Medicine"* (Montreal: Osler Library, McGill Univ., and American Osler Society, 2004); William Osler, *The Principles and Practice of Medicine, Designed for the Use of Practitioners and Students of Medicine* (New York: D. Appleton, 1892).

91. Osler, *Principles and Practice*; Golden, *History*, esp. 50–51.

92. Christopher Crenner, *Private Practice: In the Early Twentieth-Century Medical Office of Dr. Richard Cabot* (Baltimore: Johns Hopkins Univ. Press, 2005); Rubins, "Case History in Historical Perspective," 219–221.

93. Richard C. Cabot, *Case Teaching in Medicine: A Series of Graduated Exercises in the Differential Diagnosis, Prognosis and Treatment of Actual Cases of Disease* (Boston: D. C. Heath, 1906), esp. 64; Cabot, *Case Histories in Medicine, Illustrating the Diagnosis, Prognosis and Treatment of Disease*, 2nd ed. (Boston: W. M. Leonard, 1911), esp. 246–250.

94. Cabot, *Case Histories in Medicine*, 278–280.

95. Woods Hutchinson, "The Conquest of the Great Diseases," *World's Work*, 21 (1911), 13881–13883.

96. Guenter B. Risse, "From Horse and Buggy to Automobile and Telephone: Medical Practice in Wisconsin, 1848–1930," in *Wisconsin Medicine: Historical Perspectives*, ed. Ronald L. Numbers and Judith Walzer Leavitt (Madison: Univ. of Wisconsin Press, 1981), 32–34, quotation from 33.

97. Ibid., 36, 38–40.

98. Jonathan M. Liebenau, "Scientific Ambitions: The Pharmaceutical Industry, 1900–1920," *Pharmacy in History*, 27 (1985), 3–11.

99. Ramunas A. Kondratas, "Biologics Control Act of 1902," in *The Early Years of Federal Food and Drug Control*, ed. James Harvey Young (Madison, WI: American Institute of the History of Pharmacy, 1982), 8–27; Glenn Sonnedecker, "Drug Standards Become Official," in ibid., 28–39, quotation from 31; Jonathan Liebenau, *Medical Science and Medical Industry: The Formation of the American Pharmaceutical Industry* (Houndmills, UK: Macmillan, 1987), esp. chap. 6.

100. "Unemployed Medical Men," *Boston Medical and Surgical Journal*, 130 (1894), 451–452, quoting *Medical Record*.

101. John H. Rauch, quoted in ibid.

102. David G. Schuster, *Neurasthenic Nation: America's Search for Health, Happiness, and Comfort, 1869–1920* (New Brunswick, NJ: Rutgers Univ. Press, 2011); Rennie B. Schoeptlin, *Christian Science on Trial: Religious Healing in America* (Baltimore: Johns Hopkins Univ. Press, 2003); Robert C. Fuller, *Alternative Medicine and American Religious Life* (New York: Oxford Univ. Press, 1989); Heather D. Curtis, *Faith in the Great Physician: Suffering and Divine Healing in American Culture, 1860–1900* (Baltimore: Johns Hopkins Univ. Press, 2007).

103. "Medical Fees and Medical Incomes," *Columbus Medical Journal*, 11 (1892), 44–46, quoting the *Lancet*.

104. James M. Ingram, "Dr. Louis Sims Oppenheimer: Culture Among the Sandspurs," *Journal of the Florida Medical Association*, 58 (1971), 55–56.

105. Richard Harrison Shryock, *Medical Licensing in America, 1650–1965* (Baltimore: Johns Hopkins Press, 1967), chap. 2, quotation from 60.

106. Paul Starr, *The Transformation of American Medicine* (New York: Basic Books, 1982), 102–112; Charles E. Rosenberg, "Doctors and Credentials—The Roots of Uncertainty," in *The Origins of Specialization in American Medicine: An Anthology of Sources*, ed. Rosenberg (New York: Garland, 1989), 193–197.

107. Samuel L. Baker, "Physician Licensure Laws in the United States, 1865–1915," *Journal of the History of Medicine and Allied Sciences*, 39 (1984), 173–197.

108. Lynd and Lynd, *Middletown*, 442; Shryock, *Medical Licensing in America*.

109. Rosemary Stevens, "Technology and Institutions in the Twentieth Century," *Caduceus*, 12 (1996), 13.

110. Rothstein, *American Medical Schools*, chap. 5; Kenneth M. Ludmerer, *Learning to Heal: The Development of American Medical Education* (New York: Basic Books, 1985); Ludmerer, *Time to Heal: American Medical Education from the Turn of the Century to the Era of Managed Care* (New York: Oxford Univ. Press, 1999), chap. 1; Lynn E. Miller and Richard M. Weiss, "Medical Education Reform Efforts and Failures of U.S. Medical Schools, 1870–1930," *Journal of the History of Medicine and Allied Sciences*, 63 (2008), 348–387.

111. Quoted in Audrey W. Davis, *Dr. Kelly of Hopkins: Surgeon, Scientist, Christian* (Baltimore: Johns Hopkins Press, 1959), 58–59.

112. F. F. Bishop, "The Scientific Spirit in Practical Medicine," *Medical Record*, 51 (1897), 349.

113. Roger Cooter, "Medicine and Modernity," in *The Oxford Handbook of the History of Medicine*, ed. Mark Jackson (Oxford: Oxford Univ. Press, 2011), 104.

114. Hamilton Cravens, "A Modern Dilemma: Changing Notions of Truth and Expertise in 20th-Century American Medical Science," *Prospects*, 28 (2003), 245–277; Barbara Bridgman Perkins, "Shaping Institution-Based Specialism: Early Twentieth-Century Economic Organization of Medicine," *Social History of Medicine*, 10 (1997), 419–435; Rosemary Stevens, *American Medicine and the Public Interest*, 2nd ed. (Berkeley: Univ. of California Press, 1998), chap. 2; George Weisz, *Divide and Conquer: A Comparative History of Medical Specialization* (Oxford: Oxford Univ. Press, 2006), 70–83, quotation from 82.

115. Steven J. Peitzman, "'Thoroughly Practical': America's Polyclinic Medical Schools," *Bulletin of the History of Medicine*, 54 (1980), 166–187.

116. Rosenberg, *Care of Strangers*, 297.

117. Philip J. Pauly, "The Appearance of Academic Biology in Late Nineteenth-Century America," *Journal of the History of Biology*, 17 (1984), 369–397.

118. Robert E. Kohler, *From Medical Chemistry to Biochemistry: The Making of a Biomedical Discipline* (Cambridge: Cambridge Univ. Press, 1982), 109; Gerald L. Geison, "Divided We Stand: Physiologists and Clinicians in the American Context," in *The Therapeutic Revolution: Essays in the Social History of American Medicine*, ed. Morris J. Vogel and Charles E. Rosenberg (Philadelphia: Univ. of Pennsylvania Press, 1979), 67–90.

119. Russell C. Maulitz, "'Physician versus Bacteriologist': The Ideology of Science in Clinical Medicine," in Vogel and Rosenberg, *Therapeutic Revolution*, 91–107.

120. George W. Corner, *A History of the Rockefeller Institute, 1901–1953: Origins and Growth* (New York: Rockefeller Institute Press, 1964), chap. 1, esp. 23; Howard S. Berliner, *A System of Scientific Medicine: Philanthropic Foundations in the Flexner Era* (New York: Tavistock, 1985), chap. 6.

121. Corner, *History of the Rockefeller Institute*, chap. 2; *New York Evening Post*, quoted in ibid., 38.

122. *The Memorial Institute for Infectious Diseases: Brief History and Description* (Chicago: [Memorial Institute for Infectious Diseases], 1915), unpaginated.

123. Thomas N. Bonner, "German Doctors in America—1887–1914: Their Views and Impressions of American Life and Medicine," *Journal of the History of Medicine and Allied Sciences*, 14 (1959), 1–17.

Chapter 6. Physiological Medicine, 1910s to 1930s

1. Lewis Thomas, "Biostatistics in Medicine," *Science*, 198 (1977), 675.

2. Stow Persons, "The Decline of Homeopathy—The University of Iowa, 1876–1919," *Bulletin of the History of Medicine*, 65 (1991), 86.

3. B. B. Bagby, quoted in *The Structure of American Medical Practice, 1875–1941*, by George Rosen, ed. Charles E. Rosenberg (Philadelphia: Univ. of Pennsylvania Press, 1983), 56.

4. See, e.g., Joseph C. Aub and Ruth K. Hapgood, *Pioneer in Modern Medicine: David Linn Edsall of Harvard* (Cambridge, MA: Harvard Medical Alumni Association, 1970), esp. 167–169; and Russell C. Maulitz and Diana E. Long, eds., *Grand Rounds: One Hundred Years of Internal Medicine* (Philadelphia: Univ. of Pennsylvania Press, 1988).

5. William G. Rothstein, "When Did a Random Patient Benefit from a Random Physician?," *Caduceus*, Winter 1996, 3–8.

6. Fred C. Kelly, "Is Better Health Due to the Doctors?," *Current History*, 18 (1923), 48–54; Ray Lyman Wilbur, "The Doctor's Service to Humanity," ibid., 223–226.

7. Joel D. Howell, "A History of Caring in Medicine," in *The Lost Art of Caring: A Challenge to Health Professionals, Families, Communities, and Society*, ed. Leighton E. Cluff and Robert H. Binstock (Baltimore: Johns Hopkins Univ. Press, 2001), 90.

8. See, e.g., F. W. Hachtel and H. W. Stoner, "Inoculation Against Typhoid in Public Institutions and in Civil Communities: A Further Report," *JAMA*, 59 (1912), 1364–1378; Dieter Groschel, "Typhoid Vaccination—Yesterday and Today," *Pennsylvania Medicine*, July 1967, 77–81; and Terra Ziporyn, *Disease in the Popular American Press: The Case of Diphtheria, Typhoid Fever, and Syphilis, 1870–1920* (Westport, CT: Greenwood, 1988), 87–95.

9. Garland E. Allen, *Life Science in the Twentieth Century* (New York: John Wiley & Sons, 1975); John C. Burnham, "Change in the Popularization of Health in the United States," *Bulletin of the History of Medicine*, 58 (1984), 186.

10. See esp. J. Andrew Mendelsohn, "From Eradication to Equilibrium: How Epidemics Became Complex after World War I," in *Greater Than the Parts: Holism in Biomedicine, 1920–1950*, ed. Christopher Lawrence and George Weisz (New York: Oxford Univ. Press, 1998), 303–331.

11. Halbert L. Dunn, "Application of Statistical Methods in Physiology," *Physiological Reviews*, 9 (1929), 275, 288–293.

12. Walter B. Cannon, *The Wisdom of the Body*, 2nd ed. (1939; reprint, New York: W. W. Norton, 1963), esp. xiii–xiv and 303; Elin L. Wolfe, A. Clifford Barger, and Saul Benison, *Walter B. Cannon, Science and Society* (Cambridge, MA: Harvard Univ. Press, 2000), chap. 8.

13. Alfred E. Cohn, "Physiology and Medicine," *Science*, 68 (1928), 512.

14. Thomas Franklin Williams, "Cabot, Peabody, and the Care of the Patient," *Bulletin of the History of Medicine*, 24 (1950), 465.

15. Daniel M. Fox, "Health Policy and Changing Epidemiology in the United States: Chronic Disease in the Twentieth Century," *Transactions and Studies of the College of Physicians of Philadelphia*, 5th ser., 10 (1988), 11–31.

16. Alfred E. Cohn, "Purposes in Medical Research," *Journal of Clinical Investigation*, 1 (1924), 10.

17. John K. Crellin, "Internal Antisepsis or the Dawn of Chemotherapy?," *Journal of the History of Medicine and Allied Sciences*, 36 (1981), 9–18.

18. Aub and Hapgood, *Pioneer in Modern Medicine*, 39–41.

19. John H. Stokes, "Changing Causal Concepts and Investigative Methods," *Journal of Investigative Dermatology*, 3 (1940), 257–269.

20. Henry H. Donaldson, "Research Foundations in Their Relation to Medicine," *Science*, 36 (1912), 65–74, quotation from 73.

21. Esmond R. Long, *A History of American Pathology* (Springfield, IL: Charles C Thomas, 1962); W. D. Foster, *A Short History of Clinical Pathology* (Edinburgh: E. & S. Livingstone, 1961), 50.

22. Russell C. Maulitz, "Pathologists, Clinicians, and the Role of Pathophysiology," in *Physiology in the American Context, 1850–1940*, ed. Gerald L. Geison (Bethesda, MD: American Physiological Society, 1987), 209–235.

23. Robert E. Kohler Jr., "The Enzyme Theory and the Origins of Biochemistry," *Isis*, 64 (1973), 181–196; Kohler, *From Medical Chemistry to Biochemistry: The Making of a Biomedical Discipline* (Cambridge: Cambridge Univ. Press, 1982), esp. chap. 7; Philip J. Pauly, "General Physiology and the Discipline of Physiology, 1890–1935," in Geison, *Physiology in the American Context*, 195–207.

24. James Bordley III and A. McGehee Harvey, *Two Centuries of American Medicine, 1776–1976* (Philadelphia: W. B. Saunders, 1976), 256–259; W. Bruce Fye, *The Development of American Physiology: Scientific Medicine in the Nineteenth Century* (Baltimore: Johns Hopkins Univ. Press, 1987), 227–230; Hannah Landecker, *Culturing Life: How Cells Became Technologies* (Cambridge, MA: Harvard Univ. Press, 2007).

25. Gerald L. Geison, "International Relations and Domestic Elites in American Physiology, 1900–1940," in Geison, *Physiology in the American Context*, 115–154.

26. Jane Maienschein, "Physiology, Biology, and the Advent of Physiological Morphology," in ibid., 177–193.

27. Benjamin Harlow, *Glands in Health and Disease* (New York: E. P. Dutton, 1922), 9.

28. Elmer Verner McCollum, *A History of Nutrition: The Sequence of Ideas in Nutrition Investigations* (Boston: Houghton Mifflin, 1957); Percy G. Stiles, "The Vitamines," *Science Conspectus*, 4 (1914), 10–13.

29. Edward B. Vedder, *Beriberi* (New York: William Wood, 1913), iii.

30. Stiles, "Vitamines."

31. See, e.g., Leonard G. Wilson, "The Clinical Definition of Scurvy and the Discovery of Vitamin C," *Journal of the History of Medicine and Allied Sciences*, 30 (1975), 40–60; and Rima D. Apple, *Vitamania: Vitamins in American Culture* (New Brunswick, NJ: Rutgers Univ. Press, 1996), esp. 4.

32. Maxwell M. Wintrobe, *Hematology: The Blossoming of a Science—A Story of Inspiration and Effort* (Philadelphia: Lea & Febiger, 1985), esp. chap. 10.

33. Alan M. Kraut, *Goldberger's War: The Life and Work of a Public Health Crusader* (New York: Hill & Wang, 2003), esp. 99; Harry M. Marks, "Epidemiologists Explain Pellagra: Gender, Race, and Political Economy in the Work of Edgar Sydenstricker," *Journal of the History of Medicine and Allied Sciences*, 58 (2003), 34–55.

34. Kraut, *Goldberger's War*; Elizabeth W. Etheridge, *The Butterfly Caste: A Social History of Pellagra in the South* (Westport, CT: Greenwood, 1972).

35. See previous note.

36. Wintrobe, *Hematology*, 188.

37. McCollum, *History of Nutrition*; Bordley and Harvey, *Two Centuries of American Medicine*, 241–256.

38. Apple, *Vitamania*, esp. 3 and 30.

39. Diana Long Hall and Thomas F. Glick, "Endocrinology: A Brief Introduction," *Journal of the History of Biology*, 9 (1976), 229–233, esp. 231.

40. Merriley Borell, "Brown-Séquard's Organotherapy and Its Appearance in America at the End of the Nineteenth Century," *Bulletin of the History of Medicine*, 50 (1976), 309–320; Bonnie Ellen Blustein, *Preserve Your Love for Science: Life of William A. Hammond, American Neurologist* (Cambridge: Cambridge Univ. Press, 1991), chap. 12.

41. Arthur F. W. Hughes, "A History of Endocrinology," ed. Margaret Well Egar, *Journal of the History of Medicine*, 32 (1977), 292–307; R. B. Welbourn, "Endocrine Diseases," in *Companion Encyclopedia of the History of Medicine*, ed. W. F. Bynum and Roy Porter, 2 vols. (London: Routledge, 1993), 1:484–511; Dwight J. Ingle, "Edward C. Kendall, March 8, 1886–May 4, 1972," *National Academy of Sciences Biographical Memoirs*, 47 (1975), 249–290.

42. Michael Bliss, *The Discovery of Insulin* (Chicago: Univ. of Chicago Press, 1982); Chris Feudtner, *Bittersweet: Diabetes, Insulin, and the Transformation of Illness* (Chapel Hill: Univ. of North Carolina Press, 2003).

43. Saul Benison, A. Clifford Barger, and Elin L. Wolfe, *Walter B. Cannon: The Life and Times of a Young Scientist* (Cambridge, MA: Harvard Univ. Press, 1987), chap. 17; Wolfe, Barger, and Benison, *Walter B. Cannon, Science and Society*, chap. 9; Walter B. Cannon, "The Rôle of Emotion in Disease," *Annals of Internal Medicine*, 9 (1936), 1453–1465.

44. Louis Berman, *The Glands Regulating Personality: A Study of the Glands of Internal Secretion in Relation to the Types of Human Nature* (New York: Macmillan, 1921), 96–97.

45. See, e.g., Hughes, "History of Endocrinology," 292–313; and Edwards A. Park, quoted in *American Pediatrics: The Social Dynamics of Professionalism, 1880–1980*, by Sydney A. Halpern (Berkeley: Univ. of California Press, 1988), 115.

46. Maulitz, "Pathologists, Clinicians, and the Role of Pathophysiology," 225; Elliot S. Valenstein, *The War of the Soups and the Sparks: The Discovery of Neurotransmitters and the Dispute Over How Nerves Communicate* (New York: Columbia Univ. Press, 2005).

47. Robert W. Keeton, "The Advantages of the Physiological Viewpoint in Medicine," *Illinois Medical Journal*, 75 (1939), 510–513, quotation from 512.

48. Harmke Kamminga, "Vitamins and the Dynamics of Molecularization: Biochemistry, Policy and Industry in Britain, 1914–1939," in *Molecularizing Biology and Medicine: New Practices and Alliances, 1910s–1970s,* ed. Soraya de Chadarevian and Kamminga (Amsterdam: Harwood Academic, 1998), 83–105.

49. John P. Swann, *Academic Scientists and the Pharmaceutical Industry: Cooperative Research in Twentieth-Century America* (Baltimore: Johns Hopkins Univ. Press, 1988), 41; John Parascandola, "Industrial Research Comes of Age: The American Pharmaceutical Industry, 1920–1940," *Pharmacy in History,* 27 (1985), 12–21, esp. 14; Nicolas Rasmussen, "The Drug Industry and Clinical Research in Interwar America: Three Types of Physician Collaborator," *Bulletin of the History of Medicine,* 79 (2005), 50–80.

50. Joel D. Howell, "Early Perceptions of the Electrocardiogram: From Arrhythmia to Infarction," *Bulletin of the History of Medicine,* 58 (1984), 83–98.

51. Quoted in Olga Amsterdamska, "Chemistry in the Clinic: The Research Career of Donald Dexter Van Slyke," in de Chadarevian and Kamminga, *Molecularizing Biology and Medicine,* 55, 57.

52. Jonathan Liebenau, *Medical Science and Medical Industry: The Formation of the American Pharmaceutical Industry* (Houndmills, UK: Macmillan, 1987), esp. chap. 8; Dale Cooper, "The Trading with the Enemy Act of 1917 and Synthetic Drugs," *Pharmacy in History,* 47 (2005), 47–61; Swann, *Academic Scientists and the Pharmaceutical Industry,* esp. 49; Parascandola, "Industrial Research Comes of Age," esp. 13.

53. John M. Eyler, "The Fog of Research: Influenza Vaccine Trials during the 1918–19 Pandemic," *Journal of the History of Medicine and Allied Sciences,* 64 (2009), 401–428.

54. See, e.g., Gerald N. Grob, "The Rise and Decline of Tonsillectomy in Twentieth-Century America," ibid., 62 (2007), 383–421.

55. Paul B. Beeson, "Infectious Diseases (Microbiology)," in *Advances in American Medicine: Essays at the Bicentennial,* ed. John Z. Bowers and Elizabeth F. Purcell, 2 vols. (New York: Josiah Macy, Jr., Foundation, 1976), 151–152; Gerald N. Grob and Allan V. Horwitz, *Diagnosis, Therapy, and Evidence: Conundrums in Modern American Medicine* (New Brunswick, NJ: Rutgers Univ. Press, 2010), chap. 3; Andrew T. Scull, *Madhouse: A Tragic Tale of Megalomania and Modern Medicine* (New Haven, CT: Yale Univ. Press, 2005).

56. James C. Whorton, *Inner Hygiene: Constipation and the Pursuit of Health in Modern Society* (New York: Oxford Univ. Press, 2000).

57. Earl B. McKinley, "The Filterable Viruses," *Scientific Monthly,* 32 (1931); 398–403; Sally Smith Hughes, *The Virus: A History of the Concept* (New York: Science History Publications, 1977).

58. McKinley, "Filterable Viruses," 398, 400, 402–403.

59. S. B. Wolbach, "The Filterable Viruses, A Summary," *Journal of Medical Research,* 27 (1912), 1–25; Kraut, *Goldberger's War,* 86–88.

60. Saul Benison, *Tom Rivers: Reflections on a Life in Medicine and Science* (Cambridge, MA: MIT Press, 1967), 138.

61. Peter C. English, *Shock, Physiological Surgery, and George Washington Crile: Medical Innovation in the Progressive Era* (Westport, CT: Greenwood, 1980), quotation from 112.

62. See, e.g., Ralph G. Carothers, "The Advance of Surgery—Its Phases during the Past Twenty-Five Years," *International Journal of Medicine and Surgery,* 48 (1935), 69–71; and

David B. Allman, "The Newer Trends in Surgery," *Journal of the Medical Society of New Jersey*, 33 (1936), 451–453.

63. Beth Linker, *War's Waste: Rehabilitation in World War I America* (Chicago: Univ. of Chicago Press, 2011).

64. Michael Bliss, *Harvey Cushing: A Life in Surgery* (New York: Oxford Univ. Press, 2005); L. Wallace Frank, "What Is New in Surgery," *Kentucky Medical Journal*, 35 (1937), 459–462.

65. Barron H. Lerner, *The Breast Cancer Wars: Hope, Fear, and the Pursuit of a Cure in Twentieth-Century America* (Oxford: Oxford Univ. Press, 2001), chap. 2; Robert A. Aronowitz, *Unnatural History: Breast Cancer and American Society* (Cambridge: Cambridge Univ. Press, 2007), 133–139.

66. See, e.g., Charles L. Gibson, "The Educational Value of the Follow-Up," *Annals of Surgery*, 38 (1928), 772–793.

67. Sally Wilde and Geoffrey Hirst, "Learning from Mistakes: Early Twentieth-Century Surgical Practice," *Journal of the History of Medicine and Allied Sciences*, 64 (2009), 38–44.

68. Kenneth M. Ludmerer, *Learning to Heal: The Development of American Medical Education* (New York: Basic Books, 1985); Ludmerer, *Time to Heal: American Medical Education from the Turn of the Century to the Era of Managed Care* (New York: Oxford Univ. Press, 1999); William G. Rothstein, *American Medical Schools and the Practice of Medicine: A History* (New York: Oxford Univ. Press, 1987); Thomas Neville Bonner, *Becoming a Physician: Medical Education in Britain, France, Germany, and the United States, 1750–1945* (New York: Oxford Univ. Press, 1995).

69. Howard S. Berliner, *A System of Scientific Medicine: Philanthropic Foundations in the Flexner Era* (New York: Tavistock, 1985), chap. 10.

70. William C. Rappleye, "Major Changes in Medical Education during the Past Fifty Years," *Journal of Medical Education*, 34 (1959), 683–689.

71. Charles E. Rosenberg, "Social Class and Medical Care in Nineteenth-Century America: The Rise and Fall of the Dispensary," *Journal of the History of Medicine and Allied Sciences*, 29 (1974), 32–54, esp. 48–53.

72. George Rosen, "The Efficiency Criterion in Medical Care," *Bulletin of the History of Medicine*, 50 (1976), 28–44.

73. Cohn, "Purposes in Medical Research," 11; Rothstein, *American Medical Schools*, 153–159.

74. Ludmerer, *Learning to Heal*, 100–101, quotation from 101.

75. Ibid., quotation from 242; David A. Johnson and Humayun J. Chaudhry, *Medical Licensing and Discipline in America: A History of the Federation of State Medical Boards* (Lanham, MD: Lexington Books, 2012), esp. chap. 3.

76. See, e.g., "The Health Officer and the Quack," *Nation's Health*, 6 (1924), 36.

77. Abraham Flexner, *Medical Education in the United States and Canada: A Report to the Carnegie Foundation for the Advancement of Teaching* (New York: Carnegie Foundation for the Advancement of Teaching, 1910), 169, 173; Rosemary Stevens, *American Medicine and the Public Interest*, 2nd ed. (Berkeley: Univ. of California Press, 1998), 58–63; Ludmerer, *Learning to Heal*, chap. 13; Johnson and Chaudhry, *Medical Licensing and Discipline in America*, 69.

78. Ludmerer, *Learning to Heal*.

79. Thomas Neville Bonner, *Iconoclast: Abraham Flexner and a Life in Learning* (Baltimore: Johns Hopkins Univ. Press, 2002), esp. 169; Barbara Barzansky and Norman Gevitz, eds., *Beyond Flexner: Medical Education in the Twentieth Century* (New York: Greenwood, 1992).

80. Flexner, *Medical Education in the United States and Canada*, 302–303.

81. Rothstein, *American Medical Schools*, 119–120.

82. Ludmerer, *Time to Heal*.

83. Ludmerer, *Learning to Heal*, 270; Stevens, *American Medicine and the Public Interest*, 118–119.

84. Robert L. Martensen and David S. Jones, "The Emergence of the Hospital Internship," *JAMA*, 278 (1997), 963; J. A. Curran, "Internships and Residencies: Historical Backgrounds and Current Trends," *Journal of Medical Education*, 34 (1959), 873–884; Nathaniel L. Faxon, "The Problem of the Hospital Interne," *Bulletin, American College of Surgeons*, Jan. 1924, 46–49; N. P. Colwell, "The Requirement of an Intern Hospital," *JAMA*, 92 (1929), 1031–1033.

85. Judy Barrett Litoff, *American Midwives, 1860 to the Present* (Westport, CT: Greenwood, 1978), chap. 4; Faxon, "Problem of the Hospital Interne"; Colwell, "Requirement of an Intern Hospital"; Curran, "Internships and Residencies," 878.

86. Rosemary Stevens, *In Sickness and in Wealth: American Hospitals in the Twentieth Century* (New York: Basic Books, 1989), 65–67.

87. E. H. L. Corwin, *The American Hospital* (New York: Commonwealth Fund, 1946); Stevens, *In Sickness and in Wealth*, 135.

88. Rosemary A. Stevens, "The Hospital as a Social Institution, New-Fashioned for the 1990s," *Hospital and Health Services Administration*, 36 (1991), 166; Lucinda McCray Beier, *Health Culture in the Heartland, 1880–1980: An Oral History* (Urbana: Univ. of Illinois Press, 2009), 37.

89. Bordley and Harvey, *Two Centuries of American Medicine*, 291.

90. Stevens, *In Sickness and in Wealth*, chap. 5, quotation from 63; Berliner, *System of Scientific Medicine*, 135–138.

91. Guenter B. Risse, *Mending Bodies, Saving Souls: A History of Hospitals* (New York: Oxford Univ. Press, 1999), esp. 472–473; Joel D. Howell, "Hospitals," in *Medicine in the Twentieth Century*, ed. Roger Cooter and John Pickstone (Amsterdam: Harwood Academic, 2000), 505.

92. Harry F. Dowling, *City Hospitals: The Undercare of the Underprivileged* (Cambridge, MA: Harvard Univ. Press, 1982), quotation from 138.

93. Bordley and Harvey, *Two Centuries of American Medicine*, 290–291; Thomas S. Cullen, "America's Place in the Surgery of the World," *Surgery, Gynecology and Obstetrics*, 25 (1917), 387–388; John D. Porterfield, "Hospital Accreditation—Past, Present and Future," *American Journal of Hospital Pharmacy*, 27 (1970), 315.

94. Joel D. Howell, "Machines and Medicine: Technology Transforms the American Hospital," in *The American General Hospital: Communities and Social Contexts*, ed. Diana Elizabeth Long and Janet Golden (Ithaca, NY: Cornell Univ. Press, 1989), 109–110.

95. George Crile, "Modern Surgery," *Modern Hospital*, Oct. 1936, 44–45; John Hunt Shepherd, "Surgical Practice: Looking Forward," *California and Western Medicine*, 46 (1937), 378.

96. "Reliable Hospitals," *Hygeia*, 5 (1927), 364.

97. See, e.g., Dowling, *City Hospitals*, 55–56.

98. William Gerry Morgan, ed., *The American College of Physicians: Its First Quarter Century* (Philadelphia: American College of Physicians, 1940); Paul B. Beeson and Russell C. Maulitz, "The Inner History of Internal Medicine," in Maulitz and Long, *Grand Rounds*, 15–54, quotation from 41.

99. Stevens, *American Medicine and the Public Interest*, 139.

100. Rosen, *Structure of American Medical Practice, 1875–1941*, 59–60, 90–94.

101. Stevens, *American Medicine and the Public Interest*, 140, 162; Committee on the Costs of Medical Care, *Medical Care for the American People* (1932; reprint, Washington, DC: U.S. Department of Health, Education and Welfare, Public Health Service, Mental Health Administration, Community Health Service, 1970), 5.

102. See, e.g., Daniel M. Fox, *Health Policies, Health Politics: The British and American Experience, 1911–1965* (Princeton, NJ: Princeton Univ. Press, 1986), chap. 3.

103. George Weisz, *Divide and Conquer: A Comparative History of Medical Specialization* (New York: Oxford Univ. Press, 2006), 130–133.

104. Ibid., 136.

105. Stevens, *American Medicine and the Public Interest*, esp. chaps. 4 and 5; Weisz, *Divide and Conquer*, esp. 127–138; Beier, *Health Culture in the Heartland*, 107.

106. See, e.g., Maulitz and Long, *Grand Rounds*.

107. Rosemary Stevens, unpublished lecture.

108. Edward H. Ochsner, "The Future of Medicine," *Ohio Medical Journal*, 17 (1921), 767.

109. Harry M. Marks, *The Progress of Experiment: Science and Therapeutic Reform in the United States, 1900–1990* (Cambridge: Cambridge Univ. Press, 1997), esp. chaps. 1 and 2; Eyler, "Fog of Research," 401–428; Beeson and Maulitz, "Inner History of Internal Medicine," 22–24.

110. Marks, *Progress of Experiment*, chaps. 1 and 2, quotation from 32.

111. Ibid., esp. chaps. 1 and 2.

112. See, e.g., Christopher Lawrence, *Rockefeller Money, the Laboratory, and Medicine in Edinburgh, 1919–1930: New Science in an Old Country* (Rochester, NY: Univ. of Rochester Press, 2005); and Ann Westmore and David Penington, "Courting the Rockefeller Foundation and Other Attempts to Integrate Clinical Teaching, Medical Practice, and Research in Melbourne," *Health and History*, 11 (2009), 62–91.

113. Dan W. Blumhagen, "The Doctor's White Coat: The Image of the Physician in Modern America," *Annals of Internal Medicine*, 91 (1979), 111–116.

114. Helen J. Glueck, "Those Were Not the 'Good Old Days,'" *Journal of the American Medical Women's Association*, Nov./Dec. 1984, 203.

115. Edwin Post Maynard Jr., "The Practice of Medicine in 1921," *Bulletin of the New York Academy of Medicine*, 48 (1972), 807–817.

116. Guenter B. Risse, "Once on Top, Now on Tap: American Physicians View Their Relationships with Patients, 1920–1970," in *Responsibility in Health Care*, ed. George J. Agich (Dordrecht, Netherlands: D. Reidel, 1982), 28; Robert S. Lynd and Helen Merrell Lynd, *Middletown: A Study in American Culture* (New York: Harcourt, Brace, 1929), 454–455.

117. Frederick R. Green, "The Social Responsibilities of Modern Medicine," *JAMA*, 76 (1921), 1477–1478.

Chapter 7. Physicians, Public Health, and Progressivism

1. Dora L. Costa and Richard H. Steckel, "Long-Term Trends in Health, Welfare, and Economic Growth in the United States," in *Health and Welfare during Industrialization*, ed. Steckel and Roderick Floud (Chicago: Univ. of Chicago Press, 1997), 47–89; Martin V. Melosi, *The Sanitary City: Urban Infrastructure in America from Colonial Times to the Present* (Baltimore: Johns Hopkins Univ. Press, 2000), chap. 7.

2. Gretchen A. Condran and Rose A. Chaney, "Mortality Trends in Philadelphia: Age- and Cause-Specific Death Rates, 1870–1930," *Demography*, 19 (1982), 114–115; Gregory L. Armstrong, Laura A. Conn, and Robert W. Pinner, "Trends in Infectious Disease Mortality in the United States During the 20th Century," *JAMA*, 281 (1999), 61–71.

3. Alan M. Kraut, *Silent Travelers: Germs, Genes, and the "Immigrant Menace"* (New York: Basic Books, 1994).

4. Stefan Wulf and Heinz-Peter Schmiedebach, " 'Die sprachliche Verständigung ist selbstverständlich recht Schwierig': Die 'geisteskranken Rückwanderer' aus Amerika in der Hamburger Irrenstalt Friedrichsberg 1909," *Medizinhistorisches Journal*, 43 (2008), 231–263; Costa and Steckel, "Long-Term Trends," 47–89.

5. Joseph C. Aub and Ruth K. Hapgood, *Pioneer in Modern Medicine: David Linn Edsall of Harvard* (Cambridge, MA: Harvard Medical Alumni Association, 1970), 41–45; James G. Burrow, *Organized Medicine in the Progressive Era: The Move toward Monopoly* (Baltimore: Johns Hopkins Univ. Press, 1977).

6. Leon Sokoloff, "The Rise and Decline of the Jewish Quota in Medical School Admissions," *Bulletin of the New York Academy of Medicine*, 58 (1992), 497–518; Regina Markell Morantz-Sanchez, *Sympathy and Science: Women Physicians in American Medicine* (New York: Oxford Univ. Press, 1985), chap. 9.

7. Kraut, *Silent Travelers*, esp. chaps. 4 and 5; Guenter B. Risse, *Plague, Fear, and Politics in San Francisco's Chinatown* (Baltimore: Johns Hopkins Univ. Press, 2013).

8. Walter James Heimann, "The Evolution of Dermatology," *Annals of Medical History*, 1 (1917), 427–428.

9. Thomas S. Cullen, "America's Place in the Surgery of the World," *Surgery, Gynecology and Obstetrics*, 25 (1917), 386.

10. See, e.g., Gerald L. Geison, "International Relations and Domestic Elites in American Physiology, 1900–1940," in *Physiology in the American Context, 1850–1940*, ed. Geison (Bethesda, MD: American Physiological Society, 1987), 115–154; and Maxwell M. Wintrobe, *Hematology: The Blossoming of a Science—A Story of Inspiration and Effort* (Philadelphia: Lea & Febiger, 1985), 183.

11. Burrow, *Organized Medicine in the Progressive Era*, 14–15.

12. Ibid., chap. 2, quotation from 20; Donald L. Madison, "Preserving Individualism in the Organizational Society: 'Cooperation' and American Medical Practice, 1900–1920," *Bulletin of the History of Medicine*, 70 (1996), 442–483.

13. Burrow, *Organized Medicine in the Progressive Era*, chap. 2.

14. James Harvey Young, *Pure Food: Securing the Federal Food and Drugs Act of 1906* (Princeton, NJ: Princeton Univ. Press, 1989); Burrow, *Organized Medicine in the Progressive Era*, chap. 2.

15. James G. Burrow, *AMA: Voice of American Medicine* (Baltimore: Johns Hopkins Press, 1963), esp. chaps. 6 and 13, quotation from 269; Eric W. Boyle, *Quack Medicine: A History of Combating Health Fraud in Twentieth-Century America* (Santa Barbara, CA: Praeger, 2013), esp. chaps. 3 and 4.

16. Barbara Sicherman, *Alice Hamilton: A Life in Letters* (Cambridge, MA: Harvard Univ. Press, 1984); Christopher C. Sellers, *Hazards of the Job: From Industrial Disease to Environmental Health Science* (Chapel Hill: Univ. of North Carolina Press, 1997), chap. 3; Alice Hamilton, *Exploring the Dangerous Trades: The Autobiography of Alice Hamilton, M.D.* (Boston: Little, Brown, 1943), 123; Claudia Clark, *Radium Girls: Women and Industrial Health Reform, 1910–1935* (Chapel Hill: Univ. of North Carolina Press, 1997).

17. Hamilton, *Exploring the Dangerous Trades*, 4.

18. Ibid., 9–11.

19. Michael E. Teller, *The Tuberculosis Movement: A Public Health Campaign in the Progressive Era* (New York: Greenwood, 1988); John C. Burnham, *Paths into American Culture: Psychology, Medicine, and Morals* (Philadelphia: Temple Univ. Press, 1988), chap. 13.

20. See previous note; and Mark Caldwell, *The Last Crusade: The War on Consumption, 1862–1954* (New York: Atheneum, 1988).

21. Teller, *Tuberculosis Movement*, chap. 7; Sheila M. Rothman, *Living in the Shadow of Death: Tuberculosis and the Social Experience of Illness in American History* (New York: Basic Books, 1994), esp. chap. 13; Barbara Bates, *Bargaining for Life: A Social History of Tuberculosis, 1876–1938* (Philadelphia: Univ. of Pennsylvania Press, 1992), esp. 264.

22. John Parascandola, *Sex, Sin, and Science: A History of Syphilis in America* (Westport, CT: Praeger, 2008), esp. chaps. 2 and 4; Burnham, *Paths into American Culture*, esp. chaps. 9 and 10.

23. David J. Pivar, *Purity and Hygiene: Women, Prostitution, and the "American Plan," 1900–1930* (Westport, CT: Greenwood, 2002); Burnham, *Paths into American Culture*.

24. Burnham, *Paths into American Culture*, quotation from 165.

25. Parascandola, *Sex, Sin, and Science*; Pivar, *Purity and Hygiene*.

26. Patricia Spain Ward, "The American Reception of Salvarsan," *Journal of the History of Medicine and Allied Sciences*, 36 (1981), 44–62, quotations from 53.

27. Burnham, *Paths into American Culture*, esp. chap. 12; Elizabeth Lunbeck, *The Psychiatric Persuasion: Knowledge, Gender, and Power in Modern America* (Princeton, NJ: Princeton Univ. Press, 1994).

28. Nathaniel Comfort, *The Science of Human Perfection: How Genes Became the Heart of American Medicine* (New Haven, CT: Yale Univ. Press, 2012).

29. George A. Soper, "The Work of a Chronic Typhoid Germ Distributor," *JAMA*, 48 (1907), 2019–2022; Judith Walzer Leavitt, *Typhoid Mary: Captive to the Public's Health* (Boston: Beacon, 1996).

30. Robert S. Lynd and Helen Merrell Lynd, *Middletown: A Study in American Culture* (New York: Harcourt, Brace, 1929), 450–451; John Duffy, *The Sanitarians: A History of American Public Health* (Urbana: Univ. of Illinois Press, 1990), esp. chap. 13.

31. Duffy, *Sanitarians*, esp. 201.

32. John Duffy, *A History of Public Health in New York City, 1866–1966* (New York: Russell Sage Foundation, 1974), esp. 94; C.-E. A. Winslow, *The Life of Hermann M. Biggs, M.D.*,

D.Sc., LL.D., Physician and Statesman of the Public Health (Philadelphia: Lea & Febiger, 1929), 217.

33. Emery R. Hayhurst, "The Present-Day Sources of Common Salt in Relation to Health, and Especially to Iodin Scarcity and Goiter," *JAMA*, 78 (1922), 18–21; Roy D. McClure, "Thyroid Surgery as Affected by the Generalized Use of Iodized Salt in an Endemic Goitre Region—Preventive Surgery," *Annals of Surgery*, 100 (1934), 924–932; Richard D. Semba, "The Impact of Improved Nutrition on Disease Prevention," in *Silent Victories: The History and Practice of Public Health in Twentieth-Century America*, ed. John W. Ward and Christian Warren (New York: Oxford Univ. Press, 2007), 170–172.

34. Teller, *Tuberculosis Movement*, 122–123; Barbara Gutmann Rosenkrantz, "Cart before Horse: Theory, Practice and Professional Image in American Public Health, 1870–1920," *Journal of the History of Medicine and Allied Sciences*, 29 (1974), 62–73.

35. Richard A. Meckel, *Save the Babies: American Public Health Reform and the Prevention of Infant Mortality* (Baltimore: Johns Hopkins Univ. Press, 1990), 100.

36. James H. Cassedy, "The 'Germ of Laziness' in the South, 1900–1915: Charles Wardell Stiles and the Progressive Paradox," *Bulletin of the History of Medicine*, 45 (1971), 159–169; John Ettling, *The Germ of Laziness: Rockefeller Philanthropy and Public Health in the New South* (Cambridge, MA: Harvard Univ. Press, 1981); Alan I Marcus, "Physicians Open a Can of Worms: American Nationality and Hookworm in the United States, 1893–1909," *American Studies*, 30 (1989), 103–121.

37. Cassedy, "'Germ of Laziness' in the South," 159–169; Ettling, *Germ of Laziness*.

38. See previous note.

39. John Harkness, personal communication of unpublished research, 9 June 2011.

40. Teller, *Tuberculosis Movement*, chaps. 4 and 10.

41. Condran and Chaney, "Mortality Trends in Philadelphia," 114–115.

42. Samuel H. Preston and Michael R. Haines, *Fatal Years: Child Mortality in Late Nineteenth-Century America* (Princeton, NJ: Princeton Univ. Press, 1991), esp. 209.

43. Roland Marchand, *Advertising the American Dream: Making Way for Modernity, 1920–1940* (Berkeley: Univ. of California Press, 1985), esp. 218–219; Harvey Green, *The Uncertainty of Everyday Life, 1915–1945* (New York: HarperCollins, 1993), 184–185.

44. Andrew Noymer and Michael Garenne, "The 1918 Influenza Epidemic's Effects on Sex Differentials in Mortality in the United States," *Population and Development Review*, 26 (2000), 565–581; Alfred W. Crosby, *America's Forgotten Pandemic: The Influenza of 1918*, 2nd ed. (Cambridge: Cambridge Univ. Press, 1989); K. David Patterson and Gerald F. Pyle, "The Geography and Mortality of the 1918 Influenza Pandemic," *Bulletin of the History of Medicine*, 65 (1991), 4–21; Edward C. Holmes, "1918 and All That," *Science*, 303 (2004), 1787–1788. Cf. Dorothy A. Pettit and Janice Bailie, *A Cruel Wind: Pandemic Flu in America, 1918–1920* (Murfreesboro, TN: Timberlane Books, 2008); and Jocelyn Kaiser, "Resurrected Influenza Virus Yields Secrets of Deadly 1918 Pandemic," *Science*, 310 (2005), 28–29.

45. Paul Foley, "The Encephalitis Lethargica Patient as a Window on the Soul," in *The Neurological Patient in History*, ed. L. Stephen Jacyna and Stephen T. Casper (Rochester, NY: Univ. of Rochester Press, 2012), 184–211; Sherman McCall et al., "The Relationship between Encephalitis Lethargica and Influenza: A Critical Analysis," *Journal of Neurovirology*, 14 (2008), 177–185.

46. Crosby, *America's Forgotten Pandemic*, esp. 50; Rosemary Stevens, *In Sickness and in Wealth: American Hospitals in the Twentieth Century* (New York: Basic Books, 1989), 102; Janice Hume, "The 'Forgotten' 1918 Influenza Epidemic and Press Portrayal of Public Anxiety," *Journalism and Mass Communications Quarterly*, 77 (2000), 898–915; "The 1918–1919 Influenza Pandemic in the United States," *Public Health Reports*, 125, suppl. 3 (2010); Nancy K. Bristow, *American Pandemic: The Lost Worlds of the 1918 Influenza Epidemic* (New York: Oxford Univ. Press, 2012).

47. Irving Fisher, "National Vitality: Its Waste and Conservation," in *Report of the National Conservation Commission*, S. Doc. No. 676, at 636–637 (1909); Irving Fisher and Eugene Lyman Fisk, *How to Live: Rules for Healthful Living Based on Modern Science*, 15th ed. (New York: Funk & Wagnalls, 1919).

48. Meckel, *Save the Babies*, 1, 151, 154; Rosemary Stevens, *American Medicine and the Public Interest*, 2nd ed. (Berkeley: Univ. of California Press, 1998), 132n.

49. George Rosen, *Preventive Medicine in the United States, 1900–1975* (New York: Science History Publications, 1975), 59–60; Mitchell H. Charap, "The Periodic Health Examination: Genesis of a Myth," *Annals of Internal Medicine*, 95 (1981), 733–735; Paul K. J. Han, "Historical Changes in the Objectives of the Periodic Health Examination," ibid., 126 (1997), 910–913; Stanley Joel Reiser, *Technological Medicine: The Changing World of Doctors and Patients* (New York: Cambridge Univ. Press, 2009), 139–144; George Weisz, *Chronic Disease in the Twentieth Century: A History* (Baltimore: Johns Hopkins Univ. Press, 2014), 31–36 and chaps. 2 and 3.

50. Emily K. Abel, "A 'Terrible and Exhausting' Struggle: Family Caregiving During the Transformation of Medicine," *Journal of the History of Medicine and Allied Sciences*, 50 (1995), 478–506.

51. Daniel A. Dumesic, "The Physician Automobilist," ibid., 51 (1996), 208–222; Lucinda McCray Beier, *Health Culture in the Heartland, 1880–1980: An Oral History* (Urbana: Univ. of Illinois Press, 2009), 101–104; George Rosen, *The Structure of American Medical Practice, 1875–1941*, ed. Charles E. Rosenberg (Philadelphia: Univ. of Pennsylvania Press, 1983), 25–32, 49–51; Ralph E. Pumphrey, "Michael M. Davis and the Development of the Health Care Movement, 1900–1928," *Societas*, 2 (1972), 27–41.

52. Daniel M. Fox, *Health Policies, Health Politics: The British and American Experience, 1911–1965* (Princeton, NJ: Princeton Univ. Press, 1986), chap. 3; Bertha Streeter, "A New Deal in Health Education," *Hygeia*, 12 (1934), 364.

53. Bertram M. Bernheim, *Medicine at the Crossroads* (New York: William Morrow, 1939), 18–19.

54. Milton Terris, "Origins and Growth of Group Practice," *Bulletin of the New York Academy of Medicine*, 44 (1968), 1277–1281; Bernheim, *Medicine at the Crossroads*, chap. 8.

55. Committee on the Costs of Medical Care, *Medical Care for the American People* (1932; reprint, Washington, DC: U.S. Department of Health, Education and Welfare, Public Health Service, Mental Health Administration, Community Health Services, 1970).

56. Guenter B. Risse, *Mending Bodies, Saving Souls: A History of Hospitals* (New York: Oxford Univ. Press, 1999), 484–485.

57. James Harvey Young, *The Medical Messiahs: A Social History of Health Quackery in Twentieth-Century America* (Princeton, NJ: Princeton Univ. Press, 1967); Lynd and Lynd, *Middletown*, quotation from 438–439.

58. Committee on the Costs of Medical Care, *Medical Care for the American People*, esp. 5, 18, and chap. 1 generally; Jonathan Engel, *Doctors and Reformers: Discussion and Debate over Health Policy, 1925–1950* (Columbia: Univ. of South Carolina Press, 2002), esp. chap. 2.

59. David McBride, *From TB to AIDS: Epidemics among Urban Blacks since 1900* (Albany: State Univ. of New York Press, 1991), esp. 52–53.

60. See, e.g., Christopher K. Crenner, "Race and Medical Practice in Kansas City's Free Dispensary," *Bulletin of the History of Medicine*, 82 (2008), 820–847; Vanessa Northington Gamble, *Making a Place for Ourselves: The Black Hospital Movement, 1920–1945* (New York: Oxford Univ. Press, 1995), quotation from 45; and Beatrix Hoffman, *Health Care for Some: Rights and Rationing in the United States since 1930* (Chicago: Univ. of Chicago Press, 2012), xxix–xxx.

61. See esp. Keith Wailoo, *Drawing Blood: Technology and Disease Identity in Twentieth-Century America* (Baltimore: Johns Hopkins Univ. Press, 1997); and Wailoo, *Dying in the City of the Blues: Sickle Cell Anemia and the Politics of Race and Health* (Chapel Hill: Univ. of North Carolina Press, 2001).

62. Samuel Kelton Roberts Jr., *Infectious Fear: Politics, Disease, and the Health Effects of Segregation* (Chapel Hill: Univ. of North Carolina Press, 2009), esp. 20; McBride, *From TB to AIDS*, esp. chap. 3.

63. Susan L. Smith, *Sick and Tired of Being Sick and Tired: Black Women's Health Activism in America, 1890–1950* (Philadelphia: Univ. of Pennsylvania Press, 1995), esp. chap. 2; Robert B. Eleazer, quoted in *National Negro Health News*, Apr.–June 1934, suppl. p. 14; McBride, *From TB to AIDS*, esp. chaps. 3 and 4.

64. See, e.g., Alexandra Minna Stern, "Buildings, Boundaries, and Blood: Medicalization and Nation-Building on the U.S.-Mexican Border, 1910–1930," *Hispanic American Historical Review*, 79 (1999), 41–81.

65. Vanessa Northington Gamble, introduction to *Germs Have No Color Line: Blacks and American Medicine, 1900–1940*, ed. Gamble (New York: Garland, 1989), unpaginated; Stuart Galishoff, "Germs Know No Color Line: Black Health and Public Policy in Atlanta, 1900–1918," *Journal of the History of Medicine and Allied Sciences*, 49 (1985), 22–41, quotation from 29.

66. John E. Murray, *Origins of American Health Insurance: A History of Industrial Sickness Funds* (New Haven, CT: Yale Univ. Press, 2007).

67. Paul B. Bellamy, *A History of Workmen's Compensation, 1898–1915* (New York: Garland, 1997).

68. J. M. Laird, "Non-Cancellable Accident and Health Insurance Underwriting Problems," *Proceedings of the Casualty Actuarial Society*, 7 (1921), 302–307; Armand Sommer, "Development and Progress of Accident and Health Insurance during Eighty Years," *Weekly Underwriter*, 27 May 1939, 45, 139–140; Harry Dingman, *Risk Appraisal* (Cincinnati: National Underwriter, 1946), chap. 6.

69. Ronald L. Numbers, *Almost Persuaded: American Physicians and Compulsory Health Insurance, 1912–1920* (Baltimore: Johns Hopkins Univ. Press, 1978), chap. 2.

70. Ibid., esp. chap. 8; Beatrix Hoffman, *The Wages of Sickness: The Politics of Health Insurance in Progressive America* (Chapel Hill: Univ. of North Carolina Press, 2001).

71. Pumphrey, "Michael M. Davis."

72. Numbers, *Almost Persuaded*, esp. 114.

73. Lynd and Lynd, *Middletown*, 443.

74. J. Stanley Lemons, "The Sheppard-Towner Act: Progressivism in the 1920s," *Journal of American History*, 55 (1969), 776–786; Meckel, *Save the Babies*, chap. 8; Esther Everett Lape, ed., *American Medicine: Expert Testimony Out of Court*, 2 vols. (New York: American Foundation, 1937), 2:851–852; Jaap Kooijman, . . . *And the Pursuit of National Health: The Incremental Strategy toward National Health Insurance in the United States of America* (Amsterdam: Rodopi, 1999).

75. Hoffman, *Wages of Sickness*, esp. 181, citing *The Social Transformation of American Medicine*, by Paul Starr (New York: Basic Books, 1982), 258–259.

76. Charles E. Rosenberg, *No Other Gods: On Science and American Social Thought*, 2nd ed. (Baltimore: Johns Hopkins Univ. Press, 1997), chap. 7.

Chapter 8. The Era of Antibiotics, 1930s to 1950s

1. Walsh McDermott, "Social Ramifications of Control of Microbial Disease," ed. David E. Rogers, *Johns Hopkins Medical Journal*, 151 (1982), 303.

2. See, e.g., *Illness and Medical Care among 2,500,000 Persons in 83 Cities, With Special Reference to Social-Economic Factors*, Federal Security Agency, U.S. Public Health Service (Washington, DC: U.S. Government Printing Office, 1945).

3. David E. Rogers, "The Early Years: The Medical World in Which Walsh McDermott Trained," *Daedalus*, Spring 1986, 3; James Rorty, *American Medicine Mobilizes* (New York: W. W. Norton, 1939), 92.

4. Arthur Dean Bevan, "The Over-Crowding of the Medical Profession," *Journal of Association of Medical Colleges*, 11 (1936), 377–384; Beatrix Hoffman, *Health Care for Some: Rights and Rationing in the United States since 1930* (Chicago: Univ. of Chicago Press, 2012), chap. 1.

5. Robert S. Lynd and Helen Merrell Lynd, *Middletown in Transition: A Study in Cultural Conflicts* (New York: Harcourt, Brace, 1937), 391, 397–398.

6. Bertha Streeter, "A New Deal in Health Education," *Hygeia*, 12 (1934), 364.

7. Rorty, *American Medicine Mobilizes*, chap. 12.

8. Edward H. Beardsley, *A History of Neglect: Health Care for Blacks and Mill Workers in the Twentieth-Century South* (Knoxville: Univ. of Tennessee Press, 1987), esp. chap. 7.

9. Roy Lubove, "The New Deal and National Health," *Current History*, 45 (1963), 76–86, 117; Martin V. Melosi, *The Sanitary City: Urban Infrastructure in America from Colonial Times to the Present* (Baltimore: Johns Hopkins Univ. Press, 2000), chap. 12; Michael R. Grey, *New Deal Medicine: The Rural Health Programs of the Farm Security Administration* (Baltimore: Johns Hopkins Univ. Press, 1999).

10. Daniel M. Fox, *Power and Illness: The Failure and Future of American Health Policy* (Berkeley: Univ. of California Press, 1993), 49.

11. Lubove, "New Deal and National Health," 76–86, 117; Roy Lubove, *The Struggle for Social Security, 1900–1935* (Pittsburgh: Univ. of Pittsburgh Press, 1986), 89–90; Jaap Kooijman, "'Just Forget About It': FDR's Ambivalence towards National Health Insurance," in *The Roosevelt Years*, ed. Robert A. Garson and Stuart S. Kidd (Edinburgh: Edinburgh Univ. Press, 1999), 30–41; Jonathan Engel, *Doctors and Reformers: Discussion and Debate over Health Policy, 1925–1950* (Columbia: Univ. of South Carolina Press, 2002).

12. Daniel S. Hirshfield, *The Lost Reform: The Campaign for Compulsory Health Insurance in the United States from 1932 to 1943* (Cambridge, MA: Harvard Univ. Press, 1970); Colin Gordon, *Dead on Arrival: The Politics of Health Care in Twentieth-Century America* (Princeton, NJ: Princeton Univ. Press, 2003); Engel, *Doctors and Reformers*; Alan Derickson, *Health Security for All: Dreams of Universal Health Care in America* (Baltimore: Johns Hopkins Univ. Press, 2005), esp. chaps. 3 and 4.

13. M. V. Mackenzie, "Control of Medical Science," *New England Journal of Medicine*, 220 (1939), 138.

14. Monte M. Poen, *Harry S. Truman versus the Medical Lobby: The Genesis of Medicare* (Columbia: Univ. of Missouri Press, 1979), quotations from 136 and 207; James G. Burrow, *AMA: Voice of American Medicine* (Baltimore: Johns Hopkins Press, 1963), chaps. 16 and 17; Engel, *Doctors and Reformers*, chaps. 6–8. See also reports in *Medical Statistics Bulletin* (Selective Service System), 1941–46.

15. "American Medical Association Dues," *Northwest Medicine*, 49 (1950), 25.

16. Burrow, *AMA*, chaps. 16 and 17, esp. 375–376; Ronald L. Numbers, "The Specter of Socialized Medicine: American Physicians and Compulsory Health Insurance," in *Compulsory Health Insurance: The Continuing American Debate*, ed. Numbers (Westport, CT: Greenwood, 1982), 3–24; John C. Burnham, "American Medicine's Golden Age: What Happened to It?," *Science*, 215 (1982), 1474–1479.

17. Poen, *Harry S. Truman versus the Medical Lobby*, esp. chap. 6.

18. Eyewitness information from Vaughn D. Bornet, personal communication; Philip J. Hilts, *Protecting America's Health: The FDA, Business, and One Hundred Years of Regulation* (New York: Alfred A. Knopf, 2003), 126–127.

19. See, e.g., Margaret Jean Gaughan, "Effectiveness in Achieving Organizational Goals: Domain Consensus and the American Medical Association" (PhD diss., Boston University, 1977); and Frank D. Campion, *The AMA and U.S. Health Policy since 1940* (Chicago: Chicago Review Press, 1984), esp. chap. 16.

20. Grey, *New Deal Medicine*, esp. 142–144; Lloyd C. Taylor Jr., *The Medical Profession and Social Reform, 1885–1945* (New York: St. Martin's, 1974), 152–159; Hoffman, *Health Care for Some*, 47–53.

21. Burnham, "American Medicine's Golden Age"; Daniel M. Fox, *Health Policies, Health Politics: The British and American Experience, 1911–1965* (Princeton, NJ: Princeton Univ. Press, 1986), 70–74.

22. Esther Everett Lape, ed., *American Medicine: Expert Testimony Out of Court*, 2 vols. (New York: American Foundation, 1937), 1:44.

23. See, e.g., David S. Jones and Robert L. Martensen, "Human Radiation Experiments and the Formation of Medical Physics at the University of California, San Francisco and Berkeley, 1937–1962," in *Useful Bodies: Humans in the Service of Medical Science in the Twentieth Century*, ed. Jordan Goodman, Anthony McElligott, and Lara Marks (Baltimore: Johns Hopkins Univ. Press, 2003), 81–108.

24. Rosemary Stevens, *In Sickness and in Wealth: American Hospitals in the Twentieth Century* (New York: Basic Books, 1989), esp. 173; Lape, *American Medicine*, 2:811.

25. Stevens, *In Sickness and in Wealth*; E. H. L. Corwin, *The American Hospital* (New York: Commonwealth Fund, 1946), esp. 6; Odin W. Anderson, *Blue Cross Since 1929: Accountability and the Public Trust* (Cambridge, MA: Ballinger, 1975), 18.

26. Corwin, *American Hospital*, 105; Anderson, *Blue Cross Since 1929*, 18.

27. Anderson, *Blue Cross Since 1929*, 105; Sylvia A. Law, *Blue Cross: What Went Wrong?*, 2nd ed. (New Haven, CT: Yale Univ. Press, 1976), chap. 2.

28. Lewis E. Weeks and Howard J. Berman, *Shapers of American Health Care Policy: An Oral History* (Ann Arbor, MI: Health Administration Press, 1985), esp. chaps. 6 and 7; Anderson, *Blue Cross Since 1929*, 1, 42–43; Corwin, *American Hospital*, 106–108.

29. Anderson, *Blue Cross Since 1929*, 44.

30. Nathan Sinai, Odin W. Anderson, and Melvin L. Dollar, *Health Insurance in the United States* (New York: Commonwealth Fund, 1946), 69–71; Anderson, *Blue Cross Since 1929*, 44.

31. Burrow, *AMA*, chap. 12, esp. 249.

32. *California and Western Medicine*, 50–51 (1939); Anderson, *Blue Cross Since 1929*, 59–61.

33. Paul de Kruif, *Kaiser Wakes the Doctors* (New York: Harcourt, Brace, 1943); Rickey Hendricks, *A Model for National Health Care: The History of Kaiser Permanente* (New Brunswick, NJ: Rutgers Univ. Press, 1993).

34. Patricia Spain Ward, "United States versus American Medical Association et al.: The Medical Anti-Trust Case of 1938–1943," *American Studies*, 30 (1989), 123–154, quotation on 144. See also, e.g., Hendricks, *Model for National Health Care*.

35. Vannevar Bush, *Pieces of the Action* (New York: William Morrow, 1970), 43–44.

36. Fox, *Health Policies, Health Politics*; Daniel M. Fox, "Health Policy and Changing Epidemiology in the United States: Chronic Disease in the Twentieth Century," *Transactions and Studies of the College of Physicians of Philadelphia*, 5th ser., 10 (1988), 11–31.

37. Daniel M. Fox, "The Consequences of Consensus: American Health Policy in the Twentieth Century," *Milbank Quarterly*, 64 (1986), 76–99.

38. Bertram M. Bernheim, *Medicine at the Crossroads* (New York: William Morrow, 1939), 45.

39. Fox, *Health Policies, Health Politics*.

40. Lape, *American Medicine*, 1:48.

41. See, e.g., Bernheim, *Medicine at the Crossroads*, 30–34; and Roger I. Lee and Lewis Webster Jones, *The Fundamentals of Good Medical Care* (Chicago: Univ. of Chicago Press, 1933), esp. 6.

42. Fox, *Health Policies, Health Politics*, chaps. 3 and 5.

43. William G. Rothstein, *American Medical Schools and the Practice of Medicine: A History* (New York: Oxford Univ. Press, 1987), chap. 8.

44. Rosemary Stevens, *American Medicine and the Public Interest*, 2nd ed. (Berkeley: Univ. of California Press, 1998), esp. 162 and 297; George Weisz, *Divide and Conquer: A Comparative History of Medical Specialization* (Oxford: Oxford Univ. Press, 2006), esp. chap. 7.

45. Stevens, *American Medicine and the Public Interest*.

46. Ibid.; Russell C. Maulitz and Diana E. Long, eds., *Grand Rounds: One Hundred Years of Internal Medicine* (Philadelphia: Univ. of Pennsylvania Press, 1988); Weisz, *Divide and Conquer*, 131.

47. Stevens, *American Medicine and the Public Interest*; Weisz, *Divide and Conquer*, esp. 143–144; Kenneth M. Ludmerer and Michael M. E. Johns, "Reforming Graduate Medical Education," *JAMA*, 294 (2005), 1083–1087.

48. David P. Adams, "Community and Professionalization: General Practitioners and Ear, Nose, and Throat Specialists in Cincinnati, 1945–1947," *Bulletin of the History of Medicine*, 68 (1994), 664–684; Adams, *American Board of Family Practice: A History* (Lexington, KY: American Board of Family Practice, 1999).

49. Lape, *American Medicine*, 1:397.

50. Alphonse M. Schwitalla, "The Physician in the Patient-Physician Relationship," *New York State Journal of Medicine*, 46 (1946), 1464; Guenter B. Risse, "Once on Top, Now on Tap: American Physicians View Their Relationships with Patients, 1920–1970," in *Responsibility in Health Care*, ed. George J. Agich (Dordrecht, Netherlands: D. Reidel, 1982).

51. Corwin, *American Hospital*, 26.

52. John E. Lesch, *The First Miracle Drugs: How the Sulfa Drugs Transformed Medicine* (Oxford: Oxford Univ. Press, 2007), esp. 270.

53. A. McGehee Harvey, "The Story of Chemotherapy at Johns Hopkins: Perrin H. Long, Eleanor A. Bliss, and E. Kennerly Marshall, Jr.," *Johns Hopkins Medical Journal*, 138 (1976), 54; "Young Roosevelt Saved by New Drug," *New York Times*, 17 Dec. 1936.

54. Thomas E. Cone Jr., *History of American Pediatrics* (Boston: Little Brown, 1979), 212–213.

55. McDermott, "Social Ramifications," 307.

56. Lesch, *First Miracle Drugs*, esp. 6.

57. Marcel H. Bickel, "The Development of Sulfonamides (1932–1938) as a Focal Point in the History of Chemotherapy," *Gesnerus*, 45 (1988), 74; "Symposium on Sulfanilamide Therapy," *Journal of Pediatrics*, 11 (1937), 157; Lesch, *First Miracle Drugs*, 198–203, 273–274; James Bordley III and A. McGehee Harvey, *Two Centuries of American Medicine, 1776–1976* (Philadelphia: W. B. Saunders, 1976), chap. 22.

58. Bickel, "Development of Sulfonamides," 67–86, esp. 78–79; Lesch, *First Miracle Drugs*, esp. 207 and 210.

59. Cynthia Connolly, Janet Golden, and Benjamin Schneider, "'A Startling New Chemotherapeutic Agent': Pediatric Infectious Disease and the Introduction of Sulfonamides at Baltimore's Sydenham Hospital," *Bulletin of the History of Medicine*, 86 (2012), 66–93; Lesch, *First Miracle Drugs*, esp. 10, 210, and 277.

60. Edward D. Churchill, *Surgeon to Soldiers: Diary and Records of the Surgical Consultant, Allied Force Headquarters, World War II* (Philadelphia: J. B. Lippincott, 1972), 67.

61. Robert Bud, *Penicillin: Triumph and Tragedy* (Oxford: Oxford Univ. Press, 2007), esp. chaps. 1–3; Peter Neushul, "Science, Government, and the Mass Production of Penicillin," *Journal of the History of Medicine and Allied Sciences*, 48 (1993), 371–395; Neushul, "Fighting Research: Army Participation in the Clinical Testing and Mass Production of Penicillin during the Second World War," in *War, Medicine and Modernity*, ed. Roger Cooter, Mark Harrison, and Steve Sturdy (Stroud, Gloucestershire: Sutton, 1998), 203–224.

62. David P. Adams, *"The Greatest Good to the Greatest Number": Penicillin Rationing on the American Home Front, 1940–1945* (New York: Peter Lang, 1991), esp. 125n; W. H. Helfand et al., "Wartime Industrial Development of Penicillin in the United States," in *The History of Antibiotics: A Symposium*, ed. John Parascandola (Madison, WI: American Institute of the History of Pharmacy, 1980), 31–56, esp. 50–51; Bud, *Penicillin*, 58.

63. "The New Agents in Infections," *GP*, Nov. 1951, 64; Henry Welch and Félix Martí-Ibáñez, *The Antibiotic Saga* (New York: Medical Encyclopedia, 1960), 76–77; Kenneth B.

Raper, "The Progress of Antibiotics," *Scientific American*, Apr. 1952, 49–57, quotations from 49.

64. See, e.g., William B. Tucker, "The Modern Management of Tuberculosis," *Journal of the National Medical Association*, 47 (1955), 227.

65. Scott H. Podolsky, *Pneumonia before Antibiotics: Therapeutic Evolution and Evaluation in Twentieth-Century America* (Baltimore: Johns Hopkins Univ. Press, 2006).

66. Russell S. Boles, "The Press and the Patient," *JAMA*, 144 (1950), 362; Bud, *Penicillin*; Robert Bud, "From Germophobia to the Carefree Life and Back Again: The Lifecycle of the Antibiotic Brand," in *Medicating Modern America: Prescription Drugs in History*, ed. Andrea Tone and Elizabeth Siegel Watkins (New York: New York Univ. Press, 2007), 17–41; Welch and Martí-Ibáñez, *Antibiotic Saga*, 99.

67. Edward H. Kass, "History of the Subspecialty of Infectious Diseases in the United States," in Maulitz and Long, *Grand Rounds*, 91–100; Gerald N. Grob, *The Deadly Truth: A History of Disease in America* (Cambridge, MA: Harvard Univ. Press, 2002), esp. 202–204 and 216.

68. McDermott, "Social Ramifications," 305–307.

69. Louis Weinstein, quoted in *Conversations in Medicine: The Story of Twentieth-Century American Medicine in the Words of Those Who Created It*, by Allen B. Weisse (New York: New York Univ. Press, 1984), 165–166.

70. E. R. N. Grigg, quoted in *Living in the Shadow of Death: Tuberculosis and the Social Experience of Illness in American History*, by Sheila M. Rothman (New York: Basic Books, 1994), 249.

71. Tucker, "Modern Management of Tuberculosis," 227.

72. George M. Wheatley, "Mobilization against Rheumatic Fever," *Journal of Pediatrics*, 26 (1945), 237.

73. Donald G. Cooley, *The Science Book of Wonder Drugs* (New York: Franklin Watts, 1954).

74. Peter C. English, *Rheumatic Fever in America and Britain: A Biological, Epidemiological, and Medical History* (New Brunswick, NJ: Rutgers Univ. Press, 1999); Grob, *Deadly Truth*, 208–209.

75. Donald B. Armstrong, "Are They Safe at Home?," *Home Safety Review*, Aug.–Sept. 1948, 4–5, 14–15.

76. Jay M. Arena, "The Pediatrician's Role in the Poison Control Movement and Poison Prevention," *American Journal of Diseases of Children*, 137 (1983), 872–873; John C. Burnham, "Why Did the Infants and Toddlers Die? Shifts in Americans' Ideas of Responsibility for Accidents: From Blaming Mom to Engineering," *Journal of Social History*, 29 (1996), 817–837.

77. Noah Fabricant and Terry Hillel, "Antibiotics Hit the Doctor's Wallet," *Science Digest*, Apr. 1951, 21–25.

78. Harry C. Saltzstein, "Recent Advances in Surgical Care," *Harper Hospital Bulletin*, 8 (1950), 81–93, quotations from 81; Warren H. Cole, "Recent Advances in Surgery," *Northwest Medicine*, 49 (1950), 769–771, 863–866; Alfred Blalock and Helen B. Taussig, "The Surgical Treatment of Malformations of the Heart in which There Is Pulmonary Stenosis or Pulmonary Atresia," *JAMA*, 128 (1945), 189–202.

79. Saltzstein, "Recent Advances in Surgical Care," quotations from 81, 91, and 92; Cole, "Recent Advances in Surgery."

80. Bud, *Penicillin*; Helfand et al., "Wartime Industrial Development of Penicillin"; L. Ettlinger, "Wartime Research on Penicillin in Switzerland and Antibiotic Screening," in Parascandola, *History of Antibiotics*, 63; Raper, "Progress of Antibiotics," 50.

81. Arthur M. Silverstein, *A History of Immunology* (San Diego: Academic Press, 1989), esp. 225.

82. Sally Smith Hughes, *The Virus: A History of the Concept* (New York: Science History Publications, 1977), esp. 89–90; Vincent du Vigneaud, "Scientific Contributions of the Medalist," *Chemical and Engineering News*, 24 (1946), 752–755; David Dietz, "Virus Diseases," *Current History*, Feb. 1937, 97–98.

83. See, e.g., Mitchell G. Ash, "Forced Migration and Scientific Change after 1933: Steps Toward a New Overview," in *Intellectual Migration and Cultural Transformation: Refugees from National Socialism in the English-speaking World*, ed. Edward Timms and Jon Hughes (Vienna: Springer, 2003), 242–263; Herbert A. Strauss, "The Movement of People in a Time of Crisis," in *The Muses Flee Hitler: Cultural Transfer and Adaptation, 1930–1945*, ed. Jarrell C. Jackman and Barla M. Borden (Washington, DC: Smithsonian Institution Press, 1983), 54; and Atina Grossman, "German Women Doctors from Berlin to New York: Maternity and Modernity in Weimar and in Exile," *Feminist Studies*, 19 (1993), 65–88.

84. Harry M. Marks, *The Progress of Experiment: Science and Therapeutic Reform in the United States, 1900–1990* (Cambridge: Cambridge Univ. Press, 1997).

85. James Harvey Young, *The Medical Messiahs: A Social History of Health Quackery in Twentieth-Century America* (Princeton, NJ: Princeton Univ. Press, 1967), esp. chap. 8; Charles O. Jackson, *Food and Drug Legislation in the New Deal* (Princeton, NJ: Princeton Univ. Press, 1970); Marks, *Progress of Experiment*, esp. chap. 3; John P. Swann, "The 1941 Sulfathiazole Disaster and the Birth of Good Manufacturing Practices," *Pharmacy in History*, 41 (1999), 16–25; Eric W. Boyle, *Quack Medicine: A History of Combating Health Fraud in Twentieth-Century America* (Santa Barbara, CA: Praeger, 2013), chap. 5.

86. Marks, *Progress of Experiment*, esp. chap. 4.

87. Sigismund Peller, *Cancer Research since 1900: An Evaluation* (New York: Philosophical Library, 1979); *Annual Reports of the United States Public Health Service for the Fiscal Years 1941–42, 1942–43* (Washington, DC: U.S. Government Printing Office, 1943), 1941–42, xiv.

88. E. C. Andrus et al., eds., *Advances in Military Medicine Made by American Investigators Working under the Sponsorship of the Committee on Medical Research*, 2 vols. (Boston: Little, Brown, 1948); Stephen Casper, "The Origins of the Anglo-American Research Alliance and the Incidence of Civilian Neuroses in Second World War Britain," *Medical History*, 52 (2008), 327–346.

89. Churchill, *Surgeon to Soldiers*; Fox, *Health Policies, Health Politics*, 116.

90. Perrin H. Long, "Medical Progress and Medical Education during the War," *JAMA*, 130 (1946), 983–990, esp. 989; Churchill, *Surgeon to Soldiers*, esp. 19–21; Thomas A. Guglielmo, "'Red Cross, Double Cross': Race and America's World War II–Era Blood Donor Service," *Journal of American History*, 97 (2010), 63–90.

91. Virginia Kneeland Frantz, "New Surgical Plastics and Hemostatics," in Andrus et al., *Advances in Military Medicine*, 182.

92. See Long, "Medical Progress and Medical Education," in ibid.; and esp. H. L. Haller and Stanley J. Cristol, "The Development of New Insecticides," ibid., 2:622–623; and Lucille R. Farquar, "Tropical Diseases," ibid., 1:68–69.

93. Farquar, "Tropical Diseases," 66–67; Oliver Cope, "The Burn Problem," in Andrus et al., *Advances in Military Medicine*, 1:149–153.

94. John Duffy, *The Sanitarians: A History of American Public Health* (Urbana: Univ. of Illinois Press, 1990), chap. 17.

95. *Annual Report of the Surgeon General of the Public Health Service of the United States*, 1938, 1; James T. Patterson, *The Dread Disease: Cancer and Modern American Culture* (Cambridge, MA: Harvard Univ. Press, 1987), 139.

96. Patterson, *Dread Disease*, chap. 5.

97. Arthur Kallet and F. J. Schlink, *100,000,000 Guinea Pigs: Dangers in Everyday Foods, Drugs, and Cosmetics* (New York: Grosset & Dunlap, 1933), 94.

98. Jackson, *Food and Drug Legislation*; Boyle, *Quack Medicine*, chap. 5.

99. Leslie J. Reagan, Nancy Tomes, and Paula A. Treichler, eds., *Medicine's Moving Pictures: Medicine, Health, and Bodies in American Film and Television* (Rochester, NY: Univ. of Rochester Press, 2007); Nancy Tomes, "An Undesired Necessity: The Commodification of Medical Service in the Interwar United States," in *Commodifying Everything: Relationships of the Market*, ed. Susan Strasser (New York: Routledge, 2003), 97–118.

100. Burnham, "American Medicine's Golden Age," 1474–1482; Fox, *Health Policies, Health Politics*, 149–150.

101. Christopher C. Sellers, *Hazards of the Job: From Industrial Disease to Environmental Health Science* (Chapel Hill: Univ. of North Carolina Press, 1997); Diana Chapman Walsh, *Corporate Physicians: Between Medicine and Management* (New Haven, CT: Yale Univ. Press, 1987), esp. chap. 3.

102. See, e.g., Joseph C. Doane, "Why Hospital Surgical Practice Should Be Standardized," in *The Hospital in Modern Society*, ed. Arthur C. Bachmeyer (New York: Commonwealth Fund, 1943), 222–224.

103. Richard Malmsheimer, *"Doctors Only": The Evolving Image of the American Physician* (Westport, CT: Greenwood, 1988), esp. chap. 7; Peter E. Dans, *Doctors in the Movies: Boil the Water and Just Say Aah* (Bloomington, IL: Medi-Ed Press, 2000), esp. chaps. 2 and 3, quotation from 66; Philip A. Kalisch and Beatrice J. Kalisch, "When Americans Called for Dr. Kildare: Images of Physicians and Nurses in the Dr. Kildare and Dr. Gillespie Movies, 1937–1947," *Medical Heritage*, 1 (1985), 348–363.

104. Paul Witty, *The Doctor* (Boston: D. C. Heath, 1953), 12, 16, 25–27.

105. Susan M. Reverby, *Ordered to Care: The Dilemma of American Nursing, 1850–1945* (Cambridge: Cambridge Univ. Press, 1987), chaps. 8–10; James Edmonson, personal communication.

106. Reverby, *Ordered to Care*, quotation from 175.

107. Philip A. Kalisch and Beatrice J. Kalisch, *The Advance of American Nursing*, 2nd ed. (Boston: Little, Brown, 1986), esp. chaps. 13 and 14.

108. Ibid., chap. 15, quotation from 548.

109. David E. Hailman, *The Prevalence of Disabling Illness among Male and Female Workers and Housewives*, U.S. Public Health Service, Public Health Bulletin No. 260 (Washington, DC: U.S. Government Printing Office, 1941); Ernst P. Boas, *The Unseen Plague:*

Chronic Disease (New York: J. J. Augustin, 1940), 121; Daniel J. Wilson, "A Crippling Fear: Experiencing Polio in the Era of FDR," *Bulletin of the History of Medicine*, 72 (1998), 464–495.

110. Herbert Yahraes, *Something Can Be Done About Chronic Illness* (New York: Public Affairs Committee, 1951).

111. Patterson, *Dread Disease*, 141; George Weisz, *Chronic Disease in the Twentieth Century: A History* (Baltimore: Johns Hopkins Univ. Press, 2014), esp. chap. 5.

Part III. Medicine and Health in an Age of Technology

1. Michael Bliss, *The Making of Modern Medicine: Turning Points in the Treatment of Disease* (Chicago: Univ. of Chicago Press, 2011), 89.

2. Daniel M. Fox, *Health Policies, Health Politics: The British and American Experience, 1911–1965* (Princeton, NJ: Princeton Univ. Press, 1986), 149.

3. Paul Forman, "The Primacy of Science in Modernity, of Technology in Postmodernity, and of Ideology in the History of Technology," *History and Technology*, 23 (2007), 1–152; Wesley Shrum, Joel Genuth, and Ivan Chumpalov, *Structures of Scientific Collaboration* (Cambridge, MA: MIT Press, 2007); Alvin M. Weinberg, "Can Technology Replace Social Engineering?," *Bulletin of the Atomic Scientists*, 22 (1966), 4–8.

4. Jennifer Stanton, "Making Sense of Technologies in Medicine," *Social History of Medicine*, 12 (1999), 439.

5. Jonathan Simon, "The Emergence of a Risk Society: Insurance, Law, and the State," *Socialist Review*, Sept.–Oct. 1987, 61–89; William G. Rothstein, *Public Health and the Risk Factor: A History of an Uneven Medical Revolution* (Rochester, NY: Univ. of Rochester Press, 2003); Robert A. Aronowitz, *Making Sense of Illness: Science, Society, and Disease* (Cambridge: Cambridge Univ. Press, 1998).

6. Rosemary Stevens, "Health Care in the Early 1960s," *Health Care Financing Review*, 18 (1996), 11–22.

7. See, e.g., David Mechanic, *The Growth of Bureaucratic Medicine: An Inquiry into the Dynamics of Patient Behavior and the Organization of Medical Care* (New York: John Wiley & Sons, 1976), pt. 1; or John Gordon Freymann, *The American Health Care System: Its Genesis and Trajectory* (Baltimore: Williams & Wilkins, 1974), 303–304.

8. See esp. Michael L. Millman, *Politics and the Expanding Physician Supply* (Montclair, NJ: Allanheld, Osmun, 1980).

9. Freymann, *American Health Care System*, 74.

10. Sharon R. Kaufman, *The Healer's Tale: Transforming Medicine and Culture* (Madison: Univ. of Wisconsin Press, 1993), 316.

Chapter 9. The Age of Technological Medicine, 1940s to 1960s

1. Victor R. Fuchs, "The Health Sector's Share of the Gross National Product," *Science*, 247 (1990), 535.

2. *Source Book of Health Insurance Data*, 1970, 44; Barbara S. Cooper, "National Health Expenditures, 1929–72," *Social Security Bulletin*, 36 (Jan. 1973), 5; Edgar Charles, "Marketing Health Information in America," *Journal of the Alabama Academy of Science*, 44 (1973), 90.

3. Forrest E. Linder, "The Health of the American People," *Scientific American*, June 1966, 21–29, quotations from 26–27.

4. Robert Straus and John A. Clausen, "Health, Society, and Social Science," *Annals of the American Academy of Political and Social Science*, 346 (1968), 1–8.

5. Daniel M. Fox, *Health Policies, Health Politics: The British and American Experience, 1911–1965* (Princeton, NJ: Princeton Univ. Press, 1986), 150.

6. *Annual Report of the Federal Security Agency: Public Health Service*, 1949, 1–9; *Facts of Life and Death*, DHEW Publication No. (HRA) 74-1222 (Rockville, MD: National Center for Health Statistics, 1974), 3, 7, 9.

7. Monroe Lerner and Odin W. Anderson, *Health Progress in the United States, 1900–1960* (Chicago: Univ. of Chicago Press, 1963), esp. 18; *Facts of Life and Death*, 29; James C. Riley, "Why Sickness and Death Rates Do Not Move Parallel to One Another over Time," *Social History of Medicine*, 12 (1999), 101–124.

8. John E. Sutherland, Victoria W. Persky, and Jacob A. Brody, "Proportionate Mortality Trends: 1950 through 1986," *JAMA*, 264 (1990), 2178–3184; "What Ever Became of Those Quarantine Signs?," *Reader's Digest*, Nov. 1967, M2–M3.

9. H. Gordon MacLean, "Medicine on the March," *California Medicine*, 76 (1952), 316.

10. Commission on Chronic Illness, *Chronic Illness in the United States*, 2 vols. (Cambridge, MA: Harvard Univ. Press, 1957); Anselm L. Strauss and Barney G. Glaser, *Chronic Illness and the Quality of Life* (St. Louis: C. V. Mosby, 1975), 1–2; George Weisz, *Chronic Disease in the Twentieth Century: A History* (Baltimore: Johns Hopkins Univ. Press, 2014), esp. chaps. 5 and 6.

11. Daniel Fox, *Power and Illness: The Failure and Future of American Health Policy* (Berkeley: Univ. of California Press, 1993), esp. 53–55; Harry M. Marks, "Cortisone, 1949: A Year in the Political Life of a Drug," *Bulletin of the History of Medicine*, 66 (1992), 419–439; "Cortisone Output to Be Tripled," *Science News-Letter*, 59 (1951), 168.

12. "Anniversary of a Blessed Release," *New York State Journal of Medicine*, 65 (1965), 1316; Daniel J. Wilson, *Living with Polio: The Epidemic and Its Survivors* (Chicago: Univ. of Chicago Press, 2005), 3; James L. Wilson, quoted in James H. Maxwell, "The Iron Lung: Halfway Technology or Necessary Step?," *Milbank Quarterly*, 64 (1986), 7.

13. Sally Smith Hughes, *The Virus: A History of the Concept* (New York: Science History Publications, 1977), esp. 104–105.

14. Howard Markel, "The Genesis of the Iron Lung: Philip Drinker, Charles F. McKhann, James L. Wilson, and Early Attempts at Administering Artificial Respiration to Patients with Poliomyelitis," *Archives of Pediatrics*, 148 (1994), 1174–1180.

15. Naomi Rogers, *Dirt and Disease: Polio before FDR* (New Brunswick, NJ: Rutgers Univ. Press, 1992), epilogue; David M. Oshinsky, *Polio: An American Story* (New York: Oxford Univ. Press, 2005).

16. See previous note as well as Harry M. Marks, "The 1954 Salk Poliomyelitis Vaccine Field Trial," unpublished; and Paul A. Offit, *The Cutter Incident: How America's First Polio Vaccine Led to the Growing Vaccine Crisis* (New Haven, CT: Yale Univ. Press, 2005).

17. James Colegrove, *State of Immunity: The Politics of Vaccination in Twentieth-Century America* (Berkeley: Univ. of California Press, 2006); "With Polio Licked—Cold, Flu, Malaria, T.B.: Next Targets for Science," *U.S. News & World Report*, 29 Apr. 1955, 32.

18. Louis Galambos and Jane Eliot Sewell, *Networks of Innovation: Vaccine Development at Merck, Sharp & Dohme, and Mulford, 1895–1995* (Cambridge: Cambridge Univ. Press,

1995), esp. chap. 5; Louise B. Russell, *Is Prevention Better Than Cure?* (Washington, DC: Brookings Institution, 1986), 22–30.

19. R. Allan Freeze and Jay H. Lehr, *The Fluoride Wars: How a Modest Public Health Measure Became America's Longest-Running Political Melodrama* (New York: John Wiley & Sons, 2009); Alyssa Picard, *Making the American Mouth: Dentists and Public Health in the Twentieth Century* (New Brunswick, NJ: Rutgers Univ. Press, 2009), esp. chaps. 5–7; Bonnie Bullough and George Rosen, *Preventive Medicine in the United States, 1900–1990: Trends and Interpretations* (Canton, MA: Science History Publications, 1992), 113.

20. Linder, "Health of the American People," 27.

21. Alfred Goodman, quoted in "The Birth of a Medical Text," *JAMA*, 159 (1955), 25.

22. Martin L. Gross, *The Doctors* (New York: Random House, 1966), 114–116.

23. Toine Pieters, *Interferon: The Science and Selling of a Miracle Drug* (London: Routledge, 2005).

24. William A. Silverman, *Retrolental Fibroplasia: A Modern Parable* (New York: Grune & Stratton, 1986); Robert M. Jacobson and Alvan R. Feinstein, "Oxygen as a Cause of Blindness in Premature Infants: 'Autopsy' of a Decade of Errors in Clinical Epidemiologic Research," *Journal of Clinical Epidemiology*, 45 (1992), 1265–1287.

25. "The $64 Question," *New York State Journal of Medicine*, 49 (1949), 35.

26. Marc Berg, "Turning a Practice into a Science: Reconceptualizing Postwar Medical Practice," *Social Studies of Science*, 25 (1995), 437–476.

27. R. N. Wilson, quoted in *The American Health Care System: Its Genesis and Trajectory*, by John Gordon Freymann (Baltimore: Williams & Wilkins, 1974), 304.

28. MacLean, "Medicine on the March," 316; Stanley Joel Reiser, *Medicine and the Reign of Technology* (Cambridge: Cambridge Univ. Press, 1978), 159, 187.

29. Walter Alvarez, "The Need for Looking at the Patient," *Modern Medicine*, 6 Mar. 1961, 68.

30. Jeremy A. Greene and Scott H. Podolsky, "Keeping Modern in Medicine: Pharmaceutical Promotion and Physician Education in Postwar America," *Bulletin of the History of Medicine*, 83 (2009), 338.

31. See esp. Harry M. Marks, "Revisiting 'The Origins of Compulsory Drug Prescriptions,'" *American Journal of Public Health*, 85 (1995), 109–115.

32. See, e.g., James Harvey Young, "The Persistence of Medical Quackery in America," *American Scientist*, 60 (1972), 318–326; "Medical Devices: An Unhealthy Situation," *Consumer Reports*, 35 (1970), 256–259; and National Commission on Community Health Services, *Health Is a Community Affair* (Cambridge, MA: Harvard Univ. Press, 1966), 67.

33. Peter Temin, *Taking Your Medicine: Drug Regulation in the United States* (Cambridge, MA: Harvard Univ. Press, 1980), chaps. 4–6; Philip J. Hilts, *Protecting America's Health: The FDA, Business, and One Hundred Years of Regulation* (New York: Alfred A. Knopf, 2003), esp. chaps. 7–12.

34. See Scott H. Podolsky, "Antibiotics and the Social History of the Controlled Clinical Trial, 1950–1970," *Journal of the History of Medicine and Allied Sciences*, 65 (2010), 327–367; and, e.g., Hilts, *Protecting America's Health*.

35. Jordan Goodman, "Pharmaceutical Industry," in *Medicine in the Twentieth Century*, ed. Roger Cooter and John Pickstone (Amsterdam: Harwood Academic, 2000), 141–154; Andrea Tone, *The Age of Anxiety: A History of America's Turbulent Affair with*

Tranquilizers (New York: Basic Books, 2009), esp. chaps. 2 and 3; Charlotte Muller, "The Overmedicated Society: Forces in the Marketplace for Medical Care," *Science*, 176 (1972), 489.

36. Goodman, "Pharmaceutical Industry"; Fox, *Power and Illness*, 54–55.

37. See esp. James Reed, *From Private Vice to Public Virtue: The Birth Control Movement and American Society Since 1830* (New York: Basic Books, 1978).

38. See, e.g., Greene and Podolsky, "Keeping Modern in Medicine," 331–377; and Dominique A. Tobbell, "'Who's Winning the Human Race?' Cold War as Pharmaceutical Political Strategy," *Journal of the History of Medicine*, 64 (2009), 429–473.

39. M. G. Jacoby, "Medical and Associated Services in the United States," *British Medical Journal*, 8 Sept. 1956, 597.

40. Robert Bud, "From Germophobia to the Carefree Life and Back Again: The Lifecycle of the Antibiotic Brand," in *Medicating Modern America: Prescription Drugs in History*, ed. Andrea Tone and Elizabeth Siegel Watkins (New York: New York Univ. Press, 2007), 17–41.

41. See esp. Jeremy A. Greene, *Prescribing by Numbers: Drugs and the Definition of Disease* (Baltimore: Johns Hopkins Univ. Press, 2007).

42. Tone, *Age of Anxiety*; Gerald N. Grob and Allan V. Horwitz, *Diagnosis, Therapy, and Evidence: Conundrums in Modern American Medicine* (New Brunswick, NJ: Rutgers Univ. Press, 2010), chaps. 5 and 6; Greene, *Prescribing by Numbers*.

43. Alfonso Gambardella, *Science and Innovation: The US Pharmaceutical Industry during the 1980s* (Cambridge: Cambridge Univ. Press, 1985), 23–30; John P. Swann, *Academic Scientists and the Pharmaceutical Industry: Cooperative Research in Twentieth-Century America* (Baltimore: Johns Hopkins Univ. Press, 1988); Greene, *Prescribing by Numbers*; Robert M. Kaiser, "The Introduction of the Thiazides: A Case Study in Twentieth-Century Therapeutics," *Publications of the American Institute of the History of Pharmacy*, 16 (1997), 121–137; Marks, "Cortisone, 1949," esp. 438.

44. Barron H. Lerner, *When Illness Goes Public: Celebrity Patients and How We Look at Medicine* (Baltimore: Johns Hopkins Univ. Press, 2006), 66; Lerner, *The Breast Cancer Wars: Hope, Fear, and the Pursuit of a Cure in Twentieth-Century America* (New York: Oxford Univ. Press, 2001), 90.

45. David Serlin, *Replaceable You: Engineering the Body in Postwar America* (Chicago: Univ. of Chicago Press, 2004), quotation from 17.

46. Sheila M. Rothman and David J. Rothman, *The Pursuit of Perfection: The Promise and Perils of Medical Enhancement* (New York: Pantheon Books, 2003).

47. Serlin, *Replaceable You*, 16.

48. Edwin Olmos, F. Hampton Roy, and Daljit Singh, *Intraocular Lenses* (New York: Praeger, 1981), esp. chap 1; David J. Apple, "Harold Ridley, MA, MD, FRCS: A Golden Anniversary Celebration and a Golden Age," *Archives of Ophthalmology*, 117 (1999), 827–828.

49. See, e.g., Stuart S. Blume, *Insight and Industry: On the Dynamics of Technological Change in Medicine* (Cambridge, MA: MIT Press, 1992), 185; and Ira S. Brodsky, *The History and Future of Medical Technology* (St. Louis: Telescope Books, 2010), esp. 24–29.

50. Harris B. Schumacker, *The Evolution of Cardiac Surgery* (Bloomington: Indiana Univ. Press, 1992).

51. G. E. Schreiner, "The American Society for Artificial Internal Organs, Twenty Years Ago Tonight," *Transactions, American Society for Artificial Internal Organs*, 20-A (1974), 1–20, quoting L. Bluemle, 3.

52. Ulrich Tröhler, "Surgery (Modern)," in *Companion Encyclopedia of the History of Medicine*, ed. W. F. Bynum and Roy Porter, 2 vols. (London: Routledge, 1993), 2:1004–1006; "Last Ten Years Brought Major Heart Breakthroughs," *Science Digest*, June 1959, 50–51; David S. Jones, *Broken Hearts: The Tangled History of Cardiac Care* (Baltimore: Johns Hopkins Univ. Press, 2013); Francis D. Moore, *A Miracle and a Privilege: Recounting a Half Century of Surgical Advance* (Washington, DC: Joseph Henry, 1995), 109–110.

53. Susan E. Lederer, *Flesh and Blood: A Cultural History of Transplantation and Transfusion in America* (New York: Oxford Univ. Press, 2008); Harold M. Schmeck Jr., *The Semi-Artificial Man: A Dawning Revolution in Medicine* (New York: Walker, 1965), 91.

54. See, e.g., Nathan Rosenberg, Annette C. Gelijns, and Holly Dawkins, eds., *Sources of Medical Technology: Universities and Industry* (Washington, DC: National Academy Press, 1995).

55. Blume, *Insight and Industry*, esp. chap. 3; Rosenberg, Gelijns, and Dawkins, *Sources of Medical Technology*; Bettyann Holtzmann Kevles, *Naked to the Bone: Medical Imaging in the Twentieth Century* (New Brunswick, NJ: Rutgers Univ. Press, 1997), chap. 10.

56. *40 Years of People, Progress, and Patient Safety* (Arlington, VA: Association for the Advancement of Medical Instrumentation, 2007); David A. Kessler, Stuart M. Paper, and David N. Sundwall, "The Federal Regulation of Medical Devices," *New England Journal of Medicine*, 317 (1987), 357–366.

57. Harry M. Marks, "Medical Technologies: Social Contexts and Consequences," in Bynum and Porter, *Companion Encyclopedia of the History of Medicine*, 2:1603.

58. Richard A. Rettig, *Health Care Technology: Lessons Learned from the End-State Renal Disease Experience* (Santa Monica, CA: Rand Corporation, 1976); David L. Ellison, *The Bio-Medical Fix: Human Dimensions of Bio-Medical Technologies* (Westport, CT: Greenwood, 1978), chap. 5.

59. Richard A. Rettig, *End-Stage Renal Disease and the "Cost" of Medical Technology* (Santa Monica, CA: Rand Corporation, 1977); Richard A. Rettig and Ellen Marks, *The Federal Government and Social Planning for End-Stage Renal Disease: Past, Present, and Future* (Santa Monica, CA: Rand Corporation, 1983); Richard A. Rettig and Norman G. Levinsky, eds., *Kidney Failure and the Federal Government* (Washington, DC: National Academy Press, 1991); David J. Rothman, *Beginnings Count: The Technological Imperative in American Health Care* (New York: Oxford Univ. Press, 1997), chap. 4; Stanley Joel Reiser, *Technological Medicine: The Changing World of Doctors and Patients* (New York: Cambridge Univ. Press, 2009), chap. 3.

60. Marion J. Dakin, "The Psychosomatic Approach in General Practice," *Medical Clinics of North America*, 31 (1947), 213; Vincent P. Mahoney, "Evaluation of Resistance to the Psychosomatic Approach," *Journal of the Medical Society of New Jersey*, 52 (1955), 70–75; Theodore M. Brown, "The Rise and Fall of American Psychosomatic Medicine," accessed 8 Nov. 2006, http://www.human-nature.com/free-associations/riseandfall.html; Brown, "George Canby Robinson and 'The Patient as a Person,'" in *Greater Than the Parts: Holism in*

Biomedicine, 1920–1950, ed. Christopher Lawrence and George Weisz (New York: Oxford Univ. Press, 1998), 135–160.

61. David Armstrong, "The Patient's View," *Social Science and Medicine*, 18 (1984), 759.

62. Hans Selye, *The Stress of Life* (New York: McGraw-Hill, 1956); Selye, "The Stress of Life—New Focal Point for Understanding Accidents," *Industrial Medicine and Surgery*, 33 (1964), 621–625; Tone, *Age of Anxiety*.

63. John C. Burnham, "American Physicians and Tobacco Use: Two Surgeons General, 1929 and 1964," *Bulletin of the History of Medicine*, 63 (1989), 1–31; Mark Parascandola, "Skepticism, Statistical Methods, and the Cigarette: A Historical Analysis of a Methodological Debate," *Perspectives in Biology and Medicine*, 47 (2004), 244–261; Parascandola, "Epidemiology in Transition: Tobacco and Lung Cancer in the 1950s," in *Body Counts: Medical Quantification in Historical and Sociological Perspective / La quantification medicale, perspectives historiques et sociologiues*, ed. Gérard Jorland, Annick Opinel, and George Weisz (Montreal: McGill-Queen's Univ. Press, 2005), 226–248.

64. Jon Harkness, "The U.S. Public Health Service and Smoking in the 1950s: The Tale of Two More Statements," *Journal of the History of Medicine and Allied Sciences*, 62 (2007), 171–212; James T. Patterson, *The Dread Disease: Cancer and Modern American Culture* (Cambridge, MA: Harvard Univ. Press, 1987), esp. chap. 8; Allan M. Brandt, *The Cigarette Century: The Rise, Fall, and Deadly Persistence of the Product That Defined America* (New York: Basic Books, 2007).

65. See, e.g., Alessandra Parodi, David Neasham, and Paolo Vineis, "Environment, Population, and Biology," *Perspectives in Biology and Medicine*, 49 (2006), 357–368; Mervyn Susser, "Epidemiology in the United States after World War II: The Evolution of Technique," *Epidemiologic Reviews*, 7 (1985), 147–177; and William G. Rothstein, *Public Health and the Risk Factor: A History of an Uneven Medical Revolution* (Rochester, NY: Rochester Univ. Press, 2003), esp. chap. 13.

66. Harry M. Marks, *The Progress of Experiment: Science and Therapeutic Reform in the United States, 1900–1990* (Cambridge: Cambridge Univ. Press, 1997), esp. 132–133 and 155.

67. Margaret Pittman, "History of the Development of Pertussis Vaccine," *Developments in Biological Standardization*, 73 (1991), 13–29.

68. Marks, *Progress of Experiment*.

69. Rothstein, *Public Health and the Risk Factor*.

70. Bonny Kaplan, "The Computer Prescription: Medical Computing, Public Policy, and Views of History," *Science, Technology, and Human Values*, 20 (1995), 1–14, quotation from 9; Ellison, *Bio-Medical Fix*, 54.

71. John C. Burnham, "The Evolution of Editorial Peer Review," *JAMA*, 263 (1990), 1323–1329; Burnham, "How Journal Editors Came to Develop and Critique Peer Review Procedures," in *Research Ethics, Manuscript Review, and Journal Quality*, ed. H. F. Mayland and R. E. Sojka (Madison, WI: American Society of Agronomy / ACS, 1992), 55–62.

72. American Foundation, *Medical Research: A Midcentury Survey*, 2 vols. (Boston: Little, Brown, 1955), 1:xviii.

73. Donald C. Swain, "The Rise of a Research Empire: NIH, 1930 to 1950," *Science*, 138 (1962), 1233–1237; Victoria A. Harden, *Inventing the NIH: Federal Biomedical Research Policy, 1887–1937* (Baltimore: Johns Hopkins Univ. Press, 1986); Stephen P. Strickland, "Integration of Medical Research and Health Policies," *Science*, 173 (1971), 1093–1103.

74. Robert Gallo, quoted in Maxwell M. Wintrobe, *Hematology: The Blossoming of a Science—A Story of Inspiration and Effort* (Philadelphia: Lea & Febiger, 1985), 488.

75. Strickland, "Integration of Medical Research"; Stephen P. Strickland, *Politics, Science, and Dread Disease: A Short History of United States Medical Research Policy* (Cambridge, MA: Harvard Univ. Press, 1972); Fox, *Health Policies, Health Politics*, chap. 9.

76. Department of Health, Education, and Welfare, *Annual Report*, 1966, 99.

77. Rosemary Stevens, *In Sickness and in Wealth: American Hospitals in the Twentieth Century* (New York: Basic Books, 1989), 216–224; Beatrix Hoffman, *Health Care for Some: Rights and Rationing in the United States since 1930* (Chicago: Univ. of Chicago Press, 2012), chap. 4.

78. Strickland, *Politics, Science, and Dread Disease*, quotation from 213; Fox, *Health Policies, Health Politics*, 151–153.

79. Strickland, "Integration of Medical Research"; Strickland, *Politics, Science, and Dread Disease*, esp. 125–126; "Who Pays for Medical Research?," *Medical Economics*, July 1951, 64.

80. Toby A. Appel, *Shaping Biology: The National Science Foundation and American Biological Research, 1945–1975* (Baltimore: Johns Hopkins Univ. Press, 2000).

81. Strickland, *Politics, Science, and Dread Disease*, esp. chaps. 5–7.

82. Alvan R. Feinstein, Neil Koss, and John H. M. Austin, "The Changing Emphasis in Clinical Research," *Annals of Internal Medicine*, 66 (1967), 396–434; Feinstein, Koss, and Austin, "The Changing Emphasis in Clinical Research. III," ibid., 125 (1970), 885–891.

83. See esp. Daniel M. Fox, "The Consequences of Consensus: American Health Policy in the Twentieth Century," *Milbank Quarterly*, 64 (1986), 76–99.

84. Strickland, *Politics, Science, and Dread Disease*, esp. chap. 3; Kirsten E. Gardner, *Early Detection: Women, Cancer, and Awareness Campaigns in the Twentieth-Century United States* (Chapel Hill: Univ. of North Carolina Press, 2006), chap. 3; Elizabeth W. Etheridge, *Sentinel for Health: A History of the Centers for Disease Control* (Berkeley: Univ. of California Press, 1992), 108.

85. Strickland, *Politics, Science, and Dread Disease*, esp. chap. 9.

86. Thomas E. Cone Jr., *History of American Pediatrics* (Boston: Little, Brown, 1979), 232.

87. Stewart Wolf, "Disease as a Way of Life: Neural Integration in Systemic Pathology," in *Life and Disease: New Perspectives in Biology and Medicine*, ed. Dwight J. Ingle (New York: Basic Books, 1963), 367.

88. D. W. Woolley, "The Revolution in Pharmacology," in ibid., 131–154.

89. Cay Rüdiger Prüll, Andreas-Holger Maehle, and Robert Francis Halliwell, *A Short History of the Drug Receptor Concept* (Houndmills, UK: Palgrave Macmillan, 2009), esp. chap. 6 and 160–161.

90. Donald G. Cooley, *The Science Book of Modern Medicines* (New York: Pocket Books, 1963), 27.

91. Robert E. Kohler, *From Medical Chemistry to Biochemistry: The Making of a Biomedical Discipline* (Cambridge: Cambridge Univ. Press, 1982), 324–335, quotation from 332; Jean-Paul Gaudillière, "The Molecularization of Cancer Etiology in the Postwar United States: Instruments, Politics and Management," in *Molecularizing Biology and Medicine: New Practices and Alliances, 1910s–1970s*, ed. Soraya de Chadarevian and Harmke

Kamminga (Amsterdam: Harwood Academic, 1998), 139–150; Garland E. Allen, *Life Science in the Twentieth Century* (New York: John Wiley & Sons, 1975), esp. 187–189; Lily E. Kay, *The Molecular Vision of Life: Caltech, The Rockefeller Foundation, and the Rise of the New Biology* (New York: Oxford Univ. Press, 1993).

92. De Chadarevian and Kamminga, *Molecularizing Biology and Medicine*; Bruno J. Strasser, "Sickle Cell Anemia, a Molecular Disease," *Science*, 286 (1999), 1488–1490; Todd L. Savitt, *Race and Medicine in Nineteenth- and Early Twentieth-Century America* (Kent, OH: Kent State Univ. Press, 2007), chap 6.

93. Kay, *Molecular Vision of Life*, esp. 259–277.

94. Kathryn Hillier, "Babies and Bacteria: Phage Typing, Bacteriologists, and the Birth of Infection Control," *Bulletin of the History of Medicine*, 80 (2006), 733–761.

95. Maurice L. Tainter, "Medicine's Golden Age: The Triumph of the Experimental Method," *Transactions of the New York Academy of Sciences*, 18 (1956), 227; Clayton G. Loosli, "Preventive Medicine," in *Medical World Annual, 1968* (New York: McGraw-Hill, 1968), 177.

96. Moore, *Miracle and a Privilege*, 123–124.

97. "Health and Age," *Newsweek*, 14 Dec. 1959, 110; "Medical Research: New Discoveries that May Save Your Life," *Changing Times*, July 1967, 29–33.

98. See, e.g., David S. Jones, "Technologies of Compliance: Surveillance of Self-Administration of Tuberculosis Treatment, 1956–1966," *History and Technology*, 17 (2001), 279–318.

99. Robert Bud, "From Epidemic to Scandal: The Politicization of Antibiotic Resistance, 1957–1969," in *Devices and Designs: Medical Technologies in Historical Perspective*, ed. Carsten Timmerman and Julie Anderson (Basingstoke, UK: Palgrave Macmillan, 2006), 195–211; Walter Modell, "Hazards of New Drugs," *Science*, 139 (1963), 1180.

100. David M. Spain, *The Complications of Modern Medical Practices: A Treatise on Iatrogenic Diseases* (New York: Grune & Stratton, 1963), xv.

Chapter 10. Doctors, Patients, Medical Institutions, and Society in the Age of Technological Medicine

1. See, e.g., Noel Thompson, "Technology Is Getting a Bum Rap," *American Medical News*, 27 Feb. 1978; and Sidney R. Garfield, "The Delivery of Medical Care," *Scientific American*, Apr. 1970, 15–23.

2. Earl Lomon Koos, *The Health of Regionville: What the People Thought and Did About It* (New York: Columbia Univ. Press, 1954), esp. 139.

3. Ibid., esp. 32–33.

4. *Source Book of Health Insurance Data*, 1970, 45.

5. K. E. Monroe and G. A. Roback, *Reference Data on Socioeconomic Issues of Health*, 2nd ed. ([Chicago]: American Medical Association, 1971), 35–36, 42; Robert F. Rushmer, *Humanizing Health Care: Alternative Futures for Medicine* (Cambridge, MA: MIT Press, 1975), 100.

6. Monroe and Roback, *Reference Data on Socioeconomic Issues of Health*, 82.

7. Victor R. Fuchs, "The Growing Demand for Medical Care," *New England Journal of Medicine*, 279 (1968), 191.

8. Victor R. Fuchs and Maria J. Kramer, *Determinants of Expenditures for Physicians' Services in the United States, 1948–68*, DHEW Publication No. 73-3013 (Washington, DC: National Center for Health Services Research and Development, 1972), 13–15.

9. Greer Williams, "Quality versus Quantity in American Medical Education," *Science*, 153 (1966), 958.

10. Fuchs and Kramer, *Determinants of Expenditures*, 16–17.

11. David J. Rothman, "A Century of Failure: Class Barriers to Reform," in *The Politics of Health Care Reform: Lessons form the Past, Prospects for the Future*, ed. James A. Morone and Gary S. Belkin (Durham, NC: Duke Univ. Press, 1994), 17–21.

12. Milton I. Roemer, *Ambulatory Health Services in America: Past, Present, and Future* (Rockville, MD: Aspen, 1981), esp. 173; Mary E. McNamara and Clifford Todd, "A Survey of Group Practice in the United States, 1969," *American Journal of Public Health*, 60 (1970), 1303–1313.

13. Kenneth M. Ludmerer, *Time to Heal: American Medical Education from the Turn of the Century to the Era of Managed Care* (New York: Oxford Univ. Press, 1999), chaps. 8–11, esp. 187; Sydney A. Halpern, *American Pediatrics: The Social Dynamics of Professionalism, 1880–1980* (Berkeley: Univ. of California Press, 1988), 115–119; Fitzhugh Mullan, introduction to *Big Doctoring in America: Profiles in Primary Care*, ed. Mullan (Berkeley: Univ. of California Press, 2002), xii.

14. William A. Dunnagan, "Why I Stopped Being a Family Doctor," *Look*, 21 Aug. 1956, 35–37; "Incomes: Specialists vs. G.P.'s," *Medical Economics*, Sept. 1951, 59–62.

15. Donald F. Gearing and Robert L. Brenner, "They're Moving to the Suburbs," *Medical Economics*, 23 June 1958, 98–102.

16. Rosemary Stevens, *American Medicine and the Public Interest*, 2nd ed. (Berkeley: Univ. of California Press, 1998), chaps. 12 and 14 and esp. 320.

17. David P. Adams, *American Board of Family Practice: A History* (Lexington, KY: American Board of Family Practice, 1999); Stevens, *American Medicine and the Public Interest*, 305–310 and chaps. 14 and 15, esp. 342.

18. Adams, *American Board of Family Practice*; George Weisz, *Divide and Conquer: A Comparative History of Medical Specialization* (Oxford: Oxford Univ. Press, 2006), 248–256, esp. 251–252.

19. Paul E. Stepansky, *The Last Family Doctor: Remembering My Father's Medicine* (Montclair, NJ: Keynote Books, 2011), esp. chap. 11.

20. Ibid., quotation from 116.

21. Stevens, *American Medicine and the Public Interest*, 342.

22. Ibid., esp. 265 and 380; William G. Rothstein, *American Medical Schools and the Practice of Medicine: A History* (New York: Oxford Univ. Press, 1987), 275.

23. "Foreign Doctors: High Temperatures," *Newsweek*, 5 Dec. 1960, 69–70; "Foreign-Doctor 'Invasion' of U.S.," *U.S. News & World Report*, 28 Aug. 1967, 72; Paul Starr, "Too Many Doctors?," *Washington Post*, 13 Mar. 1977; Rosemary Stevens and Joan Vermeulen, *Foreign Trained Physicians and American Medicine*, DHEW Publication No. (NIH) 73-325 (Washington, DC: U.S. Government Printing Office, 1973), esp. xii and 120; Kathleen N. Williams, "Foreign Medical Graduates and Their Impact on the Quality of Medical Care in the United States," *Milbank Memorial Fund Quarterly*, 53 (1975), 549–581.

24. Dietrich C. Reitzes, *Negroes and Medicine* (Cambridge, MA: Harvard Univ. Press, 1958), esp. 5.

25. Ibid., esp. 63; Vanessa Northington Gamble, *Making a Place for Ourselves: The Black Hospital Movement, 1920–1945* (New York: Oxford Univ. Press, 1995), 182–196; Lynn Marie

Pohl, "Long Waits, Small Spaces, and Compassionate Care: Memories of Race and Medicine in a Mid-Twentieth Century Southern Community," *Bulletin of the History of Medicine*, 74 (2000), 107–137.

26. Martin L. Gross, *The Doctors* (New York: Random House, 1966), quotation from 337; Richard T. Smith, "Infectious Diseases and Immunology," *New Physician*, 11 (1962), 240.

27. John C. Burnham, "American Medicine's Golden Age: What Happened to It?," *Science*, 215 (1982), 1474–1479; Richard Carter, *The Doctor Business* (Garden City, NY: Doubleday, 1958); Barbara G. Myerhoff and William R. Larson, "The Doctor as Culture Hero: The Routinization of Charisma," *Human Organization*, 24 (1965), 188–191; Leslie J. Reagan, Nancy Tomes, and Paula A. Treichler, eds., *Medicine's Moving Pictures: Medicine, Health, and Bodies in American Film and Television* (Rochester, NY: Rochester Univ. Press, 2007).

28. Fuchs and Kramer, *Determinants of Expenditures*, 16; Burnham, "American Medicine's Golden Age," 1474–1479.

29. J. Robert Moskin, "The Challenge to Our Doctors," *Look*, 3 Nov. 1964, 27.

30. Carter, *Doctor Business*; Gross, *Doctors*.

31. Gross, *Doctors*; Eliot Freidson, *Patients' Views of Medical Practice—A Study of Subscribers to a Prepaid Medical Plan in The Bronx* (New York: Russell Sage Foundation, 1961), esp. chap. 3.

32. Gross, *Doctors*, 335.

33. Monte M. Poen, *Harry S. Truman versus the Medical Lobby: The Genesis of Medicare* (Columbia: Univ. of Missouri Press, 1979), 206.

34. Guenter B. Risse, "Once on Top, Now on Tap: American Physicians View Their Relationships with Patients, 1920–1970," in *Responsibility in Health Care*, ed. George J. Agich (Dordrecht, Netherlands: D. Reidel, 1982), 23–49.

35. "Pinpointing the Doctor Shortage," *American Journal of Public Health*, 42 (1952), 311.

36. Quentin Young, quoted in Brian J. Zink, *Anyone, Anything, Anytime: A History of Emergency Medicine* (Philadelphia: Mosby Elsevier, 2006), 4; ibid., 10.

37. Carter, *Doctor Business*, 88–90.

38. *Source Book of Health Insurance Data*, 1959, 65; Benjamin Spock, "Should You Ask Your Doctor to Make a House Call?," *Ladies' Home Journal*, Oct. 1961, 24, 28; "Doctors' Debate: Stop Home Calls?," *U.S. News & World Report*, 30 Oct. 1961, 66; "Is There Really a Doctor Shortage?," *Medical Economics*, Apr. 1951, 74–77; Evan Hill, "The American Doctor: Death of a Legend in an Era of Miracles," *Saturday Evening Post*, 15 June 1963, 30, 37.

39. Risse, "Once on Top," esp. 33–35 and 38; Rollen Waterson, "The Doctor-Patient Relationship—A Psychological Study," *GP*, Oct. 1951, 93–109; Richard H. Blum, *The Management of the Doctor-Patient Relationship* (New York: McGraw Hill, 1960).

40. Freidson, *Patients' Views of Medical Practice*; Lucinda McCray Beier, *Health Culture in the Heartland, 1880–1980: An Oral History* (Urbana: Univ. of Illinois Press, 2009).

41. Blum, *Management of the Doctor-Patient Relationship*, 284.

42. Howard S. Becker et al., *Boys in White: Student Culture in Medical School* (Chicago: Univ. of Chicago Press, 1961), esp. 10 and 191; Vernon W. Lippard, *A Half-Century of American Medical Education: 1920–1970* (New York: Josiah Macy, Jr. Foundation, 1974), esp. 30 and 33; Ludmerer, *Time to Heal*, esp. 197–198; Linda Headrick, "Seeking a Common Language in Primary Care," in Mullan, *Big Doctoring in America*, 76; Leon Sokoloff, "The Rise

and Decline of the Jewish Quota in Medical School Admissions," *Bulletin of the New York Academy of Medicine*, 58 (1992), 497–518.

43. Becker et al., *Boys in White*, quotation from 126; Rothstein, *American Medical Schools*, esp. chaps. 15 and 16.

44. Lippard, *Half-Century of American Medical Education*, chap. 2; Becker et al., *Boys in White*.

45. Norman Gevitz, *The D.O.'s: Osteopathic Medicine in America* (Baltimore: Johns Hopkins Univ. Press, 1982), esp. chaps. 8 and 9.

46. Lippard, *Half-Century of American Medical Education*, 117–120; Rothstein, *American Medical Schools*, 226.

47. Ludmerer, *Time to Heal*, chaps. 8 and 9, esp. 216–217; Rothstein, *American Medical Schools*, esp. 234.

48. Gross, *Doctors*, esp. 13 and 372–373; Ludmerer, *Time to Heal*.

49. *Health, United States*, 2009, 63; Robert E. Bulander Jr., "'The Most Important Problem in the Hospital': Nursing in the Development of the Intensive Care Unit, 1950–1965," *Social History of Medicine*, 23 (2010), 621–638.

50. Eliot Freidson and Jacob J. Felman, *The Public Looks at Hospitals* (New York: Health Information Foundation, 1958), 1, 15; Rosemary Stevens, *In Sickness and in Wealth: American Hospitals in the Twentieth Century* (New York: Basic Books, 1989), 227–228, 252.

51. Philip A. Kalisch and Beatrice J. Kalisch, *The Advance of American Nursing*, 2nd ed. (Boston: Little, Brown, 1986), 592. See also, e.g., Bernice Hotchkiss, "Nursing Homes: An Analysis of the Types of Patients and the Nursing Services," *California Medicine*, 78 (1953), 251–254. Quotation from Robert E. Burger, "Commercializing the Aged," *Nation*, 11 May 1970, 558.

52. Rosemary A. Stevens, "The Hospital as a Social Institution, New-Fashioned for the 1990s," *Hospital and Health Services Administration*, 36 (1991), 166; Karen Kruse Thomas, *Deluxe Jim Crow: Civil Rights and American Health Policy, 1935–1954* (Athens: Univ. of Georgia Press, 2011).

53. Beatrix Hoffman, "Emergency Rooms: The Reluctant Safety Net," in *History and Health Policy in the United States: Putting the Past Back In*, ed. Rosemary A. Stevens, Charles E. Rosenberg, and Lawton R. Burns (New Brunswick, NJ: Rutgers Univ. Press, 2006), 250–272, esp. 251 and 253.

54. Zink, *Anyone, Anything, Anytime*, chap. 1; Vernon D. Seifert and J. Stanley Johnstone, "Meeting the Emergency Department Crisis," *Hospitals*, 1 Nov. 1966, 55–59; Roemer, *Ambulatory Health Services*, chap. 4, esp. 49; Robert M. Cunningham Jr., "Why Our Hospitals Are in Trouble," *Medical Economics*, Apr. 1951, 78–97.

55. *Source Book of Health Insurance Data*, 1959, 44–49.

56. *Source Book of Health Insurance Data*, 1970, 48.

57. Cunningham, "Why Our Hospitals Are in Trouble"; George Kirstein, "Why Hospitals Exploit Labor," *Nation*, 4 July 1959, 3–6.

58. Leon Fink and Brian Greenberg, *Upheaval in the Quiet Zone: A History of Hospital Workers' Union, Local 1199* (Urbana: Univ. of Illinois Press, 1989); Kirstein, "Why Hospitals Exploit Labor."

59. Editorial, *New York Times*, 21 Nov. 1958, quoted in Fink and Greenberg, *Upheaval in the Quiet Zone*, 37.

60. Cunningham, "Why Our Hospitals Are in Trouble"; Fink and Greenberg, *Upheaval in the Quiet Zone.*

61. Evan M. Melhado, "Health Planning in the United States and the Decline of Public-Interest Policymaking," *Milbank Quarterly,* 84 (2006), 359–440; Stevens, *In Sickness and in Wealth,* chap. 8; Roemer, *Ambulatory Health Services;* Richard M. Magraw, *Ferment in Medicine: A Study of the Essence of Medical Practice and of Its New Dilemmas* (Philadelphia: W. B. Saunders, 1966), 97.

62. Melhado, "Health Planning in the United States"; H. Jack Geiger and Roger D. Cohen, "Trends in Health Care Delivery Systems," *Inquiry,* 8 (1971), 32–36, esp. 33.

63. Stevens, *In Sickness and in Wealth,* 236–246; Eliot Freidson, ed., *The Hospital in Modern Society* (New York: Free Press of Glencoe, 1963); James K. Skipper Jr. and Robert C. Leonard, eds., *Social Interaction and Patient Care* (Philadelphia: J. B. Lippincott, 1965), esp. pt. 4, quotation from "Alienation and the Social Structure: Case Analysis of a Hospital," by Rose Laub Coser, 237.

64. John D. Porterfield, "Hospital Accreditation—Past, Present, and Future," *American Journal of Hospital Pharmacy,* 27 (1970), 315–317; Stevens, *In Sickness and in Wealth,* esp. 244–246 and 249–250.

65. Susan Reverby, "The Search for the Hospital Yardstick: Nursing and the Rationalization of Hospital Work," in *Health Care in America: Essays in Social History,* ed. Susan Reverby and David Rosner (Philadelphia: Temple Univ. Press, 1979), esp. 216–219.

66. Hans O. Mauksch, introduction to Skipper and Leonard, *Social Interaction and Patient Care,* xiii.

67. Magraw, *Ferment in Medicine,* 85.

68. Temple Burling, Edith M. Lentz, and Robert N. Wilson, *The Give and Take in Hospitals: A Study of Human Organization in Hospitals* (New York: G. P. Putnam's Sons, 1956), esp. 89; Edith M. Lentz, "A Study of Changing Relationships in Hospitals," *American Journal of Nursing,* 56 (1956), 187–189.

69. *Source Book of Health Insurance Data,* 1970, 12.

70. "Medicare," *California Medicine,* 85 (1956), 265; Rose C. Engelman, ed., *A Decade of Progress: The United States Army Medical Department, 1959–1969* (Washington, DC: Office of the Surgeon General, Department of the Army, 1971), 57–60.

71. Beatrix Hoffman, *Health Care for Some: Rights and Rationing in the United States since 1930* (Chicago: Univ. of Chicago Press, 2012), chap. 5.

72. Philip J. Funigello, *Chronic Politics: Health Care Security from FDR to George W. Bush* (Lawrence: Univ. Press of Kansas, 2005), esp. chaps. 4 and 5; Alan Derickson, *Health Security for All: Dreams of Universal Health Care in America* (Baltimore: Johns Hopkins Univ. Press, 2005), esp. 129–130; Colin Gordon, *Dead on Arrival: The Politics of Health Care in Twentieth-Century America* (Princeton, NJ: Princeton Univ. Press, 2003), chaps. 2 and 3; Colleen M. Grogan, "A Marriage of Convenience: The Persistent and Changing Relationship between Long-Term Care and Medicaid," in Stevens, Rosenberg, and Burns, *History and Health Policy in the United States,* 210; Paul Starr, *The Social Transformation of American Medicine* (New York: Basic Books, 1982), 373–374; Colin L. Talley, *A History of Multiple Sclerosis* (Westport, CT: Praeger, 2008), chap. 6.

73. See previous note.

74. Funigello, *Chronic Politics*; Monte M. Poen, "The Truman Legacy: Retreat to Medicare," in *Compulsory Health Insurance: The Continuing Debate*, ed. Ronald L. Numbers (Westport, CT: Greenwood, 1982), 97–113; Jonathan Oberlander, *The Political Life of Medicare* (Chicago: Univ. of Chicago Press, 2003), chap. 2.

75. See previous note; and Nancy De Lew, "Medicare: 35 Years of Service," *Health Care Financing Review*, 22 (2000), 75–103, and other articles in the same issue.

76. See previous note; Daniel M. Fox, *Power and Illness: The Failure and Future of American Health Policy* (Berkeley: Univ. of California Press, 1993), 77–78; Diane Rowland, "Health Care for the Poor: Medicaid at 35," *Health Care Financing Review*, 22 (2000), 23–34; and Grogan, "Marriage of Convenience," 202–225.

77. Jonathan Engel, *Poor People's Medicine: Medicaid and American Charity Care since 1965* (Durham, NC: Duke Univ. Press, 2006).

78. Ibid.

79. Rothstein, *American Medical Schools*, 183; Lyndon Johnson, quoted in Funigello, *Chronic Politics*, 154.

80. Paul Starr, *Remedy and Reaction: The Peculiar American Struggle over Health Care Reform* (New Haven, CT: Yale Univ. Press, 2011), 73–76; Beatrix Hoffman, "Restraining the Health Care Consumer: The History of Deductibles and Co-Payments in U.S. Health Insurance," *Social Science History*, 3 (2006), 514.

81. Funigello, *Chronic Politics*, esp. 145 and 161–162; Elinor Langer, "Medical Costs: Rapid Rise Causing Government Concern," *Science*, 155 (1967), 1519–1521.

82. John Duffy, *The Sanitarians: A History of American Public Health* (Urbana: Univ. of Illinois Press, 1990), 274–275; National Commission on Community Health Services, *Health Is a Community Affair* (Cambridge, MA: Harvard Univ. Press, 1966), esp. 225–226.

83. William G. Rothstein, "The Decrease in Socioeconomic Differences in Mortality from 1920 to 2000 in the United States and England," *Journal of the History of Medicine and Allied Sciences*, 67 (2012), 515–552.

84. George Rosen, *Preventive Medicine in the United States, 1900–1975* (New York: Science History Publications, 1975), 71–72; Elizabeth W. Etheridge, *Sentinel for Health: A History of the Centers for Disease Control* (Berkeley: Univ. of California Press, 1992).

85. Duffy, *Sanitarians*, esp. 274–275.

86. Aaron Mauck, "Managing Care: The History of Diabetes Management in Twentieth Century America" (PhD diss., Harvard Univ., 2010), chap. 4.

87. Gerald M. Oppenheimer, "Becoming the Framingham Study," *American Journal of Public Health*, 95 (2005), 602–610; Daniel Levy, introduction to *50 Years of Discovery: Medical Milestones from the National Heart, Lung, and Blood Institute's Framingham Heart Study*, ed. Levy (Hackensack, NJ: Center for Biomedical Communications, 1999), 1; Daniel Levy and Susan Brink, *A Change of Heart: How the Framingham Heart Study Helped Unravel the Mysteries of Cardiovascular Disease* (New York: Alfred A. Knopf, 2005); William G. Rothstein, *Public Health and the Risk Factor: A History of an Uneven Medical Revolution* (Rochester, NY: Univ. of Rochester Press, 2003), esp. chap 15; Robert A. Aronowitz, *Making Sense of Illness: Science, Society, and Disease* (Cambridge: Cambridge Univ. Press, 1998), esp. chap. 5 and 223n; Karin Garrety, "Dietary Policy, Controversy, and Proof: Doing Something versus Waiting for the Definitive Evidence," in *Silent Victories:*

The History and Practice of Public Health in Twentieth-Century America, ed. John W. Ward and Christian Warren (New York: Oxford Univ. Press, 2007), esp. 402–405.

88. Milton Terris, "The Complex Task of the Second Epidemiologic Revolution: The Joseph W. Mountin Lecture," *Journal of Public Health Policy*, 4 (1983), 8–24; Gene Bylinsky, "The New Attack on Killer Diseases," *Fortune*, Feb. 1968, 130–132, 162, 167–168; Paul K. J. Han, "Historical Changes in the Objectives of the Periodic Health Examination," *Annals of Internal Medicine*, 126 (1997), 913–914; Stanley Joel Reiser, *Technological Medicine: The Changing World of Doctors and Patients* (New York: Cambridge Univ. Press, 2009), 146–149.

89. Louise B. Russell, *Is Prevention Better Than Cure?* (Washington, DC: Brookings Institution, 1986), chap. 4; Robert M. Kaplan, "Two Pathways to Prevention," *American Psychologist*, 55 (2000), 382–396; Rothstein, *Public Health and the Risk Factor*.

90. See, e.g., Benjamin F. Miller, "Better Health Ahead!," *Parents Magazine*, Oct. 1957, 60, 82–88; and "Where U.S. Is Winning in War against Disease," *U.S. News & World Report*, 17 Jan. 1972, 54.

91. David D. Rutstein, *The Coming Revolution in Medicine* (Cambridge, MA: MIT Press, 1967), 157–158.

92. Rosemary A. Stevens, "History and Health Policy in the United States: The Making of a Health Care Industry, 1948–2008," *Social History of Medicine*, 21 (2008), 468.

Chapter 11. Medicine in the Environmental Era, 1960s to 1980s

1. Paul B. Beeson, "Changes in Medical Therapy During the Past Half Century," *Medicine*, 59 (1980), 81.

2. Abigail Trafford et al., "Medicine's New Triumphs," *U.S. News & World Report*, 11 Nov. 1985, 46–58.

3. See, e.g., "Physicians Held in High Esteem, Survey Shows," *American Medical News*, 9 Jan. 1978, 12; Eliot Freidson, "The Medical Profession in Transition," in *Applications of Social Science to Clinical Medicine and Health Policy*, ed. Linda H. Aiken and David Mechanic (New Brunswick, NJ: Rutgers Univ. Press, 1986), 64–65; Glen T. Pearson, "Whither Now? Thoughts for a Milestone," *American Surgeon*, 49 (1983), 61; Malcolm L. Peterson, "Physicians' Forecasts of Medical Practice: Why Is the Glass Half Empty?," *Annals of Internal Medicine*, 97 (1982), 778–779; and John M. Chuck et al., "Is Being a Doctor Still Fun?," *Western Journal of Medicine*, 149 (1993), 665–669.

4. *Health, United States*, 1991, 121, 140, 141.

5. Ibid., 2009, 377; 1990, 160; 1991, 247.

6. *Seventh Report to the President and Congress on the Status of Health Personnel in the United States*, DHHS Publication No. HRS-P-OD-90-1 (Washington, DC: U.S. Department of Health and Human Services, 1990), II-5; *A Report to the President & Congress on the Status of Health Professions Personnel in the United States*, DHEW Publication No. (HRA) 78–93 (Washington, DC: U.S. Department of Health, Education, and Welfare, 1978), I-1.

7. James Michael McGinnis, "Recent Health Gains for Adults," *New England Journal of Medicine*, 306 (1982), 671–673; James F. Fries, "The Compression of Morbidity: Near or Far?," *Milbank Quarterly*, 67 (1989), 208–232; S. Jay Olshansky and A. Brian Ault, "The Fourth Stage of the Epidemiologic Transition: The Age of Delayed Degenerative Diseases," ibid., 64 (1986), 355–387.

8. Kathy J. Helzlsouer and Leon Gordis, "Risks to Health in the United States," *Daedalus*, 119 (1990), 204.

9. *Health, United States*, 1990, 172; 2010, 383.

10. Ibid., 1990, 36.

11. Ibid., 1991, 267.

12. Gregg Easterbrook, "The Revolution in Medicine," *Newsweek*, 26 Jan. 1987, 40–74.

13. George Will, foreword to *Reassessing the Sixties: Debating the Political and Cultural Legacy*, ed. Stephen Macedo (New York: W. W. Norton, 1997), 8. See also, e.g., Gail Collins, *When Everything Changed: The Amazing Journey of American Women from 1960 to the Present* (New York: Little, Brown, 2009); and Roger Kimball, *The Long March: How the Cultural Revolution of the 1960s Changed America* (San Francisco: Encounter Books, 2000).

14. See, e.g., Howard Weitzkin and Barbara Waterman, *The Exploitation of Illness in Capitalist Society* (Indianapolis: Bobbs-Merrill, 1974).

15. Roy Branson, "The Secularization of American Medicine," *Hastings Center Studies*, 1, no. 2 (1973), 18.

16. Edwin D. Kilbourne and Wilson G. Smillie, eds., *Human Ecology and Public Health*, 4th ed. (New York: Macmillan, 1969).

17. Samuel P. Hayes, *Beauty, Health, and Permanence: Environmental Politics in the United States, 1955–1985* (Cambridge: Cambridge Univ. Press, 1987), chap. 6, quotation from 172; Alan I Marcus, *Cancer from Beef: DES, Federal Food Regulation, and Consumer Confidence* (Baltimore: Johns Hopkins Univ. Press, 1994).

18. Lynne Page Snyder, "'The Death Dealing Smog over Donora, Pennsylvania': Industrial Air Pollution, Public Health Policy, and the Politics of Expertise, 1948–1949," *Environmental History Review*, 18 (1994), 117–139.

19. Christopher C. Sellers, *Hazards of the Job: From Industrial Disease to Environmental Health Science* (Chapel Hill: Univ. of North Carolina Press, 1997).

20. Linda Nash, *Inescapable Ecologies: A History of Environment, Disease, and Knowledge* (Berkeley: Univ. of California Press, 2006), 7, chap. 4, quotation from 137; Christopher C. Sellers, *Crabgrass Crucible: Suburban Nature and the Rise of Environmentalism in Twentieth-Century America* (Chapel Hill: Univ. of North Carolina Press, 2012); Adam Rome, "'Give Earth a Chance': The Environmental Movement and the Sixties," *Journal of American History*, 90 (2003), 525–554, quoting Joyce Maynard on 542.

21. Amasa B. Ford, "Casualties of Our Time," *Science*, 167 (1970), 256–263.

22. Rachel Carson, *Silent Spring* (1962; reprint, Boston: Houghton Mifflin, 1987); René J. Dubos, *Man Adapting* (New Haven, CT: Yale Univ. Press, 1965), quotation from 205.

23. See, e.g., James T. Patterson, *The Dread Disease: Cancer and Modern American Culture* (Cambridge, MA: Harvard Univ. Press, 1987), chap. 10.

24. David Price, quoted in *Silent Spring*, by Rachel Carson (1962; reprint, Boston: Houghton Mifflin, 2002), 188.

25. Christian Warren, *Brush with Death: A Social History of Lead Poisoning* (Baltimore: Johns Hopkins Univ. Press, 2000), 220–222.

26. Susan P. Baker, "Injuries in America: A National Disaster," in *Unnatural Causes: The Three Leading Killer Diseases in America*, ed. Russell C. Maulitz (New Brunswick, NJ: Rutgers Univ. Press, 1989), 135–145; John C. Burnham, *Accident Prone: A History of Technology, Psychology, and Misfits of the Machine Age* (Chicago: Univ. of Chicago Press, 2009).

27. *Healthy People: The Surgeon General's Report on Health Promotion and Disease Prevention, Background Papers 1979*, DHEW (PHS) Publication No. 79-55071A (Washington, DC: U.S. Government Printing Office, 1979), 461.

28. See, e.g., George A. Kaplan et al., "Socioeconomic Status and Health," in *Closing the Gap: The Burden of Unnecessary Illness*, ed. Robert W. Ambler and H. Bruce Dull (New York: Oxford Univ. Press, 1987), 125–129; and Lisa F. Berkman and Ichiro Kawachi, "A Historical Framework for Social Epidemiology," in *Social Epidemiology*, ed. Berkman and Kawachi (New York: Oxford Univ. Press, 2000), 4–6.

29. Robert Crawford, "Cultural Influences on Prevention and the Emergence of a New Health Consciousness," in *Taking Care: Understanding and Encouraging Self-Protective Behavior*, ed. Neil D. Weinstein (Cambridge: Cambridge Univ. Press, 1987), 95–113.

30. Nash, *Inescapable Ecologies*, 171.

31. JoAnne Brown, "The Social Construction of Invisible Danger: Two Historical Examples," in *Nothing to Fear: Risks and Hazards in American Society*, ed. Andrew W. Kirby (Tucson: Univ. of Arizona Press, 1990), 45–48.

32. John Burnham, "The Home as Environment: Changing Understandings from the History of Childhood Lead Poisoning," in *Health and the Modern Home*, ed. Mark Jackson (New York: Routledge, 2007), 285–303.

33. E. D. Palmes, "Measurement and Significance of Subclinical Effects of Chemicals," *Clinical Toxicology*, 9 (1976), 724.

34. Sellers, *Crabgrass Crucible*. See also, e.g., *Environment and Health*, ed. Norman M. Trieff (Ann Arbor, MI: Ann Arbor Science Publishers, 1980); and Kaye H. Kilburn, "Epidemics Then and Now: Chemicals Replace Microbes and Degenerations Oust Infections," *Archives of Environmental Health*, 49 (1994), 3–5.

35. Edward H. Kass, "Infectious Diseases and Social Change," *Journal of Infectious Diseases*, 123 (1971), 110–114.

36. David A. Evans and Vimla L. Patel, eds., *Cognitive Science in Medicine: Biomedical Modeling* (Cambridge, MA: MIT Press, 1989).

37. Edward J. Burger Jr., "Health as a Surrogate for the Environment," *Daedalus*, 119 (1990), 133–153, quotations from 140 and 147.

38. See, e.g., Phil Brown et al., "The Health Politics of Asthma: Environmental Justice and Collective Illness Experience in the United States," *Social Science and Medicine*, 57 (2003), 453–464.

39. Mark Jackson, *Allergy: The History of a Modern Malady* (London: Reaktion Books, 2006); Gregg Mitman, *Breathing Space: How Allergies Shape Our Lives and Landscapes* (New Haven, CT: Yale Univ. Press, 2007), esp. 246–247; Murray Dworetzky, "Changing Concepts of the Asthma Problem," *Journal of Allergy*, 43 (1969), 321.

40. Mitman, *Breathing Space*; Jackson, *Allergy*.

41. "Cigarette Smoke Pollutes Non-Smokers' Environment," *JAMA*, 219 (1972), 821–822; E.R.M., "Can Your Husband's Cigarette Give You Cancer?," *Good Housekeeping*, May 1981, 255–256.

42. See, e.g., Michelle Murphy, "The 'Elsewhere within Here' and Environmental Illness; Or, How to Build Yourself a Body in a Safe Space," *Configurations*, 8 (2000), 87–120.

43. Emmanuel Farber, "Chemical Carcinogenesis," *New England Journal of Medicine*, 303 (1981), 1379–1389, esp. 1379.

44. See, e.g., Samuel S. Epstein, "Environmental Determinants of Human Cancer," *Cancer Research*, 34 (1974), 2425–2435; and J. W. Cullen, B. H. Fox, and R. N. Isom, eds., *Cancer: The Behavioral Dimensions* (New York: Raven, 1976), esp. Peter B. Peacock, "Environmental Risks Related to Cancer," 85–92, quotations from 86–90.

45. Sigismund Peller, *Cancer Research since 1900: An Evaluation* (New York: Philosophical Library, 1979), 263; Patterson, *Dread Disease*, quotation from 257; Edith Efron, *The Apocalyptics: Cancer and the Big Lie* (New York: Simon & Schuster, 1984).

46. Patterson, *Dread Disease*.

47. Leslie J. Reagan, "Engendering the Dread Disease: Women, Men, and Cancer," *American Journal of Public Health*, 87 (1997), 1779–1787; Kirsten E. Gardner, *Early Detection: Women, Cancer, and Awareness Campaigns in the Twentieth-Century United States* (Chapel Hill: Univ. of North Carolina Press, 2006); Keith Wailoo, *How Cancer Crossed the Color Line* (New York: Oxford Univ. Press, 2011).

48. William D. Kelley, quoted in Barron H. Lerner, *When Illness Goes Public: Celebrity Patients and How We Look at Medicine* (Baltimore: Johns Hopkins Univ. Press, 2006), 146.

49. Patterson, *Dread Disease*, quotation from 170.

50. See, e.g., Samuel S. Epstein, *The Politics of Cancer* (San Francisco: Sierra Club Books, 1978), 423–428.

51. David Cantor, "Introduction: Cancer Control and Prevention in the Twentieth Century," in *Cancer in the Twentieth Century*, ed. Cantor (Baltimore: Johns Hopkins Univ. Press, 2008), 1–38; Monica J. Casper and Adele E. Clarke, "Making the Pap Smear into the 'Right Tool' for the Job: Cervical Cancer Screening in the USA, circa 1940–95," *Social Studies of Science*, 28 (1998), 255–290; Ilana Löwy, *A Woman's Disease: The History of Cervical Cancer* (Oxford: Oxford Univ. Press, 2011), esp. chap. 4.

52. Casper and Clarke, "Making the Pap Smear into the 'Right Tool' for the Job"; Barron H. Lerner, *The Breast Cancer Wars: Hope, Fear, and the Pursuit of a Cure in Twentieth-Century America* (New York: Oxford Univ. Press, 2001); Robert A. Aronowitz, *Unnatural History: Breast Cancer and American Society* (New York: Cambridge Univ. Press, 2007); Aronowitz, "Do Not Delay: Breast Cancer and Time, 1900–1970," *Milbank Quarterly*, 79 (2001), 355–386; Aronowitz, *Making Sense of Illness: Science, Society, and Disease* (Cambridge: Cambridge Univ. Press, 1998), 112–113; William G. Rothstein, *Public Health and the Risk Factor: A History of an Uneven Medical Revolution* (Rochester, NY: Univ. of Rochester Press, 2003), esp. 361–362.

53. Richard A. Rettig, *Cancer Crusade: The Story of the National Cancer Act of 1971* (Princeton, NJ: Princeton Univ. Press, 1977); Patterson, *Dread Disease*, 247–252.

54. Ilana Löwy, "Immunotherapy of Cancer from Coley's Toxins to Interferons: Molecularization of a Therapeutic Practice," in *Molecularizing Biology and Medicine: New Practices and Alliances, 1910s–1970s*, ed. Soraya de Chadarevian and Harmke Kamminga (Amsterdam: Harwood Academic, 1998), 249–271.

55. Robert A. Weinberg, "Oncogenes and Tumor Suppressor Genes," in Maulitz, *Unnatural Causes*, 83–94; Jean Marx, "Oncogenes Reach a Milestone," *Science*, 266 (1994), 1942–1944.

56. Seamus J. Martin, "Apoptosis: An Introduction," in *Apoptosis and Cancer*, ed. Martin (Basel: Karger Landes Systems, 1997), 9–10; L. David Tomei and Frederick O. Cope, introduction to *Apoptosis: The Molecular Basis of Cell Death*, ed. Tomei and Cope (Plainview,

NY: Cold Spring Harbor Laboratory Press, 1991), 3; Guido Majno and Isabelle Joris, "Apoptosis, Oncosis, and Necrosis: An Overview of Cell Death," *American Journal of Pathology*, 146 (1995), quotation from 7.

57. William H. Stewart, "Public Health," in *Medical World Annual 1968* (New York: McGraw-Hill, 1968), 184.

58. Janet Golden, *Message in a Bottle: The Making of Fetal Alcohol Syndrome* (Cambridge, MA: Harvard Univ. Press, 2005); Elizabeth M. Armstrong, *Conceiving Risk, Bearing Responsibility: Fetal Alcohol Syndrome and the Diagnosis of Moral Disorder* (Baltimore: Johns Hopkins Univ. Press, 2003).

59. Barron H. Lerner, *One for the Road: Drunk Driving since 1900* (Baltimore: Johns Hopkins Univ. Press, 2011).

60. John E. Sutherland, Victoria W. Persky, and Jacob A. Brody, "Proportionate Mortality Trends: 1950 through 1986," *JAMA*, 264 (1990), 3178–3184; Ahmedin Jemal et al., "Trends in the Leading Causes of Death in the United States, 1970–2002," ibid., 294 (2005), 1255–1259; Diana B. Dutton, *Worse than the Disease: Pitfalls of Medical Progress* (New York: Cambridge Univ. Press, 1988), 358.

61. See, e.g., Lawrence Cohen and Henry Rothschild, "The Bandwagons of Medicine," *Perspectives in Biology and Medicine*, 22 (1979), 531–538; Stephen J. Kunitz, "Some Notes on Physiologic Conditions as Social Problems," *Social Science and Medicine*, 8 (1974), 207–211; Lerner, *When Illness Goes Public*, 273; and Arthur J. Barsky, *Worried Sick: Our Troubled Quest for Wellness* (Boston: Little, Brown, 1988), 87–89.

62. Barton J. Bernstein, "The Swine Flu Immunization Program," *Medical Heritage*, 1 (1985), 236–266; Dutton, *Worse than the Disease*, chap. 5. See also, e.g., "The War Against Disease: Many Gains—But Setbacks, Too," *U.S. News & World Report*, 20 Dec. 1976, 43–46.

63. Richard C. Adelman and Lois M. Verbrugge, "Death Makes News: The Social Impact of Disease on Newspaper Coverage," *Journal of Health and Social Behavior*, 41 (2000), 347–367.

64. Lester G. Cordes and David W. Fraser, "Legionellosis: Legionnaires' Disease; Pontiac Fever," *Medical Clinics of North America*, 64 (1980), 395–397; Gary L. Lattimer and Richard A. Ormsbee, *Legionnaires' Disease* (New York: Marcel Dekker, 1981), chap. 1; David W. Fraser, "The Challenges Were Legion," *Lancet Infectious Disease*, 5 (2005), 237–242.

65. See previous note; and Thomas F. Keys, "Legionnaires' Disease: A Review of the Epidemiology and Clinical Manifestations of a Newly Recognized Infection," *Mayo Clinic Proceedings*, 55 (1980), 129–133.

66. Lattimer and Ormsbee, *Legionnaires' Disease*, 10; G. F. Mallison, "Legionellosis: Environmental Aspects," *Annals of the New York Academy of Sciences*, 353 (1980), 68–70.

67. Elisabeth Keiffer, "Mrs. Murray's Mystery Disease," *Good Housekeeping*, Mar. 1977, 80–86, quotation from 80; Jonathan A. Edlow, *Bull's Eye: Unraveling the Medical Mystery of Lyme Disease* (New Haven, CT: Yale Univ. Press, 2003); Robert A. Aronowitz, "Lyme Disease: The Social Construction of a New Disease and Its Social Consequences," *Milbank Quarterly*, 69 (1991), 79–109.

68. Edlow, *Bull's Eye*.

69. Merlin S. Bergdoll, "History," in *Toxic Shock Syndrome*, by Bergdoll and P. Joan Chesney (Boca Raton, FL: CRC, 1991), 3–7; David E. Rogers, "Infections," *Yearbook of Med-*

Lime, 1981, 17–18; Pamela Sherrid, "Tampons after the Shock Wave," *Fortune*, 10 Aug. 1991, 114–128, quotation from 128.

70. Bergdoll, "History"; Nancy Friedman, "A Major Report on Tampon Safety," reprinted from *New West* in *Working Women*, Jan. 1981, 57–58, 64, 66, 76; Tampax advertisement, *Seventeen*, Jan. 1978, 24; Karen Wright, "Bad News Bacteria," *Science*, 249 (1990), 22–24.

71. Randy Shilts, *And the Band Played On: Politics, People, and the AIDS Epidemic* (New York: St. Martin's, 1987), 12.

72. Abraham Verghese, *My Own Country: A Doctor's Story of a Town and Its People in the Age of AIDS* (New York: Simon & Schuster, 1994), 24.

73. David McBride, *From TB to AIDS: Epidemics among Urban Blacks since 1900* (Albany: State Univ. of New York Press, 1991), esp. 161; Verghese, *My Own Country*; Victoria A. Harden, *AIDS at 30: A History* (Washington, DC: Potomac Books, 2012).

74. Timeline and William Blattner transcript, 3, in "In Their Own Words: NIH Researchers Recall the Early Years of AIDS," http://nih.gov/NIHInownwords/docs.

75. Timeline, in "In Their Own Words"; Shilts, *And the Band Played On*, 267, 353; Timothy E. Cook and David C. Colby, "The Mass-Mediated Epidemic: The Politics of AIDS on the Nightly Network News," in *AIDS: The Making of a Chronic Disease*, ed. Elizabeth Fee and Daniel M. Fox (Berkeley: Univ. of California Press, 1992), 97; Jean Seligmann et al., "The AIDS Epidemic: The Search for a Cure," reprinted in *The AIDS Crisis: A Documentary History*, ed. Douglas A. Feldman and Julia Wang Miller (Westport, CT: Greenwood, 1998), 19; Stanley B. Prusiner, "Discovering the Cause of AIDS," *Science*, 298 (2002), 1726–1727.

76. Robert Yarchoan, quoted in Harden, *AIDS at 30*, 131–132.

77. Cook and Colby, "Mass-Mediated Epidemic," 84–122, quotations from 84; James Kinsella, *Covering the Plague: AIDS and the American Media* (New Brunswick, NJ: Rutgers Univ. Press, 1989).

78. John H. Gagnon, "Disease and Desire," *Daedalus*, Summer 1989, 47–77; Jonathan E. Kaplan, "A Modern-Day Plague," reprinted in Feldman and Miller, *AIDS Crisis*, 7.

79. See esp. Elizabeth Fee and Daniel M. Fox, eds., *AIDS: The Burden of History* (Berkeley: Univ. of California Press, 1988); Fee and Fox, *AIDS: The Making of a Chronic Disease*; and Harden, *AIDS at 30*.

80. Warren G. Guntheroth, *Crib Death: The Sudden Infant Death Syndrome*, 3rd ed. (Armonk, NY: Futura, 1995); Abraham B. Bergman, *The "Discovery" of Sudden Infant Death Syndrome: Lessons in the Practice of Political Medicine* (1986; reprint, Seattle: Univ. of Washington Press, 1988).

81. Arthur M. Silverstein, *A History of Immunology* (San Diego: Academic Press, 1989), esp. 327–334; Jeffrey C. Mariner et al., "Rinderpest Eradication: Appropriate Technology and Social Innovations," *Science*, 337 (2012), 1309–1310.

82. Pauline M. H. Mazumdar, ed., *Immunology 1930–1980* (Toronto: Wall & Thompson, 1989), esp. Mazumdar, "Working Out of the Theory," 1–11; Emily Martin, *Flexible Bodies: Tracking Immunity in American Culture—From the Days of Polio to the Age of AIDS* (Boston: Beacon, 1994).

83. Howard Brody and David S. Sobel, "A Systems View of Health and Disease," reprinted in *The Nation's Health: A Courses by Newspaper Reader*, ed. Philip R. Lee, Nancy

Brown, and Ida V. S. W. Red (San Francisco: Boyd & Fraser, 1981), 27–37, quotations from 30–31; Kristen L. Mueller, "Recognizing the First Responders," *Science*, 327 (2010), 283.

84. Richard M. Frankel, Timothy E. Quill, and Susan H. McDaniel, eds., *The Biopsychosocial Approach: Past, Present, Future* (Rochester, NY: Rochester Univ. Press, 2003), esp. Theodore M. Brown, "George Engel and Rochester's Biopsychosocial Tradition: Historical and Developmental Perspectives," 199–219; Sydney A. Halpern, *American Pediatrics: The Social Dynamics of Professionalism, 1880–1980* (Berkeley: Univ. of California Press, 1988), chap. 7; Martin, *Flexible Bodies*; Julius B. Richmond, *Currents in American Medicine: A Developmental View of Medical Care and Education* (Cambridge, MA: Harvard Univ. Press, 1969), chap. 5; Kenneth M. Ludmerer, *Time to Heal: American Medical Education from the Turn of the Century to the Era of Managed Care* (New York: Oxford Univ. Press, 1999), 288–291.

85. Quoted in Martin, *Flexible Bodies*, 117.

86. Ibid., esp. chaps. 7 and 9.

87. See, e.g., John W. Mason, "A Historical View of the Stress Field," *Journal of Human Stress*, 1 (1975), 21–36; Hans Selye, "Confusion and Controversy in the Stress Field," ibid., 37–44; Andrea Tone, *The Age of Anxiety: A History of America's Turbulent Affair with Tranquilizers* (New York: Basic Books, 2009); Gerry V. Stimson, "The Message of Psychotropic Drug Ads," *Journal of Communication*, 25 (1975), 153–160; Constance Holden, "Behavioral Medicine: An Emergent Field," *Science*, 209 (1980), 479–481; and Leonard I. Pearlin and Carol S. Aneshensel, "Coping and Social Supports: Their Functions and Applications," in Aiken and Mechanic, *Applications of Social Science*, 417–437.

88. Walsh McDermott, "Social Ramifications of Control of Microbial Disease," ed. David E. Rogers, *Johns Hopkins Medical Journal*, 151 (1982), 308–311.

89. Stuart S. Blume, *Insight and Industry: On the Dynamics of Technological Change in Medicine* (Cambridge, MA: MIT Press, 1992), e.g., 166.

90. *Health, United States*, 2009, 63; Russell C. Coile Jr., *The New Hospital: Future Strategies for a Changing Industry* (Rockville, MD: Aspen, 1986), 85; Bettyann Holtzmann Kevles, *Naked to the Bone: Medical Imaging in the Twentieth Century* (New Brunswick, NJ: Rutgers Univ. Press, 1997), pt. 2, quotation from 169; Blume, *Insight and Industry*; Amit Prasad, "The (Amorphous) Anatomy of an Invention: The Case of Magnetic Resonance Imaging (MRI)," *Social Studies of Science*, 37 (2007), 533–560.

91. Rosalyn S. Yalow, ed., *Radioimmunoassay* (Stroudsburg, PA: Hutchinson Ross, 1983).

92. Stanley Joel Reiser, *Medicine and the Reign of Technology* (Cambridge: Cambridge Univ. Press, 1978), 220–221.

93. David J. Rothman, *Beginnings Count: The Technological Imperative in American Health Care* (New York: Oxford Univ. Press, 1997), chap. 5.

94. E. A. Friedman, "Bionic, Organic, and Hybrid Spare Parts in Clinical Medicine," *Transplantation Proceedings*, 12 (1980), 605–611.

95. Carl Elliott, *Better than Well: American Medicine Meets the American Dream* (New York: W. W. Norton, 2003), esp. 240.

96. Verghese, *My Own Country*, 12.

97. Eric J. Cassell, "The Changing Concept of the Ideal Physician," *Daedalus*, Spring 1986, 192. See also, e.g., Gary E. Krieger, "Has a Thorough Medical History Become History?," *American Medical News*, 20 Oct. 1997, 24.

98. Bonnie L. Svarstad, "Patient-Practitioner Relationships and Compliance with Prescribed Medical Regimens," in Aiken and Mechanic, *Applications of Social Science*, 438–459, quotation from 439.

99. Herbert E. Klarman, quoted in Blume, *Insight and Industry*, 4.

100. William G. Rothstein, *American Medical Schools and the Practice of Medicine: A History* (New York: Oxford Univ. Press, 1987), 191.

101. See, e.g., "Cleaner Air for the OR," *Modern Health Care*, Nov. 1974, 82–84.

102. See, e.g., Renée C. Fox and Judith P. Swazey, *Spare Parts: Organ Replacement in American Society* (New York: Oxford Univ. Press, 1992), esp. chap. 8; Jeffrey M. Prottas, "Competition for Altruism: Bone and Organ Procurement in the United States," *Milbank Quarterly*, 70 (1992) 299–317; Mary Jo Festle, *Second Wind: Oral Histories of Lung Transplant Survivors* (New York: Palgrave Macmillan, 2012); Rothstein, *Public Health and the Risk Factor*, 290–294; Clive Handler and Michael Clemen, eds., *Classic Papers in Coronary Angioplasty* (London: Springer-Verlag, 2006); and David S. Jones, *Broken Hearts: The Tangled History of Cardiac Care* (Baltimore: Johns Hopkins Univ. Press, 2013).

103. See, e.g., Annctine C. Gelijns and Nathan Rosenberg, "From the Scalpel to the Scope: Endoscopic Innovations in Gastroenterology, Gynecology, and Surgery," in *Sources of Medical Technology: Universities and Industry,* ed. Nathan Rosenberg, Annette C. Gelijns, and Holly Dawkins (Washington, DC: National Academy Press, 1995), 67–96.

104. F. D. Moore, "Transplantation—A Perspective," *Transplantation Proceedings*, 12 (1980), 539–550, esp. 550; quotation from *Complications: A Surgeon's Notes on an Imperfect Science*, by Atul Gawande (New York: Henry Holt, 2002), 25. See also, e.g., "Equipment Salesmen in the OR? New York Eyeing Legislation to Require Patients to Be Told," *American Medical News*, 9 Jan. 1978, 27.

105. See, e.g., Gelijns and Rosenberg, "From the Scalpel to the Scope," 91–93; and Thomas Schlich, "Degrees of Control: The Spread of Operative Fracture Treatment with Metal Implants; A Comparative Perspective on Switzerland, East Germany and the USA, 1950s-1990s," in *Innovations in Health and Medicine: Diffusion and Resistance in the Twentieth Century*, ed. Jennifer Stanton (London: Routledge, 2002), 106–126.

106. See, e.g., "Tremendous Growth Predicted in Ambulatory Surgery Market," *Same-Day Surgery*, Nov. 1981, 137–139; and "Market Forces Determine Pricing Strategies for Ambulatory Surgery," ibid., June 1984, 71–73.

107. Enriqueta C. Bond and Simon Glynn, "Recent Trends in Support for Biomedical Research and Development," in Rosenberg, Gelijns, and Dawkins, *Sources of Medical Technology*, 15–38, esp. 19 and 26.

108. Robert H. Ebert, "Medical Education at the Peak of the Era of Experimental Medicine," *Daedalus*, Spring 1986, 55.

109. See, e.g., J. B. Wyngaarden, "The Clinical Investigator as an Endangered Species," *New England Journal of Medicine*, 301 (1979), 1254–1259; and Harold M. Swartz and Diane L. Gottheil, eds., *The Education of Physician-Scholars: Preparing for Leadership in the Health Care System* (Rockville, MD: Betz, 1993).

110. See, e.g., Malcolm C. Todd et al., "Orthodox Medicine, Humanistic Medicine, Holistic Health Care," *Western Journal of Medicine*, 131 (1979), 463–483; and S.V., "In Search of Identity," *JAMA*, 219 (1972), 609.

111. *Health, United States,* 1975, 286; Harry Schwartz, "Doctors Are Treated to a Dose of Competition," *Wall Street Journal,* 16 Aug. 1982; "MD Income Is Steady, Liability Costs Up: Report," *American Medical News,* 13 Feb. 1987, 2; Milton I. Roemer, *Ambulatory Health Services in America: Past, Present, and Future* (Rockville, MD: Aspen, 1981), esp. 11 and 173.

112. *Health, United States,* 1991, 98.

113. Rosemary Stevens, unpublished data, 2002.

114. George Weisz, *Divide and Conquer: A Comparative History of Medical Specialization* (Oxford: Oxford Univ. Press, 2006), 248–256.

115. Ibid.; George E. Fryer Jr., "The United States Medical Profession: An Abnormal Form of the Division of Labour," *Sociology of Health and Illness,* 13 (1991), 213–229; Gordon T. Moore, "The Case of the Disappearing Generalist: Does It Need to Be Solved?," *Milbank Quarterly,* 70 (1992), 361–379, esp. 362–363, quotation from 368; Fryer, "United States Medical Profession," 213–230.

116. David Charles Sloane and Beverlie Conant Sloane, *Medicine Moves to the Mall* (Baltimore: Johns Hopkins Univ. Press, 2003); John A. D. Cooper, "What Is Immediate Past Is Prologue—Unfortunately," *Journal of Medical Education,* 61 (1986), 114; Coile, *New Hospital,* esp. 7.

117. George J. Agich, ed., *Responsibility in Health Care* (Dordrecht, Netherlands: D. Reidel, 1982).

118. Julie Fairman, *Making Room in the Clinic: Nurse Practitioners and the Evolution of Modern Health Care* (New Brunswick, NJ: Rutgers Univ. Press, 2008); Theodor J. Litman, "Public Perceptions of the Physician's Assistant—A Survey of the Attitudes and Opinions of Rural Iowa and Minnesota Residents," *American Journal of Public Health,* 62 (1972), 343–348; Natalie Holt, "'Confusion's Masterpiece': The Development of the Physician Assistant Profession," *Bulletin of the History of Medicine,* 72 (1998), 246–278; Laura E. Ettinger, *Nurse-Midwifery: The Birth of a New American Profession* (Columbus: Ohio State Univ. Press, 2006), esp. 190.

119. Fairman, *Making Room in the Clinic.*

120. Donald Light and Sol Levine, "The Changing Character of the Medical Profession: A Theoretical Overview," *Milbank Quarterly,* 66, suppl. 2 (1988), 10–32; Marie R. Haug, "A Re-Examination of the Hypothesis of Physician Deprofessionalization," ibid., 48–56; John D. Stoeckle, "Reflections on Modern Doctoring," ibid., quotation from 81–82. See also, e.g., Diana Chapman Walsh, *Corporate Physicians: Between Medicine and Management* (New Haven, CT: Yale Univ. Press, 1987).

121. Marc Berg, "Turning a Practice into a Science: Reconceptualizing Postwar Medical Practice," *Social Studies of Science,* 25 (1995), 437–476, quotation from 452; Harvey V. Fineberg, "Medical Decision Making and the Future of Medical Practice," *Medical Decision Making,* 1 (1981), 4–6.

122. *Source Book of Health Insurance Data,* 1991, 93.

Chapter 12. Environmental-Era Health Care in a Hostile Social Climate

1. Daniel M. Fox, *Power and Illness: The Failure and Future of American Health Policy* (Berkeley: Univ. of California Press, 1993), esp. 89; Leon Kass, quoted in *The Health Care Revolution: From Medical Monopoly to Market Competition,* by Carl F. Ameringer (Berkeley: Univ. of California Press, 2008), 133–134.

2. Renée C. Fox, "The Medicalization and Demedicalization of American Society," in *Doing Better and Feeling Worse: Health in the United States*, ed. John H. Knowles (New York: W. W. Norton, 1977), 9–22; Arthur J. Barsky, *Worried Sick: Our Troubled Quest for Wellness* (Boston: Little, Brown, 1988), 9 and chap. 6; Peter Conrad, *The Medicalization of Society: On the Transformation of Human Conditions into Treatable Disorders* (Baltimore: Johns Hopkins Univ. Press, 2007); Conrad, "Medicalization and Social Control," *Annual Review of Sociology*, 18 (1992), 209–232, quotation from 213. See also, e.g., Allan V. Horwitz, *The Loss of Sadness: How Psychiatry Transformed Normal Sorrow into Depressive Disorder* (New York: Oxford Univ. Press, 2007); and Joseph E. Davis, "How Medicalization Lost Its Way," *Society*, 43 (Sept.–Oct. 2006), 51–56.

3. See, e.g., Dorothy Pawluch, *The New Pediatrics: A Profession in Transition* (New York: Aldine de Gruyter, 1996); and Paul E. Stepansky, *The Last Family Doctor: Remembering My Father's Medicine* (Montclair, NJ: Keynote Books, 2011).

4. See, e.g., Claire Tehan, "Has Success Spoiled Hospice?," *Hastings Center Report*, Oct. 1985, 10–13; Patrice C. Moore and Robert H. McCollough, "Hospice: End-of-Life Care at Home," in *Home Care Advances: Essential Research and Policy Issues*, ed. Robert H. Binstock and Leighton E. Cluff (New York: Springer, 2000), 101–116; and Joel D. Howell, "A History of Caring in Medicine," in *The Lost Art of Caring: A Challenge to Health Professionals, Families, Communities, and Society*, ed. Leighton E. Cluff and Robert H. Binstock (Baltimore: Johns Hopkins Univ. Press, 2001), 96–97.

5. Sarah A. Leavitt, "'A Private Little Revolution': The Home Pregnancy Test in American Culture," *Bulletin of the History of Medicine*, 80 (2010), 317–345. See also, e.g., Lucinda McCray Beier, *Health Culture in the Heartland, 1880–1980: An Oral History* (Urbana: Univ. of Illinois Press, 2009), 168.

6. Daniel Wikler and Norma J. Wikler, "Turkey-Baster Babies: The Demedicalization of Artificial Insemination," *Milbank Quarterly*, 69 (1991), 5–40; Carl Djerassi, "Sex in an Age of Mechanical Reproduction," *Science*, 285 (1999), 53–54.

7. Ronald Bayer, *Homosexuality and American Psychiatry: The Politics of Diagnosis* (New York: Basic Books, 1981).

8. Conrad, *Medicalization of Society*; George L. Fite, "The Practice of Medicine," *JAMA*, 230 (1974), 265.

9. "Growing Crisis in Health Care," *U.S. News & World Report*, 3 Nov. 1969, 70–73.

10. See, e.g., Harry Schwartz, *The Case for American Medicine: A Realistic Look at Our Health Care System* (New York: David McKay, 1972).

11. Edward T. Chace, "The Health Crisis," *Commonweal*, 93 (1970), 244; Peter H. Schuck, "A Consumer's View of the Health Care System," in *Ethics of Health Care*, ed. Laurence R. Tancredi (Washington, DC: National Academy of Sciences, 1974), 95–99.

12. Ernest W. Saward, "Institutional Organization, Incentives, and Change," in Knowles, *Doing Better and Feeling Worse*, 193; Marvin Henry Edwards, *Hazardous to Your Health: A New Look at the "Health Care Crisis" in America* (New Rochelle, NY: Arlington House, 1972), quotation from 33; Schwartz, *Case for American Medicine*, quotation from 34.

13. National Advisory Commission on Health Manpower, quoted in "The Health Crisis—What to Do?," *American Journal of Public Health*, 59 (1969), 2.

14. Daniel Schwartz, "Responsibility for Malpractice," in *In Failing Health: A Critical Analysis of Health Care in the United States*, ed. James J. Allen and Steven B. Hunt (Skokie,

IL: National Textbook Company, 1977), 254; Robert W. Berliner, "The Relevance of Medical Science to Medical Care," *Archives of Internal Medicine*, 125 (1970), 510; Richard Malmsheimer, *"Doctors Only": The Evolving Image of the American Physician* (Westport, CT: Greenwood, 1988), esp. chaps. 7 and 8; Joan Liebermann-Smith and Sharon L. Rosen, "The Presentation of Illness on Television," in *Deviance and Mass Media*, ed. Charles Winick (Beverly Hills: Sage, 1978), 79–93, quotation from 80–81; Nathan Maccomby, "The Effects of Mass Communication on Shaping Consumer and Provider Values in Health," in Tancredi, *Ethics of Health Care*, 130–151; Robert Michels, "Commentary," in ibid., 152–156.

15. John A. D. Cooper, quoted in "Growing Crisis in Health Care," 70.

16. Barbara J. Culliton, "Health Manpower: The Feds Are Taking Over," *Science*, 191 (1976), 447; Stephen S. Mick, "The Physician 'Surplus' and the Decline of Professional Dominance," *Journal of Health Politics, Policy and Law*, 29 (2004), 907–924.

17. See, e.g., Daniel M. Fox, "The Consequences of Consensus: American Health Policy in the Twentieth Century," *Milbank Quarterly*, 64 (1986), 76–99.

18. Jill Quadagno, *One Nation, Uninsured: Why the U.S. Has No National Health Insurance* (New York: Oxford Univ. Press, 2005). See also, e.g., Louise Lander, *National Health Insurance: He Who Pays the Piper Calls the Tune* (New York: Health Policy Advisory Center, 1975); and John Gordon Freymann, *The American Health Care System: Its Genesis and Trajectory* (Baltimore: Williams & Wilkins, 1974), chap. 8.

19. Frank D. Campion, *The AMA and U.S. Health Policy since 1940* (Chicago: Chicago Review Press, 1984), chap. 17; John K. Iglehart, "Lobbying: Cash, Counsel, Caveats," *Modern Healthcare*, May 1974, 35–39.

20. See, e.g., Knowles, *Doing Better and Feeling Worse*.

21. William Shonick, *Government and Health Services: Government's Role in the Development of U.S. Health Services, 1930–1980* (New York: Oxford Univ. Press, 1995), 482–484; Diana B. Dutton, "Social Class, Health, and Illness," in *Applications of Social Science to Clinical Medicine and Health Policy*, ed. Linda H. Aiken and David Mechanic (New Brunswick, NJ: Rutgers Univ. Press, 1986), 46–50; Jonathan Engel, *Poor People's Medicine: Medicaid and American Charity Care since 1965* (Durham, NC: Duke Univ. Press, 2006), 135–137, chap. 10; Jenna Loyd, "Where is Community Health? Racism, the Clinic and the Biopolitical State," in *Rebirth of the Clinic: Places and Agents in Contemporary Health Care*, ed. Cindy Patton (Minneapolis: Univ. of Minnesota Press, 2010), 57; Victor Sidell, "A Time for Rededication," *Nation's Health*, Dec. 1984, 2; Bonnie Lefkowitz, *Community Health Centers: A Movement and the People Who Made It Happen* (New Brunswick, NJ: Rutgers Univ. Press, 2007).

22. Hugh Tilson and Kristine M. Gebbie, "The Public Health Workforce," *Annual Review of Public Health*, 25 (2004), 345; *Health, United States*, 1991, 55.

23. See, e.g., John D. Klemm, "Medicaid Spending: A Brief History," *Health Care Financing Review*, 22 (2000), 105–112; and Christie Provost and Paul Hughes, "Medicaid: 35 Years of Service," ibid., 141–153.

24. Eleanor C. Lambertsen, "Interdisciplinary Education: 'Why and How,'" in *Educating for the Health Team: Report of the Conference on the Interrelationships of Educational Programs for Health Professionals* (Washington, DC: National Academy of Sciences, 1972), 3; Florence A. Wilson and Duncan Neuhauser, *Health Services in the United States* (Cambridge, MA: Ballinger, 1974), xiii, and ibid., 2nd ed., rev. (1985), xv; Freymann, *American*

Health Care System, xiii–xiv; Paul J. Feldstein, "The Changing Structure of the Health Care Delivery System in the United States," in *Health Care Systems and Their Patients: An International Perspective*, ed. Marilynn M. Rosenthal and Marcel Frenkel (Boulder, CO: Westview, 1992), 21–36; Natalie D. Reaves, "A Model of Effective Health Policy: The 1983 Orphan Drug Act," *Journal of Health and Social Policy*, 17 (2003), 61–71.

25. Wilson and Neuhauser, *Health Services in the United States*, 2nd ed., rev.; Marshall W. Raffel and Norma K. Raffel, *The U.S. Health System: Origins and Functions*, 3rd ed. (Media, PA: Harwal, 1989).

26. See, e.g., Kathleen M. King and Richard V. Rimkunas, *National Health Expenditures: Trends from 1960–1989*, Publication No. 91-588 EPW (Washington, DC: Congressional Research Service, Library of Congress, [1991]), esp. 72; and Lawrence D. Brown, ed., *Health Policy in Transition: A Decade of Health Politics, Policy and Law* (Durham, NC: Duke Univ. Press, 1987).

27. James E. Bryan, "View from the Hill," *American Family Physician*, Nov. 1982, 317.

28. Evan M. Melhado, "Economists, Public Provision, and the Market: Changing Values in Public Debate," *Journal of Health Politics, Policy and Law*, 23 (1998), 215–263; Fox, *Power and Illness*, chaps. 3 and 4.

29. *Source Book of Health Insurance Data*, 1990, 3.

30. Engel, *Poor People's Medicine*, esp. chaps. 10 and 11; Edward H. Beardsley, *A History of Neglect: Health Care for Blacks and Mill Workers in the Twentieth-Century South* (Knoxville: Univ. of Tennessee Press, 1987), chap. 12.

31. Irwin Miller, *American Health Care Blues: Blue Cross, HMOs, and Pragmatic Reform Since 1960* (New Brunswick, NJ: Transaction, 1996), esp. chap. 4; David J. Rothman, *Beginnings Count: The Technological Imperative in American Health Care* (New York: Oxford Univ. Press, 1997), chap. 1.

32. Sylvia Law, *Blue Cross: What Went Wrong?*, 2nd ed. (New Haven, CT: Yale Univ. Press, 1976), 115–116.

33. Alain C. Enthoven and Richard Kronick, "Universal Health Insurance through Incentive Reform," *JAMA*, 265 (1991), 2534.

34. Engel, *Poor People's Medicine*, chaps. 10 and 11; Lawrence D. Brown, "Introduction to a Decade of Transition," in Brown, *Health Policy in Transition*, 1–15; Fox, *Power and Illness*, esp. chaps. 4 and 5; "The Two Barriers to HMO Success: The Doctors and the Patients," *Modern Healthcare*, May 1974, 49–53; Bradford H. Gray, "The Rise and Decline of the HMO: A Chapter in U.S. Health-Policy History," in *History and Health Policy in the United States: Putting the Past Back In*, ed. Rosemary A. Stevens, Charles E. Rosenberg, and Lawton R. Burns (New Brunswick, NJ: Rutgers Univ. Press, 2006), 309–339; Jan Gregoire Coombs, *The Rise and Fall of HMOs: An American Health Care Revolution* (Madison: Univ. of Wisconsin Press, 2005); *The 1983 Investor's Guide to Health Maintenance Organizations* (Washington, DC: Division of Private Sector Initiatives, Office of Health Maintenance Organizations, Department of Health and Human Services, 1983).

35. Brown, "Introduction to a Decade of Transition"; Rick Mayes, "The Origins, Development, and Passage of Medicare's Revolutionary Prospective Payment System," *Journal of the History of Medicine and Allied Sciences*, 62 (2007), 21–55; Frank A. Sloan, Michael A. Morrisey, and Joseph Valvona, "Effects of the Medicare Prospective Payment System on

Hospital Cost Containment: An Early Appraisal," *Milbank Quarterly*, 66 (1988), 191–220; Linda Hughey Holt, "DRGs (Diagnostic Related Groups): The Doctors' Dilemma," *Perspectives in Biology and Medicine*, 29 (1986), 219–226.

36. Danielle A. Dolenc and Charles J. Dougherty, "DRGs: The Counterrevolution in Financing Health Care," *Hastings Center Report*, June 1985, 19–29.

37. Curtis L. Bakken and Donald S. Young, "Changing American Medicine," *British Medical Journal*, 288 (31 Mar. 1984), 956–957.

38. See, e.g., Sanford L. Weiner et al., "Economic Incentives and Organizational Realities: Managing Hospitals under DRGs," *Milbank Quarterly*, 65 (1987), 463–487.

39. Holt, "DRGs," 225–226.

40. Nancy Tomes, "Patients or Health-Care Consumers? Why the History of Contested Terms Matters," in Stevens, Rosenberg, and Burns, *History and Health Policy in the United States*, 83.

41. Social Security Amendments of 1965, Public Law 89–97, 30 July 1965, *U.S. Statutes*, 79; Tomes, "Patients or Health-Care Consumers?," 83–110.

42. See esp. Ameringer, *Health Care Revolution*; and Douglas E. Hough, *Irrationality in Health Care: What Behavioral Economics Reveals about What We Do and Why* (Stanford, CA: Stanford Univ. Press, 2013).

43. S. Lon Conner, "Professionalism as a Dirty Word," *Journal of the Medical Association of the State of Alabama*, Mar. 1979, 4; James D. Snyder, "Assaults on the Medical Profession," *Physician's Management*, Sept. 1982, 28–40; Warren Greenberg, *The Healthcare Marketplace* (New York: Springer, 1998).

44. Daniel J. McCarty et al., "Equity in Physician Compensation: The Marshfield Experiment," *Perspectives in Biology and Medicine*, 35 (1992), 261–276; Greenberg, *Health Care Marketplace*. See also, e.g., Harry Schwartz, "Doctors Are Treated to a Dose of Competition," *Wall Street Journal*, 16 Aug. 1982.

45. Ameringer, *Health Care Revolution*.

46. See, e.g., Arnold S. Relman, "The New Medical-Industrial Complex," *New England Journal of Medicine*, 303 (1980), 966.

47. Ameringer, *Health Care Revolution*; Tomes, "Patients or Health-Care Consumers?," esp. 102; Engel, *Poor People's Medicine*.

48. See, e.g., Paul Starr, *The Social Transformation of American Medicine* (New York: Basic Books, 1982), 428–449; Bradford H. Gray, ed., *For-Profit Enterprise in Health Care* (Washington, DC: National Academy Press, 1986); and Gray, *The Profit Motive and Patient Care* (Cambridge, MA: Harvard Univ. Press, 1991), chap. 12.

49. See, e.g., C. Wayne Higgins and Eugene D. Meyers, "The Economic Transformation of American Health Insurance: Implications for the Hospital Industry," *Health Care Management Review*, 11, no. 4 (1986), 21–27; Russell C. Coile Jr., *The New Hospital: Future Strategies for a Changing Industry* (Rockville, MD: Aspen, 1986); and *Source Book of Health Insurance Data*, 1991, 81.

50. Martin Kitchener, "Exploding the Merger Myth in U.S. Health Care," in *Health Policy: Crisis and Reform in the U.S. Health Care Delivery System*, ed. Charlene Harrington and Carroll L. Estes, 4th ed. (Sudbury, MA: Jones & Bartlett, 2004), 162–167; Deborah A. Savage, "Professional Sovereignty Revisited: The Network Transformation of American Medicine?," *Journal of Health Politics, Policy and Law*, 29 (2004), 666.

51. Victor R. Fuchs, *The Health Economy* (Cambridge, MA: Harvard Univ. Press, 1986), esp. 257 and 259–260.

52. T. M. Sonneborn, "Implications of the New Genetics for Biology and Man," *AIBS Bulletin*, 13 (1963), 22–26.

53. Robert Martensen, "The History of Bioethics: An Essay Review," *Journal of the History of Medicine and Allied Sciences*, 56 (2001), 168–175; Jonathan Imber, *Trusting Doctors: The Decline of Moral Authority in American Medicine* (Princeton, NJ: Princeton Univ. Press, 2007); Renée C. Fox and Judith P. Swazey, "Examining American Bioethics: Its Problems and Prospects," *Cambridge Quarterly Journal of Healthcare Ethics*, 14 (2005), 361–373.

54. David J. Rothman, "Ethics and Human Experimentation: Henry Beecher Revisited," *New England Journal of Medicine*, 317 (1987), 1195–1201; Henry K. Beecher, "Ethics and Clinical Research," ibid., 274 (1966), 1354–1360; Rothman, "Human Experimentation and the Origins of Bioethics in the United States," in *Social Science Perspectives on Medical Ethics*, ed. George Weisz (Dordrecht, Netherlands: Kluwer Academic, 1999), 185–217; Rothman, *Strangers at the Bedside: A History of How Law and Bioethics Transformed Medical Decision Making* (New York: Basic Books, 1991).

55. See esp. Daniel M. Fox, "Who We Are: The Political Origins of the Medical Humanities," *Theoretical Medicine*, 6 (1985), 327–342.

56. James H. Jones, *Bad Blood: The Tuskegee Syphilis Experiment* (New York: Free Press, 1981); Susan Reverby, *Examining Tuskegee: The Infamous Syphilis Study and Its Legacy* (Chapel Hill: Univ. of North Carolina Press, 2009).

57. Albert R. Jonsen, *A Short History of Medical Ethics* (New York: Oxford Univ. Press, 2000); Kathleen Montgomery and Amalya L. Oliver, "Shifts in Guidelines for Ethical Scientific Conduct: How Public and Private Organizations Create and Change Norms of Research Integrity," *Social Studies of Science*, 39 (2009), 137–146; Heinz Schott, "Medizingeschichte und Ethik: Zum Gedenken an Rolf Winan (1937–2006)," *Medizinhistorisches Journal*, 43 (2008), 93; introduction to *The American Medical Ethics Revolution: How the AMA's Code of Ethics Has Transformed Physicians' Relationships to Patients, Professionals, and Society*, ed. Robert B. Baker et al. (Baltimore: Johns Hopkins Univ. Press, 1999), xxxv.

58. Howard L. Kaye, "Sociology and the De-Moralization of Bioethics," in *Society and Medicine*, ed. Carla M. Messikomer, Judith P. Swazey, and Allen Glicksman (New Brunswick: Transaction, 2003), 227–241, esp. 228–229; Imber, *Trusting Doctors*.

59. See, e.g., [American Hospital Association], "Statement on a Patient's Bill of Rights," *Hospitals*, 47 (1973), 41.

60. Kenneth De Ville, "Medical Malpractice in Twentieth Century United States: The Interaction of Technology, Law, and Culture," *International Journal of Technology Assessment in Health Care*, 14 (1998), 197–211; Schwartz, "Societal Responsibility for Malpractice," 249–250; Louise Lander, *Defective Medicine: Risk, Anger, and the Malpractice Crisis* (New York: Farrar, Straus & Giroux, 1978), esp. chap. 8.

61. Marshall W. Raffel, *The U.S. Health System: Origins and Functions*, 2nd ed. (New York: John Wiley & Sons, 1984), 118; Ellen C. Annandale, "The Malpractice Crisis and the Doctor-Patient Relationship," *Sociology of Health & Illness*, 11 (1989), 1–23; Lander, *Defective Medicine*; Daniel M. Fox, "Physicians versus Lawyers: A Conflict of Cultures," in *AIDS Law Today: A New Guide for the Public*, ed. Scott Burris et al. (New Haven, CT: Yale Univ. Press, 1993), 367–376.

62. Lander, *Defective Medicine*, 224n.

63. Annandale, "Malpractice Crisis," 6–15; Lander, *Defective Medicine*, quotation from 140.

64. Lander, *Defective Medicine*, esp. 135 and 137; Leon Eisenberg, "The Search for Care," in Knowles, *Doing Better and Feeling Worse*, 243.

65. Lander, *Defective Medicine*, chap. 12, quotations from 174 and 175.

66. P. M. Strong, "Viewpoint: The Academic Encirclement of Medicine," *Sociology of Health and Illness*, 6 (1984), 339–358.

67. James L. Titchener, "A Remarkable Assemblage," unpublished memoir, 1969, copy in possession of the author; David E. Smith and John Luce, *Love Needs Care: A History of San Francisco's Haight-Ashbury Clinic and Its Pioneer Role in Treating Drug-Abuse Problems* (Boston: Little, Brown, 1971); Naomi Rogers, "'Caution: The AMA May Be Dangerous to Your Health': The Student Health Organizations (SHO) and American Medicine, 1965–1970," *Radical History Review*, 80 (2001), 5–34, quotation from 11; Fitzhugh Mullan, *White Coat, Clenched Fist: The Political Education of an American Physician* (New York: Macmillan, 1976), esp. 216.

68. Brian J. Zink, *Anyone, Anything, Anytime: A History of Emergency Medicine* (Philadelphia: Mosby Elsevier, 2006), esp. 56, 79–80, and chap. 4.

69. "Doctors Try an Image Implant," *Business Week*, 22 June 1968, 64–68; Edwards, *Hazardous to Your Health*, 31.

70. Michael G. Michaelson, "The Failure of American Medicine," *American Scholar*, 39 (1970), 702.

71. Daniel Schorr, quoted in Edwards, *Hazardous to Your Health*, 128. See also, e.g., Barbara Ehrenreich and John Ehrenreich, *The American Health Empire: Power, Profits, and Politics* (1970; reprint, New York: Vintage Books, 1971), esp. chaps. 16 and 17.

72. Russell Lynes, "After Hours," *Harper's*, Aug. 1968, 24.

73. Engel, *Poor People's Medicine*, 136. See also, e.g., Arnold S. Relman and Uwe Reinhardt, "An Exchange on For-Profit Health Care," in Gray, *For-Profit Enterprise in Health Care*, esp. 213–214; and Beatrix Hoffman, *Health Care for Some: Rights and Rationing in the United States since 1930* (Chicago: Univ. of Chicago Press, 2012), 138–142.

74. Martin Kohn and Carol Donley, eds., *Return to "The House of God": Medical Resident Education, 1978–2008* (Kent, OH: Kent State Univ. Press, 2008); Imber, *Trusting Doctors*.

75. Imber, *Trusting Doctors*.

76. See, e.g., Rosemary Stevens, "Public Roles for the Medical Profession in the United States: Beyond Theories of Decline and Fall," *Milbank Quarterly*, 79 (2001), 327–353; William Winkenwerder and John R. Ball, "Transformation of American Health Care: The Role of the Medical Profession," *New England Journal of Medicine*, 318 (1988), 317–319; John A. D. Cooper, "What Is Immediate Past Is Prologue—Unfortunately," *Journal of Medical Education*, 61 (1986), 113; and Mark Schlesinger, "A Loss of Faith: The Sources of Reduced Political Legitimacy for the American Medical Profession," *Milbank Quarterly*, 80 (2002), 185–235.

77. Ivan Illich, *Medical Nemesis: The Expropriation of Health* (London: Calder & Boyars, 1975), 11, 15; a U.S. edition followed the next year.

78. See, e.g., Paul D. Stolley, "Cultural Lag in Health Care," *Inquiry*, 8 (1971), 73; and David Halberstam, "Cures that Kill," *Harper's*, Dec. 1980, 20–24.

79. James H. Wittebols, *Watching M*A*S*H, Watching America: A Social History of the 1972–1983 Television Series* (Jefferson, NC: McFarland, 1998); Cortney Davis, "Chasing Fire Engines," in Kohn and Donley, *Return to "The House of God,"* 176.

80. Michael J. Halberstam, "Professionalism and Health Care," in Tancredi, *Ethics of Health Care*, 247.

81. See, e.g., John D. Stoeckle, "Medical Advice Books: The Search for the Healthy Body," *Social Science and Medicine*, 18 (1984), 707–712; and David W. Stepansky, in Stepansky, *Last Family Doctor*, 147.

82. Lander, *Defective Medicine*, chap. 12, quotation from 178; John P. Callan, "Holistic Health or Holistic Hoax?," *JAMA*, 241 (1979), 1156; Michael Glasser et al., "Holistic Medicine," ibid., 242 (1979), 1489–1491.

83. Rayna Rapp, *Testing Women, Testing the Fetus: The Social Impact of Amniocentesis in America* (New York: Routledge, 1999), 255.

84. Sandra Morgen, *Into Our Own Hands: The Women's Health Movement in the United States, 1969–1990* (New Brunswick, NJ: Rutgers Univ. Press, 2002); Wendy Kline, *Bodies of Knowledge: Sexuality, Reproduction, and Women's Health in the Second Wave* (Chicago: Univ. of Chicago Press, 2010); Susan Wells, *"Our Bodies, Our Selves" and the Work of Writing* (Stanford, CA: Stanford Univ. Press, 2010).

85. Morgen, *Into Our Own Hands*, esp. 7–8 and chaps. 5–8; Kline, *Bodies of Knowledge*, quotation from 11; Sara Arber and Hilary Thomas, "From Women's Health to a Gender Analysis of Health," in *The Blackwell Companion to Medical Sociology*, ed. William C. Cockerham (New York: Blackwell, 2001), 94–113.

86. Susan Gelfand Malka, *Daring to Care: American Nursing and Second-Wave Feminism* (Urbana: Univ. of Illinois Press, 2007).

87. Kenneth M. Ludmerer, *Time to Heal: American Medical Education from the Turn of the Century to the Era of Managed Care* (New York: Oxford Univ. Press, 1999), 240.

88. Wilson and Neuhauser, *Health Services in the United States*, 2nd ed., 7; Freymann, *American Health Care System*, xv.

89. Charles E. Rosenberg, "The Origins of the American Hospital System," *Bulletin of the New York Academy of Medicine*, 55 (1979), 10. See also, e.g., H. R. Nayer, "Opening Statement," ibid., 6–9.

90. Doris C. Van Doren and Alan P. Spielman, "Hospital Marketing: Strategy Reassessment in a Declining Market," *Journal of Health Care Marketing*, Mar. 1989, 15.

91. "Hospitals Break Acute-Care Mold," *American Medical News*, 16 Jan. 1987, 3; Rosemary Stevens, *In Sickness and in Wealth: American Hospitals in the Twentieth Century* (New York: Basic Books, 1989), esp. 302, quotation from 294.

92. Ludmerer, *Time to Heal*, 265–270; Stevens, *In Sickness and in Wealth*; Robert G. Petersdorf, "Managing the Revolution in Medical Care," *Journal of Medical Education*, 59 (1984), 85; Fox, "The Consequences of Consensus," 94.

93. Steven M. Shortell, Robin R. Gillies, and Kelly J. Devers, "Reinventing the American Hospital," *Milbank Quarterly*, 73 (1995), 134.

94. Stevens, *In Sickness and in Wealth*, quotation from 298.

95. M. Raffel, *U.S. Health System*, 2nd ed., 195; Eli Ginzberg, "The Grand Illusion of Competition in Health Care," *JAMA*, 249 (1983), 1857–1859; Ginzberg, "For-Profit Medicine," *New England Journal of Medicine*, 319 (1988), 757–761.

96. Stevens, *In Sickness and in Wealth*, esp. chap 11; Wilson and Neuhauser, *Health Services in the United States*, 2nd ed., rev.; Rosemary A. Stevens, "Times Past, Times Present," in *The American General Hospital: Communities and Social Contexts*, ed. Diana Elizabeth Long and Janet Golden (Ithaca, NY: Cornell Univ. Press, 1989), 191–206, quotation from 205; John E. Sauer, personal communication; Guenter B. Risse, *Mending Bodies, Saving Souls: A History of Hospitals* (New York: Oxford Univ. Press, 1999), esp. 540–546.

97. See, e.g., John D. Stoeckle, "The Citadel Cannot Hold: Technologies Go Outside the Hospital, Patients and Doctors Too," *Milbank Quarterly*, 73 (1993), 3–17.

98. Beatrix Hoffman, "Emergency Rooms: The Reluctant Safety Net," in Stevens, Rosenberg, and Burns, *History and Health Policy in the United States*, 250–272; Shelley I. White-Means and Michael C. Thornton, "Nonemergency Visits to Hospital Emergency Rooms: A Comparison of Blacks and Whites," *Milbank Quarterly*, 67 (1989), 35–57.

99. Hoffman, "Emergency Rooms"; Zink, *Anyone, Anything, Anytime*.

100. Louise B. Russell, *Technology in Hospitals: Medical Advances and Their Diffusion* (Washington, DC: Brookings Institution, 1979), esp. chap. 2, quotation from 49.

101. See, e.g., Anthony J. Sattilaro and John C. Cameron, "Change Challenges Traditional Hospital / Medical Staff Relationship," *Hospitals*, 1 Apr. 1981, 129, 142–143; and Stevens, *In Sickness and in Wealth*, esp. 234 and 323–326, quotation from 326.

102. Gordon Wolstenholme, "Medicine in the United States," *British Medical Journal*, 288 (1984), 1214.

103. See Ludmerer, *Time to Heal*.

104. Michael J. Lepore, *Death of the Clinician: Requiem or Reveille?* (Springfield, IL: Charles C Thomas, 1982), 328.

105. Vernon W. Lippard, *A Half-Century of American Medical Education: 1920–1970* (New York: Josiah Macy, Jr. Foundation, 1974), vii.

106. Renée C. Fox, "Is There a 'New' Medical Student? A Comparative View of Medical Socialization in the 1950s and 1970s," in Tancredi, *Ethics of Health Care*, 197–220; Norman Metzger, *The Health Care Supervisor's Handbook*, 2nd ed. (Rockville, MD: Aspen, 1982), chap. 6; "A New Type of Doctor Emerges," *Time*, 8 Nov. 1971, 61–62, 67; T. J. Hoff, "Professional Commitment among US Physician Executives in Managed Care," *Social Science and Medicine*, 50 (2000), 1433–1444; Alan L. Hillman et al., "Managing the Medical-Industrial Complex," *New England Journal of Medicine*, 315 (1986), 511–413.

107. See, e.g., Barbara J. Culliton, "Health Manpower Act: Aid But Not Comfort for Medical Schools," *Science*, 194 (1976), 700–704; Ludmerer, *Time to Heal*; and William G. Rothstein, *American Medical Schools and the Practice of Medicine: A History* (New York: Oxford Univ. Press, 1987).

108. Ludmerer, *Time to Heal*, 223, 264.

109. Colin L. Talley, *A History of Multiple Sclerosis* (Westport, CT: Praeger, 2008), 93.

110. Barsky, *Worried Sick*, 86.

111. C. McKeen Cowles, *Nursing Home Statistical Yearbook, 1995* (Tacoma, WA: Cowles Research Group, 1995), 128–131; Oliver E. Allen, "The State of Medical Care, 1984: An Interview with Dr. David E. Rogers," *American Heritage*, Oct./Nov. 1984, 34; Jan Shank-

roff et al., "Nursing Home Initiative," *Health Care Financing Review,* 22 (2000), 113–115; Mathy Mezey and Claire Fagin, "Caring in Institutional Settings," in *The Lost Art of Caring: A Challenge to Health Professionals, Families, Communities, and Society,* ed. Leighton E. Cluff and Robert H. Binstock (Baltimore: Johns Hopkins Univ. Press, 2001), 140–141; Engel, *Poor People's Medicine,* esp. 186–189.

112. Bruce C. Vladeck, *Unloving Care: The Nursing Home Tragedy* (New York: Basic Books, 1980); Catherine Hawes and Charles D. Phillips, "The Changing Structure of the Nursing Home Industry and the Impact of Ownership on Quality, Cost, and Access," in Gray, *For-Profit Enterprise in Health Care,* 492–541, quotation from 507; Charlene Harrington et al., "Does Investor Ownership of Nursing Homes Compromise the Quality of Care?," *American Journal of Public Health,* 91 (2001), 1452–1455; Fuchs, *Health Economy.*

113. See, e.g., Thomas N. Chirikos, "Accounting for the Historical Rise in Work-Disability Prevalence," *Milbank Quarterly,* 64 (1986), 271–301; Barsky, *Worried Sick,* 85; and Edward D. Berkowitz and Larry DeWitt, *The Other Welfare: Supplemental Security Income and U.S. Social Policy* (Ithaca, NY: Cornell Univ. Press, 2013).

114. Barbara L. Wolfe and Robert Haveman, "Trends in the Prevalence of Work Disability from 1962 to 1984, and Their Correlates," *Milbank Quarterly,* 68 (1990), 56. See also, e.g., Talley, *History of Multiple Sclerosis.*

115. John Duffy, *The Sanitarians: A History of American Public Health* (Urbana: Univ. of Illinois Press, 1990), chap. 19; Jonathan E. Fielding, "Public Health in the Twentieth Century: Advances and Challenges," *Annual Review of Public Health,* 20 (1999), esp. xiii; Mark R. Cullen, "Personal Reflections on Occupational Health in the Twentieth Century: Spiraling to the Future," ibid., 9; Phil Brown, "Environmental Health as a Core Public Health Component," in *The Contested Boundaries of American Public Health,* ed. James Colegrove, Gerald Markowitz, and David Rosner (New Brunswick, NJ: Rutgers Univ. Press, 2008), 85–109.

116. Stephen Martin and Sara Dubous (paper presented at the annual meeting of American Association for the History of Medicine, Rochester, MN, 2010).

117. Élodie Giroux, "The Framingham Study and the Constitution of a Restrictive Concept of Risk Factor," *Social History of Medicine,* 26 (2013), 94–112.

118. Daniel Wilner, Rosabelle Price Walkley, and Lenor S. Goerke, *Introduction to Public Health,* 6th ed. (New York: Macmillan, 1973), 6.

119. Gloria Ruby, "Increasing the Knowledge Base for Prevention," in *Healthy People: The Surgeon General's Report on Health Promotion and Disease Prevention, Background Papers 1979,* DHEW (PHS) Publication No. 79-55071A (Washington, DC: U.S. Government Printing Office, 1979), 461–465.

120. Lisa F. Berkman and Lester Breslow, *Health and Ways of Living: The Alameda County Study* (New York: Oxford Univ. Press, 1983), 216–222, quotation from 219.

121. U.S. Public Health Service, *Promoting Health/Preventing Disease* (Washington, DC: U.S. Government Printing Office, 1980), quotations from iii and 1.

122. Tori DeAngelis, "Should Wellness Model Replace Disease Focus?," *American Psychological Association Monitor,* Dec. 1990, 30; Nancy K. Janz, "The Health Belief Model: A Decade Later," *Health Education Quarterly,* 11 (1984), 1–47.

123. See, e.g., Lester Breslow, "Setting Objectives for Public Health," *Annual Review of Public Health,* 8 (1987), 291–292; Deborah C. Glik and Jennie J. Kronenfeld, "Well Roles:

An Approach to Reincorporate Role Theory into Medical Sociology," *Research in Sociology of Health Care*, 8 (1989), 299; Robert Crawford, "Cultural Influences on Prevention and the Emergence of a New Health Consciousness," in *Taking Care: Understanding and Encouraging Self-Protective Behavior*, ed. Neil D. Weinstein (Cambridge: Cambridge Univ. Press, 1987), 95–113; and Barsky, *Worried Sick*.

124. See, e.g., Mary Ann Scheirer, "The Life Cycle of an Innovation," *Journal of Health and Social Behavior*, 31 (1990), 203.

125. See, in general, John W. Ward and Christian Warren, eds., *Silent Victories: The History and Practice of Public Health in Twentieth-Century America* (New York: Oxford Univ. Press, 2007); and Marian Moser Jones and Ronald Bayer, "Paternalism and Its Discontents: Motorcycle Helmet Laws, Libertarian Values, and Public Health," in Colegrove, Markowitz, and Rosner, *Contested Boundaries of American Public Health*, 110–126.

126. See, e.g., Deborah A. Stone, "The Resistable Rise of Preventive Medicine," in Brown, *Health Policy in Transition*, 107–111; and Howard M. Leichter, "'Evil Habits' and 'Personal Choices': Assigning Responsibility for Health in the 20th Century," *Milbank Quarterly*, 81 (2003), 603–626.

127. Robert W. Ambler and Donald L. Eddins, "Cross-Sectional Analysis: Precursors of Premature Death in the United States," in *Closing the Gap: The Burden of Unnecessary Illness*, ed. Robert W. Ambler and H. Bruce Dull (New York: Oxford Univ. Press, 1987), 187; John H. Knowles, "The Responsibility of the Individual," in Knowles, *Doing Better and Feeling Worse*, esp. 58–60; Robert Crawford, "Individual Responsibility and Health Politics in the 1970s," in *Health Care in America: Essays in Social History*, ed. Susan Reverby and David Rosner (Philadelphia: Temple Univ. Press, 1979), 247–268; George F. Will, "A Right to Health?," *Newsweek*, 7 Aug. 1978, 88.

128. Julius B. Richmond, quoted in U.S. Senate, Select Committee on Nutrition and Human Needs, *Dietary Goals for the United States*, 2nd ed. (Washington, DC: U.S. Government Printing Office, 1977), xxxiii.

129. Barsky, *Worried Sick*, chap. 5, quotation from 89.

130. Jerrold M. Michael, "The Second Revolution in Health: Health Promotion and Its Environmental Base," *American Psychologist*, 37 (1982), 936–941; Eugene C. Nelson and Jeannette J. Simmons, "Health Promotion—The Second Public Health Revolution: Promise or Threat?," *Family and Community Health*, Feb. 1983, 1–15; Margret M. Baltes, "Health Care from a Behavioral-Ecological Viewpoint," *Health Care Dimensions*, 3 (1976), 149, 153; Deane Neubauer and Richard Pratt, "The Second Public Health Revolution: A Critical Appraisal," *Journal of Health Politics, Policy and Law*, 6 (1981), 205–228; Lawrence W. Green, *Health Promotion Planning: An Educational and Environmental Approach*, 2nd ed. (Mountain View, CA: Mayfield, 1991), esp. 12–14; Robin Bunton, Sarah Nettleton, and Roger Burrows, eds., *The Sociology of Health Promotion: Critical Analyses of Consumption, Lifestyle, and Risk* (London: Routledge, 1995).

131. John D. Loft and Phillip R. Kletke, "Human Resource Trends in the Health Field," in *Handbook of Medical Sociology*, ed. Howard E. Freeman and Sol Levine (Englewood Cliffs, NJ: Prentice Hall, 1989), 419–420; Harold M. Swartz and Ann Barry Flood, "Thirty Years of Change in Medical Education, Money, and Power: Differing Effects on Generations of Physicians and Clinical Practice," in *Money, Power, and Health Care*, ed. Evan M. Melhado et al. (Ann Arbor, MI: Health Administration Press, 1988), 287–299.

132. Eli Ginzberg, "A Non-Conforming View," *Health Management Quarterly*, 12, no. 3 (1990), 20–22, quotations from 20 and 22; Drew E. Altman, "Two Views of a Changing Health Care System," in Aiken and Mechanic, *Applications of Social Science*, 105.

133. See, e.g., Schwartz, "Doctors Are Treated to a Dose of Competition." The quotation is from David Mechanic, "Public Perceptions of Medicine," *New England Journal of Medicine*, 312 (1985), 181.

134. Alvin P. Sanoff and William Schwartz, "U.S. Medicine 'Cannot Do Everything for Everybody,' " *U.S. News & World Report*, 25 June 1984, 71–72; Richard D. Lamm, "The Brave New World of Health Care," *Annals of Thoracic Surgery*, 52 (1991), 371.

135. Madonna Harrington Meyer, ed., *Care Work: Gender, Class, and the Welfare State* (New York: Routledge, 2000); Carroll L. Estes and Donna M. Zulman, "Informalization of Long-Term Caregiving: A Gender Lens," in Harrington and Estes, *Health Policy*, 147–156; John B. McKinlay, introduction to *Milbank Quarterly*, 66, suppl. 2 (1988), 1; Charles A. Kiesler and Teru L. Morton, "Psychology and Public Policy in the 'Health Care Revolution,' " *American Psychologist*, 43 (1988), 993.

Part IV. The End of One Epoch and the Beginning of Another

1. Kaye H. Kilburn, "Epidemics Then and Now: Chemicals Replace Microbes and Degenerations Oust Infections," *Archives of Environmental Health*, 49 (1994), 3–5.

2. See, e.g., Elizabeth Haiken, *Venus Envy: A History of Cosmetic Surgery* (Baltimore: Johns Hopkins Univ. Press, 1997); Sheila M. Rothman and David J. Rothman, *The Pursuit of Perfection: The Promise and Perils of Medical Enhancement* (New York: Pantheon Books, 2003); John Hoberman, *Testosterone Dreams: Rejuvenation, Aphrodisia, Doping* (Berkeley: Univ. of California Press, 2005); and Nathaniel Comfort, *The Science of Human Perfection: How Genes Became the Heart of American Medicine* (New Haven, CT: Yale Univ. Press, 2012).

3. See, e.g., Lisa F. Berkman, "Social Epidemiology: Social Determinants of Health in the United States: Are We Losing Ground?," *Annual Review of Public Health*, 30 (2009), 27–41.

4. William G. Rothstein, "The Decrease in Socioeconomic Differences in Mortality from 1920 to 2000 in the United States and England," *Journal of the History of Medicine and Allied Sciences*, 67 (2012), 551.

5. See, e.g., Deborah A. Savage, "Professional Sovereignty Revisited: The Network Transformation of American Medicine?," *Journal of Health Politics, Policy and Law*, 29 (2004), 661–667.

6. Einer R. Elhauge, ed., *The Fragmentation of U.S. Health Care: Causes and Solutions* (New York: Oxford Univ. Press, 2010).

7. Michael S. Sparer, "Myth and Misunderstandings: Health Policy, the Devolution Revolution, and the Push for Privatization," *American Behavioral Scientist*, 43 (1999), 138–154.

8. Adele E. Clarke et al., eds., *Biomedicalization: Technoscience, Health, and Illness in the U.S.* (Durham, NC: Duke Univ. Press, 2010), chaps. 1–3.

9. Charles E. Rosenberg, "Managed Fear," *Lancet*, 373 (2009), 802–803.

10. See, e.g., John B. McKinlay, introduction to *Milbank Quarterly*, 66, suppl. 2 (1988), 1–9.

11. See, e.g., Robin Bunton and Roger Burrows, "Consumption and Health in the 'Epidemiological' Clinic of Late Modern Medicine," in *The Sociology of Health Promotion:*

Critical Analyses of Consumption, Lifestyle, and Risk, ed. Robin Bunton, Sarah Nettleton, and Roger Burrows (London: Routledge, 1995), 206–244.

Chapter 13. The Era of Genetic Medicine, Late 1980s and After

1. Mark A. Peterson, "Evidence: Its Meaning in Health Care and in Law," *Journal of Health Politics, Policy and Law*, 26 (2001), 191–192.

2. David Dranove, *The Economic Evolution of American Health Care: From Marcus Welby to Managed Care* (Princeton, NJ: Princeton Univ. Press, 2000).

3. Francis S. Collins, *The Language of Life: DNA and the Revolution in Personalized Medicine* (New York: HarperCollins, 2010), 2.

4. *Health, United States*, 1991, 247, and 2010, 346; Derek R. Smart, *Physician Characteristics and Distribution in the US* (Chicago: American Medical Association, 2010), 4.

5. *Health, United States*, 2009, 383, 374, and 2010, 351–352, 345, 350, 354; Fitzhugh Mullan, "Primary Care Roots," in *Big Doctoring in America: Profiles in Primary Care*, ed. Mullan (Berkeley: Univ. of California Press, 2002), 12–13; Carl F. Ameringer, *The Health Care Revolution: From Medical Monopoly to Market Competition* (Berkeley: Univ. of California Press, 2008), 157; Gloria J. Bazzoli et al., "The Transition from Excess Capacity to Strained Capacity in U.S. Hospitals," *Milbank Quarterly*, 84 (2006), 273–304; Robyn I. Stone, "The Direct Care Worker: The Third Rail of Home Care Policy," *Annual Review of Public Health*, 25 (2004), 521–537; V. Mor et al., "Changes in the Quality of Nursing Homes in the U.S.: A Review and Data Update," in *The Nation's Health*, ed. Leiyu Shi and Douglas A. Singh, 8th ed. (Sudbury, MA: Jones & Bartlett, 2011), 596–611.

6. *Health, United States*, 2009, 3.

7. Abner Louis Notkins, "New Predictors of Disease," in *Infectious Disease: A Scientific American Reader*, ed. Scientific American (Chicago: Univ. of Chicago Press, 2008), 233–243.

8. *Health, United States*, 2009, 3–4, and 2010, 14–15; Steven A. Schroeder, "We Can Do Better—Improving the Health of the American People," *New England Journal of Medicine*, 357 (2007), 1221–1228.

9. *Health, United States*, 2010, 19.

10. Ibid., 43.

11. See, e.g., James Colegrove, Gerald Markowitz, and David Rosner, eds., *The Contested Boundaries of American Public Health* (New Brunswick, NJ: Rutgers Univ. Press, 2008), esp. pt. 3; Roger I. Glass, "Perceived Threats and Real Killers," *Science*, 304 (2004), 927; Nancy A. Thompson and Christopher D. Van Gorder, "Healthcare Executives' Role in Preparing for the Pandemic Influenza 'Gap': A New Paradigm for Disaster Planning," *Journal of Healthcare Management*, 52 (2007), 87–90; A. Cassil, "Rising Rates of Chronic Health Conditions: What Can Be Done?," in Shi and Singh, *Nation's Health*, 797–800, esp. 797; and Gerard Anderson and Jane Horvath, "The Growing Burden of Chronic Disease in America," *Public Health Reports*, 119 (2004), 263.

12. See, e.g., Paul B. Ginsburg, "Health System Change in 1997," *Health Affairs*, 17 (1998), 165; and Kathryn McDonagh, "Interview with Kathryn McDonagh, FACHE, President and CEO, CHRISTUS Spohn Health System, Corpus Christi, Texas," by Kyle L. Grazier, *Journal of Healthcare Management*, 49 (2004), 5.

13. Randall M. Packard et al., "Introduction: Emerging Illness as Social Process," in *Emerging Illnesses and Society: Negotiating the Public Health Agenda*, ed. Randall M. Packard et al. (Baltimore: Johns Hopkins Univ. Press, 2004), 1–35.

14. Peter J. Whitehouse, Konrad Maurer, and Jesse F. Ballenger, eds., *Concepts of Alzheimer Disease: Biological, Clinical, and Cultural Perspectives* (Baltimore: Johns Hopkins Univ. Press, 2000); Matt Clark, Mary Hagar, and Dan Shapiro, "Epidemic of Senility," *Newsweek*, 5 Nov. 1979, 95. See also *Reader's Guide to Periodical Literature*, 1920–2004.

15. Robert Katzman, "The Prevalence and Malignancy of Alzheimer Disease," *Archives of Neurology*, 33 (1976), 217–218; Patrick Fox, "From Senility to Alzheimer's Disease: The Rise of the Alzheimer's Disease Movement," *Milbank Quarterly*, 67 (1989), 58–101, quotations from 58; Jesse F. Ballenger, *Self, Senility, and Alzheimer's Disease in Modern America: A History* (Baltimore: Johns Hopkins Univ. Press, 2006), quotation from 119.

16. John Hardy, "Toward Alzheimer Therapies Based on Genetic Knowledge," *Annual Review of Medicine*, 55 (2004), 15–25; Stone, "Direct Care Worker," 521–537; Madonna Harrington Meyer, ed., *Care Work: Gender, Class, and the Welfare State* (New York: Routledge, 2000).

17. Chloe Silverman, *Understanding Autism: Parents, Doctors, and the History of a Disorder* (Princeton, NJ: Princeton Univ. Press, 2012), esp. chap. 12.

18. Maryn McKenna, *Superbug: The Fatal Menace of MRSA* (New York: Free Press, 2010); John Edwards Jr., quoted in ibid.

19. Chip Chambers, quoted in ibid., 77.

20. Edward B. Hayes and Duane J. Gubler, "West Nile Virus," *Annual Review of Medicine*, 57 (2006), 181–194; A. Marm Kilpatrick, "Globalization, Land Use, and the Invasion of West Nile Virus," *Science*, 334 (2011), 323–327.

21. Thomas Abraham, *Twenty-First Century Plague: The Story of SARS* (Baltimore: Johns Hopkins Univ. Press, 2005); J. Brian Houston, Wen-yu Chao, and Sandra Ragan, "Newspaper Coverage of the 2003 SARS Outbreak," in *The Social Construction of SARS: Studies of a Health Communication Crisis*, ed. John H. Powers and Xiaosui Xiao (Amsterdam: John Benjamins, 2008), 203–221.

22. Martin Enserink, "Ground the Planes During a Flu Pandemic? Studies Disagree," *Science*, 313 (2006), 1555.

23. Nicholas B. King, "The Influence of Anxiety: September 11, Bioterrorism, and American Public Health," *Journal of the History of Medicine and Allied Sciences*, 58 (2003), 433–441; Jonathan E. Fielding, Barbara Starfield, and Ross C. Brownson, preface to *Annual Review of Public Health*, 23 (2002), v. See also, e.g., Chris L. Barrett, Stephen G. Eubank, and James F. Smith, "If Smallpox Strikes Portland . . . ," in Scientific American, *Infectious Disease: A Scientific American Reader*, 275–285.

24. Susan Lindee, *Moments of Truth in Genetic Medicine* (Baltimore: Johns Hopkins Univ. Press, 2005). See also, e.g., Susan Hockfield, "The Next Innovation Revolution," *Science*, 323 (2009), 1147.

25. Dorothy Nelkin and M. Susan Lindee, "Good Genes and Bad Genes: DNA in Popular Culture," in *The Practices of Human Genetics*, ed. Michael Fortun and Everett Mendelsohn (Dordrecht, Netherlands: Kluwer Academic, 1999), 155–167; Lindee, *Moments of Truth in Genetic Medicine*; Nathaniel Comfort, *The Science of Human Perfection:*

How Genes Became the Heart of American Medicine (New Haven, CT: Yale Univ. Press, 2012); Peter K. Vogt, "Cancer Genes," *Western Journal of Medicine*, 158 (1993), 273.

26. Lindee, *Moments of Truth in Genetic Medicine*; Diane B. Paul, "PKU Screening: Competing Agendas, Converging Stories," in Fortun and Mendelsohn, *Practices of Human Genetics*, 185–195; Rachel Grob, "A House on Fire: Newborn Screening, Parents' Advocacy, and the Discourse of Urgency," in *Patients as Policy Actors*, ed. Beatrix Hoffman et al. (New Brunswick, NJ: Rutgers Univ. Press, 2011), 231–256.

27. See, e.g., Mitchell S. Golbus et al., "Prenatal Genetic Diagnosis in 3000 Amniocenteses," *New England Journal of Medicine*, 300 (1979), 157–163; and Rayna Rapp, *Testing Women, Testing the Fetus: The Social Impact of Amniocentesis in America* (New York: Routledge, 1999).

28. Rapp, *Testing Women, Testing the Fetus*; Lindee, *Moments of Truth in Genetic Medicine*, 14–17; Robert G. Resta, "Historical Aspects of Genetic Counseling: Why Was the Maternal Age 35 Chosen as the Cut-Off for Offering Amniocentesis?," *Medicina nei Secoli*, 14 (2002), 793–811; Keith Wailoo, *Dying in the City of the Blues: Sickle Cell Anemia and the Politics of Race and Health* (Chapel Hill: Univ. of North Carolina Press, 2001).

29. Barton Childs, "Medical Genetics to Genomic Medicine," *Medicina nei Secoli*, 14 (2002), 708–710, 712.

30. Robert Williamson, "Cystic Fibrosis—A Strategy for the Future," in *The Identification of the CF (Cystic Fibrosis) Gene: Recent Progress and New Research Strategies*, ed. Lap-Chee Tsui et al. (New York: Plenum, 1991), 1–7; Keith Wailoo and Stephen Pemberton, *The Troubled Dream of Genetic Medicine: Ethnicity and Innovation in Tay-Sachs, Cystic Fibrosis, and Sickle Cell Disease* (Baltimore: Johns Hopkins Univ. Press, 2006), esp. chap. 2; Kathi E. Hanna et al., *Genetic Counseling and Cystic Fibrosis Carrier Screening: Results of a Survey*, OTA-BP-BA-97 (Washington, DC: U.S. Government Printing Office, 1992); Miriam Solomon, "Evidence-Based Medicine and Mechanistic Reasoning in the Case of Cystic Fibrosis" (paper presented at the "Logic, Methodology, and Philosophy of Science" conference, Nancy, 11 July 2011).

31. Barbara J. Culliton, "Nanomedicine—The Power of Proteins: A Conversation with Lance Liatta and Emanual Petricoin," *Health Affairs*, S27 (2008), w310–w314.

32. K. Codell Carter, *The Decline of Therapeutic Bloodletting and the Collapse of Traditional Medicine* (New Brunswick, NJ: Transaction, 2012), esp. 139–146; Huda Y. Zoghbi, "The Basics of Translation," *Science*, 339 (2013), 250.

33. Wolfgang Sadee, "Drug Therapy and Personalized Health Care: Phramacogenomics in Perspective," *Pharmaceutical Research*, 25 (2008), 2714; Earl Lane and Becky Ham, "Personalized Medicine: Successes Tempered by Significant Challenges," *Science*, 324 (2009), 1658; Mark S. Boguski, Kenneth D. Mandl, and Vikas P. Sukhatme, "Repurposing with a Difference," ibid., 1394–1395.

34. Sadee, "Drug Therapy and Personalized Health Care," 2713–2714.

35. Beverly Merz, "The Genetic Revolution: Designer Genes," in *Genetics and Society*, ed. Penelope Barker (New York: H. W. Wilson, 1995), 63–65.

36. Jennifer Couzin-Frankel, "The Promise of a Cure: 20 Years and Counting," *Science*, 324 (2009), 1504–1507; Lindee, *Moments of Truth in Genetic Medicine*, esp. 194–200. Examples of skepticism include James M. Wilson, "A History Lesson for Stem Cells," *Science*, 324 (2009), 727–728; James P. Evans, "Deflating the Genomic Bubble," ibid., 331 (2011), 861–862;

Nicola J. Marks, "Science Fiction, Cultural Knowledge and Rationality: How Stem Cell Researchers Talk About Reproductive Cloning," in *The Body Divided: Human Being and Human "Material" in Modern Medical History,* ed. Sarah Ferber and Sally Wilde (Farnham, Surrey: Ashgate, 2011), 191–222; Comfort, *Science of Human Perfection*; and Inder M. Verma, "Gene Therapy That Works," *Science,* 341 (2013), 853–855.

37. Eric T. Juengst, "Concepts of Disease after the Human Genome Project," in *Ethical Issues in Health Care on the Frontiers of the Twenty-First Century,* ed. Stephen Wear et al. (Dordrecht, Netherlands: Kluwer Academic, 2000), 127–154; Sadee, "Drug Therapy and Personalized Health Care," 2713.

38. Robert P. Erickson, "From 'Magic Bullet' to 'Specially Engineered Shotgun Loads': The New Genetics and the Need for Individualized Pharmacotherapy," *BioEssays,* 20 (1998), 683–685.

39. Collins, *Language of Life*; Greta Lorge, "You, Decoded," *Stanford Magazine,* May/ June 2009, 65–66.

40. Francis Hyslop et al., *The Toxicology of Beryllium* National Institute of Health Bulletin 181 (Washington, DC: U.S. Government Printing Office, 1943); Charles G. Wilber, *Beryllium—A Potential Environmental Contaminant* (Springfield, IL: Charles C Thomas, 1980), quotation from 117; Milton D. Rossman, Otto P. Preuss, and Martin B. Powers, *Beryllium: Biomedical and Environmental Aspects* (Baltimore: Williams & Wilkins, 1991), 4–6.

41. See, e.g., Keith C. Meyer, "Beryllium and Lung Disease," *Chest,* 106 (1994), 942–946; Richard T. Sawyer et al., "Chronic Beryllium Disease: A Model Interaction between Innate and Acquired Immunity," *International Immunopharmacology,* 2 (2002), 249–261; and Kathleen Kreiss, Gregory A. Day, and Christine R. Schuler, "Beryllium: A Modern Industrial Hazard," *Annual Review of Public Health,* 28 (2007), 259–277, quotation from 274.

42. Andrew Lakoff, *Pharmaceutical Reason: Knowledge and Value in Global Psychiatry* (Cambridge: Cambridge Univ. Press, 2005), 169–170; Adam Hedgecoe, *The Politics of Personalised Medicine: Pharmacogenetics in the Clinic* (Cambridge: Cambridge Univ. Press, 2004), chaps. 1–2.

43. T. Friedman, quoted in Juengst, "Concepts of Disease after the Human Genome Project," 129–130.

44. Christopher P. Austin, "The Impact of the Completed Human Genome Sequence on the Development of Novel Therapeutics for Human Disease," *Annual Review of Medicine,* 55 (2004), 8; Comfort, *Science of Human Perfection.*

45. See, e.g., "Cancer Genes," *Science,* 322 (2008), 1769; Lindee, *Moments of Truth in Genetic Medicine*; and Jocelyn Kaiser, "Looking for a Target on Every Tumor," *Science,* 326 (2009), 218–220.

46. Sadee, "Drug Therapy and Personalized Health Care," 2713; James M. Rippe, ed., *Lifestyle Medicine* (Malden, MA: Blackwell Science, 1999).

47. Michael McGinnis and William H. Foege, "Actual Causes of Death in the United States," *JAMA,* 270 (1993), 2207–2212; Eliot Marshall, "Public Enemy Number One: Tobacco or Obesity?," *Science,* 304 (2004), 804; Ali H. Mokdad et al., "The Spread of the Obesity Epidemic in the United States, 1991–1998," *JAMA,* 282 (1999), 1519–1522; Jay Carrington Chunn, ed., *The Health Behavioral Change Imperative: Theory, Education, and Practice in Diverse Populations* (New York: Kluwer Academic / Plenum, 2002); Brian L. Cole and Jonathan E.

Fielding, "Health Impact Assessment: A Tool to Help Policy Makers Understand Health Beyond Health Care," *Annual Review of Public Health*, 28 (2007), 393–412.

48. Peter J. Wedlund, "Techniques for Identifying Genetic Polymorphisms: An Historical Perspective," in *Genetic Polymorphisms and Susceptibility to Disease*, ed. M. S. Miller and M. T. Cronin (London: Taylor & Francis, 2000), 1.

49. Eli Kintisch, "EPA Calls for More Studies on Health Risks of Climate Change," *Science*, 321 (2008), 477; Andre Nel et al., "Toxic Potential of Materials at the Nanolevel," ibid., 311 (2005), 622–627; Stephen M. Rappaport and Martyn T. Smith, "Environment and Disease Risks," ibid., 330 (2010), 460–461; W. Sadee, "The Relevance of 'Missing Heritability' in Pharmacogenomics," *Nature*, 92 (2012), 428–430.

50. See, e.g., Richard A. Deyo, "Cascade Effects of Medical Technology," *Annual Review of Public Health*, 23 (2002), 23–44.

51. See, e.g., Hayriye V. Erkizan et al., "A Small Molecule Blocking Oncogenic Protein EWS-FLI1 Interaction with RNA Helicase A Inhibits Growth of Ewing's Sarcoma," *Nature Medicine*, 15 (2009), 750–756.

52. Robert A. Aronowitz, *Unnatural History: Breast Cancer and American Society* (New York: Cambridge Univ. Press, 2007), esp. chap. 11.

53. See, e.g., Scientific American, *Infectious Disease: A Scientific American Reader*.

54. See, e.g., Stephen S. Morse, ed., *Emerging Viruses* (New York: Oxford Univ. Press, 1993); Mart Krupovic and Dennis H. Bamford, "Revealing Virus-Host Interplay," *Science*, 333 (2011), 45; and Elizabeth Pennisi, "Going Viral: Exploring the Role of Viruses in Our Bodies," ibid., 331 (2011), 1513.

55. Keith Wailoo et al., eds., *Three Shots at Prevention: The HPV Vaccine and the Politics of Medicine's Simple Solutions* (Baltimore: Johns Hopkins Univ. Press, 2010); Ilana Löwy, *A Woman's Disease: The History of Cervical Cancer* (Oxford: Oxford Univ. Press, 2011), esp. chap. 6.

56. Gerald N. Grob, "The Rise of Peptic Ulcer, 1900–1950," *Perspectives in Biology and Medicine*, 46 (2003), 550–566; Karen J. Goodman and Myles Cockburn, "The Role of Epidemiology in Understanding the Health Effects of *Helicobacter pylori*," *Epidemiology*, 12 (2001), 266–271.

57. David Y. Graham et al., "Effect of Treatment of *Helicobacter pylori* Infection on the Long-Term Recurrence of Gastric or Duodenal Ulcer," *Annals of Internal Medicine*, 116 (1992), 705–708, quotations from 705 and 708; NIH Consensus Development Panel on *Helicobacter pylori* in Peptic Ulcer Disease, "*Helicobacter pylori* in Peptic Ulcer Disease," *JAMA*, 272 (1994), 65–69; Paul Thagard, *How Scientists Explain Disease* (Princeton, NJ: Princeton Univ. Press, 1999).

58. Kiheung Kim, *The Social Construction of Disease: From Scrapie to Prion* (London: Routledge, 2007).

59. Ibid.; Stanley B. Prusiner, "Detecting Mad Cow Disease," in Scientific American, *Infectious Disease: A Scientific American Reader*, 170–174; Warwick Anderson, *The Collectors of Lost Souls: Turning Kuru Scientists into Whitemen* (Baltimore: Johns Hopkins Univ. Press, 2008).

60. Kim, *Social Construction of Disease*; David C. Bolton, "Prions, the Protein Hypothesis, and Scientific Revolutions," in *Prions and Mad Cow Disease*, ed. Brian K. Nunnally

and Ira S. Krull (New York: Marcel Dekker, 2004), 21–60; John Collinge, "The Risk of Prion Zoonoses," *Science*, 339 (2012), 411–413.

61. See, e.g., James Harvey Young, *The Medical Messiahs: A Social History of Health Quackery in Twentieth-Century America* (Princeton, NJ: Princeton Univ. Press, 1967).

62. David Hess, "Technology, Medicine, and Modernity: The Problem of Alternatives," in *Modernity and Technology*, ed. Thomas J. Misa, Philip Brey, and Andrew Feenberg (Cambridge, MA: MIT Press, 2003), 279–302; David M. Eisenberg, Ronald C. Kessler, et al., "Unconventional Medicine in the United States: Prevalence, Costs, and Patterns of Use," *New England Journal of Medicine*, 328 (1993), 246–252; David M. Eisenberg, Roger B. Davis, et al., "Trends in Alternative Medicine Use in the United States, 1990–1997," *JAMA*, 280 (1998), 1569–1575; *NIH News*, http://nccam.nih.gov/news/2009/073009.htm; Michael S. Goldstein, "The Persistence and Resurgence of Medical Pluralism," *Journal of Health Politics, Policy and Law*, 29 (2004), 925–945; James C. Whorton, *Nature Cures: The History of Alternative Medicine in America* (New York: Oxford Univ. Press, 2002).

63. Hans A. Baer, *Biomedicine and Alternative Healing Systems in America: Issues of Class, Race, Ethnicity, and Gender* (Madison: Univ. of Wisconsin Press, 2001).

64. Hess, "Technology, Medicine, and Modernity," quotation from 282.

65. Young, *Medical Messiahs*, 401–402, 420; Gerald E. Markle and James C. Petersen, eds., *Politics, Science, and Cancer: The Laetrile Phenomenon* (Boulder, CO: Westview, 1980); Eric W. Boyle, *Quack Medicine: A History of Combating Health Fraud in Twentieth-Century America* (Santa Barbara, CA: Praeger, 2013), chap. 8.

66. See, e.g., Bailus Walker Jr., "Public Health and the New National Effort: The 1988 Presidential Address," *American Journal of Public Health*, 79 (1989), 419–421; and Tom Christoffel and Susan Scavo Gallagher, *Injury Prevention and Public Health: Practical Knowledge, Skills, and Strategies*, 2nd ed. (Sudbury, MA: Jones & Bartlett, 2006), xxiii–xxiv.

67. See, e.g., James Colegrove, *States of Immunity: The Politics of Vaccination in Twentieth-Century America* (Berkeley: Univ. of California Press, 2006).

68. Lynn Etheredge, Stanley B. Jones, and Lawrence Lewin, "What Is Driving Health System Change?," *Health Affairs*, 15 (1996), 93–101.

69. Ameringer, *Health Care Revolution*, chap. 9; Jon Gabel et al., "Job-Based Health Insurance in 2001: Inflation Hits Double Digits, Managed Care Retreats," in *Health Policy: Crisis and Reform in the U.S. Health Care Delivery System*, ed. Charlene Harrington and Carroll L. Estes, 4th ed. (Sudbury, MA: Jones & Bartlett, 2004), 318–323; Jonathan Engel, *Poor People's Medicine: Medicaid and American Charity since 1965* (Durham, NC: Duke Univ. Press, 2006), 209; Paul Starr, *Remedy and Reaction: The Peculiar American Struggle over Health Care Reform* (New Haven, CT: Yale Univ. Press, 2011), 5.

70. Phoebe Eliopoulos, ed., *Managed Care: Facts, Trends and Data: 1997–1998*, 2nd ed. (Washington, DC: Atlantic Information Services, 1997), 1.

71. "America's Health Centers," in Shi and Singh, *Nation's Health*, 549; J. H. Hibbard and P. J. Cunningham, "How Engaged Are Consumers in Their Health and Health Care, and Why Does It Matter?," in ibid., 451–458; Stephen J. Kunitz, "Changing Patterns of Mortality among American Indians," in ibid., 28–37.

72. David J. Rothman, *Beginnings Count: The Technological Imperative in American Health Care* (New York: Oxford Univ. Press, 1997), chap. 6; Vincente Navarro, "Why

Congress Did Not Enact Health Care Reform," in Harrington and Estes, *Health Policy*, 36–44; Colin Gordon, *Dead on Arrival: The Politics of Health Care in Twentieth-Century America* (Princeton, NJ: Princeton Univ. Press, 2003), 251–260; Starr, *Remedy and Reaction*, pt. 2.

73. Starr, *Remedy and Reaction*, pt. 3.

74. *The Guide to the Managed Care Industry* (Baltimore: HCIA, 1994), vii–xii, quotation from vii; *Source Book of Health Insurance Data*, 1994, 7, 24–32; Susan B. Garland and Naomi Freundlich, "Insurers vs. Doctors: Who Knows Best?," *Business Week*, 18 Feb. 1991, 64–65; Robert Kuttner, "Sick Joke," *New Republic*, 2 Dec. 1991, 20–22; Dranove, *Economic Evolution of American Health Care*; Samuel W. Bloom, *The Word as Scalpel: A History of Medical Sociology* (New York: Oxford Univ. Press, 2002), 279.

75. Robert J. Blendon et al., "Understanding the Managed Care Backlash," *Health Affairs*, 17 (1998), 80–94; David Mechanic, "The Managed Care Backlash: Perceptions and Rhetoric in Health Care Policy and the Potential for Health Care Reform," *Milbank Quarterly*, 79 (2001) 35–54, quotation from 37; Gwen Wagstrom Halaas, "Evidence-Based Doctoring," in Mullan, *Big Doctoring in America*, 175–176.

76. Mechanic, "Managed Care Backlash"; Shannon Brownlee, "The Overtreated American," *Atlantic Monthly*, Jan./Feb. 2003, 89–91; Jamie Court and Francis Smith, *Making a Killing: HMOs and the Threat to Your Health* (Monroe, ME: Common Courage, 1999).

77. Mechanic, "Managed Care Backlash"; Gail B. Agrawal and Howard R. Veit, "Back to the Future: The Managed Care Revolution," *Law and Contemporary Problems*, 11 (2002), 11–53; Thomas R. McLean and Edward P. Richards, "Health Care's 'Thirty Years War': The Origins and Dissolution of Managed Care," *N.Y.U. Annual Survey of American Law*, 60 (2004), 283–328; *Health and Health Care 2010: The Forecast, The Challenge*, 2nd ed. (San Francisco: Jossey-Bass, 2003), chap. 5; Deborah Stone, "The Doctor as Businessman: The Changing Politics of a Cultural Icon," *Journal of Health Politics, Policy and Law*, 22 (1997), 533–556.

78. See previous note, esp. Agrawal and Veit, "Back to the Future," 41–42; Gabel, "Job-Based Health Insurance in 2001," 321; Beatrix Hoffman, *Health Care for Some: Rights and Rationing in the United States since 1930* (Chicago: Univ. of Chicago Press, 2012), xiii; Steffie Woolhandler, Terry Campbell, and David U. Himmelstein, "Costs of Health Care Administration in the United States and Canada," *New England Journal of Medicine*, 349 (2003), 768–775; and Sally Sleeper et al., "Trust Me: Technical and Institutional Determinants of Health Maintenance Organizations Shifting Risk to Physicians," *Journal of Health and Social Behavior*, 39 (1998), 189–200.

79. See previous note and esp. McLean and Richards, "Health Care's 'Thirty Years War,'" 283; Eliopoulos, *Managed Care*, 25; and Donald W. Light, "Ironies of Success: A New History of the American Health Care 'System,'" *Journal of Health and Social Behavior*, 45, suppl. (2004), 2.

80. See, e.g., Helen Halpin Schauffler and Jennifer K. Mordavsky, "Consumer Reports in Health Care: Do They Make a Difference?," *Annual Review of Public Health*, 22 (2001), 69–89; Clark C. Havighurst, "How the Health Care Revolution Fell Short," *Law and Contemporary Problems*, 65 (2002), 55–101; Stephen M. Davidson, *Still Broken: Understanding the U.S. Health Care System* (Stanford, CA: Stanford Univ. Press, 2010); and Douglas E.

Hough, *Irrationality in Health Care: What Behavioral Economics Reveals about What We Do and Why* (Stanford, CA: Stanford Univ. Press, 2013).

81. Timothy Hoff, Winthrop P. Whitcomb, and John R. Nelson, "Thriving and Surviving in a New Medical Career: The Case of Hospitalist Physicians," *Journal of Health and Social Behavior*, 43 (2002), 72–91.

82. Timothy Hoff, *Practice Under Pressure: Primary Care Physicians and Their Medicine in the Twenty-First Century* (New Brunswick, NJ: Rutgers Univ. Press, 2010), quotation from 29.

83. Dranove, *Economic Evolution of American Health Care*, 104–105. See also, e.g., Ariela Royer, *Life with Chronic Illness: Social and Psychological Dimensions* (Westport, CT: Praeger, 1998).

84. Aaron Catlin et al., "National Health Spending in 2005: The Trend Down Continues," *Health Affairs*, 26 (2007), 145; P. B. Ginsburg, "High and Rising Health Care Costs: Demystifying U.S. Health Care Spending," in Shi and Singh, *Nation's Health*, 380; Alan Garber, Dana P. Goldman, and Anupam B. Jena, "The Promise of Health Care Cost Containment," *Health Affairs*, 26 (2007), 1545–1547; Kelly J. Devers, Linda R. Brewster, and Lawrence P. Sasalino, "Changes in Hospital Competitive Strategy: A New Medical Arms Race?," in Harrington and Estes, *Health Policy*, 177–178; Kevin Grumbach, "Fighting Hand to Hand over Physician Workforce Policy," in ibid., 199.

85. Davidson, *Still Broken*, 17.

86. Daniel Callahan, *Taming the Beloved Beast: How Medical Technology Costs Are Destroying Our Health Care System* (Princeton, NJ: Princeton Univ. Press, 2009).

87. Ellen Kramarow et al., "Trends in the Health of Older Americans, 1970–2005," *Health Affairs*, 26 (2007), 1417–1425. See also, e.g., Eileen M. Crimmins, "Trends in the Health of the Elderly," *Annual Review of Public Health*, 25 (2004), 79–98; and Kenneth G. Manton, "Recent Declines in Chronic Disability in the Elderly U.S. Population: Risk Factors and Future Dynamics," ibid., 29 (2008), 81–113.

88. Gordon Guyatt et al., "Evidence-Based Medicine: A New Approach to Teaching the Practice of Medicine," *JAMA*, 268 (1992), 2420–2425; Solomon, "Evidence-Based Medicine"; Bernice A. Pescosolido, Thomas W. Croghan, and Joel D. Howell, "Unexamined Discourse: The Outcomes Movement as a Shift from Internal Medical Assessment to Health Communication," in *Communicating to Manage Health and Illness*, ed. Dale E. Brashers and Daena J. Goldsmith (New York: Routledge, 2009), 41–50; Steven H. Woolf et al., "Developing Evidence-Based Clinical Practice Guidelines: Lessons Learned by the US Preventive Services Task Force," *Annual Review of Public Health*, 17 (1996), 514; Daniel M. Fox, *The Convergence of Science and Governance: Research, Health Policy, and American States* (Berkeley: Univ. of California Press, 2010), esp. chap. 1.

89. Solomon, "Evidence-Based Medicine"; David Malakoff, "Spiraling Costs Threaten Gridlock," *Science*, 322 (2008), 210–213; Floyd E. Bloom, "Science as a Way of Life: Perplexities of a Physician-Scientist," ibid., 300 (2003), 1680; Patricia Dolan Mullen and Gilbert Ramírez, "The Promise and Pitfalls of Systematic Reviews," *Annual Review of Public Health*, 27 (2006), 81.

90. George Weisz, "From Clinical Counting to Evidence-Based Medicine," in *Body Counts: Medical Quantification in Historical and Sociological Perspective/La quantification*

medicale, perspectives historiques et sociologiues, ed. Gérard Jorland, Annick Opinel, and George Weisz (Montreal: McGill-Queen's Univ. Press, 2005), 382–389.

91. Jeanne Daly, *Evidence-Based Medicine and the Search for a Science of Clinical Care* (Berkeley: Univ. of California Press, 2005), quotation from 21; Fox, *Convergence of Science and Governance*.

92. Daniel M. Fox, "Systematic Reviews and Health Policy: The Influence of a Project on Perinatal Care since 1988," *Milbank Quarterly*, 89 (2011), 425–429; Fox, *Convergence of Science and Governance*, esp. chap. 4.

93. See, e.g., Edward Shorter, *How Everyone Became Depressed: The Rise and Fall of the Nervous Breakdown* (New York: Oxford Univ. Press, 2013), esp. chap. 10; Sergio Sismondo, "Pharmaceutical Maneuvers," *Social Studies of Science*, 34 (2004), 149–159; John K. Iglehart, "An Industry Under Siege Mounts Counterattack," *Health Affairs*, 23 (2004), 7–8; Fox, *Convergence of Science and Governance*, esp. 3; Donald Kennedy, "Beauty and the Beast," *Science*, 295 (2002), 1601; Marcia Angell, *The Truth about the Drug Companies: How They Deceive Us and What to Do About It* (New York: Random House, 2004); and Marc A. Rodwin, *Medicine, Money, and Morals: Physicians' Conflicts of Interest* (New York: Oxford Univ. Press, 1993).

94. See, e.g., Kathleen Montgomery and Amalya L. Oliver, "Shifts in Guidelines for Ethical Scientific Conduct: How Public and Private Organizations Create and Change Norms of Research Integrity," *Social Studies of Science*, 39 (2009), 137–155; and Dominique A. Tobbell, *Pills, Power, and Policy: The Struggle for Drug Reform in Cold War America and Its Consequences* (Berkeley: Univ. of California Press, 2012).

95. Solomon, "Evidence-Based Medicine"; Malakoff, "Spiraling Costs Threaten Gridlock," 210–213; Bloom, "Science as a Way of Life," 1680.

96. Linda T. Kohn, Janet M. Corrigan, and Molla S. Donaldson, eds., *To Err is Human: Building a Safer Health System* (Washington, DC: National Academy Press, 2000), quotation from 1; Angela M. Wicks, "Competing Values in Healthcare: Balancing the (Un)Balanced Scorecard," *Journal of Healthcare Management*, 52 (2007), 310; William M. Sage and Rogan Kersh, eds., *Medical Malpractice and the U.S. Health Care System* (New York: Cambridge Univ. Press, 2006), esp. Peter D. Jacobson, "Medical Liability and the Culture of Technology," 115–134; David M. Studdert, Michelle M. Mello, and Troyen A. Brennan, "Health Policy Review: Medical Malpractice," in *Medical Malpractice: A Physician's Sourcebook*, ed. R. E. Anderson (Totowa, NJ: Humana, 2005), 227–245, esp. 235.

97. Davidson, *Still Broken*, 6, 11; Harold Varmus, "Proliferation of National Institutes of Health," *Science*, 291 (2001), 1903–1905; Francis S. Collins, "Opportunities for Research and NIH," ibid., 327 (2010), 37.

Chapter 14. The Recent Past as a New Epoch

1. See, e.g., William M. Sage, "The Wal-Martization of Health Care," *Journal of Legal Medicine*, 28 (2007), 503–519.

2. Huda Y. Zoghbi, "The Basics of Translation," *Science*, 339 (2013), 250.

3. Thomas J. Misa, Philip Brey, and Andrew Feenberg, eds., *Modernity and Technology* (Cambridge, MA: MIT Press, 2003), esp. David Hess, "Technology, Medicine, and Modernity: The Problem of Alternatives," 279–302, and Arthur P. J. Mol, "The Environmental Transformation of the Modern Order," 303–325.

4. Einer Elhauge, "Why We Should Care about Health Care Fragmentation and How to Fix It," in *The Fragmentation of U.S. Health Care: Causes and Solutions*, ed. Elhauge (New York: Oxford Univ. Press, 2010), 3.

5. Donald Light and Sol Levine, "The Changing Character of the Medical Profession: A Theoretical Overview," *Milbank Quarterly*, 66, suppl. 2 (1988), 10–32.

6. Ross Rubin, quoted in *The Health Care Revolution: From Medical Monopoly to Market Competition*, by Carl F. Ameringer (Berkeley: Univ. of California Press, 2008), 159.

7. See, e.g., Arnold S. Relman and Uwe Reinhardt, "An Exchange on For-Profit Health Care," in *For-Profit Enterprise in Health Care*, ed. Bradford H. Gray (Washington, DC: National Academy Press, 1986), 209–223.

8. Robert J. Sternberg and Joseph A. Horvath, eds., *Tacit Knowledge in Professional Practice: Researcher and Practitioner Perspectives* (Mahwah, NJ: Lawrence Erlbaum Associates, 1999), esp. pt. 3. See also, e.g., Henny P. A. Boshuizen and Henk G. Schmidt, "On the Role of Biomedical Knowledge in Clinical Reasoning by Experts, Intermediates and Novices," *Cognitive Science*, 16 (1992), 153–184.

9. Robert A. Aronowitz, "The Converged Experience of Risk and Disease," *Milbank Quarterly*, 8 (2009), 417–442. On another level, see Bernice A. Pescosolido, Jane McLeod, and Margarita Alegría, "Confronting the Second Social Contract: The Place of Medical Sociology in Research and Policy for the Twenty-First Century," in *Handbook of Medical Sociology*, ed. Howard E. Freeman and Sol Levine, 5th ed. (Upper Saddle River, NJ: Prentice Hall, 2000), 411–426.

10. The next several paragraphs are based on John Burnham, "The Death of the Sick Role," *Social History of Medicine*, 25 (2012), 761–776.

11. Jacqueline Hart, "Healthy Beliefs: Wellness in Action at Company X," in *Society and Medicine: Essays in Honor of Renée C. Fox*, ed. Carla M. Messikomer, Judith P. Swazey, and Allen Glicksman (New Brunswick, NJ: Transaction, 2003), 23–41.

12. See, e.g., Deborah A. Stone, "The Struggle for the Soul of Health Insurance," in *The Politics of Health Care Reform: Lessons from the Past, Prospects for the Future*, ed. James A. Morone and Gary S. Belkin (Durham, NC: Duke Univ. Press, 1994), 26–56; Élodie Giroux, "The Framingham Study and the Constitution of a Restrictive Concept of Risk Factor," *Social History of Medicine*, 26 (2013), 94–112; and Robert A. Aronowitz, *Making Sense of Illness: Science, Society, and Disease* (Cambridge: Cambridge Univ. Press, 1998), 17.

13. Carsten Timmermann, "Appropriating Risk Factors: The Reception of an American Approach to Chronic Disease in the Two German States, c. 1950–1990," *Social History of Medicine*, 25 (2012), 157–174.

14. Jeremy A. Greene, *Prescribing by Numbers: Drugs and the Definition of Disease* (Baltimore: Johns Hopkins Univ. Press, 2007), esp. 221–240, quotation from 225.

15. Jason Karlawish, "Desktop Medicine," *JAMA*, 304 (2010), 2061–2062.

16. Sergio Sismondo, "Pharmaceutical Maneuvers," *Social Studies of Science*, 34 (2004), 157–158.

17. Greene, *Prescribing by Numbers*; William G. Rothstein, *Public Health and the Risk Factor: A History of an Uneven Medical Revolution* (Rochester, NY: Rochester Univ. Press, 2003). See also, e.g., Nortin M. Hadler, *Worried Sick: A Prescription for Health in an Overtreated America* (Chapel Hill: Univ. of North Carolina Press, 2008).

18. See, e.g., Robin Bunton and Roger Burrows, "Consumption and Health in the 'Epidemiological' Clinic of Late Modern Medicine," 206–244, and Martin O'Brien, "Health and Lifestyle, A Critical Mess? Notes on the Dedifferentiation of Health," 191–205, in *The Sociology of Health Promotion: Critical Analyses of Consumption, Lifestyle, and Risk*, ed. Robin Bunton, Sarah Nettleton, and Roger Burrows (London: Routledge, 1995).

19. Matthew M. Zack et al., "Worsening Trends in Adult Health-Related Quality of Life and Self-Rated Health—United States, 1993–2001," *Public Health Reports*, 119 (2004), 493–505.